Precision Medicine in Cancers and Non-Communicable Diseases

Precision Medicine in Cancers and Non-Communicable Diseases

Edited by
Debmalya Barh

CRC Press
Taylor & Francis Group
Boca Raton London New York

CRC Press is an imprint of the
Taylor & Francis Group, an **informa** business

CRC Press
Taylor & Francis Group
6000 Broken Sound Parkway NW, Suite 300
Boca Raton, FL 33487-2742

First issued in paperback 2020

ISBN 13: 978-0-367-57103-0 (pbk)
ISBN 13: 978-1-4987-7560-1 (hbk)

Library of Congress Cataloging-in-Publication Data

Names: Barh, Debmalya, editor.
Title: Precision medicine in cancers and non-communicable diseases / [edited by] Debmalya Barh.
Description: Boca Raton, FL : CRC Press, 2019. | Includes bibliographical references and index.
Identifiers: LCCN 2018014610| ISBN 9781498775601 (hardback : alk. paper) | ISBN 9781315154749 (ebook)
Subjects: | MESH: Precision Medicine | Neoplasms--therapy | Noncommunicable Diseases--therapy
Classification: LCC RC263 | NLM WB 300 | DDC 616.99/4--dc23
LC record available at https://lccn.loc.gov/2018014610

Visit the Taylor & Francis Web site at
http://www.taylorandfrancis.com

and the CRC Press Web site at
http://www.crcpress.com

Dedication

I dedicate this book to **Dr. Candan Hizel**, one of my best friends whose philosophy of life, humanity, humbleness, and passion for science touches my heart.

Contents

Preface

Precision medicine is a holistic approach that considers variability in genetic makeup, environment, and lifestyle of an individual for personalized disease prevention and treatment. Currently, the availability of large-scale omics data, gene-environment and gene-lifestyle interaction information, as well as cutting-edge big data analytics and predictive algorithms has enabled us to develop precision medicine strategies on a case-by-case basis. However, it is the beginning and there is a long way to go!

Although the precision medicine strategies can be applied to improve every aspect of health and wellness, currently the approach is mostly restricted to non-communicable diseases (NCDs), in which cancers are the primary focus. Few books are available thus far on this topic, and those are mostly related to technical and ethical aspects. *Precision Medicine in Cancers and Non-Communicable Diseases* is probably the first book of its kind to address specific NCD-related precision medicine opportunities. In general, the book describes the prevalence, incidence, mortality rate, currently used diagnosis and treatment approaches, and explores how the precision medicine approach can benefit in predicting, preventing, managing, and treating cancers and NCDs. Further, case studies, challenges, and research opportunities are also discussed in some chapters.

A total of 17 chapters, divided between two sections, are included in this volume. Part 1 is composed of ten chapters on oncology, and the remaining seven chapters comprising Part 2 describe precision medicine aspects in a number of other NCDs.

Part 1

The book starts with a chapter (Chapter 1) by Dr. Mehmet Gunduz's group describing the overview of precision medicine in oncology and examining circulating tumor cells (CTC) and tumor DNA (CTD) as emerging tools in cancer precision medicine. In Chapter 2, Dr. Anjana Munshi and colleagues have provided a brief account on how CTC and CTD can be characterized and how they can be used in clinical decision making. Dr. Mario Pérez-Sayáns and team have focused on the importance, challenges, and opportunities of CTC in oral squamous cell carcinoma precision medicine in Chapter 3. In Chapter 4, Dr. Yusuf Izci has discussed the clinical scenario of precision medicine in brain gliomas. Dr. Candan Hizel and his colleagues have provided a detailed account of precision medicine approaches in colorectal cancer in Chapter 5. Precision medicine in prostate cancer with an emphasis on gene–environmental interactions is described by Dr. Nigel Murray in Chapter 6. In Chapter 7, the role of epigenetics in breast cancer precision medicine is discussed by Dr. Ramona G. Dumitrescu. In Chapter 8, Dr. Shailendra Dwivedi and colleagues have given an update on ovarian cancer precision medicine. Finally, in Chapters 9 and 10, Dr. Ota Fuchs presents the application of the precision medicine concept in myelodysplastic syndromes and acute myeloid leukemia, respectively.

Part 2

Part 2 deals with NCDs other than cancers. The section starts with precision medicine in coronary artery disease (Chapter 11) developed by Dr. Melvin George and team. In Chapter 12, Dr. Anjana Munshi's group has highlighted the precision medicine aspects in stroke and other neurological diseases. In Chapter 13, Dr. Fatmanur Kazanci and his colleagues have discussed the importance of lifestyle, environmental, and genetic factors in preventing and treating osteoporosis. Dr. Sandhiya Selvarajan and colleagues have demonstrated the application of precision medicine approach in type-1 and type-2 diabetes in Chapter 14. In Chapters 15 and 16, Dr. Shoaib Ahmad has reviewed the precision medicine potential in multiple sclerosis and asthma/COPD, respectively. In the final chapter (Chapter 17), Dr. Kenneth Blum and his team have provided an overview on how the DNA-directed precision nutrition benefits in managing reward deficiency syndrome and addiction.

It is my hope that this book will be useful to the readers in understanding the current status, challenges, and future aspects of various cancers and other NCD-specific precision medicine.

Debmalya Barh, PhD

About the editor

Debmalya Barh, MSc (Applied Genetics), MTech (Biotechnology), MPhil (Biotechnology), PhD (Biotechnology), PhD (Bioinformatics), Post-Doc (Bioinformatics), and PGDM (Post Graduate in Management), has more than 15 years of academic, health care, molecular diagnostic, and bioinformatics industry experience and is an expert in integrative omics-based biomarker, targeted drug discovery, molecular diagnosis, and precision medicine in cancers and various complex diseases and human traits. Dr. Barh works with more than 400 scientists from more than 100 top-ranked organizations across more than 40 countries. He has over 150 journal publications and has authored more than 30 book chapters. He is a branded editor for over 20 cutting-edge, omics-related reference books published by top-ranked publishers including Taylor & Francis. He frequently reviews articles for *Nature* publications, Elsevier, AACR journals, NAR, BMC journals, *PLOS ONE*, and Frontiers, to name a few. He has been recognized by *Who's Who in the World* and *Limca Book of Records* for his significant contributions in managing advance scientific research.

Contributors

Duygu Abbasoğlu PhD
Department of Biology
Anadolu University
Eskişehir, Turkey

Shoaib Ahmad MPharm, PhD, DBME, PGDBA
University School of Pharmaceutical Sciences
Rayat-Bahra University
Punjab, India

Arsalan Amirfallah PhD
Pathology Department/Cell Biology Unit
Landspitali University Hospital
BioMedical Center (BMC)
University of Iceland
Reykjavik, Iceland

Nishanthi Anandabaskar MD
Department of Pharmacology
Sri Manakula Vinayagar Medical College
   and Hospital
Puducherry, India

Rajendra D. Badgaiyan PhD
Department of Psychiatry
Ichan School of Medicine
New York, New York

and

Department of Psychiatry
Wright State University
and
Dayton Veterans Affairs Medical Center
Dayton, Ohio

Debmalya Barh MSc, MTech, MPhil, PhD, PGDM
Institute of Integrative Omics and Applied
   Biotechnology (IIOAB)
West Bengal, India

and

Division of Bioinformatics and Computational
   Genomics
NITTE University Center for Science Education
   and Research (NUCSER)
(NITTE Deemed to be University)
Mangalore, Karnataka, India

Yasemin Baskın MD, PhD
Department of Basic Oncology
Institute of Oncology
and
Personalized Medicine
   and Pharmacogenomics Research Center
Dokuz Eylül University
İzmir, Turkey

Kenneth Blum PhD
Western University Health Sciences
Graduate School of Biomedical Sciences
Pomona, California

and

Department of Clinical Neurology
PATH Foundation NY
New York City, New York

and

Division of Neurogenetic Research & Addiction
   Therapy
The Florida House Experience
Deerfield Beach, Florida

Eric R. Braverman PhD
Department of Clinical Neurology
Path Foundation NY
New York, New York

Satrupa Das PhD
Department of Molecular Biology
Institute of Genetics and Hospital for
   Genetic Diseases
Osmania University
Hyderabad, India
and
Dr. NTR University of Health Sciences
Andhra Pradesh, India

Zsolt Demotrovics PhD
Institute of Psychology
Eötvös Loránd University
Budapest, Hungary

William B. Downs PhD
Victory Nutrition International, LLC
Lederach, Pennsylvania

Ramona G. Dumitrescu PhD, MPH
Kelly Government Solutions
Bethesda, Maryland

Shailendra Dwivedi MSc, PhD
Department of Biochemistry
All India Institute of Medical Sciences
Jodhpur, India

Marcelo Febo PhD
Department of Psychiatry
McKnight Brain Institute
University of Florida College of Medicine
Gainesville, Florida

Lyle Fried PhD
The Shores Treatment & Recovery
Port St. Lucie, Florida

Ota Fuchs PhD
Department of Genomics
Institute of Hematology and Blood Transfusion
Prague, Czech Republic

Abel García García PhD, MD
Oral Medicine, Oral Surgery and Implantology Unit
Health Research Institute of Santiago (IDIS)
Santiago de Compostela, Spain

Melvin George DM
Department of Clinical Pharmacology
SRM Medical College Hospital & Research Centre
Kancheepuram, India

Apul Goel MBBS, MS, MCh
Department of Urology
King George Medical University
Lucknow, India

Luxitaa Goenka MSc
Department of Clinical Pharmacology
SRM Medical College Hospital & Research Centre
Kancheepuram, India

Esra Gunduz DMD, PhD
Department of Medical Genetics
Turgut Ozal University
Ankara, Turkey

Mehmet Gunduz MD, PhD
Department of Medical Genetics
Turgut Ozal University
Ankara, Turkey

Candan Hızel PhD
Opti-Thera Inc.
Montreal, Quebec, Canada

Yusuf Izci MD
Department of Neurosurgery
Gulhane Education and Research Hospital
University of Health Sciences
Ankara, Turkey

Sadishkumar Kamalanathan DM
Department of Endocrinology
Jawaharlal Institute of Postgraduate Medical
   Education and Research (JIPMER)
Puducherry, India

Fatih Kazanci MD, PhD
Department of Medical Genetics
Turgut Ozal University
Ankara, Turkey

Fatmanur Hacievliyagil Kazanci MD, PhD
Department of Medical Biochemistry
Turgut Ozal University
Ankara, Turkey

Sanjay Khattri MBBS, MD
Department of Pharmacology and Therapeutics
King George Medical University
Lucknow, India

Gizem Çalıbaşı Koçal PhD
Department of Basic Oncology
Institute of Oncology
Dokuz Eylül University
İzmir, Turkey

Mona Li PhD
Division of Neurogenetics & Addiction Therapy
Florida House
Deerfied Beach, Florida

Rafael López-López MD, PhD
Translational Medical Oncology Laboratory
Health Research Institute of Santiago (IDIS)
Santiago de Compostela, Spain

Thomas McLaughlin PhD
Center for Psychiatric Medicine
North Andover, Massachusetts

Radhieka Misra MBBS
Era's Lucknow Medical College and Hospital
Lucknow, India

Sanjeev Misra MBBS, MS, MCh
Department of Onco-Surgery
All India Institute of Medical Sciences
Jodhpur, India

Anjana Munshi MPhil, PhD
Department of Human Genetics and Molecular
    Medicine
Central University of Punjab
Bathinda, India

Nigel P. Murray MRCP BSc (Hons), MB BCh
Department of Medicine
Hospital de Carabineros de Chile
and
Faculty of Medicine
University Finis Terrae
Santiago, Chile

Haluk Onat MD
İstanbul Onkoloji Hospital Maltepe
İstanbul, Turkey

Kamlesh Kumar Pant MBBS, MD
Department of Pharmacology and Therapeutics
King George Medical University
Lucknow, India

Puneet Pareek MBBS, MD
Department of Radio-Therapy
All India Institute of Medical Sciences
Jodhpur, India

Mario Pérez-Sayáns PhD, DDS
Oral Medicine, Oral Surgery and
    Implantology Unit
Health Research Institute of Santiago (IDIS)
Santiago de Compostela, Spain

Alicia Ábalo Piñeiro PhD
Liquid Biopsy Analysis Unit
Health Research Institute of Santiago (IDIS)
Santiago de Compostela, Spain

Purvi Purohit MSc, PhD
Department of Biochemistry
All India Institute of Medical Sciences
Jodhpur, India

Laura Álvarez Rodríguez DDS
Oral Medicine, Oral Surgery
    and Implantology Unit
Health Research Institute of Santiago (IDIS)
Santiago de Compostela, Spain

Laura Muinelo Romay PhD
Liquid Biopsy Analysis Unit
Health Research Institute of Santiago (IDIS)
Santiago de Compostela, Spain

Sandhiya Selvarajan MD, DNB, DM
Department of Clinical Pharmacology
Jawaharlal Institute of Postgraduate Medical
    Education and Research (JIPMER)
Puducherry, India

Praveen Sharma MSc, PhD
Department of Biochemistry
All India Institute of Medical Sciences
Jodhpur, India

Vandana Sharma PhD
Dr. NTR University of Health Sciences
Andhra Pradesh, India
and
Department of Molecular Biology
Indraprastha Apollo Hospital
New Delhi, India

Sulena Singh DM
Department of Neurology
GGS Medical College
Faridkot, Punjab

Akila Srinivasan MBBS
Department of Pharmacology
Jawaharlal Institute of Postgraduate Medical
  Education and Research (JIPMER)
Puducherry, India

Bruce Steinberg PhD
Department of Psychology
Curry College
Milton, Massachusetts

Şükrü Tüzmen PhD
Molecular Biology and Genetics Program
Department of Biological Sciences
Eastern Mediterranean University (EMU)
Famagusta, North Cyprus

and

Arizona State University (ASU)
Phoenix, Arizona

Jeewan Ram Vishnoi MBBS, MS, MCh
Department of Onco-Surgery
All India Institute of Medical Sciences
Jodhpur, India

Roger Waite PhD
Division of Precision Behavioral Management
Geneus Health, LLC
San Antonio, Texas

M. Ramazan Yigitoglu MD, PhD
Department of Medical Biochemistry
Turgut Ozal University
Ankara, Turkey

Yeşim Yıldırım MD
Department of Internal Medicine and
  Medical Oncology
Anadolu Medical Center Hospital
Gebze, Turkey

Fazilet Yılmaz MD
Department of Medical Genetics
Turgut Ozal University
Ankara, Turkey

Sultan Ciftci Yılmaz PhD
Department of Medical Genetics
Turgut Ozal University
Ankara, Turkey

Precision medicine in oncology

PART 1

Precision medicine in oncology

Precision medicine in oncology: An overview

FAZILET YILMAZ, SULTAN CIFTCI YILMAZ, ESRA GUNDUZ, AND MEHMET GUNDUZ

EPIDEMIOLOGY OF ONCOLOGY

According to the World Health Organization (WHO), cancer is a leading cause of death worldwide. In 2012, 8.2 million people died of cancer (Stewart and Wild, 2014). The most common cancers in the United States are breast, lung, prostate, and colorectal carcinoma (American Cancer Society, 2016).

When we rank cancer types according to number of deaths, lung cancer (1.59 million deaths) is the most common, followed by liver, stomach, colorectal and breast cancer.

Beyond that, every year 14 million people receive a cancer diagnosis (Stewart and Wild, 2014). Even if it does not cause death in the short term, it causes lifestyle changes and morbidity with a heavy emotional and financial burden.

GENETIC MECHANISM OF CANCER

Cancer, the cell population with an altered genetic profile, does not fit to the normal tissue structure and proliferates without control. Morphology of the cancer cells transit from dysplasia to neoplasia (Weinberg, 2014). The genetic heterogeneity is the most common feature of the cancer cell groups. The heterogeneity increases with the cell divisions and can be due to genetic and/or epigenetic changes (Swanton, 2012).

One nucleotide change in the genome is called single nucleotide variation (SNV). Phenotypical alterations can be observed based on the location of the SNV. Generally, a change occurring in an exon is more dangerous than a change occurring in an intron. In the case of a silent mutation, no change can be observed, but missense, nonsense, or frameshift mutations usually cause dramatic changes in the protein structure. Typically, cancer is coupled with mutation in an oncogene and/or tumor suppressor gene (Klug et al., 2012).

Both proto-oncogenes and the tumor suppressor genes normally exist in a genome. Proto-oncogenes mostly code proteins related with the activation of cell division and differentiation (Todd and Wong, 1999). Some of the well-studied proto-oncogenes are RAS, WNT, MYC, ERK, and TRK. These proto-oncogenes turn into oncogenes upon a mutation or translocation to a transcriptionally active site in the genome. Oncogenes lead to formation of cancer cells by promoting cell proliferation and undifferentiation (Weinstein and Joe, 2006).

One example of a proto-oncogene turning into an oncogene is formation of a BCR-ABL fusion gene observed in chronic myeloid leukemia (CML) after reciprocal translocation between chromosome 9 and 22. Normally, an ABL gene codes a tyrosine kinase and the expression of the gene is tightly regulated. After the translocation, the newly formed fusion gene BCR-ABL starts to express tyrosine kinase independent of regulation mechanism. Excess formation of tyrosine kinase activates proteins related to cell cycle and cell division, then in turn the cell starts to proliferate without control (Druker, 2002).

Another important gene group related to cancer formation is tumor suppressor genes (TSGs). TSGs are normally responsible for inhibiting uncontrolled cell proliferation and inducing apoptosis when necessary. Genes like Rb, p53, and PTEN mutated in many cancers belong to this group. Different than oncogenes, in most cases a single allele mutation will not drive the cells to the tumor formation. Mutations in both alleles are required for the cancer development (Lodish and Zipursky, 2000).

One of the most common examples of cancer formation due to TSG mutation is BRCA1 (breast cancer 1) gene mutation, which causes familial breast and ovarian cancer. This gene is responsible for recognition of damaged DNA and can lead to cell destruction in case the damage cannot be repaired. It is mostly expressed in the breast tissue. When a mutation occurs in this gene, in spite of the DNA damage, the cell continues to divide and accumulate DNA damage. As a result of this process, an abnormal population of cells will be assembled with a potential to have cancerous formation.

Mutations in tumor suppressor genes can be inherited. However, since the mutations in oncogenes are inherited in a dominant manner and most of them are incompatible with life, they cannot be inherited. Hence, most of the familial mutations associated with the cancer formation are related with the TSGs (Lodish and Zipursky, 2000).

Upon the mutation of the proto-oncogenes and/or TSGs playing critical roles in the cell such as cell cycle, proliferation, differentiation, and apoptosis, cancerous formation begins. Anomalies accumulate in the course of time by addition of new mutations and a heterogenic tumor develops (American Cancer Society, 2014). Tumor tissue inhibits the functioning of the normal tissue and clinical symptoms arise depending on the location.

CURRENT CLINICAL APPROACHES

Conventionally, cancers are classified according to their histopathological features (Weinberg, 2014). Before we had an understanding of cancer's molecular profile, the main progression factors were cancer localization and the cancer's tissue and cell type of origin. In the past, the only curative treatment was surgery. Therefore, classification of the cancer was based on surgical treatment options and tissue localization (Brauer, 1964). Later on, supportive methods such as chemotherapy and radiotherapy are added to cancer treatment ("Cancer research," 1932).

Nowadays, besides the conventional chemotherapeutic drugs, agents targeting the precise molecules in specific cancers are being developed. These new agents produced according to molecular characterization are named as targeted therapy. Even though these new agents are basically the same as chemotherapeutics, their acting mechanisms differ from them. These agents are either in a form of monoclonal antibody or a small molecule (National Cancer Institute, 2014). Some of the agents developed for the targeted therapy are approved for the cancer treatment. The following cancers are mostly studied for the targeted therapy for the indicated reasons: breast cancer for being the most common, lung cancer for having the highest mortality rate (Stewart and Wild, 2014), and hematological cancers for having no surgical treatment alternatives. In this section, general treatment aspects of these three cancer types will be elaborated.

The most common cancer type worldwide is breast cancer (Stewart and Wild, 2014). Breast cancer can exhibit different clinical symptoms and sometimes can be silent until it has metastasized. It also shows a varying progress pattern according to histopathological classification. Infiltrating ductal carcinoma is the most common type of breast cancer. Risk factors for breast cancer include BRCA1 and BRCA2 mutations, positive family history (especially positive in mother and sisters), age, age of first menstrual cycle, age of first pregnancy and parity, age of menopause, usage of oral contraceptives, alcohol consumption, breast tissue density, bone density, previous breast mass, and previous atypical biopsies (Collaborative Group on Hormonal Factors in Breast Cancer, 1997, 2001; Modan et al., 1998).

In breast cancer, diagnosis is confirmed by histopathological existence of malignant cells. The mass is surgically removed and the hormone receptor (ER and HER2) profile is identified. Furthermore, the sample should be examined to determine whether metastasis has occurred. Considering receptor situation and metastases, chemotherapeutics are given as a single agent or combined. If there is any nodal or distance metastases with clear resection margins, surgical resection is considered as the total cure. Genetic profile (ER+, HER2+, and triple negative) is especially important for choice of chemotherapeutic agent and personalized treatment (Hutchinson, 2010). When HER2 is positive, HER2 inhibitors such as trastuzumab (Herceptin) (Haq and Gulasingam, 2016; Rugo et al., 2016) and pertuzumab (Perjeta) (Gollamudi et al., 2016) can be added to the treatment. Since there is no targetable receptor by a drug, the triple negative is the most aggressive form of breast cancer.

Among the many types of cancer, lung cancer is the deadliest worldwide with 1.8 million diagnoses and 1.6 million deaths in 2012. Because of these clinical differences, lung cancer is classified as small cell and non-small cell lung cancer (NSCLC) (Stewart and Wild, 2014). Small cell lung cancer is the most aggressive type of lung cancer and has the worst prognosis. NSCLCs include lung adenocarcinoma, squamous cell carcinoma, and the large cell carcinoma. The best treatment is surgical resection. If there is no benefit from surgery, other options are radiotherapy, radiofrequency ablation (RFA), or cryoablation. Currently, directed treatment options are available for epidermal growth factor receptor (EGFR) and anaplastic lymphoma kinase (ALK), aberrantly expressed in tumors (Popat and Yap, 2014).

Hematologic malignancies do not form solid tumors as breast and lung cancers. These malignancies originate from the circulatory and immune system–related cells. Initially, they are classified as leukemia, lymphoma, and myeloma. Subsequently, they are subclassified based on the differentiation stage of the hematopoietic cells they originate from (Harris et al., 2000).

In hematological malignancies, molecular classification and genetic-based diagnosis is preferred (Tefferi et al., 2009). Genetic-based molecular diagnosis allows us to differentiate cells that cannot be differentiated with histopathological findings. In hematological malignancies, chemotherapeutics and the targeted therapies are the treatment options. Imatinib, used for chronic myeloid leukemia, is one example of these treatment options (Zarin et al., 2015).

General information of common cancer types are summarized in Table 1.1.

Unfortunately, the current clinical approaches for the treatment of many cancer types are insufficient and the chosen chemotherapeutical methods are not specific enough for a complete treatment of the tumors. Herewith, precision treatment methods can have a higher potential in completely curing the cancer.

PERSONALIZED ONCOLOGY

The intensity of a disease varies from one individual to another. One of the most important reasons for these personal differences is single nucleotide polymorphisms (SNPs), which cause genetic variations in populations. Even a genetic disease with the same mutation exerts itself at different severities due to different penetrance and the different gene expression profile of a person (Childs et al., 2015; Ustinova et al., 2015; Zahary et al., 2015). Also, despite the same severity of a disease, response to a drug varies from person to person.

Environmental factors are other reasons for diversity in disease and drug response. Consuming certain foods at excess rates in some communities results in altered liver enzyme activities, which in turn cause rapid drug degradation, and inhibit the drug from reaching the intended blood level

Table 1.1 General cancer information

	Approved screening and prevention methods	Current treatments	Targeted therapy molecule
Oral cavity and pharynx	Visual inspection	Radiation therapy, surgery, chemotherapy in advanced disease	
Digestive system	Colorectal, colonoscopy	Surgery, chemotherapy, radiation	Colorectal: EGFR
Respiratory system	Low dose spiral computed tomography (LDCT)	Surgery, radiation therapy, chemotherapy	Lung: EGFR, ALK
Bones and joints			
Soft tissue			
Skin	Regular examination with inspection	Surgical excision, electrodissection and curettage, radiation therapy, topical agents, immunotherapy	Melanoma: BRAF
Breast	Mammography	Surgery, radiation, hormonal therapy, targeted therapy	HER2
Genital system	Cervical: Pap smear and HPV vaccine	Ovary: Surgery, chemotherapy Uterine cervix: Electrosurgical excision, cryotherapy, laser ablation, conization, surgery, radiotherapy, chemotherapy Uterine corpus: Surgery, radiation, hormone therapy, chemotherapy	
Urinary system	—	Kidney: Surgery Bladder: Surgery, immunotherapy, chemotherapy Prostate: Surgery, radiation, radioactive seed implant, hormonal therapy	Kidney: Targeted therapy studies
Eye and orbit			
Brain and other nervous system			
Endocrine system	—	Thyroid: Surgery, radioactive iodine	
Lymphoma			
Myeloma			
Leukemia	—	Chemotherapy	CML: BCR-ABL
Other and unspecified primary sites			

Source: American Cancer Society, Cancer Facts & Figures 2016, American Cancer Society, Atlanta, 2016.

(National Consumers League and the U.S. Food and Drug Administration [FDA], 2016).

Since the response of a person both to a disease and drug treatment varies, prescribing the same treatment for all patient approaches is changing. Owing to improved technologies and cumulative knowledge, specific screening and treatment for each patient are becoming available.

Wide molecular diversity in cancer emphasizes the importance of personalized medicine in this field (Boisguérin et al., 2014). Some cancer types do not have curative treatment options and urgently need personalized molecular treatment alternatives. Taking into account all of these, personalized oncology is really important and open for improvement (Sicklick et al., 2016).

Currently, most developments in personalized medicine are related to whole genome sequencing. At the beginning, a whole genome sequence cost about $2.7 billion (Weinberg, 2014). Today, whole genome/exome sequencing is easily achievable with next generation sequencing (NGS), within hours for approximately $1000. It provides high-throughput and totally personalized data in a short time with a lower price. Nevertheless, merely knowing the patients' whole sequence cannot be considered as having entire knowledge about the patient. To have precise information extracted from the sequencing for treatment purposes, the data first should be analyzed. During the analysis, other factors such as epigenetic and environmental factors such as exposomics (exposure of infections, toxic agents) should be evaluated with the obtained data.

Even though the most widely used current genetic analysis for tumor profiling is NGS, some other genetic screenings are also available. Since some of the cancer types have been well identified, instead of sequencing the entire genome, specific genes or loci related to the tumor formation can be screened. Some of the currently available tests for tumor profiling are listed in Table 1.2.

Precision medicine in oncology can be evaluated in four main sections. The first one is predicting the cancer risk in whole life and predicting the recurrence. Second is personalized prophylactic treatment in the high-risk population. The third one is personalizing the treatment according to molecular features of the cancer and the metabolic features of the patient. Fourth is predicting the drug resistance (Figure 1.1).

Table 1.2 Available genetic screens for disease profiling

Specific single gene tests
Specific gene panels
Genotyping panels
Whole genome or exome sequencing
Epigenetics arrays
Pharmacogenomics testing

Predicting the cancer risk

Prediction of the cancer risk is a relatively underdeveloped part of precision medicine. Even though some tests are available, risk assessment is extremely limited to only a couple of the cancer types (Hamilton et al., 2013). One reason for this can be that with cancer being a very genetically versatile disease, it becomes statistically hard to determine the course of the disease. A second reason can be the low level of the genetic screen performed and the absence of a well-defined data pool available to all researchers and the clinicians. Another reason can be the lack of sufficient bioinformatics tools and databases to correctly interpret the obtained data.

Cancer risk can be predicted with different genetic tests for some cancer types (Heald et al., 2012). However, performing merely a test is not enough for determining the risk and giving correct information to the patient. Pre- and posttest consultation, including epigenetic and environmental factors, are fundamental for correct evaluation and processing. Also choosing the appropriate genetic test influences accuracy and avoidance of false-negative and false-positive results. As a necessity, the results should be explained to the patient in a manner revealing the potential expectation and the potential outcomes clearly.

Recently, some companies' permission for commercial genetic tests for BRCA gene mutation in the United States has been removed. The main reason was the result obtained without expert consultation can be misleading because breast cancer does not solely depend on the BRCA mutation (Cook-Deegan and Niehaus, 2014).

Risk assessment for breast cancer is commonly practiced. There are many tools for risk assessment according to BRCA mutations, family history, and the personal risk status. Some quantitative risk assessment tools are BOADICEA (Lee et al., 2014), BRCAPRO (BRCAPRO, 2015), Tyrer–Cuzick (Boughey et al., 2010), and BCRAT (Park et al., 2011).

BRCA mutation analysis was patented from a commercial company. But, the U.S. Supreme Court invalidated the patent in 2013 (Marshall and Price, 2013). Now, there are alternative companies for BRCA mutation tests with extended test content.

More precise cancer risk prediction availability and taking the necessary precautions in the future

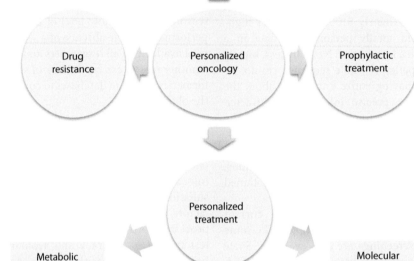

Figure 1.1 Precision medicine in oncology.

can greatly contribute to the welfare of societies. For these reasons, more platforms and agencies should promote and support research and investigations in the cancer risk prediction area.

Personalized prophylactic treatment in high-risk populations

Prophylaxis means preventing diseases before having the disease. Some trials for prophylaxis of high-risk populations were conducted for some of the most common cancer types.

An example of personalized prophylactic treatment in high-risk populations is daily aspirin usage in colorectal cancer (CRC) prevention. Aspirin (acetyl salicylic acid) is used as an analgesic, antipyretic, and antithrombolytic agent. Currently, aspirin is being used as a prophylactic treatment for colorectal cancer in high-risk populations.

The first idea about aspirin usage in colorectal cancer prevention emerged in a case control study from 1988. In 1998, another study reported no relationship with CRC and aspirin (Sturmer et al., 1998). In 2007, a study showed that aspirin is effective on PTGS2 overexpressing tumors (Harris et al., 2000). Because of some discordant results, consecutive studies investigated the molecular relation between aspirin treatment and cancer prevention. Some SNPs related with the effect of the aspirin have been identified.

Daily usage of aspirin causes many side effects, especially on the gastrointestinal system. Besides, aspirin does not exert its effect on every individual. Under certain circumstances, aspirin usage should be considered on a personal basis to obtain the most effective results with the lowest side effects. The U.S. Preventive Services Task Force (USPSTF) recommends a low dose of aspirin to only people 50–69 years old with a high-risk profile.

Familial adenomatous polyposis (FAP) is characterized with a vast number of adenomatous polyp formation in the colon and mutation in the APC gene. Individuals with FAP develop colon cancer after 40 years of age. Prophylactic colectomy is recommended to these individuals around 25 years of age (Hagen et al., 2015).

Another highly practiced preventive cancer treatment is prophylactic mastectomy. It is recommended to people with a strong positive family history and/or high mutation risk in the BRCA gene (National Cancer Institute, 2016). Bilateral prophylactic mastectomy has been shown to reduce the risk of individuals with strong family history (Rebbeck et al., 2004; Domchek et al., 2010). Two FDA-approved drugs—tamoxifen and raloxifene—have also been administered to individuals with a risk of 1.67% or greater of developing breast cancer in the next 5 years in order to reduce the risk (Cummings et al., 1999; Fisher et al., 2005).

Choosing personalized treatment

In precision medicine, the decision of drug selection for an optimal treatment should be based on the molecular features of the cancer and the metabolic features of the patients. Both aspects are elaborated upon in the next sections.

MOLECULAR FEATURES OF THE CANCER

Cancer's molecular features were first identified in hematological malignancies. The reason for such discovery was the search for alternative treatment options due to no possibility of surgical interventions. Another reason was the cases showing different prognoses with no histological distinction. Additionally, thanks to some of the solid tumors with no treatment, researchers have been directed to investigate the molecular mechanism of the malignancies for developing potential drugs for targeting the precise mechanisms.

Several FDA-approved oncology drugs with companion diagnostic tests are currently commercially available (Jørgensen, 2015). Drugs targeting oncogenes such as HER2, KRAS, BRAF, ALK, and cKi have been developed (Marrone et al., 2014). In most cancer cases, expressions of these genes are upregulated in tumors compared to normal tissues and suppressing the expression results in inhibition of cancer progression while relatively unharming the normal tissue.

The first drug used based on molecular features of the tumor is imatinib targeting the BCR-ABL fusion gene coding a tyrosine kinase always on due to translocation of the gene. Imatinib blocks the BCR-ABL protein leading to inactivation of the signaling pathway. Eventually, cell proliferation stops. An FDA-approved companion diagnostic test for imatinib is commercially available (FDA, 2016b).

Breast cancer drugs Herceptin (trastuzumab), Perjeta (pertuzumab), and Kadcyla (ado-trastuzumab emtansine) also work with the same approach. They target HER2, an oncogene activated in some breast cancer (FDA, 2016b).

Another example of drugs developed based on molecular characteristics of the tumor are epidermal growth factor receptor (EGFR) blockers that have been used for non-small cell lung cancer (NSCLC). EGFR belongs to the HER receptor family and is related to several pathways, such as RAS, Akt, and mTOR, playing a role in cell proliferation. The first generation of drugs developed for targeting EGFR are erlotinib and Iressa (gefitinib). Ten months after treatment, many patients developed resistance to these drugs, so a second-generation drug targeting pan-HER receptors, afatinib, has been developed. AZD9291 is the third-generation drug developed specifically for targeting mutant EGFR (Popat and Yap, 2014).

Until now, most of the targeted therapies for cancer treatment have been intended to affect one mutation or one pathway. Even though in most cases a driver mutation plays an essential role in cancer formation initially, in the later stages of the tumor development, cells accumulate more mutations, so usually targeting a single aberration does not completely cure the malignant formation. Today with the developing sequencing techniques, it is possible to analyze various mutations in the tumors. Treatment approaches should be radically revised based on mutations, SNPs, metabolic differences, and environmental factors (Lawrence et al., 2003).

PERSONAL METABOLIC FEATURES

Each individual possesses a different metabolism. Since drug pharmacokinetics depends on metabolism, the drugs cannot show the same effect in all individuals. Some of the drugs can be absorbed less than desired by the patients, and some of them can be degraded very rapidly based on the metabolic differences so they cannot reach the optimal level and cannot show the intended affect. All of these drug response differences due to the different genetic backgrounds of the individuals are named pharmacogenomics (Scott, 2011). Precision medicine in oncology also aims to oversee these potential differences.

The first pharmacogenomics studies were conducted on cytochrome P450 enzymes. Cytochrome P450 enzymes (CYPs) are responsible for drug metabolism in the liver. Fifty-eight identified CYPs exist in the human. Some of the SNPs have been determined for the CYPs. FDA-approved pharmacogenomic biomarkers have been used, mostly for drugs used for pulmonary, cardiac, rheumatologic, and infectious diseases (FDA, 2016a). Even though some examples exist in the oncology area, more extensive studies are needed in the oncology pharmacogenomics field.

Predicting drug sensitivity and resistance

Chemotherapeutics are either used alone or combined with a surgical treatment, depending on the cancer type. However, due to accumulated mutations over time, some of the tumors develop resistance to the chemical treatment (Szakacs et al., 2004; Restifo et al., 2016). Prediction medicine investigates the potential of the tumors to develop resistance to specific drugs.

One of the most used anticancer drugs is cis-diamminedichloridoplatinum (cisplatin). It was also the first anticancer drug containing platinum. Platinum complex binds to DNA and forms cross bands in the DNA strand. Subsequently, cells undergo apoptosis. Some of the tumor cells gain resistance after a course of the treatment. Cisplatin metabolism is explained in the following mechanism:

1. Cells take the cisplatin inside the cell by either passive transport or by an identified transporter such as CTR1, CTR2, and CTR3.
2. When it enters the cell, it binds to the DNA.
3. The damage caused by the cisplatin is fixed by the members of nucleotide excision repair (NER) pathway.
4. Cisplatin is inhibited by the glutathione s-transferase.
5. Conjugated cisplatin is exported out of the cell by either MPR2 or ATP7A/ATP7B transporters (Amable, 2016).

Based on the pharmacokinetics of cisplatin, resistance can develop due to quick inhibition of cisplatin by glutathione s-transferase or due to low levels of the drug inside the cells caused by transport of the cisplatin out of the cell. Different combinations of SNP in each molecule for each patient at this step defines the tumor's drug resistance. SNP variations among the individuals are one of the primary reasons for the development of drug resistance.

FUTURE OF PRECISION MEDICINE IN ONCOLOGY

Why future developments are extremely important in personalized oncology

Cancer is an aggressive disease with a high incidence of mortality. Besides being highly deadly, cancer seriously decreases the quality of life with high emotional and financial burden. In most cases, cancers can metastasize or relapse after a treatment and become harder to treat due to accumulated mutations. There are also several aggressive cancer types without any precise treatment.

One example is anaplastic thyroid cancer (ATC). It is an extremely aggressive cancer type in which most patients die within 6 months of diagnosis. ATC quickly invades the surrounding tissue and metastasizes, which makes surgical intervention impossible. Being also highly resistant to the known chemotherapeutics, personalized medicine can be the solution for ATC treatment.

Gallbladder cancer is another example of a cancer with a very low treatment success rate. Because of the gallbladder's anatomic location, radical surgical resection is impractical. Currently, researchers are focusing on molecular characterization and personalized medicine for gallbladder cancer treatment, thanks to genetics methods such as next-generation sequencing. Thus far, ARID1A, BRAF, CDKN2A/B, EGFR, ERBB2-4, HKN-RAS, PIK3CA, PBRM1, and TP53 mutations have been associated with this cancer type (Sicklick et al., 2016). But still, no approved treatment exists.

As described in both examples and when evaluated as a whole, the success rate in treating cancer is still extremely low. At present, cancer is the one of leading causes of disease-related deaths worldwide. Given the fact that each individual with a

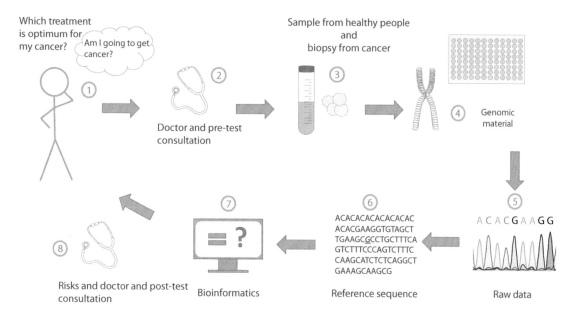

Figure 1.2 Application of precision medicine in oncology.

different genetic background responds differently to both the disease and treatment, from a cancer aspect, precise medicine with targeted molecular therapy seems extremely promising for finding an ultimate cure for cancer (Figure 1.2). Without a doubt, some of the developing fields described next bear great importance for carrying precision medicine in oncology to the next level.

Next-generation sequencing (massively parallel sequencing)

The human genome sequencing project was started using Sanger sequencing technology. However, the disadvantages of this method include high cost, labor intensiveness, and time. Later, the massively parallel sequencing method known as next-generation sequencing (NGS) was developed and is commonly used today. With this method, genomic material is digested randomly into fragments of a particular size. The method is based on the reading of many regions repeatedly (ranging from 100 to 10,000 times). This new method made it possible for the sequencing of the entire human genome (once sample prep is complete) in hours, something not possible using the Sanger sequencing method. Also, the cost is extremely low and with the spread of this technology throughout laboratories, its use has increased quite significantly. There are two or

three main manufacturers in the market today with several machines available that are based on different mechanisms and which are continuously being further developed.

While analysis was previously carried out on a specific mutation within a specific exon found within a particular gene, today in a single reaction, all of the mutations contained within a sample can be sequenced. Such analyses are referred to with terminology such as whole genome sequencing (WGS) or whole exome sequencing (WES). However, in this field one of the greatest current problems is the lack of sufficient bioinformatics software to handle the amount of data generated by this method as well as the lack of databases for analysis of incidental mutations.

When looking at NGS from the perspective of precision oncology, in the future when databases with sufficient information have been made available the following could be possible: (1) a single WGS analysis for an individual can determine genetic risk classification for all cancer types or (2) mutation analysis of an existing tumor can provide instruction regarding personalized treatment.

Single-cell genome sequencing

Cancer cells possess a wide range of intratumoral heterogeneity in terms of morphology, genotype, gene expression, metabolism, proliferative, and metastatic potential. Such heterogeneity creates a need for analysis at the single-cell level. Single-cell analysis is also important for tumors that cannot easily be cultured in an *in vitro* environment. Previously, sequencing was only able to target some specific regions. Now, whole genome/exome sequencing has become more available and single-cell analysis is being performed successfully and becoming more accessible (Van Loo and Voet, 2014). Sequencing/cell resolution is increasing each day.

For single-cell sequencing, cells should be selected individually with methods such as fluorescence-activated cell sorting (FACS), microfluidics-based cell sorting, laser capture microdissection or microdroplet technologies (Nakamura et al., 2016; Zhang et al., 2016). Normally, DNA material of a single cell is around 6 pg and not enough for analysis (Van Loo and Voet, 2014). However, this material can be amplified by commercially available genome amplification kits and the sequencing is performed thereafter (Qiagen, 2016). The processes that can be done at single cell resolution are whole genome, exome, transcriptome, (scRNA-seq), and epigenomic (methylation and chromatin structure) analyses. Since it can give detailed information about cancer's behavior and reveal the potential treatment options, single-cell analysis has great importance from the aspect of precision medicine in oncology.

Patient-derived tumor xenograft

A xenograft is the transplant of one tissue to another species. Patient-derived xenograft (PDX) is transplanting cancer tissue of one individual to another species and aims to investigate the molecular characteristics, drug response, and the aggressiveness of the tumor formation. The most common application of PDX is transplanting the cancer tissue from a patient to a nude mouse with a suppressed immune system and obtaining information about histology, genomic structure, cellular heterogeneity, and drug responsiveness of the tumor. A treatment approach can be tested on this mouse model and the potential outcome of the treatment can be anticipated. It is an important development regarding personalized medicine for being an *in vivo* study at the single-patient level (Cho et al., 2016).

Mixture of barcodes tumors PRISM

The PRISM method allows the screening of many drugs by treating several cancer cell lines contained within a single pool. Initially, cells are labeled with a 24-nucleotide barcode by using lentivirus as a carrier. After the drug administration, these barcodes make it possible to identify the dead cell lines that are considered as responsive to the drug treatment.

To test this assay's sensitivity and reliability, it was compared with currently used methods. Once the sensitivity of this method was shown, many studies were carried out using this assay. Thanks to this method, it is possible to test a host of different molecules/drugs/materials of different cell lines at once. In short, the effect of drugs on a host of different cell lines can be studied simultaneously. At the same time, by screening many molecules it affords the opportunity to identify molecules with therapeutic potential. It is thought that this method, used together with single-cell next-generation sequencing technology, can offer an even more sensitive level of resolution for the understanding of cellular mutations and cellular behavior. This high-resolution technology not only further opens the door to predicting a patient's overall tumor response to a drug but even presents the possibility of identifying the response of individual cells within that patient's tumor based on each cell's specific mutation profile (Yu et al., 2016).

With time, bioinformatics and data mining are becoming ever more important in personalized medicine.

Mutanomes

A mutanome is used to describe total mutations in tumor cells and is also known as patient centered tumor vaccination. It is a new treatment approach using the body's own defense system to eliminate tumor cells and it is specifically designed for each patient's tumors (Kuhn et al., 2011).

Via immune informatics, these mutations are analyzed and the correct mutation for using them as a vaccine is decided (Patronov and Doytchinova, 2013). Choosing the appropriate mutation is critical, since some of the mutations can be in a non-protein coding region, some of them can be shared with the patient's germ-line genome, and some of them can be nonimmunogenic (Boisguérin et al., 2014). Another consideration is choosing an epitope not targeting the other tissues enabled by bioinformatics software. In early attempts, peptides were given directly. Nowadays, modified mRNA (modRNA) technologies are improved and they are more effective than peptide vaccines. When compared to peptides, modRNAs are more stable, have a long-term effect, and are less immunogenic. Also, synthesized mRNA can be engineered as a poly-epitopic nucleotide sequence.

One example is autologous patient-specific vaccines used for advanced melanoma. Prophylactic immunization with 27mer peptides encoding specific mutations resulted in complete protection from the tumor and 40% survival of the mice, whereas all mice in the control group died within 44 days (Kreiter et al., 2015).

One major handicap of personalized vaccination is that there is a time-consuming process of approval for each vaccine. Theoretically, all new vaccines should pass the preclinical and the phase studies. Acceleration of the approval stage for personalized vaccination will greatly improve the field and can be very promising in the treatment of the cancer.

CONCLUSION

The potential of precision medicine carries great importance for several different areas of medicine, and it bears an extra significance for the oncology field for many reasons. First, cancer is a heterogenic disease that genetically varies not only among the patients but also among the cells of the same tumor requiring different clinical courses. Second is the need of precognition of the disease and taking precautions because of the lack of a definitive treatment. Third is the failure of many treatment methods due to personal differences of drug effect and resistance mechanisms. Owing to these necessities, precision medicine carries a great potential in the oncology field and some major problems should be addressed to get to a promising point for precision medicine in oncology.

Thanks to high throughput analysis, a vast amount of data is produced on a daily basis. The major problem is interpretation of the data and reflecting the outcome to the clinics. Even on a personal basis, analysis of the data obtained, determining the importance of the outcome, and

deciding how a patient should be informed regarding the results are other concerns that should be resolved.

Another major problem is the need of the bioinformatics tools to evaluate high-throughput data. Storage and the software developments should be synchronized with the data production.

Another major concern is the ethical aspect of the obtained data. The rules for which part of the patient's information can used and which part can be shared with insurance companies, relatives, or even the patients themselves should be well defined (Rothstein, 2012; Allen et al., 2014).

Even though precision medicine appears to be extremely promising for the oncology field, it still has a long journey of revision and development in the aspects of defining the targets, drug designing, improving the appropriate methods, and making new policies.

REFERENCES

Allen NL et al. Biobank participants' preferences for disclosure of genetic research results: Perspectives from the OurGenes, OurHealth, OurCommunity project. *Mayo Clin Proc* 89(6);2014:738–46.

Amable L. Cisplatin resistance and opportunities for precision medicine. *Pharmacol Res* 106;2016:27–36.

American Cancer Society. *Cancer Facts & Figures 2016*. American Cancer Society, Atlanta, 2016.

American Cancer Society. Genes and cancer. 2014. https://www.cancer.org/cancer/cancer-basics/cancer-prevalence.htm

Boisguérin V et al. Translation of genomics-guided RNA-based personalised cancer vaccines: Towards the bedside. *Br J Cancer* 111(8);2014:1469–75.

Boughey JC et al. Evaluation of the Tyrer-Cuzick (International Breast Cancer Intervention Study) model for breast cancer risk prediction in women with atypical hyperplasia. *J Clin Oncol* 28(22);2010:3591–6.

Brauer W. [The TNM System?]. *Radiobiol Radiother (Berl)* 5;1964:761–5.

BRCAPRO. 2015. Available from http://bcb.dfci.harvard.edu/bayesmendel/brcapro.php. Accessed 2016.

Cancer research: Chemotherapy and radium therapy. *Br Med J* 2(3746);1932:766–8.

Childs EJ et al. Common variation at 2p13.3, 3q29, 7p13 and 17q25.1 associated with susceptibility to pancreatic cancer. *Nat Genet* 47(8);2015:911–6.

Cho SY et al. An integrative approach to precision cancer medicine using patient-derived xenografts. *Mol Cells* 39(2);2016:77–86.

Collaborative Group on Hormonal Factors in Breast Cancer. Breast cancer and hormone replacement therapy: Collaborative reanalysis of data from 51 epidemiological studies of 52,705 women with breast cancer and 108,411 women without breast cancer. *Lancet* 350(9084);1997:1047–59.

Collaborative Group on Hormonal Factors in Breast Cancer. Familial breast cancer: Collaborative reanalysis of individual data from 52 epidemiological studies including 58,209 women with breast cancer and 101,986 women without the disease. *Lancet* 358(9291);2001:1389–99.

Cook-Deegan R, and Niehaus A. After Myriad: Genetic testing in the wake of recent Supreme Court decisions about gene patents. *Curr Genet Med Rep* 2(4);2014:223–41.

Cummings SR et al. The effect of raloxifene on risk of breast cancer in postmenopausal women: Results from the MORE randomized trial. Multiple Outcomes of Raloxifene Evaluation. *JAMA* 281(23);1999:2189–97.

Domchek SM et al. Association of risk-reducing surgery in BRCA1 or BRCA2 mutation carriers with cancer risk and mortality. *JAMA* 304(9);2010:967–75.

Druker BJ. Perspectives on the development of a molecularly targeted agent. *Cancer Cell* 1(1);2002:31–6.

Fisher B et al. Tamoxifen for the prevention of breast cancer: Current status of the National Surgical Adjuvant Breast and Bowel Project P-1 study. *J Natl Cancer Inst* 97(22);2005:1652–62.

Gollamudi J et al. Neoadjuvant therapy for early-stage breast cancer: The clinical utility of pertuzumab. *Cancer Manag Res* 8;2016:21–31.

Hagen CE, Setia N, and Lauwers GY. Familial adenomatous polyposis: A review of gastrointestinal manifestations. *Diagn Histopathol* 21(4);2015:152–60.

Hamilton W et al. Evaluation of risk assessment tools for suspected cancer in general practice: A cohort study. *Br J Gen Pract* 63(606);2013:e30–6.

Haq R, and Gulasingam P. Duration of trastuzumab in patients with HER2-positive metastatic breast cancer in prolonged remission. *Curr Oncol* 23(2);2016:91–5.

Harris NL et al. The World Health Organization classification of hematological malignancies report of the Clinical Advisory Committee Meeting, Airlie House, Virginia, November 1997. *Mod Pathol* 13(2);2000:193–207.

Heald B, Edelman E, and Eng C. Prospective comparison of family medical history with personal genome screening for risk assessment of common cancers. *Eur J Hum Genet* 20(5);2012:547–51.

Hutchinson L. Breast cancer: Challenges, controversies, breakthroughs. *Nat Rev Clin Oncol* 7(12);2010:669–70.

Jørgensen JT. Companion diagnostics: The key to personalized medicine. *Expert Rev Mol Diagn* 15(2);2015:153–6.

Klug WS et al. *Concepts of Genetics.* 10th edition. Pearson, Boston, 2012.

Kreiter S et al. Mutant MHC class II epitopes drive therapeutic immune responses to cancer. *Nature* 520(7549);2015:692–6.

Kuhn AN et al. Determinants of intracellular RNA pharmacokinetics: Implications for RNA-based immunotherapeutics. *RNA Biol* 8(1);2011:35–43.

Lawrence A, Loeb KRL, and Anderson JP. Multiple mutations and cancer. *PNAS* 100;2003:776–81.

Lee AJ et al. BOADICEA breast cancer risk prediction model: Updates to cancer incidences, tumour pathology and web interface. *Br J Cancer* 110(2);2014:535–45.

Lodish H et al. Section 24.2, Proto-oncogenes and tumor-suppressor genes. In: *Molecular Cell Biology.* 4th edition. W. H. Freeman, New York, 2000. Available from: https://www.ncbi.nlm.nih.gov/books/NBK21662/.

Marrone M et al. Multi-marker solid tumor panels using next-generation sequencing to direct molecularly targeted therapies. *PLoS Curr* 2014. http://currents.plos.org/genomictests/article/multi-marker-solid-tumor-panels-using-next-generation-sequencing-to-direct-molecularly-targeted-therapies/.

Marshall E, and Price M. U.S. Supreme Court strikes down human gene patents. *Science*. June 13, 2013.

Modan B et al. Cancer incidence in a cohort of infertile women. *Am J Epidemiol* 147(11);1998:1038–42.

Nakamura K et al. Culture-independent method for identification of microbial enzyme-encoding genes by activity-based single-cell sequencing using a water-in-oil microdroplet platform. *Sci Rep* 6;2016:22259.

National Cancer Institute. Surgery to reduce the risk of breast cancer. https://www.cancer.gov/types/breast/risk-reducing-surgery-fact-sheet. Accessed April 8, 2016.

National Cancer Institute. Targeted cancer therapies. April 25, 2014. Available from http://www.cancer.gov/about-cancer/treatment/types/targeted-therapies/targeted-therapies-fact-sheet#q1. Accessed 2016.

National Consumers League and the U.S. Food and Drug Administration (FDA). Avoid food-drug interactions. 2016. Publication no. CDER 10-1933.

Park S et al. Comparison of breast cancer risk estimations, risk perception, and screening behaviors in obese rural Korean women. *Oncol Nurs Forum* 38(6);2011:E394–401.

Patronov A, and Doytchinova I. T-cell epitope vaccine design by immunoinformatics. *Open Biol* 3(1);2013:120139. http://doi.org/10.1098/rsob.120139

Popat S, and Yap T. Toward precision medicine with next-generation EGFR inhibitors in non-small-cell lung cancer. *Pharmacogenomics Pers Med* 7;2014:285–92.

Qiagen. Overview of whole genome amplification. Available from: https://www.qiagen.com/de/resources/technologies/wga/overview-on-wga/. Accessed 2016.

Rebbeck TR et al. Bilateral prophylactic mastectomy reduces breast cancer risk in BRCA1 and BRCA2 mutation carriers: The PROSE Study Group. *J Clin Oncol* 22(6);2004:1055–62.

Restifo NP, Smyth MJ, and Snyder A. Acquired resistance to immunotherapy and future challenges. *Nat Rev Cancer* 16(2);2016:121–6.

Rothstein MA. Disclosing decedents' research results to relatives violates the HIPAA Privacy Rule. *Am J Bioeth* 12(10);2012:16–7.

Rugo HS et al. A clinician's guide to biosimilars in oncology. *Cancer Treat Rev* 46;2016:73–9.

Scott SA. Personalizing medicine with clinical pharmacogenetics. *Genet Med* 13(12);2011:987–95.

Sicklick JK et al. Genomics of gallbladder cancer: The case for biomarker-driven clinical trial design. *Cancer Metastasis Rev* 35(2);2016:263–75.

Stewart BW, and Wild CP. (Eds). *World Cancer Report 2014*. World Health Organization, Geneva, 2014.

Sturmer T et al. Aspirin use and colorectal cancer: Post-trial follow-up data from the Physicians' Health Study. *Ann Intern Med* 128(9);1998:713–20.

Swanton C. Intratumor heterogeneity: Evolution through space and time. *Cancer Res* 72(19);2012:4875–82.

Szakacs G et al. Predicting drug sensitivity and resistance: Profiling ABC transporter genes in cancer cells. *Cancer Cell* 6(2);2004:129–37.

Tefferi A, Thiele J, and Vardiman JW. The 2008 World Health Organization classification system for myeloproliferative neoplasms: Order out of chaos. *Cancer* 115(17);2009:3842–7.

Todd R, and Wong DT. Oncogenes. *Anticancer Res* 19(6a);1999:4729–46.

Ustinova M et al. Impact of KRAS variant rs61764370 on breast cancer morbidity. *Exp Oncol* 37(4);2015:292–4.

U.S. Food and Drug Administration (FDA). Homepage. 2016a. Available from http://www.fda.gov.

U.S. Food and Drug Administration (FDA). List of cleared or approved companion diagnostic devices (in nitro and imaging tools). 2016b. Available from http://www.fda.gov/MedicalDevices/ProductsandMedicalProcedures/InVitroDiagnostics/ucm301431.htm.

Van Loo P, and Voet T. Single cell analysis of cancer genomes. *Curr Opin Genet Dev* 24;2014:82–91.

Weinberg RA. *The Biology of Cancer*. Second edition. Garland Science, New York, 2014.

Weinstein IB, and Joe AK. Mechanisms of disease: Oncogene addiction—A rationale for molecular targeting in cancer therapy. *Nat Clin Pract Oncol* 3(8);2006:448–57.

Yu C et al. High-throughput identification of genotype-specific cancer vulnerabilities in mixtures of barcoded tumor cell lines. *Nat Biotechnol* 34(3);2016:419–23.

Zahary MN et al. Polymorphisms of cell cycle regulator genes G870A and C215G: Association with colorectal cancer susceptibility risk in a Malaysian population. *Oncol Lett* 10(5);2015:3216–22.

Zarin DA, Tse T, and Ross JS. Trial-results reporting and academic medical centers. *N Engl J Med* 372(24);2015:2371–2.

Zhang X et al. Single-cell sequencing for precise cancer research: Progress and prospects. *Cancer Res* 76(6);2016:1305–12.

Circulating tumor cells and circulating tumor DNA in precision medicine

ANJANA MUNSHI AND SATRUPA DAS

INTRODUCTION

Oncology is a unique area of therapeutics owing to the plethora of cancers that have emerged over several years. It gets more interesting because the mutations that drive progression of tumors also serve as the specific targets for treating cancer. Significant progress has been made in cancer research, to identify critical targets for its treatments, but unfortunately there is no universal answer that can help reduce or eliminate the burden of cancer. Several modes of treatment (chemotherapy, surgery, radiation therapy) have been developed, but procedures get complicated with advancement in stage and development of resistant tumor cells due to genetic heterogeneity. Although tumor biopsies (solid biopsy), traditional biomarkers, and imaging techniques have yielded valuable information for years, they are backed by several drawbacks due to their invasiveness and inability to capture variable gene expressions (Batth et al., 2017).

This has led researchers to look into alternative biological samples, one being "liquid biopsy," which involves screening of peripheral blood by collecting and enriching the circulating tumor cells (CTCs) and circulating tumor DNA (ctDNA). This is increasingly being considered due to its higher sensitivity, noninvasive nature, early tumor diagnosis ability, better recurrence monitoring, and therapeutic guidance (Sestini et al., 2015). The primary cause for cancer mortality is its ability of tissue invasion and metastasis, and thus the most effective therapy would be to delay or prevent this process. Therefore, an early and accurate detection of tumor status is highly advantageous, and liquid biopsy offers the potential "real-time" scenario diagnosis (Batth et al., 2017; Zhang et al., 2017).

However, despite the numerous benefits as a potential diagnostic resource, clinical utility of liquid biopsy suffers due to certain limitations. In general there is a lack of consensus on detection methods and also difficulty in analysis of overwhelming sequencing data and lack of strong proof for application in evidence-based medicine. Discovery of this biological specimen has revealed the presence of a unique cancer cell protein in CTCs, oncogenic mutations, and also epigenetic changes in CTCs and ctDNA that cause "genetic evolution of cells" to adapt to various treatment strategies (Batth et al., 2017; Zhang et al., 2017). Therefore, it is necessary to understand the role of CTCs and ctDNA to further knowledge of oncology precision medicine for diagnosis, recurrence monitoring, prognosis assessment, and medication planning of various kinds of cancers.

CHARACTERISTICS AND UTILITY OF CIRCULATING TUMOR CELLS (CTCs)

General features of CTCs

Circulating tumor cells were discovered in the 1860s by Thomas Ashworth during his microscopic examination of peripheral blood. Later, studies suggested that solid tumor cells could break and enter the bloodstream by passive and active approaches (Thiery, 2002; Tsuji et al., 2008, 2009; Joosse and Pantel, 2013; Thiery and Lim, 2013). CTCs serve as a significant prognostic factor and it has been estimated that approximately 10^6 separating tumor cells/gm of tumor mass make contact with blood, and this strongly correlates with metastasis and secondary tumor foci (Chang et al., 2000). Thus, it is rightly said that CTCs act as ambassadors of a localized disease that may remain dormant or become activated at a later time causing a relapse. As such thousands of tumor cells leak into vasculature but most get eliminated from the bloodstream and have obstacles to their survival due to shearing forces of blood flow, immune cell attack, and anoikis (Douma et al., 2004; Mitchell and King, 2013; Steinert et al., 2014). Most of them are accidental CTCs, as they are passively pushed by external forces like tumor growth, mechanical force due to surgical operation, or other kinds of friction (McDonald and Baluk, 2002). Once in circulation they persist for a short time and their morphology is highly heterogeneous. Many may be apoptotic or damaged following simple isolation techniques, whereas others appear similar in appearance to cells from matched tumor biopsies. They may also travel in clusters, ranging from two CTCs undergoing mitosis or a large microemboli having >50 cells. Proliferative index of CTCs is defined by Ki67 staining that is highly variable among different patients (Fidler, 1973; Liotta et al., 1976; Duda et al., 2010; Cho et al., 2012). Similarly, single-cell analyses have revealed heterogeneity in signaling pathways among CTCs derived from individual patients (Miyamoto et al., 2012; Powell et al., 2012; Heitzer et al., 2013a; Ozkumur et al., 2013). On average in a metastatic carcinoma patient approximately 5 to 50 CTCs for every 7.5 mL of blood can be detected, which undoubtedly leads to grave technical difficulties (Allard et al., 2004; Cristofanilli et al., 2007; Ross and Slodkowska, 2009; Hou et al., 2013).

CTC selection approaches

The primary challenge in clinical use of CTCs is their detection due to rarity (1 cell per 1×10^9 normal blood cells in patients with metastatic cancer), which makes them difficult to identify and isolate (Pantel et al., 2008). The simplest method of isolation is based on size-based membrane filters, but studies have revealed that there is considerable overlap between CTCs and leukocytes, so this method would miss a large proportion of CTCs. As a result there are multiple selection and capture methods that have been recently developed for their enrichment. Depending on detection principle, the methods have been categorized into two types: cell surface marker-dependent and marker-independent approaches.

The cell surface marker-dependent approach utilizes positive selection that relies on epithelial cell adhesion markers like EpCAM and cytokeratin (CK) (Parkinson et al., 2012). The CellSearch system (Veridex, New Jersey), a technique approved by the U.S. Food and Drug Administration (FDA), is the only method that has demonstrated the prognostic value of determining CTC numbers and has set the benchmark for CTC isolation technologies. Further, CTC-chips, having a microfluidic platform containing antibody-coated microposts (against EpCAM or MUC1) have been used to improve the capturing ability. This technology has been regarded as highly sensitive, with good yield accompanied by simplification of prelabeling processes (Nagrath et al., 2007; Maheswaran et al., 2008; Thege et al., 2014). However, the one limitation of this technology is in subsequent single-CTC analysis, which was overcome by replacing the microposts with an advanced chip that uses surface ridge or herringbone grooves in the ceiling of the channel. This was called the herringbone or [HB]CTC-chip (Stott et al., 2010a).

Similarly, there are tumors that lack epithelial markers that can be enriched by use of negative selection (CD45) and size-based methods like the ISET system or the ScreenCell approach based on density centrifugation or filtration (Powell et al., 2012; Wu et al., 2014; Kallergi et al., 2016). However, in cases of CTC metastasis there is a need of the EMT (epithelial to mesenchymal transition) process. In this, transformed tumor cells can lose epithelial markers like EpCAM and CKs, and those with epithelial markers may essentially not be the reason for cancers. Further, it is also suggested that rare primary tumor cells simultaneously expressed mesenchymal and epithelial markers, but mesenchymal cells were highly enriched in CTCs. Therefore, it is essential that CTCs be detected and allocated for different phenotypes, for example, epithelial (epithelial+/mesenchymal), complete EMT (epithelial/mesenchymal+), and intermediate EMT (epithelial+/mesenchymal+) (Yu et al., 2013; Zhang et al., 2017). Additionally, it is also found that there is heterogeneity of EpCAM expression on the surface of CTCs that can contribute to variation in detection, and EpCAM methods cannot detect nonepithelial cancers such as sarcomas (Allard et al., 2004).

To distinguish epithelial from mesenchymal cancer cells, a sophisticated method is RNA *in situ* hybridization (RNA-ISH), which differentially stains cells according to the expression levels of epithelial and mesenchymal genes (Yu et al., 2013). Yet another capture platform called CTC-iChip is virtually applicable to all cancers because it helps in isolation of EpCAM+ and EpCAM− CTCs using a series of steps. The only limitation in its widespread application is its lack of validation as compared to the CellSearch system with regard to its specificity, reproducibility, and clinical relevance.

CTC molecular characterization and cell analysis

After the capturing of CTCs they need to be characterized and analyzed for molecular properties. Baseline enumeration involves counting of cells, which barely makes use of potential information that exists within these cells and is in no way useful for oncologists. Therefore, we need methods to study

the molecular properties of CTCs by analyzing the protein, and RNA and DNA contents (Pantel et al., 2008; Yu et al., 2011). Traditional methods include use of fluorescent-conjugated antibodies for identifying cells based on positive results for pan-cytokeratin and negative for the common leukocyte antigen CD45 (Allard et al., 2004). More recently, the use of microfluidic isolation technologies has enabled the use of high-resolution light microscopy to study various cytopathological protocols along with standardized immunohistochemistry applications (Ozkumur et al., 2013). Recently, an antibody-based method for quantification of live CTCs is EPISPOT (Epithelial ImmunoSpot) assay that captures the secreting proteins such as cathepsin D, MUC1, and CK19. Isolated CTCs via negative selection using anti-CD45 immunomagnetic beads are cultured in tissue culture plates precoated with antibodies that capture the secreted protein of interest. It can be rightly called "CTC protein fingerprinting," because it differentiates between apoptotic and viable CTCs, and also identifies and differentiates between different proteins within CTCs (Alix-Panabieres, 2012; Alix-Panabieres and Pantel, 2013). The CTCs are also characterized for their expression of protein markers by fluorescence microscopy (immunofluorescence to simultaneously visualize differently labeled targets within CTCs). Promising protein-based identification involves dual Ki67/PSA staining in prostate cancer CTCs (Stott et al., 2010b). Also dual staining for androgen-induced PSA and androgen-suppressed prostate-specific membrane antigen (PSMA) markers quantitate heterogeneity in androgen signaling status of prostate CTCs before and after hormonal therapy (Miyamoto et al., 2012).

Another technology is RNA-based expression monitoring of CTCs (successful where isolation techniques do not involve formaldehyde fixation). Studies have demonstrated successful reverse-transcription polymerase chain reaction (RT-PCR) amplification of lineage-specific transcripts in CTC-enriched cell populations, with readily detectable tumor-specific translocations too (e.g., *TMPRSS2-ERG* in prostate cancer and *EML4-ALK* in non-small cell lung cancer) (Seiden et al., 1994; Xi et al., 2007; Attard et al., 2009; Stott et al., 2010a; Ozkumur et al., 2013). Apart from all these methods, the most recent application of the next-generation sequencing (NGS) in whole-genome expression profiling has been used (Yu et al., 2012, 2013). Isolation of single CTCs and their transcriptome profile has shed light on heterogeneity of CTCs. Further, the development of highly sensitive, quantifiable dual-colorimetric RNA-ISH has also contributed in understanding the epithelial versus mesenchymal transcripts (by direct visualization of the hybridization pattern within cells) within individual CTCs (Yu et al., 2012, 2013). Genotyping of CTCs is yet another explored avenue where allele-specific PCR-based assays for CTC-enriched cell populations have been demonstrated, for example, EGFR-mutant non-small cell lung cancer having high concordance between tumor biopsies at presentation and CTC-derived genotypes (Maheswaran et al., 2008; Attard et al., 2009; Heitzer et al., 2013a; Ozkumur et al., 2013).

The cytogenetic composition of CTCs can also be assessed with interphase fluorescence *in situ* hybridization (FISH) that detects copy number changes only for genomic regions. For a more genome-wide level, whole genome amplification by array-comparative genomic hybridization (array-CGH) for single or pooled CTCs can be performed (Yu et al., 2012; Heitzer et al., 2013a; Magbanua et al., 2012, 2013). Recent studies also reveal that CTC lines can be developed and kept for long-term culture; these can be used to study functional properties for invasiveness or metastases when xenografted into nude mice (Ameri et al., 2010; Zhang et al., 2013).

CTC clinical and research utility

Clinical trials for CTC count as a predictor of overall survival (OS) and progression-free survival (PFS) have been conducted in various tissues. Although studies do support the prognostic potential, there is no substantial evidence of correlation found between CTC count and pathological status. However, one study suggests CTC counts to be viable predictors of disease relapse alone and not as potential indicators of tumor characteristics, lymph node positivity, therapeutic response, or OS (Pierga et al., 2008). In contrast to this, the SWOG S0500 study confirms CTC counts to have strong prognostic value for OS (Raimondi et al., 2014; Smerage et al., 2014). Moving further from simple counting of cells, presence of CTC in peripheral blood can also be used as a measure of tracking therapeutic response (cancer cells may

develop resistance against a therapy and may recur or spread so there is a need to identify secondary mutations) in cancer patients (Allard et al., 2004; Cristofanilli et al., 2004, 2005; de Bono et al., 2008; Olmos et al., 2009; Danila et al., 2011). This can help patients who are unlikely to benefit from initial therapy to spare them from side effects and loss of time (Heitzer et al., 2013b).

The use of a CTC-chip in patients with metastatic small-cell lung cancer to detect for *EGFR* mutation states has also been reported. Monitoring of CTC cells revealed that the attainment of recurrent T790M-EGFR drug resistance mutation coincided with development of clinically refractory disease (Maheswaran et al., 2008). Similarly, use of a CTC-iChip for RNA sequencing on a breast CTC cluster led to the discovery of plakoglobin that helps in the maintenance of CTC clusters (Aceto et al., 2014). Now with the availability of genome-wide analysis strategies, use of array-CGH and NGS can reveal all possible mechanisms of resistance instead of analyzing only specific previously known mutations (Heitzer et al., 2013a). In a whole exome sequencing of two prostate cancer patients, CTCs revealed 70% overlap with mutations found in lymph node metastasis and primary tissue (Lohr et al., 2014). CTCs also showed heterogeneity in the *TP53*, *PIK3CA*, *ESR1*, and *KRAS* genes. Further, it is also suggested that we can study other relevant features of the tumor genome not present or observed during earlier initial diagnosis. This was observed in a patient where CTC examination at 34 and 24 months after diagnosis of primary tumor and liver metastasis revealed high amplification of *CDK8* not found earlier (Heitzer et al., 2013a). Such critical identification has led to identification of viable targets for therapy (in the aforementioned case use of CDK inhibitors) (Dickson et al., 2010; Wang and Ren, 2010; Ramaswamy et al., 2012).

Another use of CTCs is to understand the process of metastasis and understanding EMT process in tumor metastasis. Using an endogenous mouse pancreatic cancer model, single-molecule RNA sequencing from CTCs revealed enriched expression of Wnt2. WNT2 resulted in increased metastatic propensity also in human pancreatic cancer cells (Yu et al., 2012). Another study using xenograft assay demonstrated that primary human luminal breast cancer CTCs contain cells that give rise to metastasis in mice in various organs. The other interesting study is that of measuring the expression of mesenchymal and epithelial markers in CTCs from breast cancer patients. Monitoring of serial CTCs suggested an association of mesenchymal CTCs with disease progression. Additionally, in one patient reversible shifts between cell fates for mesenchymal and epithelial cells were associated with response to therapy and disease progression (Baccelli et al., 2013).

CHARACTERISTICS AND UTILITY OF CIRCULATING TUMOR DNA (ctDNA)

General features of ctDNA

Cell-free DNA (cfDNA) is released into circulation through various pathologic and normal physiologic mechanisms. Fragments of DNA are released into the bloodstream from dying cells during cellular turnover or from apoptotic or necrotic cells (Jahr et al., 2001; Stroun et al., 2001b). Normal physiological pathways clear these cells through infiltrating phagocytes and this leads to low levels of cfDNA. Similarly, from solid tumors, cfDNA can be released through necrosis, autophagy, and other physiologic events induced by microenvironmental stress and treatment pressure. However, unlike apoptosis, necrosis generates larger DNA fragments due to an incomplete and random digestion of genomic DNA (Wang et al., 2003). Nevertheless, not all cfDNA originate from cell death and it is observed that live cells spontaneously release newly synthesized DNA as part of a homeostatically regulated system (Anker et al., 1975; Stroun et al., 2000, 2001a, 2001b). Further, stimulation of lymphocytes also results in the release of large amounts of cfDNA in the absence of cell death (Rogers et al., 1972; Rogers, 1976; Stroun et al., 2000).

In cancer patients, a fraction of cfDNA is tumor derived and is termed ctDNA (derived from a primary tumor, metastatic lesions or CTCs) (Jen et al., 2000). Cancer patients normally have higher levels of ctDNA than healthy individuals, but the levels vary widely, from 0.01% to more than 90% (Diehl et al., 2008; Delgado et al., 2013). It is found in fragment lengths in range of 140–170 bp due to its original format as a histone-packaged nucleosome (Pixberg et al., 2015). The variability of ctDNA levels in cancer patients can be associated with tumor burden, stage, vascularity, cellular turnover, and response to therapy (Diehl et al., 2008; Kohler

et al., 2011). Somatic genetic mutations and tumor-specific alterations of cancers can be detected in ctDNA. Therefore, ctDNA carries genomic and epigenomic alterations concordant to tumor mutational spectrum, such as point mutations, degree of integrity, rearranged genomic sequences, copy number variation (CNV), microsatellite instability (MSI), loss of heterozygosity (LOH), and DNA methylation (Marzese et al., 2013). These biological characteristics differentiate ctDNA from cfDNA and qualify ctDNA as a biomarker that can help in precision medicine to detect residual disease or help in monitoring of tumor progression during therapy.

ctDNA selection approaches

Unlike CTC capture, ctDNA isolation does not depend on specialized equipment and is analyzed from plasma. Plasma is preferable to serum, and should be processed and stored promptly after whole blood collection in EDTA tubes to prevent increase in cfDNA levels due to cell lysis of normal blood cells. cfDNA can then be extracted from plasma using commercially available kits, and the analysis of ctDNA can proceed using assays designed to detect somatic genomic aberrations. ctDNA analysis is independent of EpCAM markers and reflects an average of all tumor cells releasing DNA into circulation (Heitzer et al., 2013).

ctDNA molecular characterization and cell analysis

Recent advances in genomics technologies are paving the way for the analysis of ctDNA. NGS technologies are now being used for plasma DNA analysis to allow more comprehensive detection of mutations across wider genomic regions. Types of tumor-specific aberrations that have been detected in ctDNA include somatic single nucleotide variants (SNVs), chromosomal rearrangements, and epigenetic alterations (Leon et al., 1977; Esteller et al., 1999; Silva et al., 1999; Wong et al., 1999; Lecomte et al., 2002; Hanley et al., 2006; Umetani et al., 2006; Warton and Samimi, 2015). The detection of SNVs in plasma DNA has been achieved by the use of a variety of PCR approaches (Nawroz et al., 1996; Tomita et al., 2007; Zhai et al., 2012), but digital PCR has now emerged as a sensitive analytical tool for the detection of mutations at low allele fractions (Diaz and Bardelli, 2014). Methods involving the use of digital PCR include droplet-based systems; microfluidic platforms; and the use of beads, emulsions, amplification, and magnetics (BEAMing) (Yoshimasu et al., 1999; Ito et al., 2002; Hanley et al., 2006; Heitzer et al., 2013b; Lipson et al., 2014); Targeted deep sequencing using PCR-based (e.g., TAm-Seq, Safe-Seq, Ion AmpliSeq™) or capture-based (e.g., CAPP-seq) approaches have been used to sequence specified genomic regions in plasma DNA (Leon et al., 1977; Vasioukhin et al., 1994; Shinozaki et al., 2007; Heyn and Esteller, 2012; Schwarzenbach et al., 2012). Additionally, whole-exome analysis of plasma DNA has opened up new opportunities to carefully characterize mutation profiles, without the need to focus on predefined or existing mutations (Esteller et al., 1999). Chromosomal rearrangements such as translocations or gains/losses of chromosomal regions can also be detected in ctDNA, providing excellent sensitivity and specificity as tumor biomarkers (Umetani et al., 2006; Bidard et al., 2014). Personalized analysis of rearrangement ends (PARE) is a method that involves the identification of specific somatic rearrangements in tumor tissue and the subsequent design of PCR-based assays to detect these alterations in plasma DNA (Leon et al., 1977). Moreover, in selected cases, whole-genome sequencing has now been directly applied to plasma DNA analysis to view somatic chromosomal alterations and copy number aberrations in ctDNA genome-wide (Silva et al., 1999; Wong et al., 1999; Kawakami et al., 2000).

ctDNA clinical utility

ctDNA analysis involves ease of collection, higher levels as compared to CTCs, and better medium for high-throughput analysis. Owing to this, genotyping of ctDNA can be said to be rapid, economical and reliable for clinical applications (Sausen et al., 2013; Bettegowda et al., 2014). Use of ctDNA in clinical care has been tested in cases where the presence of mutations is associated with response to targeted therapy. In Europe, regulatory labeling of Gefitinib was updated to allow use of ctDNA for assessment of activating EGFR mutations in patients where a tumor sample is not available for tests. Similarly, ctDNA use in KRAS or NRAS mutations for EGFR inhibitors in colorectal cancer proved slower to be adopted in clinical care due

to concerns over false-negative tests (Batth et al., 2017). However, efforts are underway to make use of ctDNA testing for reliable results due to its higher turnaround time and specimen availability. Tumor-specific DNA levels can be used as a surrogate of treatment response. Therefore, ctDNA analysis can act as a biomarker to reveal the tumor burden in a patient by measuring the amount of DNA released, as dying tumor cells will increase the DNA release into circulation during treatment (McBride et al., 2010; Haber and Velculescu, 2014). Additionally, it can also assist in detection of early relapse, that is, the primary tumor can be sequenced for identifying tumor-specific "driver" or "passenger" translocations that can help in sensitive monitoring for early tumor recurrence (Leary et al., 2010, 2012; Sausen et al., 2013).

DISCUSSION AND CONCLUSION

CTCs and ctDNA are capable of providing snapshots of genomic alterations of tumors. Both specimens have revolutionized the blood-based diagnostics in clinical oncology owing to their multiple applications through different stages of cancers. CTCs use highly selective approaches and represent pure tumor cell population and combined with WGA and NGS provide unique insights into tumors and their evolution at various stages of progressive cancers. Similarly, ctDNA offers a more easily obtainable specimen and analysis of real-time DNA content from tumors. Therefore, it can be said that liquid biopsy has an important contribution in the field of precision medicine by offering a non-invasive cancer monitoring technology. Although with certain limitations in clinical care as of now, it is more sensitive, accurate, and provides substantially more information when compared with traditional detection and monitoring approaches.

Focus now should be on using the combined knowledge from these specimens and the need for larger research studies. Although many studies have been carried out to examine their clinical utility most of them were retrospective. Therefore, to validate their potential, there is a need for rigorous clinical research. As these technologies mature over the years, there is hope of better diagnosis, treatment, and management of cancers.

In conclusion we can say that both biological specimens need to be studied independently and simultaneously in their applications to clinical oncology and precision medicine. Tumors evolve with time, and, hence, there is a need for repeated sampling to view the real-time tumor status for patient management. Both specimens offer their particular capabilities and limitations and with time they might become essential tools for better cancer management.

REFERENCES

Aceto N, Bardia A, Miyamoto DT et al. Circulating tumor cell clusters are oligoclonal precursors of breast cancer metastasis. *Cell* 158;2014:1110–22.

Alix-Panabieres C. EPISPOT assay: Detection of viable DTCs/CTCs in solid tumor patients. *Recent Results Cancer Res* 195;2012:69–76.

Alix-Panabieres C, and Pantel K. Circulating tumor cells: Liquid biopsy of cancer. *Clin Chem* 59;2013:110–8.

Allard WJ, Matera J, Miller MC, Repollet M, Connelly MC, Rao C, Tibbe AG, Uhr JW, and Terstappen LW. Tumor cells circulate in the peripheral blood of all major carcinomas but not in healthy subjects or patients with nonmalignant diseases. *Clin Cancer Res* 10;2004:6897–904.

Ameri K, Luong R, Zhang H, Powell AA, Montgomery KD, Espinosa I, Bouley DM, Harris AL, and Jeffrey SS. Circulating tumour cells demonstrate an altered response to hypoxia and an aggressive phenotype. *Br J Cancer* 102;2010:561–9.

Anker P, Stroun M, and Maurice PA. Spontaneous release of DNA by human blood lymphocytes as shown in an in vitro system. *Cancer Res* 35;1975:2375–82.

Attard G, Swennenhuis JF, Olmos D et al. Characterization of ERG, AR and PTEN gene status in circulating tumor cells from patients with castration-resistant prostate cancer. *Cancer Res* 69;2009:2912–8.

Baccelli I, Schneeweiss A, Riethdorf S et al. Identification of a population of blood circulating tumor cells from breast cancer patients that initiates metastasis in a xenograft assay. *Nat Biotechnol* 31;2013:539–44.

Batth IS, Mitra A, Manier S, Ghobrial IM, Menter D, Kopetz S, and Li S. Circulating tumor markers: Harmonizing the Yin and Yang of CTCs and ctDNA for precision medicine. *Ann Oncol* 28;2017:468–77.

Bettegowda C, Sausen M, Leary RJ et al. Detection of circulating tumor DNA in early- and late-stage human malignancies. *Sci Transl Med* 6;2014:224ra24.

Bidard FC, Madic J, Mariani P et al. Detection rate and prognostic value of circulating tumor cells and circulating tumor DNA in metastatic uveal melanoma. *Int J Cancer* 134;2014:1207–13.

Chang YS, di Tomaso E, McDonald DM, Jones R, Jain RK, and Munn LL. Mosaic blood vessels in tumors: Frequency of cancer cells in contact with flowing blood. *Proc Natl Acad Sci USA* 97;2000:14608–13.

Cho EH, Wendel M, Luttgen M et al. Characterization of circulating tumor cell aggregates identified in patients with epithelial tumors. *Phys Biol* 9;2012:016001.

Cristofanilli M, Broglio KR, Guarneri V et al. Circulating tumor cells in metastatic breast cancer: Biologic staging beyond tumor burden. *Clin Breast Cancer* 7;2007:471–9.

Cristofanilli M, Budd GT, Ellis MJ et al. Circulating tumor cells, disease progression, and survival in metastatic breast cancer. *N Engl J Med* 351;2004:781–91.

Cristofanilli M, Hayes DF, Budd GT et al. Circulating tumor cells: A novel prognostic factor for newly diagnosed metastatic breast cancer. *J Clin Oncol* 23;2005:1420–30.

Danila DC, Fleisher M, and Scher HI. Circulating tumor cells as biomarkers in prostate cancer. *Clin Cancer Res* 17;2011:3903–12.

de Bono JS, Scher HI, Montgomery RB, Parker C, Miller MC, Tissing H, Doyle GV, Terstappen LW, Pienta KJ, and Raghavan D. Circulating tumor cells predict survival benefit from treatment in metastatic castration-resistant prostate cancer. *Clin Cancer Res* 14;2008:6302–9.

Delgado PO, Alves BC, Gehrke Fde S et al. Characterization of cell-free circulating DNA in plasma in patients with prostate cancer. *Tumour Biol* 34;2013:983–6.

Diaz LA Jr, and Bardelli A. Liquid biopsies: Genotyping circulating tumor DNA. *J Clin Oncol* 32;2014:579–86.

Dickson MA, Shah MA, Rathkopf D et al. A phase I clinical trial of FOLFIRI in combination with the pan-cyclin-dependent kinase (CDK) inhibitor flavopiridol. *Cancer Chemother Pharmacol* 66;2010:1113–21.

Diehl F, Schmidt K, Choti MA et al. Circulating mutant DNA to assess tumor dynamics. *Nat Med* 14;2008:985–90.

Douma S, Van Laar T, Zevenhoven J, Meuwissen R, Van Garderen E, and Peeper DS. Suppression of anoikis and induction of metastasis by the neurotrophic receptor trkb. *Nature* 430;2004:1034–9.

Duda DG, Duyverman AM, Kohno M et al. Malignant cells facilitate lung metastasis by bringing their own soil. *Proc Natl Acad Sci USA* 107;2010:21677–82.

Esteller M, Sanchez-Cespedes M, Rosell R, Sidransky D, Baylin SB, and Herman JG. Detection of aberrant promoter hypermethylation of tumor suppressor genes in serum DNA from non-small cell lung cancer patients. *Cancer Res* 59;1999:67–70.

Fidler IJ. The relationship of embolic homogeneity, number, size and viability to the incidence of experimental metastasis. *Eur J Cancer* 9;1973:223–7.

Haber DA, and Velculescu VE. Blood-based analysis of cancer: Circulating tumor cells and circulating tumor DNA. *Cancer Discovery* 4;2014:650–61.

Hanley R, Rieger-Christ KM, Canes D et al. DNA integrity assay: A plasma-based screening tool for the detection of prostate cancer. *Clin Cancer Res* 12;2006:4569–74.

Heitzer E, Auer M, Gasch C et al. Complex tumor genomes inferred from single circulating tumor cells by array-CGH and next-generation sequencing. *Cancer Res* 73;2013a:2965–75.

Heitzer E, Auer M, Ulz P, Geigl JB, and Speicher MR. Circulating tumor cells and DNA as liquid biopsies. *Genome Med* 5;2013:73.

Heitzer E, Ulz P, Belic J et al. Tumor-associated copy number changes in the circulation of patients with prostate cancer identified through whole-genome sequencing. *Genome Med* 5;2013b:30.

Heyn H, and Esteller M. DNA methylation profiling in the clinic: Applications and challenges. *Nat Rev Genet* 13;2012:679–92.

Hou HW, Warkiani ME, Khoo BL, Li ZR, Soo RA, Tan DS, Lim WT, Han J, Bhagat AA, and Lim CT. Isolation and retrieval of circulating tumor cells using centrifugal forces. *Sci Rep* 3;2013:1259.

Ito K, Hibi K, Ando H et al. Usefulness of analytical CEA doubling time and half-life time for overlooked synchronous metastases in colorectal carcinoma. *Jpn J Clin Oncol* 32;2002:54–8.

Jahr S, Hentze H, Englisch S et al. DNA fragments in the blood plasma of cancer patients: Quantitations and evidence for their origin from apoptotic and necrotic cells. *Cancer Res* 61;2001:1659–65.

Jen J, Wu L, and Sidransky D. An overview on the isolation and analysis of circulating tumor DNA in plasma and serum. *Ann NY Acad Sci* 906;2000:8–12.

Joosse SA, and Pantel K. Biologic challenges in the detection of circulating tumor cells. *Cancer Res* 73;2013:8–11.

Kallergi G, Politaki E, Alkahtani S, Stournaras C, and Georgoulias V. Evaluation of isolation methods for circulating tumor cells (ctcs). *Cell Physiol Biochem* 40;2016:411–9.

Kawakami K, Brabender J, Lord RV et al. Hypermethylated APC DNA in plasma and prognosis of patients with esophageal adenocarcinoma. *J Natl Cancer Inst* 92;2000:1805–11.

Kohler C, Barekati Z, Radpour R, and Zhong XY. Cell-free DNA in the circulation as a potential cancer biomarker. *Anticancer Res* 31;2011:2623–8.

Leary RJ, Kinde I, Diehl F et al. Development of personalized tumor biomarkers using massively parallel sequencing. *Sci Transl Med* 2;2010:20ra14.

Leary RJ, Sausen M, Kinde I et al. Detection of chromosomal alterations in the circulation of cancer patients with whole-genome sequencing. *Sci Transl Med* 4;2012:162ra54.

Lecomte T, Berger A, Zinzindohoue F et al. Detection of free-circulating tumor-associated DNA in plasma of colorectal cancer patients and its association with prognosis. *Int J Cancer* 100;2002:542–8.

Leon SA, Shapiro B, Sklaroff DM, and Yaros MJ. Free DNA in the serum of cancer patients and the effect of therapy. *Cancer Res* 37;1977:646–50.

Liotta LA, Saidel MG, and Kleinerman J. The significance of hematogenous tumor cell clumps in the metastatic process. *Cancer Res* 36;1976:889–94.

Lipson EJ, Velculescu VE, Pritchard TS et al. Circulating tumor DNA analysis as a real-time method for monitoring tumor burden in mela-noma patients undergoing treatment with immune checkpoint blockade. *J Immunother Cancer* 2;2014:42.

Lohr JG, Adalsteinsson VA, Cibulskis K et al. Whole-exome sequencing of circulat-ing tumor cells provides a window into metastatic prostate cancer. *Nat Biotechnol* 32;2014:479–84.

Magbanua MJ, Sosa EV, Roy R, Eisenbud LE, Scott JH, Olshen A, Pinkel D, Rugo HS, and Park JW. Genomic profiling of isolated circulating tumor cells from metastatic breast cancer patients. *Cancer Res* 73;2013:30–40.

Magbanua MJ, Sosa EV, Scott JH, Simko J, Collins C, Pinkel D, Ryan CJ, and Park JW. Isolation and genomic analysis of circulat-ing tumor cells from castration resistant metastatic prostate cancer. *BMC Cancer* 12;2012:78.

Maheswaran S, Sequis LV, Nagrath S et al. Detection of mutations in EGFR in circu-lating lung-cancer cells. *N Engl J Med* 359;2008:366–77.

Marzese DM, Hirose H, and Hoon DS. Diagnostic and prognostic value of circulating tumor-related DNA in cancer patients. *Expert Rev Mol Diagn* 13;2013:827–44.

McBride DJ, Orpana AK, Sotiriou C et al. Use of cancer-specific genomic rearrange-ments to quantify disease burden in plasma from patients with solid tumors. *Genes Chromosomes Cancer* 49;2010:1062–9.

McDonald DM, and Baluk P. Significance of blood vessel leakiness in cancer. *Cancer Res* 62;2002:5381–5.

Mitchell MJ, and King MR. Computational and experimental models of cancer cell response to fluid shear stress. *Front Oncol* 3;2013:44.

Miyamoto DT, Lee RJ, Stott SL et al. Androgen receptor signaling in circulating tumor cells as a marker of hormonally responsive prostate cancer. *Cancer Discov* 2;2012:995–1003.

Nagrath S, Sequist LV, Maheswaran S et al. Isolation of rare circulating tumour cells in cancer patients by microchip technology. *Nature* 450;2007:1235–9.

Nawroz H, Koch W, Anker P, Stroun M, and Sidransky D. Microsatellite alterations in serum DNA of head and neck cancer patients. *Nat Med* 2;1996:1035–7.

Olmos D, Arkenau HT, Ang JE et al. Circulating tumour cell (CTC) counts as intermediate end points in castration-resistant prostate cancer (CRPC): A single-centre experience. *Ann Oncol* 20;2009:27–33.

Ozkumur E, Shah AM, Ciciliano JC et al. Inertial focusing for tumor antigen-dependent and -independent sorting of rare circulating tumor cells. *Sci Transl Med* 5;2013:179ra47.

Pantel K, Brakenhoff RH, and Brandt B. Detection, clinical relevance and specific biological properties of disseminating tumour cells. *Nat Rev Cancer* 8;2008:329–40.

Parkinson DR, Dracopoli N, Petty BG et al. Considerations in the development of circu-lating tumor cell technology for clinical use. *J Transl Med* 10;2012:138.

Pierga J-Y, Bidard F-C, Mathiot C, Brain E, Delaloge S, Giachetti S, de Cremoux P, Salmon R, Vincent-Salomon A, and Marty M. Circulating tumor cell detection predicts early metastatic relapse after neoadjuvant chemotherapy in large operable and locally advanced breast cancer in a phase II random-ized trial. *Clin Cancer Res* 14;2008:7004–10.

Pixberg C, Schulz W, Stoecklein N, and Neves R. Characterization of DNA methylation in circulat-ing tumor cells. *Genes (Basel)* 6;2015:1053–75.

Powell AA, Talasaz AH, Zhang H et al. Single cell profiling of circulating tumor cells: Transcriptional heterogeneity and diversity from breast cancer cell lines. *PLOS ONE* 7;2012:e33788.

Raimondi C, Gradilone A, Naso G et al. Clinical utility of circulating tumor cell counting through CellSearch (VR): The dilemma of a concept suspended in limbo. *Onco Targets Ther* 7;2014:619–25.

Ramaswamy B, Phelps MA, Baiocchi R et al. A dose-finding, pharmacokinetic and phar-macodynamic study of a novel schedule of flavopiridol in patients with advanced solid tumors. *Invest New Drugs* 30;2012:629–38.

Rogers JC. Identification of an intracellular pre-cursor to DNA excreted by human lympho-cytes. *Proc Natl Acad Sci USA* 73;1976:3211.

Rogers JC, Boldt D, Kornfeld S, Skinner A, and Valeri CR. Excretion of deoxyribonucleic acid by lymphocytes stimulated with phytohemag-glutinin or antigen. *Proc Natl Acad Sci USA* 69;1972:1685–9.

Ross JS, and Slodkowska EA. Circulating and disseminated tumor cells in the management of breast cancer. *Am J Clin Pathol* 132;2009:237–45.

Sausen M, Leary RJ, Jones S et al. Integrated genomic analyses identify ARID1A and ARID1B alterations in the childhood cancer neuroblastoma. *Nat Genet* 45;2013:12–7.

Schwarzenbach H, Eichelser C, Kropidlowski J, Janni W, Rack B, and Pantel K. Loss of heterozygosity at tumor suppressor genes detectable on fractionated circulating cell-free tumor DNA as indicator of breast cancer progression. *Clin Cancer Res* 18;2012:5719–30.

Seiden MV, Kantoff PW, Krithivas K et al. Detection of circulating tumor cells in men with localized prostate cancer. *J Clin Oncol* 12;1994:2634–9.

Sestini S, Boeri M, Marchiano A et al. Circulating micro RNA signature as liquid-biopsy to monitor lung cancer in low-dose computed tomography screening. *Oncotarget* 6;2015:32868–77.

Shinozaki M, O'Day SJ, Kitago M, Amersi F, Kuo C, and Kim J. Utility of circulating B-RAF DNA mutation in serum for monitoring melanoma patients receiving biochemotherapy. *Clin Cancer Res* 13;2007:2068–74.

Silva JM, Dominguez G, Villanueva MJ, Gonzalez R, Garcia JM, Corbacho C, Provencio M, España P, and Bonilla F. Aberrant DNA methylation of the p16INK4a gene in plasma DNA of breast cancer patients. *Br J Cancer* 80;1999:1262–4.

Smerage JB, Barlow WE, Hortobagyi GN et al. Circulating tumor cells and response to chemotherapy in metastatic breast cancer: SWOG S0500. *J Clin Oncol* 32(31);2014:1–8.

Steinert G, Scholch S, Niemietz T et al. Immune escape and survival mechanisms in circulating tumor cells of colorectal cancer. *Cancer Res* 74;2014:1694–704.

Stott SL, Hsu CH, Tsukrov DI et al. Isolation of circulating tumor cells using a microvortex-generating herringbone-chip. *Proc Natl Acad Sci USA* 107;2010a:18392–7.

Stott SL, Lee RJ, Nagrath S et al. Isolation and characterization of circulating tumor cells from patients with localized and metastatic prostate cancer. *Sci Transl Med* 2;2010b:25ra23.

Stroun M, Lyautey J, Lederrey C, Mulcahy HE, and Anker P. Alu repeat sequences are present in increased proportions compared to a unique gene in plasma/serum DNA: Evidence for a preferential release from viable cells? *Ann NY Acad Sci* 945;2001a:258–64.

Stroun M, Lyautey J, Lederrey C, Olson-Sand A, and Anker P. About the possible origin and mechanism of circulating DNA apoptosis and active DNA release. *Clin Chim Acta* 313;2001b:139–42.

Stroun M, Maurice P, Vasioukhin V et al. The origin and mechanism of circulating DNA. *Ann NY Acad Sci* 906;2000:161–8.

Thege FI, Lannin TB, Saha TN, Tsai S, Kochman ML, Hollingsworth MA, Rhim AD, and Kirby BJ. Microfluidic immunocapture of circulating pancreatic cells using parallel epcam and muc1 capture: Characterization, optimization and downstream analysis. *Lab Chip* 14;2014:1775–84.

Thiery JP. Epithelial-mesenchymal transitions in tumour progression. *Nat Rev Cancer* 2;2002:442–54.

Thiery JP, and Lim CT. Tumor dissemination: An EMT affair. *Cancer Cell* 23;2013:272–3.

Tomita H, Ichikawa D, Ikoma D et al. Quantification of circulating plasma DNA fragments as tumor markers in patients with esophageal cancer. *Anticancer Res* 27;2007:2737–41.

Tsuji T, Ibaragi S, and Hu GF. Epithelial-mesenchymal transition and cell cooperativity in metastasis. *Cancer Res* 69;2009:7135–9.

Tsuji T, Ibaragi S, Shima K, Hu MG, Katsurano M, Sasaki A, and Hu GF. Epithelial-mesenchymal transition induced by growth suppressor p12cdk2-ap1 promotes tumor cell local invasion but suppresses distant colony growth. *Cancer Res* 68;2008:10377–86.

Umetani N, Giuliano AE, Hiramatsu SH et al. Prediction of breast tumor progression by integrity of free circulating DNA in serum. *J Clin Oncol* 24;2006:4270–6.

Vasioukhin V, Anker P, Maurice P, Lyautey J, Lederrey C, and Stroun M. Point mutations of the N-ras gene in the blood plasma DNA of patients with myelodysplastic syndrome or acute myelogenous leukaemia. *Br J Haematol* 86;1994:774–9.

Wang BG, Huang HY, Chen YC et al. Increased plasma DNA integrity in cancer patients. *Cancer Res* 63;2003:3966–8.

Wang LM, and Ren DM. Flavopiridol, the first cyclin-dependent kinase inhibitor: Recent advances in combination chemotherapy. *Mini Rev Med Chem* 10;2010:1058–70.

Warton K, and Samimi G. Methylation of cell-free circulating DNA in the diagnosis of cancer. *Front Mol Biosci* 2;2015:13.

Wong IH, Lo YM, Zhang J et al. Detection of aberrant p16 methylation in the plasma and serum of liver cancer patients. *Cancer Res* 59;1999:71–3.

Wu S, Liu Z, Liu S, Lin L, Yang W, and Xu J. Enrichment and enumeration of circulating tumor cells by efficient depletion of leukocyte fractions. *Clin Chem Lab Med* 52;2014:243–51.

Xi L, Nicastri DG, El-Hefnawy T, Hughes SJ, Luketich JD, and Godfrey TE. Optimal markers for real-time quantitative reverse transcription PCR detection of circulating tumor cells from melanoma, breast, colon, esophageal, head and neck, and lung cancers. *Clin Chem* 53;2007:1206–15.

Yoshimasu T, Maebeya S, Suzuma T et al. Disappearance curves for tumor markers after resection of intrathoracic malignancies. *Int J Biol Markers* 14;1999:99–105.

Yu M, Bardia A, Wittner BS et al. Circulating breast tumor cells exhibit dynamic changes in epithelial and mesenchymal composition. *Science* 339;2013:580–4.

Yu M, Stott S, Toner M, Maheswaran S, and Haber DA. Circulating tumor cells: Approaches to isolation and characterization. *J Cell Biol* 192;2011:373–82.

Yu M, Ting DT, Stott SL et al. RNA sequencing of pancreatic circulating tumour cells implicates WNT signalling in metastasis. *Nature* 487;2012:510–3.

Zhai R, Zhao Y, Su L, Cassidy L, Liu G, and Christiani DC. Genome-wide DNA methylation profiling of cell-free serum DNA in esophageal adenocarcinoma and Barrett esophagus. *Neoplasia* 14;2012:29–33.

Zhang L, Ridgway LD, Wetzel MD, Ngo J, Yin W, Kumar D, Goodman JC, Groves MD, and Marchetti D. The identification and characterization of breast cancer CTCs competent for brain metastasis. *Sci Transl Med* 5;2013:180ra148.

Zhang W, Xia W, Lv Z, Xin Y, Ni C, and Yang L. Liquid biopsy for cancer: Circulating tumor cells, circulating free DNA or exosomes? *Cell Physiol Biochem* 41;2017:755–68.

3

Oral squamous cell carcinoma and liquid biopsies: A new tool for precision oncology

LAURA ÁLVAREZ RODRÍGUEZ, LAURA MUINELO ROMAY,
ABEL GARCÍA GARCÍA, RAFAEL LÓPEZ-LÓPEZ, ALICIA ÁBALO PIÑEIRO,
AND MARIO PÉREZ-SAYÁNS

ORAL SQUAMOUS CELL CARCINOMA (OSCC)

Head and neck cancers are those malignant tumors located in the area of the paranasal sinus, the nasopharynx, oropharynx (tonsils, soft palate, base of the tongue), hypopharynx, larynx, oral cavity (oral mucosa, gums, hard palate, tongue, and floor of the mouth), and salivary glands (Barnes, 2005).

The term *oral cancer* is used as a synonym for oral squamous cell carcinoma (OSCC), which accounts for 90% of all head and neck cancers and 3%–4% of malignancies (Bagan et al., 2010). It is the sixth cancer in the world in terms of incidence (approximately half a million people per year). The disease is located among the main types of solid cancers and is responsible for over 65,000 deaths in Europe every year (Wikner et al., 2014). The incidence increases with age and affects middle- to

advanced-aged adults in a higher frequency (from the fourth decade onward, with a maximum peek in the 60s). However, a preoccupying number of neoplasms have been detected in young adults (under 45 years). This disease is more prevalent in men, the proportion being 2:1 (men:women), although these figures are matching up since women are acquiring toxic habits similar to men (Ragin et al., 2007).

The etiology of this type of cancer is multifactor; the main risk factors are smoking and alcohol. In some cases, vitamin-poor diets (A and C, mainly), poor oral hygiene, infections such as Epstein-Barr or HPV 16 and 18, radiation, and the presence of genetic factors have proven to be relevant (Lifelong Learning Programme, 2015).

The most commonly used system to discover the extension of this type of head and neck cancer is TNM, which refers to the tumor size (T), the extension of the propagation to the regional lymph nodes (N), and the development of metastasis (M) (Barnes, 2005).

Patients diagnosed with this type of cancer are treated by surgery, radiotherapy, chemotherapy, or a combination of these three treatments. The choice of treatment depends on the location of the tumor, its extension, histological subtype, tumoral stage, and the patient's general condition. Surgery and radiotherapy are used independently to treat cases of nonmetastatic disease (stages I and II), but cancers in advanced stages (III and IV) require concomitant radiotherapy and chemotherapy. One third of patients have tumors in their initial phases (stages I and II), while the other two thirds show advanced stage tumors (stages III and IV). Over 50% of cancer patients suffer local recurrence and 25% develop long-distance metastasis (Figure 3.1a–d).

However, prognosis of these patients still remains poor (Zbären and Lehmann, 1987; Slootweg et al., 1992; Wheeler et al., 2014). Although the most important prognostic indicator of recurrence is metastasis of the lymph glands in the neck, the incidence of distant metastasis has dramatically increased (Shah, 1990; Forastiere et al., 2001). Despite the significant advances in diagnostics (Yongkui et al., 2013) and the therapeutic options (Teymoortash and Werner, 2012), the 5-year survival rate has remained stable at approximately 50% throughout the last decades and in all tumor phases (Cohen et al., 2004; Cooper et al., 2004). Although

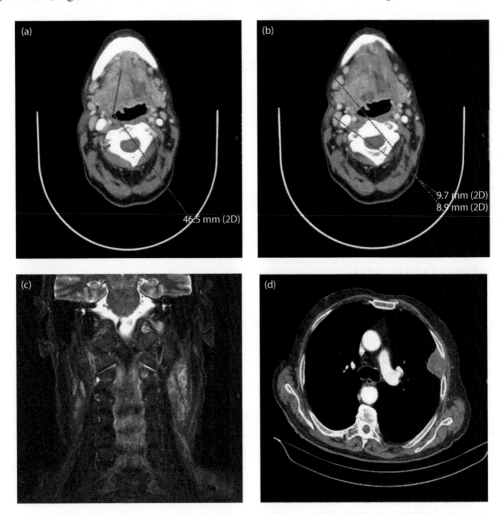

Figure 3.1 (a) Woman, 60 years old, with an OSCC located on the base of the tongue. (b) Same woman with cervical metastasis some months later. (c and d) Male patients (71 and 97 years, respectively) who, after presenting an OSCC in the floor of the mouth and lower gum, developed metastasis, cervical for the first one and thoracic for the second one.

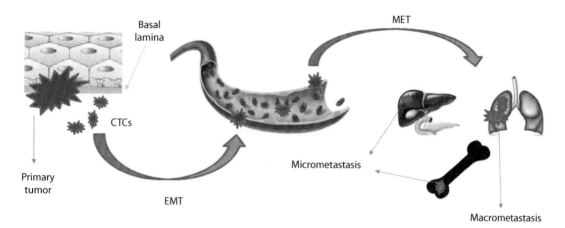

Figure 3.2 Process by which tumor cells leave the primary tumor, invade blood vessels, and are then able to extravasate to target organs and form metastases. In this process the cells begin to show the phenomenon of epithelial–mesenchymal transition (EMT) by which they lose their adhesion properties and acquire a migratory phenotype. Once they are in the blood, these cells must be able to survive and be implanted into a new tissue, so a process called mesenchymal–epithelial transition (MET) is believed necessary.

currently fewer patients suffer from locoregional recurrence, several have developed distant disease due to hidden cervical metastasis. These micrometastasis seem to contribute to an increase in mortality and morbidity (Partridge, 1999; Psychogios et al., 2013). Therefore, better methods are needed for early detection of metastatic spreading and residual tumors, and for the decision-making process in terms of individual therapeutic interventions.

CIRCULATING TUMOR CELLS (CTCs): BIOLOGICAL RELEVANCE AND TECHNOLOGICAL DEVELOPMENT

Distant metastasis is mainly the result of the hematogenous spreading of tumor cells from a primary tumor. It is known that primary tumor cells spread to other areas of the organism through the bloodstream and the lymphatic system, which are called disseminated tumor cells (DTCs), and more specifically those traveling through the bloodstream, circulating tumor cells (CTCs). When these cells reach other organs they may settle and develop a new tumor, depending on the conditions and the microenvironment of these new areas.

This process is considered very inefficient. It is estimated that only 1 of each 10^5–10^6 CTCs in peripheral blood enter into tissues at a distance from the primary tumor and only a small percentage of these cells develop a metastasis (Ring et al.,

2004) (Figure 3.2). But the appearance of the disseminated disease has dramatic consequences in cancer evolution.

The presence of CTCs was first described by Ashworth in 1869 (Ashworth, 1869; Cristofanilli et al., 2005) for women with breast cancer, but due to the lack of sensitive cytology techniques to detect these cells, their analysis had not shown clinical impact until the past two decades. One of the main problems for CTC analysis in blood is the low numbers compared to other cell types. The frequency is estimated in approximately 1 CTC per 10^6 nucleated cells in the blood of a metastatic patient; therefore we require highly sensitive isolation techniques for their analysis (Alix-Panabières, 2012). To solve this problem, different enrichment strategies have been developed to recover the CTCs present in the blood, based on the differential properties of these cells compared to the other cells present in the bloodstream (Figure 3.3). To date, there are more than 40 published technologies for the isolation of these cells, and numerous research groups are working to increase the list of devices every day (Parkinson et al., 2012). The techniques for CTC isolation can be divided into two groups: those based on physical parameters such as cell size, density, deformability, or electric properties; and those based on biological characteristics, such as invasion capacity or the expression of surface antigens (Partridge, 1999).

From among the methods based on biologic properties, CellSearch System (Janssen

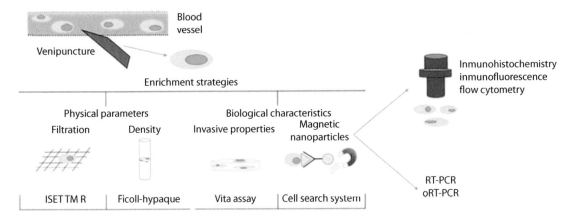

Figure 3.3 Different preanalytic enrichment strategies for further analysis of CTCs.

Diagnostics) is the most used. This system captures CTCs using magnetic nanoparticles conjugated with antibodies targeting the EpCAM surface antigen (immunomagnetic separation of EpCAM-positive tumor cells). Upon isolation, immunofluorescence is developed to detect the expression of epithelial cytokeratins (CK8, 18 and 19) and CD45 (which is only expressed in leukocytes) (Villegas et al., 2011). Following this step, an automated system analyzes the data through florescence microscopy. Nucleotic cells >4 μm in size, with positive marking for cytokeratins and negative for CD45, are considered as CTCs, always taking into account that the detection limit of the CellSearch is about 1 CTC for every 105–107 mononuclear cells (Figure 3.4).

This is the only system approved for clinical use by the U.S. Food and Drug Administration (FDA),

based on the results obtained in metastatic patients with colon, breast, and prostate cancer (Cristofanilli et al., 2004; de Bono et al., 2008; Cohen et al., 2009). The highest limitation to this technology is the low efficiency in certain tumors that do not express sufficient EpCAM levels and the low purity of the obtained CTCs, which condition the performance of postisolation molecular studies.

Another commercial system based on immuno-isolation is CellCollector (GILUPI). This medical device has shown high specificity and sensitivity for *in vivo* isolation of CTCs from patients with breast and microcytic lung cancer. CellCollector consists of a medical wire, functionalized with anti-EpCAM antibodies, that is inserted for 30 minutes into a peripheral vein, allowing the isolation of CTCs from a higher volume of blood than other conventional methods (Saucedo-Zeni et al., 2012).

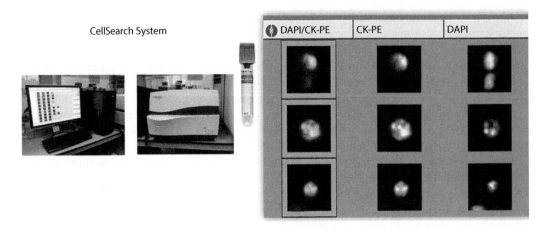

Figure 3.4 Images of the CellSearch System and the CTCs isolated from a metastatic cancer patient using this technology.

The advantage of the methods based on physical properties to isolate CTCs compared to immunological techniques is that these may be applied regardless of the antigenic tumor phenotype. Different systems are based on the size of the CTCs for their isolation, such as ISET, ScreenCell, or Metacell (Vona et al., 2000; Paterlini-Brechot and Benali, 2007; Kolostova et al., 2015). These systems are based on the use of porous membranes (pores normally <8 μm) in which tumor cells larger than 10 μm are retained. The downside of this strategy, once again, is the low purity of the CTCs after their isolation. There are also methods based on centrifugation with density gradients such as Oncoquick, using Ficoll to retain cells because of their high density (Rosenberg et al., 2002).

More recently, numerous systems based on microfluids have been developed. Such systems can combine both physical and biological strategies. These systems improve the purity of the isolation of systems such as CellSearch. CTC-chip, for example, isolated EpCAM-positive cells with higher efficiency than CellSearch passing the sample through a chip with microcolumns functionalized with anti-EpCAM antibodies (Nagrath et al., 2007; Maheswaran et al., 2008).

Once these cells are isolated, they can be detected with two strategies: using direct methods, based on immunocytochemistry, immunoflourescense (as those developed by CellSearch) and flow cytometry; and indirect methods, mainly based on the study of nucleotic acid. The most commonly used methods for indirect detection are the quantitative polymerase chain reaction (PCR) and the quantitative reverse transcription (qRT-PCR). These methods depend on the expression patterns of certain genes, the detection of known genetic mutations, amplifications, genomic aberrations, or methylation patterns in tumor cells (Lacroix, 2006). The main problem for the detection of CTCs is that there is no universally applicable marker for all tumors and all patients due to tumoral intra- and interheterogeneity.

WHY SHOULD THE STUDY OF CTCs BE INCLUDED IN PRECISION MEDICINE FOR OSCC PATIENTS?

As in most tumors, in OSCCs there is an important clinical need to be covered, which consists of the identification of early screening markers and especially prognostic and therapy-monitoring markers that allow the development of a personalized and more efficient treatment. Quantification and characterization of CTCs in patients with this tumor could be an invaluable tool to evaluate the state of the disease in a dynamic and noninvasive way.

CTC analysis, with its potential to inform us of the tumor burden, is a prognostic factor that has been validated for multiple tumors (Cristofanilli et al., 2004; de Bono et al., 2008; Cohen et al., 2008; Hou et al., 2009). In advanced head and neck tumors, a study developed using the CellSearch system by Bozec et al. (2013) detected CTCs in 16% of patients.

In locally advanced head and neck squamous carcinomas, Tinhofer et al. (2014) detected CTCs in 29% of the 144 patients under study. The number of CTCs was higher in cases with lymph node affectation and location in the tonsils or the base of the tongue, regardless of the papillomavirus status or smoking habits. The presence of lymphatic metastasis is currently the most important prognostic factor affecting the survival of head and neck cancer patients (survival can be reduced by 50% in the N1 stages) (Snow et al., 1982; Leemans et al., 1993; Jones et al., 1994); a significant percentage develop distant disease due to hidden cervical metastasis. There are differing opinions in regard to the impact of CTC presence on the development of regional or distant metastasis. The lymph node affectation (the most determining factor in the prognosis and survival of head and neck cancer patients) has been related to the presence of CTCs in peripheral blood by some authors. Hristozova et al. (2011) defend that there is a significant relationship between the number of CTCs and the development of locorregional metastasis in head and neck carcinomas, despite CTC-positive cases observed in patients with N0/N1. Their frequency was significantly higher in patients with stage N2b or higher. However, studies by Grobe et al. (2014) and Jatana et al. (2010) found no association between the presence of CTCs and locorregional metastasis in OSCC. In addition, Grobe et al. (2014) and Nichols et al. (2012) established that the presence of distant metastasis is directly correlated to the presence of CTCs in OSCC and head and neck carcinomas. In Nichols's study, 40% of patients with pulmonary nodes affected had CTCs. Supporting the value of monitoring the CTC levels as a prognostic marker, Jatana et al. (2010) determined in a study with 48 head and neck squamous cell carcinoma (HNSCC) patients that those without CTCs presented higher progression-free survival (PFS).

In a recent study using CellSearch, Bozec et al. (2013) found that CTCs were detected in 3/9 patients (33%) and 8/49 (16%), respectively, in advanced-stage HNSCC patients. He et al. (2013) reported that CTCs were present in patients presenting with nodal stage 2 and 3 rather than 0 or 1. In addition, Nichols et al. (2012), reported that CTCs were detected in 6/15 (40%) of HNSCC patients with suspected metastasis. In regard to the value of these cells as a prognostic marker, in the study of oral tumors developed by Tinhofer et al. (2014) they found no association between CTCs and PFS, but they found this association in tumors not located in the oral cavity, being the rate of PFS was lower in CTC-positive cases. In another study analyzing head and neck tumors, the correlation between the presence of CTCs and the risk of local and distant recurrence and even the patients' response to treatment has been described (Hristozova et al., 2011; Buglione et al., 2012).

All of these results show the potential of CTC quantification to help oncologists detect tumor dissemination and predict a patient's prognosis in a more realistic way. Despite this, more research is needed to clarify the entire clinical potential of CTCs in these types of tumors.

WHY IS PERSONALIZED AND PRECISION MEDICINE THROUGH CTC STUDY NECESSARY? DOES IT HAVE THE POTENTIAL TO IMPROVE THE HEALTH OF OSCC PATIENTS?

The ability to detect CTCs has a clear impact on the prognosis and treatment of cancer patients, providing clear evidence of the dissemination of the primary cancer and being a risk factor for future development of metastasis in addition to a peripheral marker to predict treatment susceptibility and disease surveillance.

The chance to isolate OSCC patients' CTCs at different moments throughout the evolution of the disease allows for the perspective of the tumor in real time that would avoid inefficient over-treatments and secondary effects as well as the expenses derived from these treatments. In this sense, the possibility to molecularly characterize OSCC tumors is especially interesting using CTCs, especially when patients undergo surgery for the

primary tumor and generate metastasis that cannot be biopsied in many cases. In this sense, the group of Tinhofer et al. (2012) determined using flow cytometry that the number of CTCs of head and neck cancer patients increased after radiotherapy, although they were unable to determine if this increase had clinical impact. In CTC-positive patients, the presence of epidermal growth factor receptor (EGFR) was detected in 100% of cases and the p-EGFR form in 55%. Additionally, combined treatment of the anti-EGFR antibody Cetuximab in combination with radiotherapy reduced the receptor-active CTC levels, in comparison with the treatment based on cisplatin/5-fluorouracil. This study showed the potential to monitor the molecular alterations in CTCs to guide OSCC treatments. For example, in monitoring p-EGFR levels in CTCs, we could say that Cetuximab is reaching its therapeutic target efficiently. Although currently Cetuximab is the only target therapy approved for use on head and neck cancer patients, other therapies are emerging as good alternatives. Such is the case of Rapamycin to block the PI3 K pathway. The status of this gene has been analyzed in the CTC population in other tumors (Schneck et al., 2013) and this strategy could also be applied to OSCC.

Another example of how molecular characterization of a CTC can help in the diagnosis and therapies selection is the possibility of analyzing the viral origin of oral carcinomas. The group of Weismann et al. (2009) determined the presence of E6/E7 HR-HPV 16 or HR-HPV 18 variants in isolated CTCs of patients with cervical carcinoma. HPV-negative head and neck squamous carcinomas are related to p53 mutations and drops in p16 levels, whereas those virus-induced tumors are characterized by a p53 wild type with an increase of both p16 and pRb levels, which makes these tumors two different entities from a molecular point of view (zur Hausen, 2000; Beachler et al., 2012). There are also multiple studies demonstrating that HPV-induced tumors have a better prognosis than those that are not HPV-induced being in a similar disease stage (Vermorken et al., 2014).

One of the main characteristics of high aggressive head and neck carcinomas are the expression of mesenchymal markers. For epithelial tumor cells, for migration through the basal lamina and invasion into other tissues, an epithelial to mesenchymal transition or EMT is necessary. This process is a dynamic mechanism through which epithelial cells increase their mobility and cell plasticity due to the loss of epithelial markers such as E-cadherin or EpCAM and the increase of mesenchymal markers' expression, such as N-cadherin and vimentin. It is important to identify those tumors with a more mesenchymal phenotype, since these will be less sensitive to different therapies and will have higher chances of relapse after surgery. In this sense, different studies have characterized the EMT phenomenon in CTCs (Alonso-Alconada et al., 2014; Barriere et al., 2014; Lim et al., 2014) and described the expression of mesenchymal markers in CTCs.

The only study developed on OSCC to characterize the EMT process in CTCs was developed on a cohort of 20 patients diagnosed with oral squamous cell carcinoma by the Maxillofacial Surgery Service of the Complejo Hospitalario Universitario de Santiago de Compostela. An amount of 7.5 mL of peripheral blood was extracted from each patient before surgery to extract the tumor. Collected samples were processed using the CELLection™ Epithelial Enrich kit (Invitrogen, Dynal, Oslo, Norway) to isolate CTCs based on EpCAM expression. The CTC population was analyzed for the expression of 12 markers related to the process of EMT (E-cadherin, HIFT1A, KRT 8, KRT 19, KRT 20, SNAIL1, SNAIL2, TGF-β1, TIMP1, VIM, ZEB1 and ZEB 2). Our study's data, which have not been previously published, demonstrated that a combination of immunoisolation based on the presence of EpCAM, and a gene expression analysis by RT-qPCR is a valid strategy for the detection and molecular characterization of CTCs in OSCC patients. Additionally, TGFβ 1, VIM, and ZEB 1 genes (related to the EMT) were identified as potential CTC markers in OSCC patients, being their expression levels are statistically different than ones obtained in healthy controls. In regard to this result, Buglione et al. (2012) related the low CTC frequency in head and neck tumors located in the oral cavity to the highest expression of mesenchymal markers in these locations, making CTC isolation based on EpCAM inefficient.

CHALLENGES AND RESEARCH OPPORTUNITIES FOR THE CLINICAL INCLUSION OF CTC ANALYSIS TO MANAGE OSCC

Nowadays it is completely accepted that the most direct path to develop precision medicine is based on liquid biopsy study. Despite this evidence, we must face several challenges before CTC analysis could be a clinical reality for OSCC patient management.

First, it is necessary to validate the results obtained with systems such as CellSearch in larger and more homogeneous groups of patients, providing more robust results in terms of prognostic value of the CTC monitoring. Furthermore, there is a scarce amount of studies that monitor the CTC levels through the different stages of the disease, at a basal level, after surgery or after the treatment with chemo- or radiotherapy. Undoubtedly, in addition to the possibility of determining tumor dissemination in a more precise way than imaging tests, the option of using CTC levels to monitor the disease would allow for a qualitative leap in the management of metastatic OSCC.

On the other hand, as mentioned earlier, there is a technical limitation to studying CTCs in OSCC residing in the heterogeneous expression of epithelial markers, such as EpCAM, which are not always sufficient to allow the CTCs isolation in an efficient manner. Therefore, it is very important to validate and develop new isolation and detection techniques that provide higher detection sensitivity. In this sense, new microfluidics-based technologies for the enrichment of CTCs and new detection methods, such as biosensors, offer a promising opportunity to make CTC detection systems of small dimensions and provide ease of use that can be applied at health centers and hospitals without requiring large facilities and significant financial costs.

In parallel, it is important to start applying high-resolution molecular techniques such as massive sequencing of genomes and exomes to characterize the CTCs of OSCC patients. These techniques are being used in other types of tumors to identify the molecular bases that are behind resistance to therapies and are enabling the development of single-cell studies that provide a better understanding of the phenomenon of tumor heterogeneity.

Last, the greatest challenge in the field of CTCs consists of their isolation and *ex vivo* culture. Due to the reduced number of CTCs in circulation and the mechanical damage suffered during the isolation process and their limited proliferation capacity, there are very few groups that have been able to obtain stable *in vitro* cultures. The generation of these cultures from OSCC patients would allow us to develop functional studies to identify and validate new therapeutic targets making *in vitro* screening of drugs the most efficient therapy for each patient in each stage of the disease.

CONCLUSION

Despite the great advances in terms of therapeutic alternatives and the surge of new monitoring and predictive markers in the past two decades, OSCC tumors still have an important void to be filled in terms of clinical tools to develop precision oncology. The analysis and characterization of the CTC population that is escaping from the primary tumor and metastasis are a true opportunity to learn more about the molecular events behind carcinogenesis and aggressiveness of OSCC tumors, as well as the best way to monitor patients with systemic disease. Although different studies support the clinical usefulness of the CTC study to reach precision oncology in OSCC, there is a higher commitment to the development of a promising work field that may provide knowledge and technology to bring clinicians a real tool for personalized medicine.

REFERENCES

Alix-Panabières C. EPISPOT assay: Detection of viable DTCs/CTCs in solid tumor patients. In M Ignatious, C Sotiriou, and K Pantel (Eds), *Minimal Residual Disease and Circulating Tumor Cells in Breast Cancer*, pp. 69–76. Berlin, Springer; 2012.

Alonso-Alconada L, Muinelo-Romay L, Madissoo K et al. Molecular profiling of circulating tumor cells links plasticity to the metastatic process in endometrial cancer. *Mol Cancer* 13;2014:223.

Ashworth T. A case of cancer in which cells similar to those in the tumours were seen in the blood after death. *Aust Med J* 14(3);1869:146–9.

Bagan J, Sarrion G, and Jimenez Y. Oral cancer: Clinical features. *Oral Oncol* 46(6);2010:414–7.

Barnes L. *Pathology and Genetics of Head and Neck Tumours.* Lyon, IARC Press; 2005.

Barriere G, Fici P, Gallerani G, Fabbri F, Zoli W, and Rigaud M. Circulating tumor cells and epithelial, mesenchymal and stemness markers: Characterization of cell subpopulations. *Ann Transl Med* 2(11);2014:109.

Beachler DC, Weber KM, Margolick JB et al. Risk factors for oral HPV infection among a high prevalence population of HIV-positive and at-risk HIV-negative adults. *Cancer Epidemiol Biomarkers Prev* 21(1);2012:122–33.

Bozec A, Peyrade F, Hebert C et al. Significance of circulating tumor cell detection using the CellSearch system in patients with locally advanced head and neck squamous cell carcinoma. *Eur Arch Otorhinolaryngol* 270(10);2013:2745–9.

Buglione M, Grisanti S, Almici C et al. Circulating tumour cells in locally advanced head and neck cancer: Preliminary report about their possible role in predicting response to non-surgical treatment and survival. *Eur J Cancer* 48(16);2012:3019–26.

Cohen EE, Lingen MW, and Vokes EE. The expanding role of systemic therapy in head and neck cancer. *J Clin Oncol* 22(9);2004:1743–52.

Cohen SJ, Punt CJ, Iannotti N et al. Relationship of circulating tumor cells to tumor response, progression-free survival, and overall survival in patients with metastatic colorectal cancer. *J Clin Oncol* 26(19);2008:3213–21.

Cohen SJ, Punt CJ, Iannotti N et al. Prognostic significance of circulating tumor cells in patients with metastatic colorectal cancer. *Ann Oncol* 20(7);2009:1223–9.

Cooper JS, Pajak TF, Forastiere AA et al. Postoperative concurrent radiotherapy and chemotherapy for high-risk squamous-cell carcinoma of the head and neck. *N Engl J Med* 350(19);2004:1937–44.

Cristofanilli M, Budd GT, Ellis MJ et al. Circulating tumor cells, disease progression, and survival in metastatic breast cancer. *N Engl J Med* 351(8);2004:781–91.

Cristofanilli M, Hayes DF, Budd GT et al. Circulating tumor cells: A novel prognostic factor for newly diagnosed metastatic breast cancer. *J Clin Oncol* 23(7);2005:1420–30.

de Bono JS, Scher HI, Montgomery RB et al. Circulating tumor cells predict survival benefit from treatment in metastatic castration-resistant prostate cancer. *Clin Cancer Res* 14(19);2008:6302–9.

Forastiere A, Koch W, Trotti A, and Sidransky D. Head and neck cancer. *N Engl J Med* 345(26);2001:1890–1900.

Grobe A, Blessmann M, Hanken H et al. Prognostic relevance of circulating tumor cells in blood and disseminated tumor cells in bone marrow of patients with squamous cell carcinoma of the oral cavity. *Clin Cancer Res* 20(2);2014:425–33.

He S, Li P, Long T et al. Detection of circulating tumour cells with the CellSearch system in patients with advanced-stage head and neck cancer: Preliminary results. *J Laryngol Otol* 127;2013:788.

Hou J, Greystoke A, Lancashire L et al. Evaluation of circulating tumor cells and serological cell death biomarkers in small cell lung cancer patients undergoing chemotherapy. *Am J Pathol* 175(2);2009:808–16.

Hristozova T, Konschak R, Stromberger C et al. The presence of circulating tumor cells (CTCs) correlates with lymph node metastasis in non-resectable squamous cell carcinoma of the head and neck region (SCCHN). *Ann Oncol* 22(8);2011:1878–85.

Jatana KR, Balasubramanian P, Lang JC et al. Significance of circulating tumor cells in patients with squamous cell carcinoma of the head and neck: Initial results. *Arch Otolaryngol Head Neck Surg* 136(12);2010:1274–9.

Jones A, Roland N, Field J, and Phillips D. The level of cervical lymph node metastases: Their prognostic relevance and relationship with head and neck squamous carcinoma primary sites. *Clin Otolaryngol Allied Sci* 19(1);1994:63–69.

Kolostova K, Spicka J, Matkowski R, and Bobek V. Isolation, primary culture, morphological and molecular characterization of circulating tumor cells in gynecological cancers. *Am J Transl Res* 7(7);2015:1203.

Lacroix M. Significance, detection and markers of disseminated breast cancer cells. *Endocr Relat Cancer* 13(4);2006:1033–67.

Leemans CR, Tiwari R, Nauta J, Van der Waal I, and Snow GB. Regional lymph node involvement and its significance in the development of distant metastases in head and neck carcinoma. *Cancer* 71(2);1993:452–6.

Lifelong Learning Programme. Cáncer Oral. *Prevención y diagnóstico precoz.* 2015.

Available at http://oralcancerldv.org/es/. Accessed October 6, 2015.

Lim SH, Becker TM, Chua W, Ng WL, de Souza P, and Spring KJ. Circulating tumour cells and the epithelial mesenchymal transition in colorectal cancer. *J Clin Pathol* 67(10);2014:848–53.

Maheswaran S, Sequist LV, Nagrath S et al. Detection of mutations in EGFR in circulating lung-cancer cells. *N Engl J Med* 359(4);2008:366–77.

Nagrath S, Sequist LV, Maheswaran S et al. Isolation of rare circulating tumour cells in cancer patients by microchip technology. *Nature* 450(7173);2007:1235–9.

Nichols AC, Hammond A, Palma DA et al. Detection of circulating tumor cells in advanced head and neck cancer using the CellSearch system. *Head Neck* 34(10);2012:1440–4.

Parkinson DR, Dracopoli N, Petty BG et al. Considerations in the development of circulating tumor cell technology for clinical use. *J Transl Med* 10(1);2012:138.

Partridge M. Head and neck cancer and precancer: Can we use molecular genetics to make better predictions? *Ann R Coll Surg Engl* 81(1);1999:1–11.

Paterlini-Brechot P, and Benali NL. Circulating tumor cells (CTC) detection: Clinical impact and future directions. *Cancer Lett* 253(2);2007:180–204.

Psychogios G, Mantsopoulos K, Bohr C, Koch M, Zenk J, and Iro H. Incidence of occult cervical metastasis in head and neck carcinomas: Development over time. *J Surg Oncol* 107(4);2013:384–7.

Ragin CC, Modugno F, and Gollin SM. The epidemiology and risk factors of head and neck cancer: A focus on human papillomavirus. *J Dent Res* 86(2);2007:104–14.

Ring A, Smith IE, and Dowsett M. Circulating tumour cells in breast cancer. *The Lancet Oncology* 5(2);2004:79–88.

Rosenberg R, Gertler R, Friederichs J et al. Comparison of two density gradient centrifugation systems for the enrichment of disseminated tumor cells in blood. *Cytometry* 49(4);2002:150–8.

Saucedo-Zeni N, Mewes S, Niestroj R et al. A novel method for the *in vivo* isolation of circulating tumor cells from peripheral blood of cancer patients using a functionalized and structured medical wire. *Int J Oncol* 41(4);2012:1241–50.

Schneck H, Blassl C, Meier-Stiegen F et al. Analysing the mutational status of PIK3CA in circulating tumor cells from metastatic breast cancer patients. *Molr Oncol* 7(5);2013:976–86.

Shah JP. Patterns of cervical lymph node metastasis from squamous carcinomas of the upper aerodigestive tract. *Am J Surg* 160(4);1990:405–9.

Slootweg PJ, Bolle CW, Koole R, and Hordijk G. Cause of death in squamous cell carcinoma of the head and neck: An autopsy study on 31 patients. *J Craniomaxillofac Surg* 20(5);1992:225–7.

Snow G, Annyas A, Ev S, Bartelink H, and Hart A. Prognostic factors of neck node metastasis. *Clin Otolaryngol Allied Sci* 7(3);1982:185–92.

Teymoortash A, and Werner J. Current advances in diagnosis and surgical treatment of lymph node metastasis in head and neck cancer. *GMS Curr Top Otorhinolaryngol Head Neck Surg* 11;2012:Doc04.

Tinhofer I, Hristozova T, Stromberger C, Keilholz U, and Budach V. Monitoring of circulating tumor cells and their expression of EGFR/phospho-EGFR during combined radiotherapy regimens in locally advanced squamous cell carcinoma of the head and neck. *Int J Radiat Oncol Biol Phys* 83(5);2012:e685–90.

Tinhofer I, Konschak R, Stromberger C et al. Detection of circulating tumor cells for prediction of recurrence after adjuvant chemoradiation in locally advanced squamous cell carcinoma of the head and neck. *Ann Oncol* 25(10);2014:2042–7.

Vermorken JB, Psyrri A, Mesia R et al. Impact of tumor HPV status on outcome in patients with recurrent and/or metastatic squamous cell carcinoma of the head and neck receiving chemotherapy with or without cetuximab: Retrospective analysis of the phase III EXTREME trial. *Ann Oncol* 25(4);2014:801–7.

Villegas LFS, Pérez NMS, Lázaro MV, de las C, María Luisa M, Fernández SR, and de Castro SV. Actualidad y futuro en las técnicas de cuantificiación de células tumorales circulantes: su importancia en tumores sólidos epiteliales [Current and future techniques in the quantification of circulating tumor cells: Their importance in solid epithelial tumors]. *Revista del Laboratorio Clínico* [*Journal of the Clinical Laboratory*] 4(3);2011:163–9.

Vona G, Sabile A, Louha M et al. Isolation by size of epithelial tumor cells: A new method for the immunomorphological and molecular characterization of circulating tumor cells. *Am J Pathol* 156(1);2000:57–63.

Weismann P, Weismanova E, Masak L et al. The detection of circulating tumor cells expressing E6/E7 HR-HPV oncogenes in peripheral blood in cervical cancer patients after radical hysterectomy. *Neoplasma* 56(3);2009:230–8.

Wheeler SE, Shi H, Lin F et al. Enhancement of head and neck squamous cell carcinoma proliferation, invasion, and metastasis by tumor-associated fibroblasts in preclinical models. *Head Neck* 36(3);2014:385–92.

Wikner J, Gröbe A, Pantel K, and Riethdorf S. Squamous cell carcinoma of the oral cavity and circulating tumour cells. *World J Clin Oncol* 52;2014:114.

Yongkui L, Jian L, and Jingui L. 18 FDG-PET/CT for the detection of regional nodal metastasis in patients with primary head and neck cancer before treatment: A meta-analysis. *Surg Oncol* 22(2);2013:e11–16.

Zbären P, and Lehmann W. Frequency and sites of distant metastases in head and neck squamous cell carcinoma: An analysis of 101 cases at autopsy. *Arch Otolaryngol Head Neck Surg* 113(7);1987:762–4.

zur Hausen H. Papillomaviruses causing cancer: Evasion from host-cell control in early events in carcinogenesis. *J Natl Cancer Inst* 92(9);2000:690–8.

4

Precision medicine for brain gliomas

YUSUF IZCI

BACKGROUND

Introduction

Gliomas are the most common primary intracranial tumors, representing about 80% of malignant brain tumors (Zong et al., 2012; Rajesh et al., 2017). Three different types of gliomas have been described in the brain as astrocytoma, oligodendroglioma, and ependymoma (Izci, 2014). Astrocytoma is the most frequent histological type of glioma and arises from the astrocytes (Figures 4.1 and 4.2) (Zong et al., 2012). About 10% of the gliomas are oligodendrogliomas (Figure 4.3). Mixed gliomas, primarily oligoastrocytomas, account for about 5%–10% of all gliomas. Ependymomas arise from the ependymal cells, which lie in the ventricular system and the central canal of the spinal cord. It is relatively rare in adults, accounting for 2%–3% of primary brain tumors, but it is frequent in children. Today, biological markers help pathologists separate oligodendrogliomas from other types of gliomas. Glioma has a poor prognosis (Perry and Wesseling, 2016). Although relatively rare, it causes significant mortality and morbidity.

Glioblastoma is the most common and malignant histological type of glioma (approximately 45% of all gliomas) (Akay et al., 2002; Zong et al., 2012). It may be solitary or multicentric in the brain (Izci et al., 2005). Glioblastoma may present with different clinical and radiological characteristics and has a 5-year relative survival of about 5% (Akay et al., 2002; Izci et al., 2005; Bondy et al., 2008; de Robles et al., 2015).

Prevalence

Complete prevalence proportions of brain gliomas estimate the total number of people living with glioma and include both newly diagnosed cases within the year and patients surviving during previous years (Porter et al., 2010). Complete prevalence of glioma reflects the complex relationships among incidence, survival, and population demographics of this disease (Ostrom et al., 2016; Zhang et al., 2016). Therefore, it provides important information about the total burden of brain gliomas in a population. So, the estimation of complete prevalence provides critical information about gliomas for public health and health care planning,

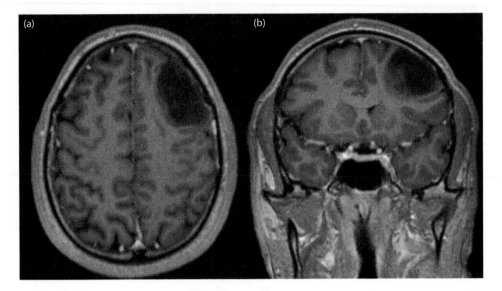

Figure 4.1 (a) Axial and (b) coronal magnetic resonance images of a patient with grade II astrocytoma (low grade).

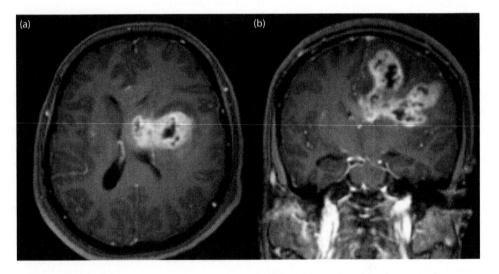

Figure 4.2 (a) Axial and (b) coronal magnetic resonance images of a patient with grade IV astrocytoma (glioblastoma).

including treatment protocols, decision making, and funding research for gliomas (de Robles et al., 2015). With increasing focus on precision medicine and the incorporation of molecular parameters into the World Health Organization (WHO) 2016 central nervous system (CNS) tumor classification, it is critical that the data are complete, and factors collected at the population level are fully integrated into brain glioma reporting (Louis et al., 2016). It is

also critical that glioma registries continue to collect established and emerging prognostic and predictive factors for this important disease (Ostrom et al., 2016).

The most prevalent types of brain glioma varied by age at prevalence. The most prevalent histological types are glioblastoma (6.46 per 100,000), diffuse astrocytoma (3.76 per 100,000), and pilocytic astrocytoma (3.71 per 100,000). The most

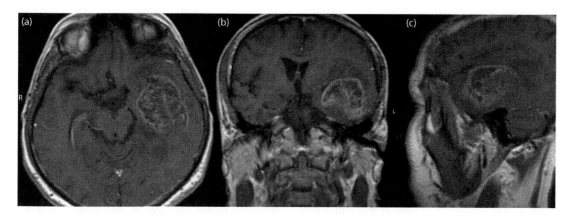

Figure 4.3 (a) Axial, (b) coronal, and (c) sagittal magnetic resonance images of a patient with oligodendroglioma grade II.

prevalent tumor types by age are childhood pilocytic astrocytoma (30.6% of cases, prevalence 6.82 per 100,000), adolescent and young adult pilocytic astrocytoma (12.2% of cases, prevalence 5.92 per 100,000), and adult glioblastoma (22.1% of cases, prevalence 12.76 per 100,000) (Ostrom et al., 2016). Even though brain tumors are not the most prevalent cancer in all age groups, these tumors are consistently prevalent throughout life and therefore merit documentation (Ostrom et al., 2014).

Incidence

The incidence rate of a disease can be defined as the number of at-risk cases per population in a specific time period. The average annual age-adjusted incidence rate of brain tumors in the United States between 2007 and 2011 was 7.25 per 100,000 persons (Ostrom et al., 2014). Many different organizations track the incidence of brain gliomas. This can be found using data collected through government cancer surveillance (i.e., statewide or countrywide cancer registries) or through the use of health system records. Incidence rates of brain glioma vary significantly by histological type, age at diagnosis, gender, and country (Bondy et al., 2008; Porter et al., 2010). It is difficult to compare the incidence rates of brain gliomas from different sources because of the lack of consistent definition of glioma and various glioma histological types as well as differences in data collection techniques. Overall age-adjusted incidence rates (adjusted to the national population of each respective study) for all types of gliomas range from 4.67 to 5.73 per 100,000 persons. Age-adjusted incidence of glioblastoma (WHO grade IV), the most common and most deadly glioma subtype in adults, ranges from 0.59 to 3.69 per 100,000 persons (Porter et al., 2010; Ostrom et al., 2014).

Low-grade glioma accounts for 20% of all gliomas, with most patients presenting in the fourth decade (Hayhurst, 2017). Men are more often affected than women, and approximately 80% of patients present with seizures (Pouratian and Schiff, 2010; Perry and Wesseling, 2016; Bush and Butowski, 2017). The incidence of low-grade glioma is uncertain as International Statistical Classification of Diseases codes do not differentiate grade of tumor. However, the RARECARE European registry project estimates the overall incidence across Europe of astrocytoma to be 4.8 per 100,000 and of oligodendroglioma to be 0.4 per 100,000 (Crocetti et al., 2012).

Survival and mortality

Standard treatment (surgery, radiotherapy, and chemotherapy) in glioblastoma has very limited effectiveness, with median overall survival of patients no longer than 15 months (Akay et al., 2002; Izci, 2014). The most important prognostic factors for glioblastoma are extent of tumor resection, age at diagnosis, and Karnofsky performance status of the patient (Walid, 2008). Survival also varies significantly by grade across all glioma subtypes. Many groups that track the incidence of glioma also track the proportion of persons who survive set periods of time after their diagnoses.

Pilocytic astrocytoma (grade I) has the highest relative survival. Glioblastoma has the poorest overall survival, with only 0.05%–4.7% of patients surviving 5 years after the diagnosis. In general, gliomas with an oligodendroglial component have increased survival, as opposed to those with an astrocytic component. Age is significantly associated with survival after diagnosis for all types of gliomas, but the effect is most pronounced for glioblastoma. Recently, it was shown in population-based parallel cohorts of diffuse low-grade gliomas that early surgical resection was associated with better overall survival than were biopsy and watchful waiting. The 22981/26981 trial by the European Organisation for Research and Treatment of Cancer/National Cancer Institute of Canada demonstrated a survival benefit for glioblastoma patients who received concurrent temozolomide with postoperative radiation, with median survival of 14.6 months for those receiving concurrent therapy versus 12.1 months for those who received radiotherapy alone (Stupp et al., 2005). This treatment regimen, known as the "Stupp protocol," was the result of this trial and was first presented in 2004. In the years since this trial was completed, it has been established as the standard of care for primary glioblastoma. For various reasons—including tolerance of chemotherapy, access to chemotherapeutic agents, and overall performance status—not all persons with glioblastoma receive this regimen. This result was then confirmed in a large study of glioblastoma patients, and several analyses found statistically significant trends in increasing survival for glioblastoma after this development, especially in those who received surgery followed by radiation. There has been an increasing trend in survival from oligodendroglioma, which is also attributed to improvements in diagnosis and treatment.

Precision medicine for brain gliomas means matching the right patient to the right treatment at the right time (Halkin, 2015). This requires gaining a deeper understanding of the specific molecular characteristics that are driving a patient's tumor growth and finding the right treatments to target those specific molecular abnormalities that are responsible for the disease. So, sequencing the patient's genome and tracking the changes in a billion biomarkers, including RNA, proteins, metabolites, and antibodies, to predict and validate disease risk are the basis of precision medicine for brain gliomas (Izci, 2014).

CURRENT APPROACHES FOR BRAIN GLIOMAS

Current management techniques for low- and high-grade brain gliomas are different. The traditional approach in low-grade gliomas was to "watch, wait and rescan," with intervention at the time of tumor progression, which often heralds change to a malignant glioma (Hayhurst, 2017). However, recent advances in our knowledge of the biology of low-grade gliomas, advances in surgical techniques, and learning the long-term impact of adjuvant treatments have caused a shift toward aggressive early treatment of low-grade glioma, aiming to delay the time to malignant transformation, to increase overall survival, and to improve neurological functional outcome.

The role of surgery in the management of low-grade glioma is still controversial. Because they have a predilection for eloquent regions, the risk of developing new neurological deficits had tempered enthusiasm for radical resection of these tumors (Pouratian and Schiff, 2010). However, there is increasing evidence that resection of low-grade glioma improves overall survival of the patient and importantly delays the time to malignant progression of the disease. Contemporary data supporting early surgery and the widespread adoption of awake craniotomy with intraoperative functional imaging and mapping have significantly changed the management of low-grade gliomas.

The standard treatment for high-grade gliomas starts with surgical procedures aiming to remove the tumor as much as possible without injuring the normal brain tissue (Akay et al., 2002). Then, radiotherapy and chemotherapy protocols including temozolomide follow the surgical treatment (Stupp et al., 2005). Although this protocol is effective at shrinking the tumors, it may cause many side effects that significantly affect the quality of life of these patients. At this point, it is critical to identify genetic and molecular markers of the tumor in order to determine the targets for personalized treatments (Ataman et al., 2004; Tan et al., 2016). Despite various treatment modalities for high-grade gliomas, the patients have a dismal prognosis (Walid, 2008).

IMPACT OF PRECISION MEDICINE ON THE TREATMENT OF BRAIN GLIOMAS

The synonyms of precision medicine are individualized medicine, personalized medicine, and targeted therapy. This approach is particularly pertinent in the area of brain tumors, especially with the most common and deadliest form of brain cancer, glioblastoma (Olar and Aldape, 2014; Duffau and Taillandier, 2015; Chen et al., 2016). DNA and RNA sequencing technology and the sprouting fields of proteomics and metabolomics are transforming our understanding of glioma biology. Researchers are now rapidly discovering biomarkers to probe the underlying pathology and define the dynamic changes that occur during development of glioma (Izci, 2014).

Progress in glioma therapy could be attained by improved comprehension of glioma biology, identification of relevant targets and signaling pathways for treatment interventions, development of personalized medicine, optimization of surgery and radiotherapy, and innovative neuroimaging modalities (Grant et al., 2014; Rajesh et al., 2017). Proteogenomic characterization is a potential strategy that could lead to identification of molecular drivers, molecular classification of disease subgroups, and glioma treatment (Grant et al.,

2014). The ultimate goal of targeted therapy should be "selectivity" inhibiting only tumor cells (Adair et al., 2014; Oh et al., 2014; Morokoff et al., 2015). The targeted approaches currently in clinical trials or in laboratory development include drugs, monoclonal antibodies, immunotherapy, small molecules inhibiting specific proteins, and specific targeting of glioma stem cells (Oh et al., 2014; Prados et al., 2015; Miller and Wen, 2016). Some of these targets for glial tumors are summarized in Table 4.1. Thus, there is a need for unconventional treatment strategies to curb glioma. Strategies such as gene therapy, microRNA (miRNA) therapy, stem cells, and immunotherapy may potentially lead to effective treatments (Kouri et al., 2015; Morokoff et al., 2015; Hodges et al., 2016; Chandran et al., 2017).

Integrative personal omics profiles (iPOPs) is the analysis of personal transcriptomes, proteomes, cytokines, metabolomes, and autoantibodies from a single individual over a 14-month period (Chen et al., 2012). It is important to learn how iPOPs can provide critical insight into the causes and development of disease. By learning this profile, it will be easy to understand

- How "omics" profiling of transcriptomes, proteomes, cytokines, metabolomes, and autoantibodies has revealed dynamic and wide-ranging

Table 4.1 Some of the potential targets for different types of brain gliomas

Tumor type	Target
Low-grade astrocytoma (Grades I and II)	RAS/RAF/mitogen-activated protein kinase pathway, phosphatidylinositol 3-kinase/mammalian target of rapamycin pathway, tyrosine kinase receptor, EGFR, PDGFR, BRAF/MAPK signaling pathway, PI3 K/mTOR signaling pathway
Oligodendroglioma (Grades II and III)	1p19q deletion, p130Cas, VEGF receptor, tyrosine kinase
Oligoastrocytoma (Grades II and III)	VEGF receptor, tyrosine kinase
Glioblastoma (Grade IV)	mTOR/PI3 K/Akt pathway, e-MET pathway, ZEB1 pathway, Wnt family, Sonic hedgehog, Notch2, transforming growth factor beta (TGF-β), bone morphogenetic protein (BMP) signaling, 27 Homeobox family, B lymphoma Mo-MLV insertion region 1 homolog (Bmi-1), PTEN (phosphatase and tensin homolog), telomerase, efflux transporters, epidermal growth factor (EGF), microRNA, and VEGF receptors, L1 CAM, p38MAPK signaling
Ependymoma (Grades I, II, III)	ErbB2, farnesyltransferase, p53, histone deacetylase

variation in the different molecular components that occur during the progression of a glioma.

- How analysis of heteroallelic expression suggests that significant changes in differential allele expression occur in healthy and diseased states.
- How integrated analysis of multiomic data on coding and gene regulation can lead to enhanced prediction of glioma risk.

Glioblastoma has four well-characterized subtypes, which can all respond differently to different treatments. These subtypes are proneural, neural, classical, and mesenchymal. The subtypes were found to differ by the type of genetic abnormalities they carried and by the patients' clinical characteristics (Verhaak et al., 2010).

Classical glioblastomas are characterized by abnormally high levels of epidermal growth factor receptor (EGFR). EGFR is a protein found on the surface of some cells that, when bound by epidermal growth factor, sends signals for the cell to keep growing. The EGFR abnormalities occur at a much lower rate in the three other glioblastoma subtypes. However, TP53, the most frequently mutated gene in glioblastoma, is not mutated in any of the classical glioblastoma. TP53 is the gene for a protein that normally suppresses tumor growth. Clinically, the classical group survived the longest of the subgroups in response to aggressive treatment (Verhaak et al., 2010).

Different than classical glioblastomas, TP53 is significantly mutated in proneural glioblastomas (54%). These glioblastomas are also characterized by having the most frequent mutations in the IDH1 gene. IDH1, when mutated, codes for a protein that can contribute to abnormal cell growth (Verhaak et al., 2010). Another gene, platelet derived growth factor receptor alpha (PDGFRA), was mutated and expressed in abnormally high amounts only in the proneural glioblastomas and not in any other subgroups. When PDGFRA is altered, too much of its protein can be produced, leading to uncontrolled tumor growth (Müller et al., 2016). Unlike the other groups, whose patients were similar in age on average, the proneural subgroup was significantly younger. They also tended to survive longer. However, patients in the proneural group who received aggressive treatment did not survive significantly longer than proneural patients who did

not receive aggressive treatment. This is important for precision medicine in glioblastoma and clinicians may be able to use this information in the future to avoid unnecessary treatment regimens for patients in the proneural glioblastomas (Verma et al., 2016).

The mesenchymal glioblastoma subgroup contains the most frequent number of mutations in the NF1 tumor suppressor gene (37%) (Verhaak et al., 2010). Frequent mutations in the PTEN and TP53 tumor suppressor genes also occurred in the group. Patients in the mesenchymal group had significant increases in survival after aggressive treatment, unlike those in the proneural, and, to an extent, in the neural subgroups (Verhaak et al., 2010; Evans, 2011).

Although the neural subgroup had mutations in many of the same genes as the other groups, the group did not stand out from the others as having significantly higher or lower rates of mutations. The neural group was characterized by the expression of several gene types that are also typical of the brain's normal, noncancerous nerve cells, or neurons. Patients in the neural group were the oldest, on average. They also had some improvement in survival after aggressive treatment, but not as much as the classical and mesenchymal groups (Verhaak et al., 2010; Evans, 2011).

Furthermore, glioblastoma is one of the most heterogeneous tumors—meaning individual tumors might be made up of cancer cells, which have a wide variety of molecular alterations and/or mutations (Halkin, 2015). As such, it is not likely that one treatment will be able to serve as a silver bullet for all glioblastoma patients (Halkin, 2015). This also helps explain why chemotherapy and radiotherapy have been largely ineffective in the treatment of glioblastoma patients. What is more likely is that treatments, or combinations of treatments, will need to be administered to distinct subgroups (or subpopulations) of patients whose tumors harbor the specific molecular alteration(s) and/or mutation(s) that the medicines can specifically affect (Lathia et al., 2015).

Defeat Glioblastoma Research Collaborative (www.defeatgbm.org) is a research group that was founded in 2013 for the development of targeted treatment of glioblastoma. This group is made up of four "synergistic research cores," or project teams. Three of these cores are working simultaneously and the fourth core is focusing on adaptive clinical trial designs. The aim of Core 1 (Target Discovery) is to identify targets (the molecular alterations and mutations) within glioblastoma cells that are driving the growth of these tumors. The aim of Core 2 (Drug Discovery) is to find specific drugs, or combinations of drugs, that attack the targets that are being discovered. The aim of Core 3 (Biomarker Development) is to identify specific biological processes that occur in glioblastoma cells that can predict whether a particular drug, or combination of drugs, will be effective for a patient or group(s) of patients. After researchers identify a promising target for treatment (for example, a mutation in a specific gene), they find a drug(s) that is designed to attack the mutation, and finally identify the predictive markers that let scientists and doctors know how patients may benefit from the potential new drug. The next step, before a new treatment regimen is approved for use by patients, is to test it through the clinical trial process (www.defeat-gbm.org).

Johnston et al. (2013) developed a patient-specific mathematical model for glioblastoma, yet the incidence is low, posing difficulty in creating statistically powered clinical trials, delaying advances in treatment. This model for disease progression coupled with clinical imaging data may provide a tool for quantifying each individual tumor's response relative to its inherent growth kinetics rather than a universal approach, and, potentially, predicting treatment efficacy. They used this model to evaluate therapeutic effect of a novel treatment for newly diagnosed glioblastoma compared to standard care treatment. Patient-specific mathematical modeling can be used to evaluate treatment gains in small clinical-trial design. The addition of O^6-benzylguanine to temozolomide in the postradiotherapy milieu can provide greater treatment gains per milligram of temozolomide dose than temozolomide alone (Adair et al., 2014).

WHY PRECISION MEDICINE IS NEEDED IN BRAIN GLIOMAS

Repeated surgeries, followed by many rounds of chemo-/radiotherapy in different doses make the glioma patients unhappy and depressed. New therapeutic approaches are needed to overcome this problem. As we enter an era that anticipates more tailored therapies for brain gliomas, many brain tumor research groups seek to address several

challenges facing precision medicine. In particular, it is necessary to understand the significance of whether targeting specific molecular alterations by therapies will be sufficient to have clinical benefit as not all molecular alterations are closely tied to tumor growth. Moreover, because of the histological and molecular complexity of brain gliomas, it is likely that combinations of targeted therapies will be required to affect the course of the disease. Therapies themselves can produce changes in the molecular profile of tumors and cause drug resistance, so we need to be able to identify and block resistance pathways. In our field, we need to build a solid base of clinical evidence through prospective clinical trials to define the best therapies for specific types of molecular "signatures" found in brain glioma cells.

CLINICAL SCENARIO FOR PRECISION MEDICINE IN BRAIN GLIOMAS

A 10-year-old female patient presented with epilepsy and weakness in her left arm and leg lasting for 2 days. Magnetic resonance imaging revealed a right parietal contrast enhancing mass lesion at the right parietal lobe with mild brain edema and midline shift. In her ophthalmological examination, grade 2 papilledema was present. She underwent surgery and the mass lesion was subtotally removed. The histopathological diagnosis was glioblastoma, which is a rare malignant brain glioma in childhood. She received chemotherapy and whole brain radiotherapy after surgery. But the tumor reoccurred in the same site 2 years after surgery and the neurological condition of the patient deteriorated. She was reoperated for the tumor and the histopathological diagnosis was again glioblastoma. But the molecular analysis of the tumor by PCR amplification and subsequent sequencing revealed a *BRAF V600E* mutation and she received a *BRAF* inhibitor "vemurafenib." The vemurafenib therapy was started at 720 mg orally twice daily for 28 days (about 600 mg/m² per dose). A complete response was observed after 4 months of therapy, and there was no residual or recurrent tumor in the MRI of the patient 1 year after the second surgery.

This patient was treated with targeted *BRAF* inhibitor therapy after the second surgery. If the molecular analysis of the tumor was performed after the first surgery, she would be better and might not undergo a second surgery for recurrence of glioblastoma. In addition, this mutation is not frequent in glioblastoma, but it is well known that *BRAF* inhibitors extend survival and improve the quality of life in patients with *BRAF V600E*-mutated melanoma. Vemurafenib is a *BRAF* inhibitor. It was recently approved by the United States Food and Drug Administration (FDA) for the treatment of melanoma with V600E mutation. But, it should not be forgotten that this response is unlikely to be uniform for all brain gliomas. This case also showed that the identification of the potential responders through careful histological and mutational analysis is critical for targeted therapy in brain gliomas.

FUTURE DIRECTIONS

Despite the continuous development of new chemotherapeutic agents, the ability to bypass the blood-brain barrier, and acquired drug resistance are still constraints (Ataman et al., 2004; Chen et al., 2012; Verma et al., 2016). Rather than attempting to control the migration of diffuse glioma, interventions that specifically target the invasive phenotype should be developed. However, recent advancements in neuroimaging and genetics will contribute to early diagnosis and initiation of glioma management. Advancements in noninvasive imaging protocols and a better understanding of glioma biology will enable neuro-oncologists to decipher the molecular, cellular, genetic, and epigenetic makeup of tumors. This information might pave the way to personalized glioma therapies (Levin, 2016).

Stem cell transplantation, nanotechnology, drug dosing or timing variations, moderation of immunosuppressive mechanisms, and a better understanding of miRNA and mRNA interactions are some of the strategies that may facilitate the stratification of brain gliomas (Lathia et al., 2015).

Proteogenomic characterization may identify molecular drivers and lead to molecular classification of glioma subgroups. The insights provided by omics studies will facilitate identification of glioma protein signatures and biomarkers as well as the design of potential treatment regimens for brain gliomas (Izci, 2014).

Future therapies comprising progressive neurointervention techniques, termination of DNA synthesis by prodrug-activating genes, silencing gliomagenesis genes, targeting miRNA oncogenic activity, sensitizing cancer stem cells by HedgehogGli/Akt inhibitors+radiation, and employing tumor lysates as antigen sources for efficient depletion of tumor-specific cancer stem cells by cytotoxic T lymphocytes along with conventional strategies will provide a new paradigm in glioma therapeutics with a focus on early diagnosis and successful management (Rajesh et al., 2017).

Finally, a number of elements should be blended for big data to deliver on precision medicine. First, appropriate data sources must be available at scale, whether it is from research cohorts, health care systems, or dedicated biobanks. Second, the method of data collection must be robust, with technologies that extract viable measurements with sufficient precision and accuracy at high throughput. Third, a multidisciplinary team with complementary expertise in bioinformatics, statistics, software engineering, quality control, biological sample processing, and clinical medicine must be engaged in a collaborative framework, preferably across jurisdictional and institutional boundaries. Fourth, expert panels that collect and assimilate the information generated by these efforts must be able to distill it into actionable interventions for patient care. Finally, a body of educators must be convened to train both their peers and the next generation of health care practitioners to disseminate and implement these interventions, which can then be iteratively evaluated for public health impact.

CONCLUSION

Precision medicine for brain glioma is still under construction. Many studies on molecular and genetic basis of gliomas are needed to better understand the development of these malignant tumors. In order to perform personalized treatment for these patients, the etiology and pathophysiology should be well known. In addition, definitive and clear targets should be determined for precise treatment.

REFERENCES

Adair JE, Johnston SK, Mrugala MM et al. Gene therapy enhances chemotherapy tolerance and efficacy in glioblastoma patients. *J Clin Invest* 124(9);2014:4082–92.

Akay KM, Baysefer A, Kayali H, Izci Y, and Timurkaynak E. Glioblastoma multiforme: Correlation of radiological findings surgery and prognosis. *Gulhane Med J* 44(2);2002:142–8.

Ataman F, Poortmans P, Stupp R, Fisher B, and Mirimanoff RO. Quality assurance of the EORTC 26981/22981; NCIC CE3 intergroup trial on radiotherapy with or without temozolomide for newly-diagnosed glioblastoma multiforme: The individual case review. *Eur J Cancer* 40(11);2004:1724–30.

Bondy ML, Scheurer ME, Malmer B et al. Brain tumor epidemiology: Consensus from the brain tumor epidemiology consortium (BTEC). *Cancer* 113(7 Suppl.);2008:1953–68.

Bush NA, and Butowski N. The effect of molecular diagnostics on the treatment of glioma. *Curr Oncol Rep* 19(4);2017:26.

Chandran M, Candolfi M, Shah D, Mineharu Y, Yadav VN, Koschmann C, Asad AS, Lowenstein PR, and Castro MG. Single vs. combination immunotherapeutic strategies for glioma. *Expert Opin Biol Ther* 17(5);2017:543–54.

Chen R, Mias GI, Li-Pook-Than J et al. Personal omics profiling reveals dynamic molecular and medical phenotypes. *Cell* 148(6);2012:1293–307.

Chen Y, Gao Z, Wang B, and Xu R. Towards precision medicine-based therapies for glioblastoma: Interrogating human disease genomics and mouse phenotypes. *BMC Genomics* 17(Suppl. 7);2016:516.

Crocetti E, Trama A, Stiller C, Caldarella A, Soffietti R, Jaal J, Weber DC, Ricardi U, Slowinski J, and Brandes A, RARECARE Working Group. Epidemiology of glial and non-glial brain tumours in Europe. *Eur J Cancer* 48(10);2012:1532–42.

de Robles P, Fiest KM, Frolkis AD, Pringsheim T, Atta C, St Germaine-Smith C, Day L, Lam D, and Jette N. The worldwide incidence and prevalence of primary brain tumors: A systematic review and meta-analysis. *Neuro Oncol* 17(6);2015:776–83.

Duffau H, and Taillandier L. New concepts in the management of diffuse low-grade glioma: Proposal of a multistage and individualized therapeutic approach. *Neuro Oncol* 17(3);2015:332–42.

Evans C. TCGA scientists discover four distinct subtypes of glioblastoma distinguished by gene expression patterns and clinical characteristics. The Cancer Genome Atlas. https://cancergenome.nih.gov/researchhighlights/research-briefs/foursubtypes. Posted February 1, 2011. Accessed online March 31, 2017.

Grant R, Kolb L, and Moliterno J. Molecular and genetic pathways in gliomas: The future of personalized therapeutics. *CNS Oncol* 3(2);2014:123–36.

Halkin T. Precision medicine, adaptive clinical trials, and defeat GBM. National Brain Tumor Society. http://blog.braintumor.org/precision-medicine-adaptive-clinical-trials-defeat-gbm. Published online February 9, 2015. Accessed on March 31, 2017.

Hayhurst C. Contemporary management of low grade glioma: A paradigm shift in neuro-oncology. *Pract Neurol* 17(3);2017:183–90.

Hodges TR, Ferguson SD, and Heimberger AB. Immunotherapy in glioblastoma: Emerging options in precision medicine. *CNS Oncol* 5(3);2016:175–86.

Izci Y. Biomarkers for brain gliomas. In: *Cancer Biomarkers: Minimal and Noninvasive Early Diagnosis and Prognosis*. Barh D, Carpi A, Verma M, and Gunduz M (Eds), CRC Press/Taylor & Francis Group, Boca Raton, FL, 2014, pp. 199–218.

Izci Y, Gurkanlar D, and Timurkaynak E. Multicentric gliomas: Still remains a controversial issue. Report of three cases and literature review. *Turk Neurosurg* 15(2);2005:71–5.

Johnston SK, Adair JE, Bridge C et al. Patient-specific mathematical modeling as a precision-medicine approach to evaluating therapeutic gains of a novel chemoprotection treatment in newly diagnosed glioblastoma. Abstract. *J Clin Oncol* 31(15; suppl);2013. doi:10.1200/jco.2013.31.15_suppl.e13041.

Kouri FM, Ritner C, and Stegh AH. miRNA-182 and the regulation of the glioblastoma phenotype-toward miRNA-based precision therapeutics. *Cell Cycle* 14(24);2015:3794–800.

Lathia JD, Mack SC, Mulkearns-Hubert EE, Valentim CL, and Rich JN. Cancer stem cells in glioblastoma. *Genes Dev* 29(12);2015:1203–17.

Levin VA. Personalized medicine in neuro-oncology. *CNS Oncol* 5(2);2016:55–8.

Louis DN, Perry A, Reifenberger G, von Deimling A, Figarella-Branger D, Cavenee WK, Ohgaki H, Wiestler OD, Kleihues P, and Ellison DW. The 2016 World Health Organization classification of tumors of the central nervous system: A summary. *Acta Neuropathol* 131(6);2016:803–20.

Miller JJ, and Wen PY. Emerging targeted therapies for glioma. *Expert Opin Emerg Drugs* 21(4);2016:441–52.

Morokoff A, Ng W, Gogos A, and Kaye AH. Molecular subtypes, stem cells and heterogeneity: Implications for personalised therapy in glioma. *J Clin Neurosci* 22(8);2015:1219–26.

Müller S, Liu SJ, Di Lullo E et al. Single-cell sequencing maps gene expression to mutational phylogenies in PDGF- and EGF-driven gliomas. *Mol Syst Biol* 12(11);2016:889.

Oh YT, Cho HJ, Kim J et al. Translational validation of personalized treatment strategy based on genetic characteristics of glioblastoma. *PLOS ONE* 9(8);2014:e103327.

Olar A, and Aldape KD. Using the molecular classification of glioblastoma to inform personalized treatment. *J Pathol* 232(2);2014:165–77.

Ostrom QT, Bauchet L, Davis FG et al. The epidemiology of glioma in adults: A "state of the science" review. *Neuro Oncol* 16(7);2014:896–913.

Ostrom QT, Gittleman H, Kruchko C, Louis DN, Brat DJ, Gilbert MR, Petkov VI, and Barnholtz-Sloan JS. Completeness of required site-specific factors for brain and CNS tumors in the Surveillance, Epidemiology and End Results (SEER) 18 database (2004-2012, varying). *J Neurooncol* 130(1);2016:31–42.

Perry A, and Wesseling P. Histologic classification of gliomas. *Handb Clin Neurol* 134;2016:71–95.

Porter KR, McCarthy BJ, Freels S, Kim Y, and Davis FG. Prevalence estimates for primary brain tumors in the United States by age, gender, behavior, and histology. *Neuro Oncol* 12(6);2010:520–7.

Pouratian N, and Schiff D. Management of low-grade glioma. *Curr Neurol Neurosci Rep* 10(3);2010:224–31.

Prados MD, Byron SA, Tran NL et al. Toward precision medicine in glioblastoma: The promise and the challenges. *Neuro Oncol* 17(8);2015:1051–63.

Rajesh Y, Pal I, Banik P, Chakraborty S, Borkar SA, Dey G, Mukherjee A, and Mandal M. Insights into molecular therapy of glioma: Current challenges and next generation blueprint. *Acta Pharmacol Sin* 38(5);2017 Mar 20:591–613, doi: 10.1038/aps.2016.167, Epub ahead of print.

Stupp R, Mason WP, van den Bent MJ et al. Radiotherapy plus concomitant and adjuvant temozolomide for glioblastoma. *N Engl J Med* 352(10);2005:987–96.

Tan SY, Sandanaraj E, Tang C, and Ang BT. Biobanking: An important resource for precision medicine in glioblastoma. *Adv Exp Med Biol* 951;2016:47–56.

Verhaak RG, Hoadley KA, Purdom E et al. Integrated genomic analysis identifies clinically relevant subtypes of glioblastoma characterized by abnormalities in PDGFRA, IDH1, EGFR, and NF1. *Cancer Cell* 17(1);2010:98–110.

Verma A, Gunasekar S, Goel V et al. A molecular approach to glioblastoma multiforme. *Int J Mol & Immuno Oncology* 1(1);2016:35–44.

Walid MS. Prognostic factors for long-term survival after glioblastoma. *Perm J* 12(4);2008 Fall:45–8.

www.defeatgbm.org. Accessed on March 21, 2017.

Zhang AS, Ostrom QT, Kruchko C, Rogers L, Peereboom DM, and Barnholtz-Sloan JS. Complete prevalence of malignant primary brain tumors registry data in the United States compared with other common cancers, 2010. *Neuro Oncol* 19(5);2016:726–35.

Zong H, Verhaak RG, and Canoll P. The cellular origin for malignant glioma and prospects for clinical advancements. *Expert Rev Mol Diagn* 12(4);2012:383–94.

5

Precision medicine for colorectal cancer

CANDAN HIZEL, ŞÜKRÜ TÜZMEN, ARSALAN AMIRFALLAH,
GIZEM ÇALIBAŞI KOÇAL, DUYGU ABBASOĞLU, HALUK ONAT,
YEŞIM YILDIRIM, AND YASEMIN BASKIN

INTRODUCTION: OVERVIEW OF EPIDEMIOLOGY/INCIDENCE, ETIOLOGY, AND PRECISION MEDICINE FOR COLORECTAL CANCER (CRC) GENOMICS IN "POSTGENOMIC" ERA

CRC epidemiology

Colorectal cancer (CRC) is a multifactorial polygenic complex disorder that represents the supremacy of health care leading to the third most commonly diagnosed malignancy among both gender groups (behind lung and breast cancer) with 1.4 million cases in 2012 (Ferlay et al., 2015) and the fourth leading cause of cancer deaths worldwide. It accounts for approximately 700,000 deaths per year, affecting men and women almost equally (third most commonly diagnosed malignancy for men behind lung and prostate cancer, and the second most commonly diagnosed malignancy for women behind breast cancer) (Ferlay et al., 2015; Siegel et al., 2018) with the highest incidence rates in high-income countries (Ferlay et al., 2015; Arnold et al., 2017; Bhandari et al., 2017). However, CRC incidence is also rapidly increasing in low- and middle-income countries (LMICs) (Bhandari et al., 2017). It is expected that its burden will be increased by 60% with more than 2.2 million new cases and 1.1 million cancer deaths by 2030 (Arnold et al., 2017). Survival of CRC patients mostly depends on the disease stage at diagnosis, and approximately 20%–25% of patients with CRC already have liver metastases at the time of diagnosis and 20%–30% of patients will develop metastases subsequently during disease progression with stage II/III (Loree and Kopetz, 2017).

Despite these bleak statistics, due to progress in high-throughput techniques in molecular biology from genetic sequencing to targeted therapies has led to the availability of more precise and personalized screening practices leading to a gradual decrease in both CRC incidence and mortality over the past few decades particularly for late-onset CRC in adults aged between 50 and 75 (Siegel et al., 2018). On the flip side of this, most research publications have reported a steady increase in incidence rates in both hereditary and sporadic forms of young-onset CRC (diagnosed at ≤40 years old) diagnosed with advanced-stage tumors having distinctive heterogeneous molecular profile in the distal colon and rectum (Ballester et al., 2016; Connell et al., 2017). It is anticipated that increased screening for CRC has been associated with a decreased incidence in the past two decades (Stracci et al., 2014), albeit so far no evidence supports that screening of average-risk individuals less than 50 years old will lead to early detection and increased survival. Hence, together with clinicopathologic features, understanding of the distinctive heterogeneous and poorly clarified molecular profile, including genetic and epigenetic mechanism of young-onset CRC, will be crucial to tailor specific screening and management strategies (Stigliano et al., 2014; Ballester et al., 2016; Tezcan et al., 2016).

Although several screening tests are now available, at present none of them has been proven the best one (Pox, 2014). Although colonoscopy is usually used in high-risk subjects with family history of either CRC or adenomas, fecal occult blood test (FOBT) is preferred (Health Quality Ontario, 2009) in intermediate-risk subjects. Colonoscopy is performed only when FOBT results are positive (Stracci et al., 2014). For comprehensive reviews of CRC screening, the reader is referred to articles by Issa and Noureddine (2017) and Bhurgri and Samiullah (2017).

Although CRC occurs in all ethnic groups and races, variation in occurrence and survival rate (stage-specific survival) in different ethnic groups and races is underlined (Wallace et al., 2016; Fedewa et al., 2017). The American Cancer Society (ACS) documented that racial and ethnic minorities are more likely to develop cancer, particularly CRC and die from it when compared to the general population of the United States (Jackson et al., 2016). Accordingly, the high incidence and mortality rate in African American compared to white Americans in the United States is reported (Jackson et al., 2016; Fedewa et al., 2017). However, Hispanic Americans and Native Americans (Alaska Natives) have the lowest incidence rate.

Genetic and nongenetic paradigms of CRC

The vast majority of tumors such as CRC encountered in human pathology are carcinomas, that is, tumors of epithelial origin that generally develop in several phases (Vogelstein et al., 2013). Additionally, cancer such as CRC is a genetic disease under the influence of the environmental/lifestyle factors that is developed in a multistep process resulting from an accumulation of a series of genetic changes leading to gain/loss of gene function, such as mutations in DNA repairing genes, in oncogenes and tumor suppressor genes, which regulate normal cellular function in the genome of cells (Vogelstein et al., 2013). Moreover, once a cancer has begun its anarchic growth it will pick up more and more mutations (Hanahan and Weinberg, 2011; De Palma and Hanahan, 2012). Thus, understanding the molecular basis of tumor-associated genetic and epigenetic modifications that alter gene expression patterns is crucial (De Palma and Hanahan, 2012; Pierotti, 2017; Inamura, 2018).

CRC is one of the best molecularly characterized neoplasms and it is important to note that most of the important theoretical frameworks for understanding the basis of carcinogenesis have grown from genomic research in CRC (Vogelstein et al., 2013). As formulated in the 1990s by Vogelstein and Fearon (Vogelstein et al., 1988, 1989; Fearon and Vogelstein, 1990), it is widely accepted that in most cases carcinoma arises from preexisting adenomas and every step from normal mucosa toward the carcinoma involves specific well-known genetic (conversion of proto-oncogene to oncogenes, and/or inactivation of tumor suppressor genes; Wnt, P13K, and RAS-MAPK pathways with *KRAS*, *NRAS*, and *BRAF* genes) (Smith et al., 2002) and epigenetic alterations (DNA methylation, histone modifications, noncoding RNAs/miRNAs) (Bardhan and Liu, 2013; Danese and Montagnana, 2017). Hence, all these alterations are driving forces of CRC carcinogenesis, which are associated with the stages of CRC progression. Moreover, the connection between inflammation (through immune cells, cytokines) and colorectal carcinogenesis, including initiation, promotion, progression, and metastasis, has been established (Terzić et al., 2010).

It is anticipated that CRC can arise from one or a combination of genetic alterations of chromosomal instability (CIN), DNA microsatellite instability (MSI), and CpG island methylator phenotype (CIMP). Accordingly, a majority of CRCs develop via CIN, which accounts for about 80%–85% of sporadic CRC and it is also associated with familial adenomatous polyposis (Wiesner et al., 2010; Roper and Hung, 2013). MSI phenotype is associated with approximately 15%–20% of all sporadic CRC cases (Issa, 2008; Wiesner et al., 2010). As for CIMP, besides genomic instability CIN and MSI, a broad spectrum of studies underline importance of a crosstalk between genetic and epigenetic mechanisms in cancer formation, including CRC, indicating that genetic aberrancies such as CIMP that are inherited or acquired during a lifetime have the potential of disrupting epigenetic patterns, which in turn epigenetic modifications can drive genome instability and mutagenesis (Pancione et al., 2012; You and Jones, 2012; Hughes et al., 2013; Nazemalhosseini Mojarad et al., 2013). Additionally, other numerous factors are important in the development of CRC, including colonocyte metabolism, high-risk luminal environment, inflammation, and environmental and lifestyle factors such as such as diet (Aguirre-Portolés et al., 2017), physical inactivity (Namasivayam and Lim, 2017), cigarette smoking (Nishihara et al., 2013; Drew et al., 2017), and alcohol consumption (may be relative to MTHFR genotype status) (Kim et al., 2012; Svensson et al., 2016). All these potentially modifiable/behavioral risk factors, which are a valuable tool for setting priorities for CRC prevention and control, could affect epigenetic process (Aleksandrova et al., 2014; Rattray et al., 2017). Likewise, disease conditions, such as inflammatory bowel disease (IBD) (Du et al., 2017), existing of previous polyps (Jelsig, 2016), diabetes (González et al., 2017; Yang et al., 2017), obesity (Heo et al., 2004), and nonalcoholic fatty liver disease (NAFLD) (Mikolasevic et al., 2017). On the other hand, nonmodifiable risk factors, such as gender, increasing age, personal history of adenomatous polyps and inflammatory bowel disease, ethnicity/race, and genetic inheritance, are associated with the incidence of colorectal cancer. Contrary to modifiable risk factors that could theoretically be avoided, nonmodifiable risk factors are not considered part of the "environmental nature" of this disease. Therefore, they are not controllable; however, they play a crucial role in screening and identifying susceptible individuals for CRC (Thélin and Sikka, 2015).

The majority of CRC cases is sporadic (approximately 70%–75% of all CRCs), with no known genetic predisposing factors for disease and about 80% are caused by somatic defects in the CIN pathway through the *KRAS*, *APC*, and *TP53* mutations (Wiesner et al., 2010; Roper and Hung, 2013).

The microsatellite instability (MSI) pathway, which is caused by DNA mismatch repair (MMR), represents a major pathway for inherited and familial form (Boland and Goel, 2010; Pino and Chung, 2011). The inherited form, which represents approximately 15%–20% of all CRCs, is more likely to be the result of low-penetrance alleles (with the possibility of some undiscovered high- and moderate-penetrance gene) at several genetic loci (Mishra and Hall, 2012; Lynch and Shaw, 2013; Hahn et al., 2016). Well-described forms of hereditary CRC include familial adenomatous polyposis (FAP) (Jaiswal et al., 2005; Half et al., 2009; Lynch and Shaw, 2013; Hahn et al., 2016) and another autosomal dominant disorder, hereditary nonpolyposis colorectal cancer (HNPCC) or Lynch syndrome (approximately 5%–10% of the total CRC), which is highly penetrant with an 80% lifetime risk for CRC (Lynch and Shaw, 2013). Certain other rare polyposis syndromes such as hamartomatous polyposis syndromes comprise less than 1% of all hereditary colorectal cancer (Campos et al., 2015; Jelsig, 2016). Being located predominantly in the right colon (Iacopetta, 2002), FAB is an autosomal dominant disorder due to the presence of a mutated copy of the adenomatous polyposis gene (*APC*) located on chromosome 5 (Nishisho et al., 1991). Tumor suppressor *APC* is considered a "gatekeeper" gene that is responsible for controlling, or inhibiting, cell growth (Deininger, 1999; Eshghifar et al., 2017). Whereas, being an autosomal dominant disorder, HNPCC, which occurs predominantly in the right colon and cannot be differentiated endoscopically from sporadic colon polyps (Iacopetta, 2002; Rijcken et al., 2002; Jang and Chung, 2010), is caused by inheritance of defective DNA mismatch repair genes *MLH1*, *MSH2*, *PMS2*, and *MSH6* (Jaiswal et al., 2005; Lynch and Shaw, 2013; Hahn et al., 2016). They are "caretaker" genes involved in the maintenance of the genome stability including mismatch repair of DNA (Deininger, 1999).

Taken as a whole, even if genomic studies are one of the principal components for identifying the "Achilles' heel" for the subject affected by the disease, it is only one piece of the puzzle (Hizel et al., 2017). Thereby, the creation of robust data from genomic risk to the "exposome," together with medical histories, social factors, and lifestyle factors, is pivotal and they cannot act in isolation. But they can act in concert in order to fulfill the prior engagement of precision medicine in

susceptibility to common multifactorial complex diseases such as cancer as well as in drug effectiveness and safety for more precise personalized health care (Hamet, 2012; Kittles, 2012; Hizel et al., 2017). To this end, the interplay between genetic and environmental factors in CRC pathology should be taken into consideration. Together with recent knowledge of the human genome, conventional trio therapies—radiotherapy, chemotherapy, surgery—in CRC and all other types of cancer are all still improving and have become more efficient in recent years. Accordingly, the tools of molecular biology have moved out of the lab and into the clinic, and recent progress in high-throughput technologies such as next-generation sequencing, as well as "omics" technologies, such as transcriptomics, proteomics, metabolomics, epigenomics, and metagenomics, have untangled the genomic architecture and the molecular heterogeneity of CRC and a new branch of therapy has sprung up for more personalized and precision medicine in cancer (Bhati et al., 2012).

IMPLEMENTATION OF BIOMARKER-DRIVEN COLON CANCER THERAPY: PROGNOSTIC AND PREDICTIVE BIOMARKERS

In the era of precision medicine, with the progress of molecular genomics and high-throughput omics-based technologies, biomarker studies hold promise for early and effective prediction of susceptibility of disease and treatment efficacy/toxicity, diagnosis, and prognosis for a tailored health care approach to an individual (Barh and Azevedo, 2017; Kamel and Al-Amodi, 2017; Wilson and Altman, 2017).

The term *biomarker* relates to a measurement variable of biological molecules that is associated with disease and treatment outcome (Ballman, 2015). Respectively, exploitation of precision medicine in oncology demands a precise understanding of disease progression as well as proper application of molecular targeting of therapies (Hizel et al., 2017). However, since not all biomarkers are the same, there is noteworthy confusion about the distinction between a predictive biomarker and a prognostic biomarker (Ballman, 2015). To this end, while prognostic biomarkers are associated with the clinical disease outcome that is independent of treatment, predictive biomarkers are measures of the possibility of response (responder/nonresponder) of a particular therapy as well as identification of patients most likely to benefit from a given treatment without toxicity (Oldenhuis et al., 2008; Carethers and Jung, 2015; Verdaguer et al., 2017). It is also anticipated that measurements coming from the signature of combined multiple biomarkers have stronger predictive power compared to a single measurement (Ballman, 2015).

The following will review recent knowledge in the application of molecular biomarker in CRC pathogenesis and treatment.

Microsatellite instability (MI) and chromosomal instability (CIN)

Being important in the carcinogenesis pathway in many cancers including CRC, microsatellite instability (MSI) is the state of genetic hypermutability (highly predisposed to mutation) (Strauss, 1998) due to impaired DNA base MMR error such as frameshift mutations and base-pair substitutions during DNA replication (S phase of the cell cycle) in the DNA MMR allowing accumulation of mutations in the DNA (Popat et al., 2005). The presence of MSI is an indication of phenotypic evidence that MMR is not functioning properly. Under normal conditions, dMMR corrects spontaneously occurring DNA polymerase errors during DNA replication; however, cells having an impaired dMMR system are unable to correct errors leading to accumulation of blunders resulting in creation of novel microsatellite fragments that are repeated sequences of DNA (ranging in length from 1 to 6 or more base pairs) (Boland and Goel, 2010; Kawakami et al., 2015; Inamura, 2018).

The MMR and CIN (accounting for about 80%–85% of sporadic CRCs) have little overlap in CRC tumorigenesis (Kawakami et al., 2015). CIN results from defects in the chromosomal segregation process characterized by an imbalance in chromosomal number (aneuploidy), subchromosomal genomic amplifications, and a high frequency of loss of heterozygosity (LOH) and changes in chromosomal copy number through classic adenoma-to-carcinoma progression (alterations in the Wnt and TGF-β signaling pathways, activation of KRAS, and inactivation of APC and TP53) (Vogelstein et al., 1989; Fearon and Vogelstein, 1990; Lengauer et al., 1997; Dienstmann et al., 2017).

Having few or no somatic copy number alterations (CNAs) tumors with dMMR/MSI phenotype occur in approximately 15%–20% of CRC patients of which 12% in sporadic CRC tumors and 3% are associated with Lynch syndrome (Boland and Goel, 2010). Contrary to microsatellite stable MSS tumors (also termed chromosomal instability), MSI is attributable to DNA mismatch repair genes in sporadic and germ line tumors (Boland and Goel, 2010; Lee and Chan, 2011; Sinicrope and Sargent, 2012; Kawakami et al., 2015).

Most of the sporadic CRCs with MSI tumors occur generally in older individuals who carry somatic mutations in the BRAF oncogene (V600E) in approximately half of cases (Boland and Goel, 2010; Kawakami et al., 2015). Whereas, while MSI tumors in sporadic CRCs are due to transcriptional silencing/epigenetic inactivation (hypermethylation) of the DNA mismatch repair MLH1 gene promoter together with the CpG island methylator phenotype (Issa, 2004; Grady and Carethers, 2008; Barault et al., 2008a; Nosho et al., 2008), MSI tumors in Lynch syndrome, which occurs in younger individuals carrying KRAS mutations (but never BRAF mutations), are caused by germ line mutations in DNA mismatch repair genes: MLH1, MSH2, MSH6, and PMS2 (Boland and Goel, 2010; Lee and Chan, 2011; Sinicrope and Sargent, 2012; Kawakami et al., 2015). Respectively, this distinction concerning methylation status suggests that most of the CRCs with sporadic MSI tumors come from CIMP status creating an important distinction from Lynch syndrome (Barault et al., 2008a; Boland and Goel, 2010); however, inheritance of a cancer-associated MLH1 germ line epimutation (abnormal gene silencing) is demonstrated (Hitchins et al., 2007). Consequently, presence of the BRAF V600E mutation in sporadic CRCs with MSI tumors essentially excludes Lynch syndrome (Funkhouser et al., 2012), with the exception of rare cases related to germ line mutation in the PMS2 gene that encodes mismatch repair endonuclease PMS2 (Nicolaides et al., 1994; Senter et al., 2008; Gatalica et al., 2016).

It anticipated that in CRC, tumor location to the splenic flexure, left side (distal) (LC) versus right side (proximal) (RC) of the colon affects colorectal carcinogenesis (Sugai et al., 2006), and incidence of location change according to geographic region, age, and gender (Iacopetta, 2002). Accordingly, recent studies have revealed that CRCs with an

MSI phenotype with frequent BRAF mutations, infrequent TP53 mutation, poorly differentiated histology, abundant tumor-infiltrating lymphocytes (TILs), and CIMP-high status preferentially occur on the RC (proximal portion) of the colon. Whereas CRC with an MSS phenotype characterized by frequent copy number alterations, CIMP-low status, TP53 mutation, and is commonly found on the LC (distal portion) of the colon (Sugai et al., 2006; Takahashi et al., 2016), suggesting that tumor location may be associated with the development of CRC. Moreover, compelling evidence from published studies suggest that besides poorer survival characteristics of RC compared to LC, patients with an RC tumor may respond poorly to antiepidermal growth factor receptor (EGFR) antibodies, such as cetuximab or panitumumab (Meguid et al., 2008; Brule et al., 2015; Kim et al., 2017).

Though not all studies have confirmed the prognostic value of MSI, most studies have emphasized that CRC patients with an MSI tumor with greater numbers of cytotoxic TILs (Michael-Robinson et al., 2001; Phillips et al., 2004) have been strongly associated to longer survival as compared with proficient-MMR tumors including microsatellite stable (MSS) and MSI-low (MSI-L) CRCs (Sankila et al., 1996; Popat et al., 2005; Sinicrope and Sargent, 2012; Cortes-Ciriano et al., 2017) particularly among young patients.

According to an international consensus meeting in 1997 (Boland et al., 1998) the definition of MSI was standardized as MSI-high (MSI-H) (resulting from MSI of greater than 30% of unstable MSI biomarkers) and MSI-low (MSI-L) or microsatellite stable (MSS) (resulting from less than 30% of unstable MSI biomarkers) (Boland et al., 1998; Pawlik et al., 2004). In CRC, recent studies revealed another type of MSI called "elevated microsatellite alterations at selected tetranucleotide repeats" (EMAST) (Lee et al., 2012; Venderbosch et al., 2015; Watson et al., 2016). Although scientific data from the available literature is not consistent and still, additional work needs to be done, MSI-H status is also associated with a different response to classic chemotherapeutic treatment modalities (Devaud and Gallinger, 2013; Kawakami et al., 2015; Fujiyoshi et al., 2017). Several studies highlighted MSI-H status as a strong predictive factor for nonresponse (resistant) to adjuvant 5-fluorouracil (5-FU) chemotherapy (Ribic et al., 2003; Des Guetz et al., 2009; Jensen et al., 2009; Buecher et al., 2013; Webber et al., 2015; Kawakami et al., 2015; Fujiyoshi et al., 2017) but may be sensitive to irinotecan (Fallik et al., 2003; Pino and Chung, 2011) and mitomycin C (Devaud and Gallinger, 2013). Also, there is evidence to suggest that MSI-H patients respond to surgery alone better than combined surgery and chemotherapy and, therefore preventing patients from needless chemotherapy exposure (Buecher et al., 2013; Kawakami et al., 2015). Additionally, according to recent studies, colorectal tumors with MSI-H have increased immunogenicity leading to positive response to immunotherapy compared to MSS (Xiao and Freeman, 2015; Bashir and Snook, 2017).

In summary, the MSI-H phenotype is an important feature of the molecular pathogenesis of cancer that may provide an important basis for novel diagnostic and personalized therapeutic approaches (Zhang et al., 2016).

18q loss of heterozygosity (LOH)

Loss of heterozygosity (LOH) is a cross-chromosomal event due to loss of one parent's contribution to the cell, which may be caused through a variety of pathways, such as mitotic recombination, inappropriate repair of DNA, deletion, gene conversion, or chromosomal loss leading to loss of the entire gene along with the surrounding chromosomal region (Ryland et al., 2015). During LOH the loss of one allele leaves the other allele hemizygote (possession of only one allele at a given locus) (Ryland et al., 2015). Clinical manifestations of LOH and MSI are different in colorectal cancer patients. High-frequency LOH is associated with high metastatic potential of colorectal cancers (Roper and Hung, 2013). As it was first reported in 1989 by Vogelstein and his colleagues, LOH of tumor suppressor genes (TSGs) is believed to be one of the crucial steps to the carcinogenesis of colorectal cancer. Respectively, the interrelationship between LOH, loss of TSG initiation, and progression of CRC is demonstrated (Ozaslan and Aytekin, 2009). As proven in several studies, mitotic crossover occurs more frequently in tumor cells and an increased mitotic crossover index in colorectal carcinoma cells have been also confirmed (Rovcanin et al., 2014). Thus, mitotic crossover is underlined as one of the principal sources of increased LOH in the tumor cells of patients with CRC (LaFave and Sekelsky, 2009).

As is well known, inactivation of TSGs is one of the most crucial steps in both sporadic and familial cancer development (Payne and Kemp, 2005). Equally, LOH, which is the most common molecular genetic alteration in a variety of human cancers including CRC, indicates the presence of a TSG. Since TSGs are recessive genes (an exception is the X-linked tumor suppressor such as Wilms tumor gene on the X chromosome), two alleles must be inactivated in order for a TSG to lose its inhibitor effect on uncontrolled cell proliferation (Payne and Kemp, 2005). Therefore, a "first hit" such as physical loss of one copy of an allele is not a trigger for the loss of the second allele, yet it renders one allele inactive. Thus, in most of the cancer as proposed in Knudson's "two-hit" hypothesis (Devilee et al., 2001), the second copy (the remaining wild type allele) of a tumor suppressor gene has been inactivated by a "second hit" such as point mutation or hypermethylation (Paige, 2003; Payne and Kemp, 2005). To this end, in cancer, LOH analysis is the common method to identify genomic potential regions harboring tumor suppressor genes by noting the presence of heterozygosity at a genetic locus in an organism's germ line DNA, as well as to characterize different tumor types, pathological stages, and progression (Zheng et al., 2005; Cacev et al., 2006).

Since LOH is a pivotal characteristic of CIN-positive tumors, determination of LOH distinguishes the tumors arising from the CIN pathway from tumors with the dMMR/MSI. Additionally, in CRC the association of a worse disease prognosis with small and large allelic loss is indicated (Brosens et al., 2010; Roper and Hung, 2013). The interrelationship between CRC tumorigenesis and LOH at various loci on different chromosomes, such as 1p, 1q, 4q, 5q, 8p, 9q, 10p, 11q, 12p, 14q, 15q, 17p (loss of p53), 17q, 18p, 18q, and 22q, is emphasized in a large number of such studies suggesting that LOH on those chromosome regions anchoring TSGs is essential for colorectal tumorigenesis mechanisms (Shima et al., 2005; Wan et al., 2006; Ozaslan and Aytekin, 2009; Mayrhofer et al., 2014). In particular, several studies have confirmed that chromosome 18q is commonly deleted in most cancers, such as prostate, esophageal, bladder carcinomas, and sporadic CRC (Jen et al., 1994; Wang et al., 2010). Since the chromosome 18q contains several potentially important genes implicated in CRCs, and although the outcomes

of studies are controversial, deletion of 18q has been demonstrated in most studies as an important step in the development and the progression of CRC tumorigenesis (Jen et al., 1994; Wang et al., 2010). For example, among the genes located on 18q, of particular importance is for DCC (deleted in colorectal carcinoma) encoding for a neutrin-1 receptor that is important in apoptosis, cell adhesion and tumor suppression (Mazelin et al., 2004; Kefeli et al., 2017). The other one is SMAD-4, which encodes for a nuclear transcription factor in transforming growth factor-β1 (TGF-beta) signaling involved in tumor suppression (Zauber et al., 2008). Given the higher prevalence of *Smad4* mutations in CRCs, specifically in those with distant metastases compared to locally advanced tumors, a strong prognostic value of chromosome 18q deletion and SMAD-4 protein inactivation is underscored (Tanaka et al., 2008; Zauber et al., 2008; Wang et al., 2010; Jia et al., 2017).

Taken together although evidence from most published studies suggest that high-frequency 18q LOH with high metastatic potential of CRC could be prognostic genetic markers, large-scale prospective randomized control trials are needed to confirm the final verdict.

TP53 in CRCs

The tumor suppressor *TP53* gene is located on the short arm of chromosome 17 (17p13.1). It is composed of 11 exons and encodes a 53 kD nuclear phosphoprotein p53, which is a transcription factor interacting with as many as 106 other proteins within the cell (Isobe et al., 1986; McBride et al., 1986). Even though TP53 gene germ-line mutations cause inherited syndrome de Li–Fraumeni (LFS) (OMIM #151623) (60%–80% of "classic" LFS families) (Varley, 2003; Malkin, 2011; Merino and Malkin, 2014), acquired mutations occur in a wide variety of cancer types including CRC for which acquired *TP53* mutations is significant as a late event from adenoma to carcinoma (Fearon and Vogelstein, 1990). The p53 protein plays a crucial role in tumor suppression and the loss of its function is required for cancer progression (Surget et al., 2013). Though the p53 protein is encoded by wild-type *TP53* tumor suppressor gene, which is in an inert and unstable state in normal cells in the absence of stresses, its transcriptional activity and stability are significantly induced in response to a variety of signals, such as oncogenic stresses and genotoxicity (Ko and Prives, 1996; Vogelstein et al., 2000), leading to control of transcription of many genes through senescence of DNA damage, and G1/S cell cycle arrest by stimulating p21^{Wafl} (cyclin-dependent kinase inhibitor 1) protein (Taylor and Stark, 2001). Accordingly, once DNA repair is successfully complete, the p53 reactivates the cell cycle; yet if repair fails, programmed cell death apoptosis is triggered to eliminate damaged cells (Amelio et al., 2016). The TP53 gene is among the most frequently mutated in cancer having mostly missense mutations (>75% of all p53 alterations) including the hot spot mutations (R175H, R248W and R273H) (Muller and Vousden, 2013; Leroy et al., 2014) which cause loss of p53-dependent tumor suppressor function, but also gains new oncogenic function (gain-of-function mutations) in order to promote cancer as well as drug resistance suggesting an important implication on cancer therapy (Xu, 2008; Liu et al., 2010; Liu et al., 2014; Alexandrova et al., 2017).

Tp53 is an enigmatic cancer gene due to its caretaker/gatekeeper character (Rubbi and Milner, 2007). Its caretaker function is related to its capacity to arrest cell cycle and/or apoptosis in response to DNA damage that points to P53 as the "guardian of the genome" for maintaining genomic stability, and its gatekeeper function is related to its role in controlling cell proliferation during which mutations in the Tp53 gene result in disruption of cell cycle control mechanisms (Levine, 1997; Deininger, 1999; Rubbi and Milner, 2007). Moreover, a large number of such studies reported that a guanine (G)/cytosine (C) common single SNP (rs1042522) at the second position of codon 72 in exon 4 of TP53 gene (Arg72Pro amino acidic substitution) may modulate individual cancer susceptibility.

Since mutation of the TP53 gene is thought to increase the protein half-life that is associated with overexpression in the nucleus and cytoplasm, most translational studies directed their efforts at determining whether p53 mutation and overexpression have prognostic value in CRC (Remvikos et al., 1990). The majority of translational studies carried out in the 1990s were aimed at determining whether p53 mutation and overexpression have prognostic value in CRC. Though p53 research is one of the controversial areas and still remains unclear with inconsistent

and inconclusive results related to its role in clinical practice (Lee and Chan, 2011), the importance of abnormal p53 protein expression as well as somatic mutation with poor survival prognosis in late stage CRC or lack of response to therapy are underlined by several studies (Russo et al., 2005; Iacopetta et al., 2006; Robles and Harris, 2010; Li XL et al., 2015). A recent meta-analysis has underlined biologically different molecular pathways between bowel disease-associated colorectal cancer (IBD-CRC) and sporadic colorectal cancer (S-CRC), with IBD patients having TP53 mutation more likely to develop IBD-CRC (Du et al., 2017). It has also been demonstrated that TP53 mutation was associated with more advanced stage cancer (Dukes class A, B, C) in patients with IBD-CRC suggesting that TP53 status may be an important biomarker to improve cancer and dysplasia screening among patients with IBD (Du et al., 2017). It is anticipated that the standardized immunohistochemical (IHC) procedure to evaluate p53 overexpression could be beneficial for CRC patients under a standard chemotherapy regime to get optimal response (Iacopetta, 2003; Akshatha et al., 2016; Vandana et al., 2017).

In summary, although TP53 targeted therapies are still in the early phases of testing, research in this field is expanding rapidly. Respectively, Tp53 having dual caretaker–gatekeeper functions, deserves additional prospective investigations in CRC for the development of personalized therapeutics (Surget et al., 2013; Sorrell et al., 2013).

PIK3CA and PTEN in CRC

Phosphatidylinositol-3-kinase (PI3K) (also called phosphatidylinositol-4,5-bisphosphate 3-kinase) is one of the pivotal kinase enzymes in the PI3K/AKT1/MTOR intracellular signaling pathway, playing a crucial role in cell cycle regulation, such as cellular growth, and proliferation as well as survival of various solid tumors (Shaw and Cantley, 2006; Manning and Cantley, 2007). It has regulatory (p85) and a catalytic (p110) subunit PI3K phosphorylate phosphatidylinositol, which is a phospholipid and an important cell membrane constituent in cell signaling as second messenger (Manning and Cantley, 2007). The catalytic subunit of PI3K p110α is encoded by a PICK3CA gene that is localized on chromosome 3q26.32 (Volinia et al., 1994).

Having oncogenic character, the PIK3CA gene is mutated in many different tumors, including CRC (Samuels et al., 2004; Cathomas, 2014). PIK3CA is mutated in approximately 10%–30% of CRCs and 80% of mutations found in two hot spots in exon 9 (E532K, E545K) and exon 20 (H1047R) (Samuels et al., 2004; Velho et al., 2005; Velho et al., 2008; Cathomas, 2014), which makes PIK3CA one of the most frequently mutated genes in CRC. Additionally, in the majority of studies, PIK3CA mutations have shown to be significantly associated with KRAS mutation and proximal tumor location (Barault et al., 2008a; Rosty et al., 2013). Though it has not been consistently reported (Mei et al., 2016), association of PIK3CA mutations (particularly for exon 20 mutation) with a worse clinical outcome and with a negative prediction of a response to targeted therapy by anti-EGFR monoclonal antibodies (mAbs), such as cetuximab and panitumumab, have been demonstrated (Ogino et al., 2009a; Souglakos et al., 2009; Whitehall et al., 2012; Mao C et al., 2012; Day et al., 2013; Cathomas, 2014). Besides lack of response to anti-EGFR mAbs, particularly in KRAS wild-type patients, a promising predictive value of PIK3CA mutations for poor survival in metastatic colorectal cancer (mCRC) patients treated with anti-EGFR mAbs has been suggested (Wu et al., 2013). Furthermore, some in vitro and in vivo studies demonstrated the benefit of aspirin in CRC pathogenesis harboring PIK3CA mutation, suggesting that PIK3CA mutation may predict response to aspirin therapy for colorectal cancer (Liao et al., 2012; Paleari et al., 2016; Gu et al., 2017; Zumwalt et al., 2017).

PIK3CA gene mutation, in many instances, occurs simultaneously with *PTEN* mutation (Chalhoub and Baker, 2009; Day et al., 2013). Acting as tumor suppressor gene, *PTEN*, which is localized on chromosome 10q23, encodes for a protein phosphatase and tensin homolog (PTEN) implicated in cell cycle regulation to prevent cells from growing and dividing too rapidly (Steck et al., 1997). Despite the significance of PTEN mutations in sporadic CRC remains still unclear some recent studies emphasized that PTEN loss together with BRAF mutations, PIK3CA exon 20 mutations could be predictive of worse outcomes in KRAS wild-type mCRC treated with anti-EGFR MAbs (Sood et al., 2012; Yang et al., 2013). Accordingly, this suggests the power of combined multiple biomarkers' clinical benefit from anti-EGFR antibodies in metastatic colorectal cancer (mCRC) (Sood et al., 2012; Yang et al., 2013; Therkildsen et al., 2014).

BRAF in CRC

The BRAF gene is located on chromosome 7 (7q34) and encodes the BRAF protein, which mainly functions in the MAP kinase/ERK signaling pathway. This intracellular pathway located downstream to the EGFR signaling cascade regulates cellular growth, differentiation, proliferation, senescence, and apoptosis of tumor cells (Peyssonnoux and Eychene, 2001). BRAF mutations have been described in approximately 9%–14% of CRC with wild-type KRAS mutation (Di Nicolantanio et al., 2008; Bokemeyer et al., 2012; Therkildsen et al., 2014). A point mutation, which leads to a valine substitution by glutamic amino acid, is the most common mutation in the BRAF gene (Ikenoue et al., 2003). It has been well demonstrated that KRAS and BRAF mutations are almost mutually exclusive in CRC (Rajagopalan et al., 2002; Fransen et al., 2004). In clinical aspects, BRAF mutations have been observed more frequently in colon cancer versus rectal cancer and also in proximal tumors compared to distal tumors (Domingo et al., 2005; Deng et al., 2008). Moreover it has been stated that BRAF mutation can be specific for sporadic cancer rather than Lynch syndrome–associated hereditary nonpolyposis colorectal cancer (Domingo et al., 2005; Funkhouser et al., 2012). In a recent study evaluating the correlation between KRAS/BRAF mutational status and clinicopathological features in advanced and recurrent CRC found that in nearly 60% of patients with CRCs with BRAF mutations, the tumor metastasized to the peritoneum compared with nearly 15% of patients with other subtypes (Souglakos et al., 2009). Furthermore it has been demonstrated that at least two-thirds of the poorly differentiated or mucinous carcinomas had BRAF mutation (Yokota et al., 2011). In mucinous adenocarcinoma that accounts for 5%–15% of all primary colorectal cancer (Symonds and Vickery, 1976; Lungulescu et al., 2017), a poor prognosis due to a poor response to oxaliplatin and/or irinotecan-based chemotherapy was reported (Catalano et al., 2009). Poor prognosis of BRAF mutant tumors may be partially explained with these histological properties. In a recent study including 2328 patients, BRAF mutation status and patient and tumor characteristics have been evaluated (Gonsalves et al., 2014). In this study BRAF mutated tumors were more likely found in patients aged 70 years or more, women, and non-Hispanic white people. Additionally, these tumors were also more likely to have four or more positive lymph nodes and T4 stages, which lead to more aggressive phenotype and poor prognosis (Gonsalves et al., 2014). BRAF-mutated and wild-type tumors can be differentiated by means of some molecular features, namely, decreased expression of CDX2, loss of P16, and positivity for DNA methyltransferase-3B, a marker of de novo CpG island methylation phenotype (CIMP) or SIRT1 histone deacylase expression (Baba et al., 2009). Molecularly BRAF mutations have been linked to high levels of MSI (MSI-H), MLH1 hypermetilation, and CIMP-high status (Ahlquist et al., 2008; Barault et al., 2008a, 2008b; Nosho et al., 2008; Funkhouser et al., 2012). Colon cancers that arise from serrated polyps are typically proximal tumors, dMMR status, and high CIMP and BRAF mutant phenotypes (Leggett and Whitehall, 2010). Velho et al. (2008) postulated that BRAF mutations are likely to precede MSI carcinomas because the frequency of mutation in serrated polyps is similar to that of MSI cancers but statistically different from MSS tumors. All these findings support that BRAF mutation may be an early event in colorectal carcinogenesis. Zlobec et al. (2009) by using a representative cohort of 404 sporadic colorectal cancers, described BRAF mutation as a significant molecular biomarker of poor outcome in patients with right-sided disease. This poor outcome was independent from tumor size, lymph node status, and MSI status.

The prognostic role of BRAF was tested in patients with resected stage II and stage II CRC. In this translational study, including three phase III trials (PETTAC-3, EORTC 40993, SAKK), 1564 paraffin blocks were collected and tested (Roth, 2010). It was noticed that BRAF tumor mutation status was prognostic for OS in all patients (both stages II and III combined) and in stage III alone. This is especially prominent in microsatellite stable (MSS) CRC with BRAF mutation (Roth, 2010). Lochhead et al. (2013) recently demonstrated that patients with BRAF wild/MSI high tumor had significantly

better survival compared to patients with BRAF mutant/MSS tumors. Moreover, it was stated that V600E BRAF mutations were a negative prognostic feature in MSS early stage resected CRC (Roth, 2010; Popovici et al., 2013). Patients with advanced and recurrent CRC treated with systemic chemotherapy (n = 229) were analyzed for KRAS/BRAF genotypes. The median overall survival (OS) for BRAF mutation-positive was 11.0 months, which was significantly worse than that for patients with wild-type BRAF (40.6 months, P < 0.001). This finding is consistent with other studies including both early stages and advanced stages of CRC (Roth, 2010; Ogino et al., 2009b; Fariña-Sarasqueta, 2010). Many reports indicated that BRAF mutations were associated with lack of response to anti-EGFR targeting agents (Di Nicolantanio et al., 2008; Laurent-Puig et al., 2009; Zlobec et al., 2009; Loupakis et al., 2009; Bokemeyer et al., 2012). One of these studies was conducted by Di Nicolantanio et al. (2008) and in this trial patients with BRAF mutation had significantly shorter overall and progression-free survival when treated with cetuximab- or panitumumab-based chemotherapy, compared with wild-type patients. Another study addressed the predictive value of BRAF in patients treated with cetuximab in first line, second line or further setting in combination with chemotherapy (Souglakos et al., 2009). No patients with BRAF mutation responded to cetuximab. The response rate to cetuximab-based chemotherapy was significantly lower in case of BRAF mutation than wild type (8.3% versus 38%) (De Rook et al., 2010). An analysis of CRYSTAL and OPUS studies revealed that BRAF mutation was detected in 9% of KRAS wild-type tumors. Although some benefit was observed in KRAS-wild/BRAF-mutant patients after adding cetuximab to the chemotherapy, this was not statistically significant (Tejpar et al., 2014). Overall findings may suggest that BRAF mutation appears not to be predictive for resistance to cetuximab plus chemotherapy (Van Cutsem et al., 2009). In the PRIME study, mCRC patients were treated with panitumumab plus FOLFOX4 and BRAF mutation was detected in 8% of the patients (Di Nicolantanio et al., 2008). In patients without RAS and BRAF mutations, 28.3 months overall survival was observed in the panitumumab–FOLFOX4 group. However, OS in patients with RAS wild-type BRAF mutant was 10.5 months. This study also confirmed that BRAF mutations appeared to confer a poor prognosis, regardless of the treatment group.

Recently, the predictive role of BRAF mutations in randomized trials that compared cetuximab or panitumumab plus chemotherapy (or monotherapy) with standard therapy or best supportive care (BSC) were included. Cetuximab and panitumumab were assessed in a meta-analysis (Doullard et al., 2013). Nine phase III trials and one phase II trial (six first-line and two second-line trials, plus two trials involving chemorefractory patients), including 463 RAS-wild/BRAF-mutant CRC patients, were analyzed. The addition of anti-EFGR treatment in the BRAF-mutant subgroup did not show significantly improved PFS (HR, 0.88; $p = 0.33$), OS (HR, 0.91; $p = 0.63$), and ORR (relative risk, 1.31; $p = 0.25$) compared with control regimens (Pietrantonio et al., 2015).

Taken together, CRC with BRAF mutation seems to originate from the serrated pathway of carcinogenesis and has aggressive features like poor differentiated or mucinous histology, four or more lymph nodes involvement, and have a tendency to peritoneal spread. Moreover, association of BRAF mutations with the efficacy of anti-EGFR therapy remains controversial (van Brummelen et al., 2017), but BRAF mutation appears to be a strong negative prognostic marker for patients with KRAS wild-type colorectal cancer treated with anti-EGFR therapies. All these findings support the idea that BRAF mutations are negative prognostic biomarkers.

KRAS in CRC

Ras is a family of genes that encode a class of 21 kD membrane-bound proteins that bind guanine nucleotides and have intrinsic GTPase activity. This family consists of three functional genes—H-ras, K-ras, and N-ras—which encode highly similar proteins. These are the most common oncogenes in human cancer. KRAS is the most frequently altered gene and is located on the short arm of chromosome 12. The action of this gene manifests in the RAS-RAF-MAPK pathway, downstream of epidermal growth factor receptor (EGFR) (Moerkerk et al., 1994). Of the KRAS mutations, 90% are located in exon 2 and 10% in exons 3 and 4. Activating mutations in exons 2 and 3 have been suggested to have similar effects

on RAS GTPase activity, whereas exon 4 mutations have been suggested to increase GDP to GTP exchange (Malumbres and Barbacid, 2003).

KRAS mutations are believed to be an early event in colorectal tumorigenesis. In fact, it is not required to initiate adenomas but mainly contributes to their progression. Mutations of the KRAS gene have been identified in tissues from both adenoma and carcinoma cases, but at much lower frequencies in colon adenoma tissues than in carcinoma tissues. Many have hypothesized that development of KRAS mutation is an important role in the multistep process early in the carcinogenesis. Nearly 30%–40% of CRCs have KRAS mutations and the point mutations are the most common mutation in KRAS gene (Capella et al., 1991; Oudejans et al., 1991). Almost 90% of KRAS mutations occur in codons 12 and 13 in the phosphate-binding loop, and mutations in either codon have transforming capacity (Andreyev et al., 1998). In a recent study, seven of the most common KRAS mutations in codon 12 and codon 13 were examined in 2478 colorectal tumor samples (Yoon et al., 2014). Mutations in these codons were detected in 35.4% of tumors (27.6% in codon 12 and 7.8% in codon 13). Within codon 12, 82% of mutations were found in the second base position (G12D). In vitro studies illustrated that when compared to codon 13 mutations, Kras codon 12 mutations have greater transforming ability characterized by inhibition of apoptosis, enhanced loss of contact inhibition, and increased predisposition to anchorage-independent growth (Moerkerk et al., 1994).

Prognostic value of KRAS has been investigated in several studies in both early and advanced stages; however, inconsistent results had been released in early trials (Oudejans et al., 1991; Dix et al., 1994; Tanaka et al., 1994). Therefore RASCAL (Kirsten Ras In Colorectal Cancer Collaborative Group) investigators aimed to definitively determine whether the presence of a KRAS mutation is of prognostic significance. In this study 2721 patients from 22 groups in 13 countries were recruited (Andreyev et al., 1998). The study group included patients of all stages but only 10% were Dukes stage D. KRAS codon 12 or codon 13 was detected in 37.7% of the tumors. Mutations were not associated with sex, age, tumor site or Dukes stage. Multivariate analysis suggested that the presence of a mutation increased the risk of recurrence ($p < 0.001$) and death ($p = 0.004$). When comparing all of the specific point mutations, overall survival was adversely affected by the presence of a glycine to valine amino acid substitution on codon 12 (C12V).

In a similar study conducted in 404 sporadic and 94 hereditary CRC patients, KRAS G12D mutation had poorer prognosis compared to patients with other KRAS mutations (Zlobec et al., 2010). In the RASCAL II study, association of Kras mutations with different stages of colorectal cancer was evaluated. The results confirmed that C12V mutation of KRAS was associated with a 50% increase of relapse or death in patients with Dukes stage C cancer but not the other stages of CRC (Andreyev et al., 2001).

The prognostic role of KRAS/BRAF was tested in patients with resected stage II and stage II CRC. In this translational study, paraffin blocks from PETTAC-3, EORTC 40993, and SAKK trials were evaluated (Ogino et al., 2009b). KRAS and BRAF tumor mutation rates were 37% and 7.9%, respectively, and were not significantly different according to tumor stage. In a multivariate analysis, KRAS mutation was associated with grade of the tumor. In univariate and multivariate analysis, KRAS mutations did not have a major prognostic value regarding relapse-free survival (RFS) or OS. Other recent trials, NCCTG N0147 and PETACC-8, have evaluated chemotherapy with and without cetuximab for patients with stage III disease (Blons et al., 2014; Gonsalves et al., 2014). In the first trial, primary tumors were assessed for KRAS and BRAF mutations and defective MMR status (Gonsalves et al., 2014). The results showed that three-year disease-free survival in patients with wild-type KRAS was significantly better than that in patients with KRAS mutants (72.3% versus 64.2% HR = 0.7 p = 0.004). It is suggested that KRAS mutations are independent prognostic factors. Furthermore, tumors with KRAS mutations were less likely to have dMMR and high-grade histology but more often right-sided. Consistent with this study, in the PETTAC8 trial, the prognostic impact of KRAS exon 2 mutations in stage III colon cancer were analyzed (Blons et al., 2014). Frequency KRAS mutations were 38.5% and presence of the mutation was linked to shorter TTR (p < 0.001). In this study it was also indicated that KRAS mutation was more frequent in women, in proximal tumors, and was associated with age, and no vascular and lymphatic invasion. Furthermore,

codon 12 mutations were detected in 79% of mutated cases and significantly associated shorter TTR. A similar trend was also noted for the pG13D mutations suggesting that KRAS exon 2 mutations are related to a worse prognosis.

Both RASCAL and PETTAC8 showed that a tumor-carrying codon 12 mutation had a statistically significant impact on worse PFS (Oudejans et al., 1991; Malumbres and Barbacid, 2003). KRAS 12 mutations may confer a more aggressive behavior than KRAS 13 mutations (Guerrero et al., 2000). However, a similar trend was shown also for codon 13 mutations in different studies. In 2002 Bazan et al. found that codon 13 mutations had a stronger predictive value than any K-ras mutations. Multivariate analysis showed that this specific codon 13 K-ras mutation was an independent prognostic factor for DFS and OS in 160 untreated patients who underwent surgery for a primary tumor. The prognostic role of KRAS/BRAF mutation was analyzed in the CRYSTAL and OPUS studies including mCRC patients who received FOLFIRI/FOLFOX with or without cetuximab as the first-line treatment (Bokemeyer et al., 2012). Subgroup analyses in this study showed that patients with KRAS codon 13 D mutations showed poor prognosis. Another study conducted by Yokota et al. (2011) also supports the idea that OS for patients with KRAS codon 13 mutation was significantly worse than for those with wild-type tumors.

KRAS mutation status was assessed in a series of resected or sampled CRC liver metastases (Nash et al., 2010). Nash et al. (2010) observed that KRAS mutations have also been associated with more rapid and aggressive metastatic behavior of CRC liver metastases. KRAS mutation was independently associated with poor prognosis after liver resection. Both codon 12 and codon 13 mutations in the KRAS gene may result in different biological properties that might lead to poor prognosis in colorectal cancer.

In conclusion, presence of KRAS mutations both in codon 12 and 13 are associated with inferior survival both in early stage and mCRC patients.

Immune-related markers in CRC

In the last few years years increasing proof from genetically engineered mice and clinical epidemiology have shown the significance of the immune barrier in formation and progression of tumors (Hanahan and Weinberg, 2011). The development of cancer in immunodeficient mice and the existence of antitumor immune responses in colon and ovarian cancers has been supported through clinical epidemiology studies (Pagès et al., 2010). During growth of cancer cells, the major histocompatibility complex (MHC) interacts with T-cell receptor (TCR) and expresses CD4+ or CD8+ lineage markers (Sukari et al., 2016). Both innate immunity (natural killer [NK] cells, unconventional T lymphocytes, and tumor infiltrating macrophages [TIM]) and adoptive immunity (intratumoural memory CD8T cells, and tumor-infiltrating CD45RO$^+$ memory T lymphocytes [CD45RO$^+$ T cells]) have an important role in the development as well as in prevention and recurrence of CRC tumors leading to better prognosis of colorectal cancers (Sandel et al., 2005; Galon et al., 2006; Sanchez-Castañón et al., 2016). Expression of tumor-associated antigens (TAAs), such as Wilms' tumor gene 1 (WT1), carcinoembryonic antigen (CEA), mucin 1 (MUC1), and melanoma-associated antigen gene (MAGE) in human CRC cells make it an appropriate and potential target for immunological-based therapies (Kajihara et al., 2016). In CRC genetic and epigenetic aberrations due to CIN pathway, MSI phenotype, and CIMP could lead to manifold oncogenes mutations, resulting in immunogenic CRC (i.e., capable of inducing an immune response). Thus, since dendritic cell–based cancer immunotherapy induces TAAs (CEA, WT1, MAGE, and MUC1) in CRC cells, it is anticipated that a combination of immunotherapy with conventional chemotherapy and/or radiotherapy, surgery, and even with immune-checkpoint inhibitors could mediate a potent antitumor effect (Koido et al., 2012; Kajihara et al., 2016). Furthermore, since TILs interact most closely with the tumor cells and reflect tumor host interactions more accurately (Holmes, 1985), it is demonstrated that CRC patients with MSI tumors owing to high levels of TILs have better prognosis (Michael-Robinson et al., 2001; Phillips et al., 2004).

PD-L1 is the ligand for transmembrane programmed cell death protein 1 (PD1) (also known as cluster of differentiation 279-CD279), which is expressed on T cells and B cells and is a critical immune checkpoint pathway responsible for mediating tumor-induced immune suppression (McDermott and Atkins, 2013). PD-L1 is highly expressed in several cancers, including CR. It's well established that this leads to cancer immune evasion, and consequently, this is often associated with a worse prognosis (McDermott and Atkins, 2013). Accordingly, interaction of PD-1 with its ligand PD-L1 which promotes the exhaustion of peripheral T-effector cells (CD4+, CD8+, Treg cells), induction of regulatory T cells (Treg) (Zhang X et al., 2015; Francisco et al., 2009), have been shown to associate with poor prognosis in CRC. Furthermore, the PD-1/PD-L1 interaction directly inhibits apoptosis of the tumor cell (Francisco et al., 2009). To this end, large pharmaceutical companies have developed drugs targeting cytotoxic T cell-associated antigen 4 (CTLA-4), programmed cell death receptor 1(PD-1), and its ligands PD-L1 and PD-L2 in mCRC (Sanchez-Castañón et al., 2016). The association between the MSI status and the efficacy of the programmed cell death 1 (PD-1) receptor inhibitor pembrolizumab has been also demonstrated (Pardoll, 2012; Boland and Ma, 2017; Gong et al., 2017). Respectively, the U.S. Food and Drug Administration (FDA) has granted accelerated approval to checkpoint inhibitor pembrolizumab (Keytruda) for pediatric and adult patients with unrespectable, metastatic, or MSI-H that has progressed following chemotherapy (Bilgin et al., 2017; Chang et al., 2017; Lemery et al., 2017).

Cytokines are small, secreted proteins providing interactions and communications between cells through cell signaling. They are other immune-related markers that have an important role in innate and specific immune response (Van der Meide and Schellekens, 1996). Despite some cytokines approved in cancer therapy, such as IL-2 in the treatment of melanoma and renal cell carcinoma and IFN-α (also called Multiferon) in the treatment of multiple hematological malignancies, cervical cancer, carcinoid syndrome, medullary thyroid cancer, basal cell, and squamous cell carcinoma, studies involving cytokine therapies in CRC are limited so far (Lynch and Murphy, 2016). However, a phase I study of 33 patients with advanced solid tumors, including four CRC patients, demonstrated that daily subcutaneous administration recombinant IL-10 (AM0010) resulted in stable systemic Th1 immune stimulation. Yet, even though cytokine therapy represents an important approach, it needs more investigation for being used in CRC as a therapy target (Lynch and Murphy, 2016). Major histocompatibility complex class I (MHC-I) low expression is associated with poor prognosis in colorectal patients.

Protease related markers in CRC: Focus on cathepsin B (CATSB)

An important property of metastatic cells is their ability to dissolve the extracellular matrix (Liotta et al., 1986; Hart and Saini, 1992). In fact, an important level of enzymes, degrading the extracellular matrix, is often found in malignant tumors. Most of these enzymes may be involved in the activation, inhibition, or regulation of other enzymes released by normal or tumor cells. The degradation of the extracellular matrix depends on a series of extracellular neutral proteases; a protease (also called peptidase, proteinase or peptide hydrolase) is an enzyme that participates in the catabolism (degradation) of proteins by cleaving peptide bonds. Intracellular proteolysis is an extremely controlled process that takes place in all cell compartments. Proteolytic activity is involved in different processes: protein renewal especially during cell growth as well as protumorigenic processes, such as cell transformation, proliferation, invasion, and malignant progression (Sloane and Berquin, 1993; Fennelly and Amaravadi, 2017).

Several independent studies underlined the important role of lysosomal cysteine proteases cathepsin B (CATSB) (Herszènyi et al., 1999; Tan et al., 2013; Bian et al., 2016) and L (CATSL) (Herszènyi et al., 1999; Tan et al., 2013; Sudhan and Siemann, 2015), and the serine protease urokinase type plasminogen activator (UPA) with its inhibitor type-1 (PAI-1) (Herszènyi et al., 1999; Sakakibara et al., 2005) in colorectal cancer invasion and metastasis, and they appear to be synthesized in an increased amount and exaggeratedly secreted (Sloane and Berquin, 1993). The human cathepsin B gene is a "housekeeping" gene having 12 exons and is located on chromosome 8 at p 22. Being very variable in 5′ untranslated region (UTR) the expression of human cathepsin B is regulated through alternative splicing of the mRNAs producing several transcript species (Gong et al., 1993; Berquin et al., 1995).

Close relationship between the intensity of angiogenesis and overexpression of the CATSB mRNA in resected colon adenocarcinoma is demonstrated (Kruszewski et al., 2004), which could lead to a poor prognosis (Khan et al., 2004). Moreover, in human tumors, such as melanoma, breast cancer, and colon cancer, absence of exon 2 and exon 3 are also demonstrated (Gong et al., 1993). Likewise, it is reported that CATSB enzyme activity levels are inversely correlated with the Dukes stages; levels of expression of CATSB mRNA in colon carcinomas were higher in tumors belonging to the Dukes A1 and B1 stages (Dukes, 1932) than in Duke stage B3 carcinomas and in the normal mucosa (Murnane et al., 1991). In addition, the increase in CATSB activity is associated with increased CATSB mRNAs in the earliest stages, suggesting that CATSB elevation may be important in tumors that are about to invade the intestinal wall (Tan et al., 2013). In agreement with other studies, a comparative study of CATSB gene transcripts in colon tumors and the corresponding normal tissues established a relationship between the spliced form for exon 2 and the complete form of the mRNA of cathepsin B comprising exons 1–5. Accordingly, authors demonstrated that the ratio form spliced for exon 2/complete form was significantly increased in colorectal tumors compared with matched normal samples, suggesting that alternative splicing of CATSB mRNA at the 5′ UTR may be an indicator of cellular transformation during colorectal carcinogenesis (Hizel et al., 1998). Furthermore, a recent study in Middle East populations diagnosed with CRC confirmed that there was a significant increase in the CATSB expression at the mRNA and protein levels in tumor tissue compared to normal adjacent mucosa in late stage CRC suggesting that CATSB may be an important prognostic biomarker for late stage CRC with lymph node metastases in this population (Abdulla et al., 2017). Hence, characterization of the aberrant splice variants may improve understanding of malignant transformation in CRC.

In light of the aforementioned studies, since proteolytic enzymes play a crucial role in tumor development and progression due to their abilities to degrade extracellular matrix, validated proteinase markers may be of interest as independent, predictive, and prognostic factors of malignant progression of neoplastic disease. Furthermore, high expression of proteinases such as CATSB in advanced stages of cancer can offer significant therapeutic benefit when using proteinase inhibitors in the treatment of CRC leading to new methods for individualized cancer therapy in the precision medicine approach (Guzińska-Ustymowicz, 2006; Duffy and Crown, 2008; Herszényi et al., 2014).

Telomere length in CRC

Telomeres are DNA sequences containing noncoding arrays of TG-rich/TTAGGG tandem repetitive nucleotide sequences (approximately 1000–2000 TTAGGG tandem base pair) at each end of a chromosome (Moyzis et al., 1988) playing a crucial role in facilitating chromosome replication and protecting the end of the chromosome from fusion with neighboring chromosomes or from DNA repairing activities that reseal chromosome internal DNA breaks occurring during DNA damage (De Lange, 2009). During DNA replication, telomere shortening occurs due to incomplete replication of telomere, and continuous shortening leads to chromosomal degradation and cell death (O'Sullivan and Karlseder, 2010; Shammas, 2011). In order to maintain telomere elongation, at the end of each cell division telomere is replenished by "ribonucleoprotein complex," which is reverse transcriptase enzyme telomerase, and it is actively expressed in more than 80% of tumors, particularly in tumors with metastatic potential allowing to escape from the inhibition of cell proliferation because of shortened telomeres (Aschacher et al., 2016; Ivancich et al., 2017). In humans, while detectable levels of telomerase activity is limited to stem cells, germ line cells, and also cardiovascular cells, its activity is absent in most normal cells (Collins and Mitchell, 2002; Shay and Wright, 2010; Pech et al., 2015; Zurek et al., 2016; Ivancich et al., 2017). Since cancer cells necessitate a mechanism to maintain their telomeric DNA in order to continue dividing indefinitely, shortening of telomeres and the induction of cell senescence could be a potential cancer biomarker and therapeutic tool as anticancer drugs (Philippi et al., 2010; Ruden and Puri, 2013; Piñol-Felis et al., 2017).

It is widely acknowledged that telomere dysfunction and telomere length (TL) is at the origin of various degenerative disorders as well as predisposition to cancer (Shin et al., 2006; Shammas, 2011; Shay and Wright, 2011; Armanios and Blackburn, 2012). Although telomerase activity is associated with poor prognosis (Garcia-Aranda et al., 2006; Bertorelle et al., 2013), the prognostic role of telomere length in CRCs is still unclear (Wang W et al., 2017). Although some scientific data from the available research publication is controversial (Wang W et al., 2017), a broad spectrum of studies emphasize the critical association between TL and CRC for improved molecular diagnosis. Accordingly, TL was smaller in CRC tumors than in the healthy adjacent mucosa, suggesting that longer telomeres could be indicative of poor prognosis in CRC (Mzahma et al., 2015; Zhang C et al., 2015; Fernández-Marcelo et al., 2016; Balc'h et al., 2017). A correlation between TL and tumor stage is also reported, demonstrating the presence of the longer TL in advanced CRC tumors (Engelhardt et al., 1997; Gertler et al., 2004). A significant association between length and KRAS mutation status has also been reported (Balc'h et al., 2017). Furthermore, recently, the clinical utility of TL has been demonstrated as a potential biomarker for mCRC patients with nonmutated KRAS status treated with anti-EGFR indicating an elevated telomere length and a good response (inhibition of proliferation) to anti-EGFR (Augustine et al., 2015; Augustine et al., 2017).

Though some current evidence supports the use of telomere status as a prognostic factor for overall survival and treatment in CRC patients, additional large well-designed prospective cohort studies are needed to further assess the telomere status as an independent prognostic factor for personalized approach in CRC patients.

CRC EPIGENETICS

Epigenetic alterations in cancer include abnormal DNA methylations, histone modifications, and expressions of miRNAs. According to the literature, these alterations in colorectal cancers are more frequent than genetic alterations (Puccini et al., 2017). Epigenetic changes in cancers were unfamiliar to scientists until 30 years ago when the first results were published about DNA methylation alteration in human cancers. Accordingly, these results comparing human colorectal tumors with normal mucosa of the same patient demonstrated the existence of high hypomethylation of almost one-third of single copy genes (Feinberg and Vogelstein, 1983).

DNA methylation

In human cancers, DNA methylation occurs at cytosine residues in regions with a high frequency of CpG sites (CpG dinucleotides), which constitute CpG islands. Almost 70% of the human gene promoter region harbors CpG islands with high CpG content (Saxonov et al., 2006), which are methylated by DNA methyltransferases (DNMTs) (Puccini et al., 2017). Although hypomethylation of CpG islands in the promoter region results in overexpression of the genes or set of genes (Ehrlich, 2009), hypermethylation could be an important cause of loss of expression of genes (Jin et al., 2011). DNA hypermethylation in colorectal cancers is due to CIMP, which strongly correlates with clinicopathological and molecular characteristics of CRC (Issa, 2004; Nazemalhosseini Mojarad et al., 2013). Besides MLH1, MSH2, MSH6, and PMS2 gene mutations, DNA MMR system hypermethylation could be another source for MSI of colorectal tumors (Grady and Carethers, 2008) On the other hand, global DNA hypomethylation has a crucial role in tumor formation (Antelo et al., 2012; Pavicic et al., 2012). LINE-1 or short interspersed nucleotide elements (SINE, or Alu) are mostly hypomethylation sites in DNA in most cancers including colorectal cancers, and some studies showed LINE-1 hypomethylation correlated with worse survival in colorectal patients (Antelo et al., 2012; Pavicic et al., 2012). Overall, it is obvious that epigenetic deregulations caused by DNA hypermethylation and hypomethylation contribute separately to the process of CRC carcinogenesis Hence, acknowledging epigenetic data–related methylation status might represent a promising tool for a more precise diagnostic, prognostic, and personalized therapeutic approach (Lam et al., 2016; Puccini et al., 2017; Werner et al., 2017).

Histone modifications

The chromatin in the mammalian genome includes tightly packed DNA inside the nucleus. The smallest unit of chromatin is nucleosomes formed by wrapping of 147 bp of DNA around eight histone proteins, (including H3/H4 tetramer and two H2A/H2B dimers). Due to the numerous direct protein–DNA interactions, nucleosome has a highly stable unit. During the conducting of genomic functions, processive enzymes (with the ability to catalyze consecutive reactions without dissociating from its substrate) reach the coding regions of genes (Das and Tyler, 2013). Except for DNA methylation-based transcriptional

regulation of gene expression, posttranslational modifications at N-terminal tails of the histones control chromatin structure as well as gene expression (Das and Tyler, 2013). Histone proteins expose various posttranslational modifications, which include methylation, acetylation, phosphorylation, deamination, and ribosylation. These modifications have a regulatory role in gene expression due to determination of chromatin structure as tight or loose (Sachan and Kaur, 2015). The type of modification and the specific amino acid that is involved in modification control the effect of histone modification. Acetylation/deacetylation and methylation/demethylation of lysine and arginine residues within histone tails are the most extensively characterized modifications in CRC. Most of the acetylation locations were determined on the N-terminal tail of histones due to the easy access for modifications (Goel and Boland, 2012).

The acetylation reactions of lysine amino acids on histone tails are reversible modifications and serve as activators or repressors of transcription function. For example, hypoacetylation silences gene expression, whereas hyperacetylation permits active gene transcription due to the destabilization of chromatin fibers and increasing the mobility of nucleosomes (Das and Tyler, 2013). These acetylation reactions are carried out by removing acetyl groups from lysine amino acids by histone acetyltransferases (HATs) and histone deacetylases (HDACs), as coactivators and corepressors of transcription. The balance between HATs and HDACs controls the transcriptional inhibition of tumor suppressor genes. The opposite of acetylation associates with transcriptional repression of genes (Bardhan and Liu, 2013). HDACs play a crucial role in CRC development. Up to now, 18 HDACs have been identified as corepressor multiprotein complexes and they are divided into four classes: Class I (HDAC1, HDAC2, HDAC3, and HDAC8), Class II (HDAC4, HDAC5, HDAC6, HDAC7, HDAC9, and HDAC10), Class III (Sirt1, Sirt2, Sirt3, Sirt4, Sirt5, Sirt6, and Sirt7), and Class IV HDACs (only HDAC11). Classes I, II, and IV HDACs have similar features due to their structure and function, but class III does not show similarity (Bolden et al., 2006). Increased expression of several HDACs has been determined in CRC. In addition, upregulation of Class I HDACs (HDAC1, HDAC2, HDAC3) has been associated with reduced patient survival in CRC (Ashktorab et al., 2009). The elevated levels of HDAC2 are accompanied with the hypoacetylation of H4K12 and H3K18 histones on the multistep carcinogenesis process of CRC (Ishihama et al., 2007). The overexpression of HDAC3 was recorded in duodenal adenomas of Apc1638N/+ mice and human colon cancers. Silencing of HDAC3 in colon cancer cell lines induced growth inhibition, and shortened survival and apoptosis (Wilson et al., 2006). Silencing of HDAC4 expression in HCT116 colorectal cancer cells resulted in growth inhibition and increased apoptosis and p21 transcription (Wilson et al., 2008).

Like histone acetylation, methylation of lysine and arginine amino acids on N-terminal histone tails is a reversible posttranslational modification. It is regulated by histone methyltransferases (HMTs) and histone demethylases (HDMs) (Klose and Zhang, 2007). Each of the HMTs have a role either alone or in a complex with other HMTs. Histone H3 lysine 4 (H3K4) is methylated by SET1 and MLL, and demethylated by the LSD1 and JARID1 family of HDMs (Barski et al., 2007). In histone H3, multiple-methylation can be observed at multiple lysine sites (including histone H3 lysine 4 [H3K4], histone H3 lysine 9 [H3K9], histone H3 lysine 27 [H3K27], histone H3 lysine 36 [H3K36], and histone H3 lysine 79 [H3K79]) or at a single lysine (up to three methyl groups) (Bardhan and Liu, 2013). H3K27 trimethylation (H3K27me3), controlled by the RAS pathway, is frequently correlated with gene silencing and affects cyclin D1 and E-cadherin expression. Increased expression of RAS oncoprotein could influence gene-specific histone modification during the epithelial mesenchymal transition of the CRC cell line, Caco2 (Peláez et al., 2010).

microRNAs

MicroRNAs (miRNAs), small noncoding RNAs 18–25 nucleotides long, are involved in the regulation of proliferation, cellular growth, differentiation, and apoptosis by controlling the translation of target mRNAs (Huang et al., 2011). miRNA biosynthesis includes a complex process. Initially, primary miRNAs (pri- miRNAs, up to 1 kb) are transcribed by the RNA polymerase II, second precursor miRNAs (pre-miRNAs, 60–70 base pairs long) are generated from pri-miRNAs by the cleavage of RNase III Drosha. Finally, formed mature miRNA is integrated into the RNA-induced silencing complex to induce the degradation and suppression of the target mRNA (Corté et al., 2012). The initial data, obtained by Michael et al.'s research, suggested that miRNA-143 and miRNA-145 were steadily downregulated at different stages of CRC (Michael et al., 2003). Then, accumulation of information about CRC-associated miRNAs supported that miRNAs have a role on the initiation, progression, and development of CRC (Strubberg and Madison, 2017). They can have tumor suppressors or oncogene function according to the target gene (Zhu and Fang, 2016). Different miRNA profiles have been assessed in several studies: hsa-miR-17-3p, hsa-miR-20a, hsa-miR-21, hsa-miR-25, hsa-miR-31, hsa-miR-92, hsa-miR-93, hsa-miR-96, hsa-miR-106 (includes 106a and 106b), hsa-miR-135b, hsa-miR-183, and hsa-miR-203 were detected as upregulated in CRC; and hsa-miR-1, hsa-miR-30a, hsa-miR-126, hsa-miR-133b, hsa-miR-143, hsa-miR-145, hsa-miR-191, and hsa-miR-192 were detected as downregulated in CRCs when compared with healthy individuals (Corté et al., 2012; Ng et al., 2009; Bandrés et al., 2006).

Circulating miRNAs have been found in the plasma or serum samples of CRC patients. Therefore plasma miRNAs can be used as noninvasive early detection biomarkers. hsa-miR-17-3p and hsa-miR-92a-1 have been found to be associated with CRC in Ng et al.'s study (Ng et al., 2009). Especially, the function of hsa-miRNA-92a-1 was found and it was validated that it is associated with the discrimination of advanced adenoma patients from healthy individuals. Compared to healthy individuals, hsa-miR-29a and hsa-miR-221 have been detected as overexpressed in the plasma samples of all stages of CRC patients (Huang et al., 2010; Pu et al., 2010). As screening biomarkers, miRNAs from fecal specimens of CRC patients have been evaluated. In a study, it was found that hsa-miR-17 cluster and hsa-miR-135 expression levels were significantly elevated in CRC patients than healthy individuals (Koga et al., 2010). In another study from Link and his colleagues hsa-miR-21 and hsa-miR-106a have been evaluated as diagnostic biomarkers due to their overexpression in CRC patients compared with healthy controls (Link et al., 2010).

Dysregulation of miRNA synthesis and function have a role in the development and progression of CRC due to their effect on the translation of oncogene or tumor suppressor genes. In two independent studies, over-expression of hsa-miR-21 was associated with poor prognosis (Corté et al., 2012). In a group comprised of stage II CRC patients, low expression levels of hsa-miR-320 or hsa-miR-498 were correlated with poor progression free survival compared to tumors with high expression (Schepeler et al., 2008). In another study that was conducted by Cheng and his colleagues, hsa-miR-141 was found to be related with stage IV colon cancer and poor survival (Cheng et al., 2011). High plasma levels of hsa-miR-221 were associated with poor survival by Pu and his colleagues (Pu et al., 2010).

In literature it was reported that miRNAs have a role in drug resistance or sensitivity. Therefore miRNAs can have the potential to predict anti-cancer agents' response. In a study conducted by Nakajima et al. (2006), hsa-let-7g and hsa-miR-181b were found related with 5-fluorouracil-based treatment response in colon cancer. The miRNA let-7 family controls the translation of KRAS protein and has a role in the resistance to EGFR-targeting monoclonal antibodies. A polymorphism on the let-7 complementary site of KRAS was correlated with increased KRAS expression, which is associated with resistance to anti-EGFR treatment (Johnson et al., 2005; Chin et al., 2008). Different studies confirmed that miRNA profiling plays an important role in the understanding of mechanisms underlying colorectal carcinogenesis, cancer progression, and metastases. Application of miRNAs as cancer biomarkers for early diagnosis, prognosis, and treatment response prediction are promising, but profiling studies need to be validated due to various differences between studies that can be explained by several components such as sample type, tumor location, stage of disease, genetic background, and methodology (Slattery et al., 2011).

PHARMACOGENETICS (PGX) FOR PRECISION MEDICINE IN CRC CHEMOTHERAPY

In many countries, lack of response, that is, resistance and adverse drug reactions (ADRs) are a significant public health problem. Among individuals, genetic differences in pharmacogenes, that is, any gene that interacts with a pharmaceutical drug, significantly affects drug clearance and responses. Recent breakthroughs in genomics and improvements in bioinformatics have uncovered the leading role of genes involved in drug metabolism and disposition. Variation in drug-response phenotypes, in the forms of resistance and ADRs as well as the motivation to administer medications more efficiently, are one of important major components of precision medicine for personalized treatment (Aydin Son et al., 2013).

Accordingly, pharmacogenetics (PGx), which studies genetic variations among individuals to predict responses to therapeutic agents, has gained considerable momentum for both prescribed and over-the-counter medications. Dating back to the 1950s (Kalow, 1962), PGx, which may uncover the associations between genetic variation and ADRs, can be regarded as the 21st century's answer for rational drug use by selecting "the right drug" to "the right patient" at "the right dosage" to perform personalized drug dosing in clinic (Ingelman-Sundberg, 2001; Becquemont, 2009; Ferraldeschi and Newman, 2011; Aydin Son et al., 2013). In contrast to the current "one size fits all" approach for drug prescription and dosing, pharmacogenetics/genomics being among one of the first clinical applications in the field of genomics medicine represents the near-term payoff for the pioneering area of precision medicine for personalized health care (Ingelman-Sundberg, 2001; Becquemont, 2009; Ferraldeschi and Newman, 2011; Aydin Son et al., 2013).

Since the completion of the Human Genome Project in 2003 our world is today going through a new era that is commonly referred to as the "postgenomic era." Meanwhile pharmacogenetics has gradually evolved into pharmacogenomics through the intensive application of genome-wide association studies (GWAS), and today these two terms are used interchangeably. (Pirmohamed, 2001; Lindpaintner, 2002; Moraes and Góes, 2016). However, while pharmacogenetics aims to distinguish responders from nonresponders (i.e., starts with an unexpected drug response result and looks for a genetic cause), pharmacogenomics as the whole-genome application takes a much more global approach to the impact of variation in genetic information on drug response, by looking for genetic differences within a population that explains certain observed responses to a drug or susceptibility to a health problem (Pirmohamed, 2001). Moreover, much of the work in the area of pharmacogenomics is focused on identification of disease susceptibility genes representing potential new molecular drug targets for therapeutic intervention by correlating gene expression or single nucleotide polymorphisms (SNPs) with a drug's efficacy or toxicity (Pirmohamed, 2001; Sheffield and Phillimore, 2009; Milos and Seymour, 2004). However, this article does not distinguish between those terms, but the abbreviation "PGx" will be used to cover both pharmacogenetics and pharmacogenomics throughout this chapter. Though application of PGx testing is not widely accepted in routine medical practice, the scenario is likely to be changing in the future for pharmacogenetics based diagnostics (PGDx) (Cecchin et al., 2017; Lemay et al., 2017; Manson et al., 2017). In PGx research, numerous studies have used candidate gene approaches to identify toxicity biomarkers. Since the mechanisms of action of cancer drugs through which a drug molecule produces its pharmacological effect are well demonstrated, these candidate gene studies are relatively prosperous compared to the identification of cancer susceptibility genes (Rodríguez-Antona and Taron, 2015). Furthermore, starting with the determination of phenotypes (in the case of PGx studies, patients responding or developing drug-induced ADRs/toxicities, and patients who respond to the treatment without ADRs as controls), GWAS, which is a very useful method for identifying common SNPs with a minor allele frequency (MAF) >5%, became a very powerful tool for PGx study to identify SNPs associated with ADRs/toxicities or drug resistance phenotype. (Daly, 2010; Ritchie, 2012).

Being an important backbone treatment of cancer patients with lymph node positive and metastatic disease, cytotoxic chemotherapy is becoming increasingly personalized because of large differences among treatment outcomes (resistance/toxicity) (Hammond et al., 2016). Together with nonmodifiable genetic factors other nonmodifiable factors (ethnicity, age, gender, organ function) and modifiable factors (smoking, alcohol consumption, comorbidity, pregnancy, drug–drug interactions, and drug–food interactions) have a critical impact on drug efficacy and toxicity (Marques and Ikediobi, 2010). While chemotherapeutic agents act through cytotoxic effects, targeted drugs are directed toward a tumor-specific alteration exhibiting cytostatic effects (Kummar et al., 2006).

In cancer PGx predictive biomarkers for drug responses due to genetic variations can originate from the host (individual's germ-line mutations) and tumor cells (somatic/de novo mutations) (Pesenti et al., 2015). The former affects pharmacokinetic (PK) (i.e., absorption, distribution, metabolism, and excretion [ADME]) profile of the drug and are more likely to be related to prediction of ADRs through alteration of drug plasma levels such as UGT1A1 variants and irinotecan-induced neutropenia (Rodríguez-Antona and Taron, 2015; Bertholee et al., 2017), whereas the latter affects the selection of chemotherapeutics related to effectiveness of the response of a tumor to the treatment, such as cetuximab and EGFR/KRAS status (Deenen et al., 2011b; Rodríguez-Antona and Taron, 2015). Polymorphisms in the human genome have been implicated in the expression and functioning of enzymes that is involved in the distribution and metabolism of anticancer drugs affecting drug efficacy and toxicity, and consequently the outcome of treatment of patients (risk/benefit) (Weng et al., 2013). Surgical resection, particularly for non-mCRC as the first choice treatment, and chemotherapy are extensively used in the treatment of CRC patients (Pozzo et al., 2008; Chua and Morris, 2012). In spite of complete resection, approximately half of CRC patients could still develop metastases (Nordlinger et al., 2009; Power and Kemeny, 2010). Fluoropyrimidine (e.g., 5-fluorouracil, capecitabine), irinotecan, and oxaliplatin are the most common anticancer drugs used to treat colorectal cancers (Aparicio et al., 2005; Braun and Seymour, 2011). Moreover, for more effective therapy in conjunction with chemotherapy and the addition of the monoclonal antibodies, such as cetuximab, bevacizumab, panitumumab, or vascular endothelial growth factor aflibercept, have ameliorated the median survival of mCRC approximately from 8 to 24 months (Gallagher and Kemeny, 2010; Van Cutsem et al., 2012). However, important interpatient variability in drug resistance and toxicity has been reported in many mCRC patients receiving combination therapy FOLFIRI (5-fluorouracil infusion leucovorin, and irinotecan) or FOLFOX (fluorouracil, leucovorin, and oxaliplatin). Chemotherapy side effects hamper the efficient treatment of mCRC due to somatic (acquired) mutation in tumors and/or individual's germ-line DNA (intrinsic) mutation affecting the activity of metabolic enzymes, transporters, and receptors (Braun and Seymour, 2011). To this end, since drug resistance and toxicity have become more significant public health problems, in the context of precision medicine there is an unmet and urgent need to develop predictive biomarkers for a personalized approach in cancer pharmacotherapy (Derks and Diosdado, 2015).

Toxicity-related biomarkers: PGx for germ-line variations in anticancer agent

This section focuses on the anticancer drugs 5-FU and irinotecan as well as related genetic polymorphisms, including UGT1A1, DPYD, TS, and MTHF that alter significantly the pharmacokinetics of these drugs.

IRINOTECAN PGX (IRINOGENETICS): IRINOTECAN AND UGT1A1

The prodrug irinotecan (Camptosar, CPT-11), which is a semisynthetic analog of the natural alkaloid camptothecin, was introduced in the early 1990s into clinical applications as an important DNA topoisomerase I inhibitor leading to cell death (Xu and Villalona-Calero, 2002). It has been established as an effective treatment in several solid tumor malignancies such as lung cancer (Sevinc et al., 2011) and CRC either as a single agent or in combination with 5-FU (Stucky-Marshall, 1999; Pizzolato and Saltz, 2003; Ferraldeschi, 2010; Marcuello et al., 2011; Fujita et al., 2015).

Having a complex metabolism, irinotecan is metabolized by carboxylesterases principally in the liver, and also in various tissues, into an active cytotoxic metabolite SN-38 (7-ethyl-10-hydroxy-camptothecin). SN-38, with poor aqueous solubility that is at least 100-fold more potent than irinotecan (Kawato et al., 1991; Gupta et al., 1994; Khanna et al., 2000; Mathijssen et al., 2001; Gagne et al., 2002), binds the nuclear enzyme topoisomerase I, which is essential for DNA synthesis and replication (Innocenti and Ratain, 2003; Ferraldeschi, 2010). SN-38 is further catabolized (glucuronidated) principally by hepatic phase II biotransformation enzymes uridine 5′-diphospho-glucuronosyltransferase (UDP-glucuronosyl transferase)—UGT 1A enzymes (1A1, UGT1A7, and UGT1A9) to the inactive compound SN-38 glucuronide (SN-38G) (Gagne et al., 2002; Innocenti

and Ratain, 2004; Guillemette et al., 2014), which is then excreted via bile into the intestine (Innocenti and Ratain, 2003; Ferraldeschi, 2010). However, in the intestine, SN-38G is deconjugated back to active SN-38 by bacterial β-glucuronidases and then SN-38 is reabsorbed in the systemic circulation called enterohepatic circulation of SN-38 (Ferraldeschi, 2010). Consequently, glucuronosyltransferases expressed in the gut play a role in the reglucuronidation and detoxification of SN-38 in the intestine (Chen et al., 2013). Most UGTs, including UGT1A1 (also small intestine and colon, but not in the kidney) are predominantly expressed in the liver (Strassburg et al., 1998). Human UGTs show tissue-specific expression with UGT1A7, UGT1A8, and UGT1A10, which are extrahepatic and utterly expressed in the gastrointestinal tract (Strassburg et al., 1997; Hazama et al., 2013). Being essential for glucuronidation of hydrophobic endobiotic and xenobiotic compounds (Tukey and Strassburg, 2000), glucuronidation leads to the formation of water-soluble metabolites for different therapeutic drugs, including irinotecan, morphine (Coffman et al., 1997), acetaminophen (Court and Greenblatt, 2000), immunosuppressants (tacrolimus, cyclosporine) (Strassburg et al., 2001), anti-HIV drugs (raltegravir), chloramphenicol, and oral contraceptive agents (a-ethinylestradiol), as well as for an array of compounds, such as bilirubin, bile acids, and steroid hormones (de Wildt et al., 1999). Glucuronidation of all these compounds is mediated primarily by phase II biotransformation enzymes UGT1A isoforms (1A1, 1A7 and 1A9) (Innocenti and Ratain, 2004; Guillemette et al., 2014), which is then excreted in the bile through the small intestine (Innocenti and Ratain, 2003; Ferraldeschi, 2010). UGT1A is encoded by the UGT1A gene located on chromosome 2q37 (van Es et al., 1993). Irinotecan-related toxicities, such as diarrhea and myelosuppression, which have been associated with an increased level of SN-38, are the most important dose-limiting toxicities interfering with its optimal utilization (Innocenti and Ratain, 2006; Ramchandani et al., 2007; Etienne-Grimaldi et al., 2015).

Increased or decreased UGT1A1 enzyme expression affects the glucuronidation activity and also the metabolic clearance of drugs (Sugatani, 2013; Guillemette et al., 2014). Respectively, it is well known that mutations in the UGT1A1 gene have been implicated in Gilbert's syndrome and

Crigler–Najjar type 2 syndrome leading to mild hyperbilirubinemia as well as the more aggressive childhood subtype, Crigler–Najjar type 1 (Bosma et al., 1995; Costa, 2006; Strassburg, 2008). TATA box polymorphism in the promoter region of UGT1A1, which can vary between 5 and 8 thymine–adenine (TA) repeats, is the most thoroughly studied polymorphism for irinotecan metabolism (Lankisch et al., 2005). Respectively, UGT1A1*28 or TA7 (rs8175347) polymorphism is characterized by seven TA repeats in the promoter region of UGT1A1) as opposed to six TA repeats representing the wild-type allele UGT1A1*1 or TA6 (Guillemette, 2003; Lankisch et al., 2005; Barbarino et al., 2014). The length of this TA repeat sequence is inversely correlated with UGT1A1 enzyme activity; thus, the *28 polymorphism results in a decreased rate of transcription initiation/expression of UGT1A1 affecting the elimination of substrate drugs. Extra TA repeats impair proper UGT1A1 gene transcription leading to a decrease in the transcriptional activity of the gene by ≈70% in homozygous UGT1A1*28 (TA7/7) and ≈25% in heterozygous UGT1A1*28 (TA6/7) (Bosma et al., 1995; Tukey et al., 2002; Barbarino et al., 2014). Furthermore, individuals who are carriers of homozygous UGT1A1*28 genotype (TA7/7) have higher levels of serum bilirubin compared with those who have heterozygous UGT1A1*28 (TA7/6) or the wild-type allele (TA6/6) genotype (Barbarino et al., 2014; Guillemette et al., 2014). UGT1A1*28 allelic frequencies vary among ethnic groups, which are commonly found in African-American (≈43%) and Caucasian (≈36%) populations; yet is the lowest in Asian populations (≈16%) (Ando et al., 1998; Beutler et al., 1998; Premawardhena et al., 2003). Although some studies are less compelling based on several scientific findings, it has been acknowledged that patients with homozygotes (TA7/7) or heterozygous (TA6/7) for UGT1A1*28 have a lower irinotecan glucuronidation capacity together with increased accumulation of cytotoxic active metabolite SN-38 compared to wild-type UGT1A1*1 (TA6/6) (Innocenti et al., 2004; Toffoli et al., 2006; Côté et al., 2007; Parodi et al., 2008; Braun et al., 2009; Takano and Sugiyama, 2017). Patients with TA7/7 genotype are exposed to a higher risk of toxicity including diarrhea (though there are few data regarding the relationship with diarrhea) (Gupta et al., 1994; Iyer et al., 2002; Carlini et al., 2005; Massacesi et al., 2006; Tziotou et al., 2014; Peng

et al., 2017) and neutropenia, which can be life-threatening (Marcuello et al., 2004; Toffoli et al., 2006; Stewart et al., 2007; Ramchandani et al., 2007; Hoskins et al., 2007; Kweekel et al., 2008; Ruzzo et al., 2008; Palomaki et al., 2009; Hu et al., 2010; Toffoli et al., 2006; Glimelius et al., 2011; Tziotou et al., 2014; Emami et al., 2017; Liu XH et al., 2017), of which the incidence depends also in part on patient characteristics such as age and lifestyle factors (Marques and Ikediobi, 2010). A meta-analysis study of 821 patients assessing the association between irinotecan dose and risk of irinotecan-related hematologic toxicities (grade III–IV neutropenia) in patients with a UGT1A1*28/*28 genotype demonstrated that the risk of experiencing irinotecan-induced hematologic toxicity in homozygous UGT1A1*28/*28 carrier patients was the function of the administered irinotecan dose, but not only UGT1A1*28/*28 status. Respectively, UGT1A1*28/*28 patients had a greater risk of hematologic toxicities at high (>250 mg/m^2) and moderate (150–250 mg/m^2) irinotecan dosage. However, no association was established in UGT1A1*28/*28 carrier patients at low irinotecan dosage (80–125 mg/m^2) (Hoskins et al., 2007). Moreover, the same study found no association between UGT1A1*28 status and irinotecan-induced diarrhea. If these aforementioned results are bona fide, then the susceptibility of UGT1A1*28/*28 carrier patients to irinotecan toxicity (severe neutropenia) is dose-dependent and increment or reduction could be indicative of the possible involvement of additional novel target genes and pathways (Lévesque et al., 2013; Makondi et al., 2017) Additionally, nongenetic lifestyle factors (de Jong et al., 2008b; Marques and Ikediobi, 2010), such as smoking habits (Dumez et al., 2005), medical conditions, including renal function (de Jong et al., 2008a), and comedication (Kiang et al., 2005) may lead to risk of toxicity. Additionally, the frequencies of the genetic variations in genes encoding drug metabolizing enzymes (DMEs) as well as drug targets associated with therapeutic phenotypes with known toxicity within a racially and ethnically defined population is important to investigate and to take into consideration for dosing guidelines (Innocenti et al., 2002; Nguyen et al., 2007; Shimoyama, 2010). Ethnic differences related to UGT enzymes should be considered in order to determine gene polymorphisms as a predictor of treatment outcome in patients receiving irinotecan-based chemotherapy (Nguyen et al., 2007; Shimoyama et al., 2010). Respectively, in addition to UGT1A1*28 polymorphism in Asians, UGT1A1*6 polymorphisms are an important predictor of irinotecan-induced hematologic toxicity and severe diarrhea compared with that in Caucasians (Gao et al., 2013; Hazama et al., 2013; Han et al., 2014; Yang et al., 2015; Cheng et al., 2014; Zhang et al., 2017).

Based on the aforementioned studies related to UGT1A1*28 status and irinotecan-based toxicity, in 2005 the FDA (https://www.fda.gov/Drugs/ScienceResearch/ucm572698.htm) approved a UGT1A1*28 genotyping test based on individual patient tolerance to treatment without any emphasis or clear recommendation on the necessity of screening patients for UGT1A1*28 mutation before the administration of irinotecan (preemptive PGx test) (Innocenti and Ratain, 2006; Perera et al., 2008; Vivot et al., 2015). However, leaving this preemptive PGx test initiative for UGT1A1*28 to the discretion of the treating physician brings up the issue of the importance of PGx education for the physicians (Frueh and Gurwitz, 2004; Perera et al., 2008; Zgheib et al., 2011).

Given that irinotecan doses are usually lower in Europe (180 mg/m^2, biweekly, combination) and Japan (150 mg/m^2, biweekly, combination) than in the United States (350 mg/m^2, triweekly, monotherapy) (Takano and Sugiyama, 2017), a guideline from the European Society for Medical Oncology (ESMO) considers UGT1A1 testing only if potentially severe toxicity related to irinotecan treatment occurs underlying the importance of UGT1A1 PGx testing when irinotecan is used at high doses (300–350 mg/m^2) (Schmoll et al., 2012). A guideline from the Dutch Pharmacogenetics Working Group (KNMP) and Japanese Society for Cancer of the Colon and Rectum (JSCCR) both consider UGT1A1 PGx testing. However, while KNMP considers a reduction by 30% in the initial irinotecan dose for TA7/7 carriers if the regimen contains >250 mg/m^2 of irinotecan but not for heterozygous (TA6/7) carriers that is in agreement with the FDA (Swen et al., 2011), the guideline from JSCCR considers UGT1A1 PGx testing as preemptive, before administering irinotecan to patients, particularly if the patient has a high serum bilirubin level (Watanabe et al., 2015). On the other hand, in 2009 a guideline from the Evaluation of Genomic Applications in Practice

and Prevention (EGAPP™) Working Group did not find enough evidence indicating benefit of routine UGT1A1 genotyping for metastatic CRC cases being treated with irinotecan, whereas it did not either conclude that it should never be used (EGAPP, 2009; https://www.cdc.gov/genomics/gtesting/egapp/recommend/ugt1a1.htm). Considering different recommendations from different organizations related to UGT1A1*28 and irinotecan response there is obvious and urgent need to find additional genetic variants contributing to variability in irinotecan pharmacokinetics (Rosner et al., 2008).

Even though interindividual variability of irinotecan response (dose/toxicity) has been attributed essentially to inherited UGT1A1*28 genetic variation, mounting of scientific evidence underlined the critical role of different UGT1 haplotypes in the 3' UTR and central region of the gene (Biason et al., 2008; Lévesque et al., 2013) in different ethnic populations other than UGT1A1*28. Accordingly, UGT1A1*6 (211G > A or G71R, rs4148323), particularly prevalent in the Asian population with frequency of 0.13 to 0.25, UGT1A1*60 (−3279T > G, rs4124874), UGT1A1*93 (−3156G > A, rs10929302), UGT1A11A7*2 and*3 (387T > G, 622T > C), UGT1A9*22 contribute to irinotecan response (Innocenti et al., 2002; Carlini et al., 2005; Saito et al., 2009; Cecchin et al., 2009; Xu et al., 2013; Hazama et al., 2013; Crona et al., 2016; Cui et al., 2016; Bai et al., 2017; Campbell et al., 2017; Takano and Sugiyama, 2017). Additionally, genes encoding drug DMEs and transporters, such as CYP3A4/3A5 (cytochrome P450) (Santos et al., 2000; Sai et al., 2008; van der Bol, 2010), P-glycoprotein or multidrug resistance proteins MDR1 (ATP-binding cassette transporter ABCB1) (Smith et al., 2006; Li W et al., 2016), ABC transporter (ABCC5 ABCG1) (Di Martino et al., 2011; Chen et al., 2015), SLCO1B1 (solute carrier organic anion transporter family member 1B1) (Xiang et al., 2006; Teft et al., 2015; Crona et al., 2016) as well as gene encoding carboxylesterase 2 (activates the prodrug irinotecan into SN-38) (CES2) (Khanna et al., 2000; Cecchin et al., 2005; Capello et al., 2015) are all implicated in irinotecan metabolism, and genetic variations in all these genes could affect irinotecan response. Related to inhibitor/inducers effect, CYP3A4 enzyme could be particularly important in drug–drug (e.g., ketoconazole) (Haaz et al.,

1998; Charasson et al., 2002; Kehrer et al., 2002; Sasaki et al., 2013) and drug–lifestyle/herbal interactions (Goey et al., 2013). Respectively, clinical outcome of irinotecan could probably be due to the result of complex interplay between gene encoding DMEs in metabolic detoxification pathways phase I (CYP3A4, CYP3A5) and in phase II (UGT) as well as transporters, suggesting that a combined signature of the UGT1 haplotypes and other genes might provide more precise information about irinotecan pharmacokinetics, pharmacodynamics affecting increased systemic exposure to SN-38 on a cellular level that lead to increased irinotecan-based toxicity, such as neutropenia and diarrhea, as well as efficacy (Mathijssen et al., 2001; Innocenti et al., 2009; Crona et al., 2016). To this end, cognizance of these genomic variations probably improve the personalized approach for irinotecan treatment. Though genotyping for the UGT1A1 polymorphisms could be important to prevent severe toxicity such as neutropenia, prediction for resistance (high levels of UGT activity) to irinotecan is not clear (Panczyk, 2014). Since the intratumoral SN-38 level can be altered by increased efflux through active transport out of cells, some in vivo and in vitro studies have been investigated with the role of polymorphisms of ATP-binding cassette (ABC) transporter proteins, such as ABCC1/MRP1, ABCC2/MRP2, and ABCG2/BCRP (Sun et al., 2012; Li W et al., 2016; Tuy et al., 2016; Nielsen et al., 2017) as well as polymorphisms of CYP3A4 and CYP3A5 on treatment resistance (Hammond et al., 2016). However, they are inconclusive and need to be discovered more. Additionally, epigenetic changes, essentially hypermethylation, have also been implicated in development of resistance processes (Hammond et al., 2016).

Even if published clinical data are limited in relation to the role of epigenetics in the regulation of UGTs, and for the DMEs in general, preclinical studies in colon cancer cell lines and clinical studies in colon tumors underlined the crucial role of DNA methylation at CpG sites in the promoter region of the UGT1A1 gene in the silencing of UGT1A1 expression in colon cancer and on cellular SN-38 detoxification influencing clinical response to irinotecan (Gagnon et al., 2006; Bélanger et al., 2010; Xie et al., 2014) suggesting that differential methylation of the CpG site may explain interindividual variability in hepatic glucuronidation by UGT1A1, which could influence treatment toxicity and efficacy (Yasar et al., 2013).

5-FU PGX

5-FU and dihydropyrimidine dehydrogenase (DPD)

The fluoropyrimidine anticancer drug 5-fluorouracil (5-FU) and its oral prodrug capecitabine are the backbone of chemotherapy for colorectal cancer (Tanaka et al., 2000). Even though 5-FU was reported in the 1950s to have anticancer activity (Heidelberger et al., 1957), its efficacy was first recognized in 1990 (Moertel et al., 1990) before it became the mainstay of therapy for advanced or mCRC. There are three methods using FU: FOLFOX contains infusional fluorouracil, leucovorin, and oxaliplatin; FOLFIRI contains fluorouracil, leucovorin, and irinotecan; and XELOX contains capecitabine plus oxaliplatin (Aparicio et al., 2005).

Some of the patients (10%–30%) using these regimens have grade 3 toxicity such as diarrhea, nausea, mucositis/stomatitis myelosuppression, hand-foot syndrome, and occasionally cardiac toxicity. Along with this, FU leads to 0.5%–1.0% mortality (grade 5). Therefore, it is important to identify biomarkers that predict 5-FU toxicity (Rosmarin et al., 2014). 5-FU metabolism comprises a plurality of enzyme reactions. After the use of parenteral 5-FU, 70%–90% of the drug is metabolized with dihydropyrimidine dehydrogenase (DPD). The dihydropyrimidine dehydrogenase DPD, which plays a key role in the metabolism of the fluoropyrimidines, is encoded by the DPYD gene. DPD is the rate-controlling enzyme for inactivation of 5-FU and more than 80% of 5-FU is metabolized by DPD in the liver to the inactive metabolite 5,6-dihydro-5-fluorouracil (van Staveren et al., 2013).

The DPYD gene contains 23 exons and is located on chromosome 1p22. More than 30 genetic polymorphisms in DPYD caused reduced function or nonfunctional DPD enzyme leading to reduced clearance of 5-FU resulting in increased 5-FU toxicity in CRC patients (Del Re et al., 2010). DPD deficiency occurs in 4%–5% of the population (Deenen et al., 2011b). The most common genetic variant of the DPYD gene with partial or complete DPD deficiency is due to a G to A point mutation within the 5'-splicing site of intron 14

(IVS14 + 1G > A) called *DPYD*2A* (rs3918290) polymorphism. *DPYD*2A* leads to catalytically inactive enzyme with a frequency of approximately 1% in Caucasians. DPD activity is reduced by 50% in heterozygous genotype resulting in increased 5-FU exposure. However, DPD activity in patients with homozygous *DPYD*2A* is about 0% (van Kuilenburg et al., 2001; Meulendijks et al., 2016a, 2016b; Deenen et al., 2016). In patients with complete DPD, nonfloropyrimidin-based treatment is recommended instead of fluoropyrimidines, or if fluoropyrimidines treatment is imperative for attentive monitoring, a starting dose reduction of approximately 10% is recommended (Caudle et al., 2013). However, some variants do not lead to completely inactive enzymes; the variant c. 1129-5923C > G (rs75017182), known as haplotype B3, is a deeply intronic variant encoding partially nonfunctional protein expression with enzyme activity 50% lower in homozygous individuals (Meulendijks et al., 2016a, 2016b).

Although enzyme activities can be measured for toxicity estimation, these tests can be cumbersome and expensive for routine use. Following initial reports linked to the use of lethal FU in the severe DPD deficiency, several common genetic polymorphisms in genes involved in FU metabolism and rare variants have been reported to affect the risk of adverse events (Henricks et al., 2015).

Theoretically, prior to dosage adjustments, FU toxicity can be estimated by testing a panel of polymorphisms. However, the important limitations are the presence of different polymorphisms in the same gene, as well as the involvement of patients in different FU programs. At the same time, several polymorphisms with lack of validation may have been included in FU toxicity kits (Boisdron-Celle et al., 2007; Afzal et al., 2011).

In order to accelerate clinical uptake of DPD testing, the following points are recommended: (a) a consensus definition of DPD deficiency should be decided internationally to determine the incidence of DPD deficiency by comparing different study results; (b) sensitivity and specificity for the prevention of fluoropyrimidine-induced toxicity should be determined in each test; (c) optimization of a genotyping and phenotyping strategy should be explored to determine the most appropriate test for detection of patients with DPD deficiency; and finally (d) the test should focus specifically on cost effectiveness. To this end, there is an obvious need

to optimize and to apply quick *DPYD* PGx tests for screening patients with DPD deficiency before being treated with fluoropyrimidines for the first time (van Staveren et al., 2013; Etienne-Grimaldi et al., 2017). Hence, the recommended 5-FU starting doses should be adjusted according to the enzyme activity of DPD (Henricks et al., 2015; Etienne-Grimaldi et al., 2017).

5-FU and thymidylate synthase (TS)

Thymidylate synthase (TYMS) is crucial enzyme for providing essential nucleotide precursors (de novo pyrimidine synthesis) in order to maintain DNA synthesis and repair. Respectively, TYMS is implicated in the conversion of deoxyuridine monophosphate (dUMP) to deoxythymidine monophosphate (dTMP) (Carreras and Santi, 1995). Hence, TYMS expression is a rate-limiting step for cell proliferation as well as for cancer growth (Rustum, 2004). Compelling evidence from several clinical studies have demonstrated that TYMS protein and mRNA levels were higher in different cancers (Berger and Berger, 2006), including breast cancer, lung cancer, gastric cancer, and CRC (Yamada et al., 2001; Kamoshida et al., 2004; Popat et al., 2006), and this increase in TYMS levels has been associated with poor clinical outcome in these cancers. It is also shown that TYMS expression correlates closely with transcription factor E2F1 expression in colon cancer specimens. In accordance with this TYMS—E2F1 correlation TYMS could be considered as an E2F1-regulated enzyme, which is essential for DNA synthesis and repair (Kasahara et al., 2000; Rahman et al., 2004). Additionally, an in vitro study in immortalized NIH/3T3 mouse fibroblast cells underlined oncogene-like activity of TYMS that suggests a connection between TYMS-regulated DNA synthesis and the induction of a neoplastic phenotype (Rahman et al., 2004; Bertino and Banerjee, 2004).

TYMS has a potential therapeutic interest as critical target of chemotherapeutic agents 5-FUl, and its prodrug uracil-tegafur with leucovorin (UFT/LV), capecitabine, and methotrexate, which exert their anticancer effects by inhibiting TYMS (Longley et al., 2003; Rustum, 2004). Moreover, overexpression of TYMS is linked to chemotherapy resistance (Panczyk, 2014; Hammond et al., 2016).

TYMS is encoded by the *TYMS* gene located on human chromosome 18 *TYMS* gene, which contains seven exons, and various types of polymorphisms of the *TYMS* gene have been linked to variable TYMS protein levels and a therapeutic outcome related to 5-FU treatment (Ferraldeschi, 2010). A common 28-bp variable number tandem repeat (VNTR) polymorphism (rs34743033) is found in the 5′ untranslated regions (5′UTR) of TYMS (Horie et al., 1995) that occurs in a different population with a variable number of reiterations leading to occurrence of two alleles, two tandem repeats (2R), and three tandem repeats (3R). While 3R represents the wild-type form, 2R represents the variant form with three different genotypes: 2R/2R, 2R/3R, and 3R/3R (Marsh et al., 2001; Kawakami and Watanabe, 2003). The number of tandem repeats affects TYMS activity levels mediated through effects of the repeats on translation efficiency. Subjects with two tandem repeats, three tandem repeats, or a heterozygous genotype were observed (Marsh and McLeod, 2001; Ferraldeschi, 2010). A triple repeat presence (TSER*3) results in 2.6 times more mRNA expression than double repeats (TSER*2) (Marsh and McLeod, 2001; Ferraldeschi, 2010; Lima et al., 2013). TSER*3, which leads to higher TYMS mRNA expression, is significantly higher in the Asian population compared to other ethnic groups (67% in Chinese and about 40% in Caucasians) (Marsh et al., 1999). On the other hand, according to world population studies, alleles with 4, 5, and 9 TSER repeats have been observed mostly in Asian and African populations (Marsh et al., 1999; Marsh et al., 2000) and in particular TSER*5 is found mostly in Asians (0.18%). Although the outcome of published studies are conflicting, it is anticipated that the TSER genotype is associated with 5-fluorouracil toxicity and efficacy (Ferraldeschi, 2010; Schwarzenbach, 2010; Panczyk, 2014). According to the results of different studies the 3R allele is responsible for an approximately four times higher mRNA level of the *TYMS* gene, suggesting that individuals homozygous for TSER*3 (TSER*3/TSER*3) have much less favorable response to 5-FU treatment as compared to those who had the 2R/2R with low TYMS mRNA (Marsh et al., 2001; Pullarkat et al., 2001; Gosens et al., 2008). However, the 2R/2R genotype is associated with an increased risk of 5-FU toxicity (Marsh et al., 2001; Pullarkat et al., 2001; Gosens et al., 2008). An additional SNP (rs2853542) of guanine instead of cytosine (G > C) at the 12th nucleotide of 3R alleles has been described as two different alleles (3RC/3G)

(Mandola et al., 2003). The presence of this polymorphism leads to the tri-allelic locus of 2R, 3RG, and 3RC with two different levels of TYMS activity such as a high expression group (carriers of 2R/3G, 3C/3G, and 3G/3G genotype) and a low expression group (carriers of 2R/2R, 2R/3C, and 3C/3C genotypes). This SNP occurs within upstream stimulatory factor 1 (USF-1), leading to a decreased transcriptional activity of *TYMS* gene by converting the transcriptional activity from a 3R to a 2R, thereby decreasing TS levels (approximately three- to four-fold) of the wild-type triple repeat (3RG/3RG) variant, which is comparable with the 2R/2R or 2R/3RG genotypes (Kawakami and Watanabe, 2003; Gusella and Padrini, 2007). The presence of this SNP (rs2853542) and double polymorphism in the TYSM gene could explain low TS expression level together with good response to 5-FU chemotherapy (Panczyk, 2014). In other words, testing for all these polymorphisms may help stratify patients at risk for 5-FU treatment. It is underscored that patients with TSER*3 and TSER*3 G > C SNPs may be at greatest risk for toxicity and decreased response. Furthermore, an important ins/del polymorphism of the hexanucleotide TTAAAG sequence (6-bp insertion/deletion) at 1494 position (1494del6, rs34489327, also referred to as rs16430) within the untranslated transcriptional region (3′UTR) of the TYMS gene is described (Ulrich et al., 2000). Consequently, this ins/del polymorphism yields three different genotypes: ins/ins (homozygous for insertion of 6 bp), del/del (homozygous for deletion of the 6 bp), and ins/del (heterozygous). Hence, it was demonstrated that individuals with homozygous deletion (del/del) had significantly lower mRNA levels of the *TYMS* gene, which was also associated with greater sensitivity to 5-FU–based therapy as compared to individuals with homozygous insertion (ins/ins) (P = 0.017) (Kawakami and Watanabe, 2003; Mandola et al., 2003; Stoehlmacher et al., 2008).

Related to 5-FU chemotherapy resistance and TS, higher TS expression in MSI-H is both sporadic (86%) and hereditary (100%). Tumors compared to MSI-negative tumors had been demonstrated as a strong predictive factor for nonresponse (resistant) to adjuvant 5-FU chemotherapy (Gatalica, 2014; Gatalica et al., 2015, 2016). In a poorly differentiated gastric cancer cell line MKN45, the important role of decreased activity of orotate phosphoribosyltransferase (OPRT), which is involved in phosphoribosylation of 5-FU leading to tumor growth inhibition, is also suggested (Tsutani et al., 2008). However, a recent in vitro study in the same cell line (MKN45) hypothesized that irrespective of decreased OPRT levels, a decrease in the intracellular FdUMP level could be a probable mechanism involved in the resistance to 5-FU (Mori et al., 2017). Though most retrospective clinical trials have shown that TYMS genotyping may help predict the 5-FU response, insufficiency of TYMS genotyping alone in accurate prediction of outcome response to 5-FU is underlined (Vignoli et al., 2011). To this end, a comprehensive evaluation of each TYMS polymorphism in large-scale prospective randomized control trials is needed to guarantee the efficacy of PGx testing for TYMS in patients prior to 5-FU chemotherapy treatment.

5-FU and MTHFR

MTHFR (5,10-methylenetetrahydrofolate reductase) is a pivotal enzyme in folate metabolism catalyzing irreversible conversion of 5,10-methylenetetrahydrofolate (CH_2THF) to 5-methyltetrahydrofolate (CH_3THF), a cosubstrate for homocysteine remethylation to methionine (Goyette et al., 1994). This enzyme is encoded by *MTHFR* gene and is located on human chromosome 1 (Goyette et al., 1994). Two nonsynonymous variants, C677T in exon 4 (Ala222Val, rs1801133) and A1298C in exon 7 (Glu 429Ala, rs1801131), have been identified for the MTHFR gene and have been associated with decreased enzymatic activity and altered intracellular folate distribution (Frosst et al., 1995). Despite contradictory data, several clinical studies have demonstrated the potential predictive role of MTHFR genetic variants in toxicity and efficacy of 5-FU (De Mattia and Toffoli, 2009). Since a reduction in MTHFR enzymatic activity could lead to a decrease in intracellular CH_3THF concentrations, it is hypothesized that tumors exhibiting the rare MTHFR variants may be more sensitive to 5-FU cytotoxicity compared to patients with a wild-type genotype (Etienne-Grimaldi et al., 2007).

The critical point of 5-FU activity is the formation of an inhibitor triple complex; the active metabolite consists of 5-fluoro-2-deoxyuridine-5-monophosphate (5-FdUMP), TS, and 5,10 methylenetetrahydrofolate, hence inhibition of TS activity. Accordingly, it has been hypothesized that MTHFR polymorphisms may increase the cytotoxic activity of 5-FU by increasing the formation and stability of the triple inhibitory complex by increasing intracellular concentrations of CH_2THF (Longley et al., 2003). While some studies demonstrated only the significant association of MTHFR C677T polymorphism with increased 5-FU response but not for toxicity prediction (Jakobsen et al., 2005; Etienne-Grimaldi et al., 2010), the others also demonstrated the increased risk of undesirable side effects together with increased 5-FU sensitivity survival in stage III and stage IV CRC (Derwinger et al., 2009). However, in some studies of advanced CRC, only populations with the 1298CC genotype were associated with the risk of developing serious adverse events after 5-FU–based chemotherapy.

In conclusion, considering contradictory results on *MTHFR* polymorphisms and 5-FU response and toxicity, it still remains controversial whether *MTHFR* polymorphisms can possibly predict toxicity or 5-FU response in patients treated with 5-FU (Panczyk, 2014; Ab Mutalib et al., 2017).

Targeted therapies: PGx for somatic mutations in anticancer agents

While most of the standard cytotoxic chemotherapies act on cancerous and all rapidly dividing normal cells, cytostatic targeted therapies act on specific molecular targets that are associated with cancer (Calvo et al., 2016). To this end, in the era of personalized medicine targeted therapies (also referred to as biologic treatments/molecularly targeted drugs) involving tumor growth and progression is an attractive topic of oncology in patient stratification (Papadatos-Pastos et al., 2015), hence the introduction of EGFR inhibitors has provided new treatment options for metastatic CRC patients (Haddad et al., 2017).

TARGETING GENES

KRAS

RAS mutational status is a pivotal factor when using these targeted therapies. EGFR activation shows its effect via the RAS-RAF-MAPK and PI3K-AKT-mTOR pathways. Mutations in these signaling pathways may result in receptor-independent continuous activation that results in unresponsiveness to the therapy. The EGFR antibodies cetuximab, a chimeric IgG1 antibody, and panitumumab, a humanized IgG2 antibody, both bind to and block the EGFR and have proven effective in all lines of mCRC treatment (Siddiqui and Piperdi, 2010).

The association of KRAS gene mutation and response to therapy was first reported in 2006 in patients with metastatic colorectal cancer who were treated with an EGFR agent (Lievre et al., 2006). In this study it was also noted that EGFR overexpression has also been linked to poor prognosis and increased risk of metastasis in colorectal cancer (Lievre et al., 2006). Studies have shown that tumors with a mutation in exon 2 (codon 12 or 13 mutation) of the KRAS gene are unlikely to get benefit from cetuximab or panitumumab treatment (Peeters et al., 2013). Recent evidence has suggested that other RAS mutations in exons 3 and 4 of KRAS and NRAS genes may also be predictive of unresponsiveness to anti-EGFR therapies (Doullard et al., 2013). A meta-analysis including nine randomized controlled trials and a total of 5948 patients demonstrated tumors with all RAS wild-type had significantly superior survival with the anti-EGFR therapy compared to tumors with KRAS exon 2 mutation along with other mutations, including KRAS exons 3 and 4 and NRAS exons 2, 3, and 4. Accordingly, these results suggest that extended RAS mutation testing in addition to KRAS exon 2 (KRAS exon 3 and 4 and NRAS exon 2, 3, and 4) should be undertaken before the administration of an anti-EGFR mAb. A study evaluating addition of cetuximab to chemotherapy (FOLFIRI) for metastatic CRC in the CRYSTAL trial indicated that first-line treatment with cetuximab plus FOLFIRI reduced the risk of progression of metastatic colorectal cancer compared to FOLFIRI alone. Additionally, this benefit of cetuximab was limited to KRAS wild-type tumors (Van Cutsem et al., 2009). Follow-up and post hoc analysis showed significant improvements in RR, PFS, and OS when treated with FOLFIRI/cetuximab compared with FOLFIRI alone in the RAS wild-type population compared to RAS mutations (KRAS exons 2, 3, and 4, and NRAS exons 2, 3, and 4) (Van Cutsem et al., 2011, 2015).

Moreover, efficacy of cetuximab in combination with FOLFOX as a first-line treatment for mCRC was tested in the OPUS study (Bokemeyer et al., 2011). A retrospective evaluation of the OPUS study showed that in patients with KRAS exon 2 wild-type tumors, the addition of cetuximab to FOLFOX was associated with an increased objective response rate (61% versus 37%; p = 0.011) and improved PFS (7.7 versus 7.2 months HR, 0.57; 95% CI, 0.36–0.91; p = 0.016) compared with chemotherapy alone (Bokemeyer et al., 2009). New results from the OPUS study revealed that patients with any activating mutation of KRAS or NRAS are unlikely to benefit from the addition of cetuximab to FOLFOX4 (Tejpar et al., 2014).

Pooled individual patient data from OPUS and CRYSTAL studies were analyzed and 845 patients with KRAS wild-type tumors with the addition of cetuximab to chemotherapy improved OS (hazard ratio [HR] 0.81; p = 0.0062), PFS (HR 0.66; p < 0.001) and ORR (odds ratio 2.16; p < 0.0001). Consequently, prognosis was worse in each treatment arm for patients with BRAF tumor mutations compared to those with BRAF wild-type tumors (Bokemeyer et al., 2012). However, in contrast to other studies in phase III MRC COIN trial, no benefit in OS or PFS was observed with the addition of anti-EGFR agents to FOLFOX or CapeOx as first-line treatments for patients with locally advanced or metastatic CRC with wild-type KRAS exon 2 (Maughan et al., 2011). Even patients with wild-type tumors for all three genes (KRAS, BRAF, NRAS) did not show any evidence of a benefit from the addition of cetuximab. This was explained by a more advanced stage of disease at the first presentation in the COIN trial than other similar trials.

Panitumumab was also tested in the first-line setting in metastatic CRC patients. In the PRIME study, RAS mutations were assessed in patients treated with FOLFOX4 with and without panitumumab. Patients with additional RAS mutation other than at KRAS exon 2 showed lack of response to treatment with FOLFOX4/panitumumab versus FOLFOX4 alone. This particular group showed a significantly shorter median PFS and OS with the addition of anti-EGFR therapy. Patients with no RAS mutations treated with FOLFOX4/panitumumab conferred a longer median PFS and OS compared with FOLFOX4 alone. Addition of panitumumab to FOLFIRI in the second-line setting was assessed by Peeters et al. (2015). Consistent with the PRIME study, among all RAS wild-type (KRAS at exons 2, 3, and 4, and NRAS at exons 2, 3, and 4) patients, improvements in outcome were observed in PFS when panitumumab was added to chemotherapy as a second-line treatment (Peeters et al., 2014).

BRAF

BRAF is a targetable mutation in many tumors including malign melanoma. Recently, improvement in both overall survival and response rate was achieved in metastatic malignant melanoma with the use of new BRAF inhibitors, namely, vemurafenib and dabrafenib. Unlike the results in melanoma, the response of BRAF inhibitors was not successful in metastatic CRC. In a small phase I study in patients with BRAFV600E mutant metastatic disease, only 1 of 19 patients had a partial response with single-agent vemurafenib (Kopetz et al., 2015). Similarly, a phase I trial of the BRAF inhibitor dabrafenib also failed in BRAF-mutated colorectal cancer (Kefford et al., 2010).

One reason for unresponsiveness of BRAF inhibition in BRAF-mutated colorectal cancer could be explained by Prahallad et al.'s study that suggested that BRAF (V600E) inhibition causes a rapid feedback activation of EGFR and this results in continuous proliferation in the presence of BRAF (V600E) inhibition. Because of low levels of EGFR expression in melanoma cells, they are not subject to this feedback activation (Prahallad et al., 2012). In light of this data, Prahallad et al. could show that the combination of vemurafenib and the EGFR-blocking antibody cetuximab was significantly more effective in mice bearing BRAF V600E mutated colorectal cancer than each of the drugs alone. Connolly et al. (2014) presented a case report of cetuximab and vemurafenib combination for a refractory patient with BRAF mutant metastatic colon cancer.

The next step in developing targeted therapy for BRAF-mutated colorectal cancer was based on the hypotheses that inhibiting BRAF and its downstream signaling partner MEK would be more effective by increasing the level of inhibition of the MAPK pathway and blocking some mechanisms of acquired resistance. Recently tested was the strategy of combined BRAF and MEK inhibition with dabrafenib and trametinib in BRAF V600–mutant CRC, which showed that 5 out of 43 patients (12%) achieved a partial response or better, including one complete response, with duration of response >36 months; 24 patients achieved stable disease as best confirmed

response (Corcoran et al., 2015). Although combined BRAF/MEK inhibition did show more activity than single-agent BRAF inhibition, the effectiveness of combined BRAF/MEK inhibition was still limited. Analysis of tumor biopsies demonstrated that the level of phosphorylated ERK, which could be a marker of inhibition, was decreased in the posttreatment biopsies analyzed (Corcoran et al., 2015). Triplet therapy that inhibits the BRAF, MEK, and EGFR pathways appears promising in BRAF-mutated colorectal cancer that does not respond to BRAF inhibition alone. A phase I/II trial evaluated the combination of dabrafenib (150 mg twice daily) plus panitumumab (6 mg/kg every 2 weeks) in 20 patients with BRAF V600E–mutated metastatic colorectal cancer (Atreya et al., 2015). The study included 35 patients on the triplet arm, with a dose-escalation scheme. Of these patients, 24 received the phase II regimen of dabrafenib (150 mg twice daily) plus panitumumab (6 mg/kg every 2 weeks) and trametinib (2 mg daily). The primary endpoint was overall response rate, and progression-free survival was a secondary endpoint. A total of 2 of 20 patients (10%) responded to the doublet of dabrafenib and panitumumab; 1 had a confirmed complete response. With triplet therapy, however, 9 of 35 patients (26%) responded, including 1 complete response. The unconfirmed response rate was 34%, and the stable disease rate was 60%. Median progression-free survival was 3.4 months with the doublet and 4.1 months with the triplet. Median duration of response for all patients receiving the triplet was 5.4 months. Tumor biopsies before and after 2 weeks of treatment revealed that phosphorylated ERK staining intensity was consistently reduced by triplet therapy, but not doublet therapy (Atreya et al., 2015). The other reason for unresponsiveness of BRAF inhibitors in colon cancer could be explained by PI3K/AKT pathway activation in BRAF-mutated colorectal cancers. It was shown that the phosphoinositide 3-kinase (PI3K)/AKT pathway was activated to a greater extent in BRAF-mutated colorectal cancers than in melanoma. As a result of PI3K/AKT pathway activation de novo an acquired resistance to BRAF inhibition was observed in cell lines and murine models. This activation occurs through PIK3CA mutation or PTEN loss and is associated with the inherent CpG island phenotypes associated with BRAF-mutated CRC through epigenetic silencing. In vivo, vemurafenib combined with either inhibitors of AKT or methyltransferase showed greater tumor growth inhibition than vemurafenib alone. Clones with acquired resistance to vemurafenib in vitro showed PI3K/AKT activation with EGFR or KRAS amplification (Mao M et al., 2012).

As a conclusion, the existence of the specific mutation in colon cancer does not confer sufficient sensitivity to a single agent targeting the particular mutation. Combination of inhibitors in the RAS/RAF/MEK/ERK pathway seem to be promising in BRAF-mutated colorectal cancers.

TARGETING THE PATHWAYS

RAF kinases

RAF is the most characterized downstream effector kinase of RAS. The Raf serine threonine kinase family consists of three isoforms: RAF-1 (C-RAF), A-RAF, and B-RAF. Mutations in RAF-1 (C-RAF) or A-RAF have not been detected in human cancers (Chong et al., 2003).

B-RAF is the strongest RAF kinase that activates MEK. Due to many missense mutations in the BRAF gene, it could be a potential target for various types of cancers. Activation of the upstream signaling pathway of RAS results in RAF activation, which induces a downstream signal transduction cascade beginning with MEK (Pearson et al., 2001). Increased MEK activity promotes activation of two kinases, namely, ERK1 and Erk2. Signaling through the ERK pathway may lead to increased growth factors and cytokines expression so that this pathway is further stimulated in an autocrine fashion (Steelman et al., 2004). Through interaction with the RB gene p27kip, ERK pathways may also increase cyclin D and E expression in the cell cycle (Steelman et al., 2004). Besides this, ERK signaling regulates cell motility, extracellular matrix remodeling, and induce the production of VEGF and other antiangiogenic factors together with inactivation of caspase and BAD. ERK also downregulates p53, which is a pivotal tumor suppressor protein. ERK could be a potential target in many cancer types. Because RAF is the only activator of ERK, drugs targeting the ERK pathway at the level of RAF may be particularly useful to control tumor development and progression. In colon cancer, mutations of B-RAF and K-RAS are often found in a mutually exclusive fashion in the same tumor.

Furthermore, dominant-negative mutants of RAF can impair RAS transforming activity, confirming that inhibition of RAF is a viable therapeutic approach.

Sorafenib is one of the most promising agents of RAF kinase inhibitors. It targets the ERK pathway and inhibits angiogenesis through VEGFR-2 and PDGFR tyrosine kinases and their associated signaling cascades. In a preclinical model, sorafenib demonstrated inhibition of the MAPK pathway in colon cancer cell lines expressing mutant KRAS or wild-type or mutant BRAF (Wilhelm et al., 2004; Samalin et al., 2014).

Vemurafenib is a specific BRAF inhibitor and has proven effects in BRAF-mutated malignant melanoma; however, it is ineffective in BRAF-mutant colorectal cancer (Prahallad et al., 2012). Another BRAF inhibitor dabrafenib also failed in BRAF-mutant colorectal cancer (Holderfield et al., 2014). One possible explanation of resistance of BRAF inhibitor therapy in colon cancer could be EGFR-mediated reactivation of MAPK signaling (Samalin et al., 2014). A blockade of BRAF causes rapid feedback activation of EGFR and triggers sustained MAPK signaling and cell proliferation via activation of RAS and CRAF. Besides this, transactivation of BRAF–CRAF heterodimers in the presence of vemurafenib may result in resistance (Corcoran et al., 2015).

To overcome EGFR-activated resistance of Braf inhibitors, Connolly et al. (2014) presented a case report of cetuximab and vemurafenib combination for a refractory patient with BRAF-mutant metastatic colon cancer. Recently Hong et al. (2014) presented a phase 1B study of vemurafenib in combination with irinotecan and cetuximab in patients with BRAF-mutated advanced cancers and metastatic colorectal cancer. The results showed that the combination of vemurafenib with irinotecan and cetuximab was well tolerated in patients with BRAF-mutated mCRC. Even with a low vemurafenib dose, PRs were seen in 4 of 5 evaluable mCRC patients in the first cohort (Hong et al., 2014).

Combined BRAF and MEK inhibition with dabrafenib and trametinib in BRAF V600–mutant colorectal cancer was recently tested (Corcoran et al., 2015). The results showed that 5 out of 43 patients (12%) achieved a partial response or better, including 1 (2%) complete response, with duration of response >36 months; 24 patients (56%)

achieved stable disease as the best confirmed response. Although combined BRAF/MEK inhibition did show more activity than single-agent BRAF inhibition, the effectiveness of combined BRAF/MEK inhibition was still limited.

Another reason for unresponsiveness was shown by Mao et al. that the phosphoinositide 3-kinase (PI3K)/AKT pathway in colon cell lines is activated to a greater extent in BRAF-mutated CRC tumors than in melanoma (Mao M et al., 2012) PI3K/AKT pathway activation results in both de novo and acquired resistance to BRAF inhibition.

Combination therapy that inhibits the BRAF, MEK, and EGFR pathways appears promising in BRAF-mutated colorectal cancer. A phase I/II trial evaluated the combination of dabrafenib (150 mg twice daily) plus panitumumab (6 mg/kg every 2 weeks) in 20 patients with BRAF-mutated metastatic colorectal cancer (Atreya et al., 2015). With this triple therapy, the unconfirmed response rate was 34%, and the stable disease rate was 60%. One patient had complete response. Median progression-free survival was 4.1 months with the triple therapy. Median duration of response was 5.4 months.

Regorafenib is a novel oral multikinase inhibitor that targets protein kinases involved in tumor angiogenesis (VEGFR1–3 and tyrosine kinase with immunoglobulin and epidermal growth factor homology domain 2 [TIE2]), oncogenesis (KIT, RET, and RAF) and the tumor microenvironment (platelet-derived growth factor receptor-b and fibroblast growth factor receptor 1 [FGFR1]). Recently, in the CORRECT trial, regorafenib demonstrated a significant improvement in overall survival (6.4 months in the regorafenib group versus 5 months in the placebo group ($p = 0.0052$)) in a phase III study in patients with refractory metastatic CRC (Grothey et al., 2013). Regorafenib was also tested in an Asian population in the phase III CONCUR trial, which also supports overall survival benefit in previously treated metastatic colorectal cancer patients (Li J et al., 2015). Although regorafenib is an RAF inhibitor, its antitumor effect mostly depends on inhibition of angiogenesis in tumor cells (Tampellini et al., 2016). Preclinical studies demonstrate that regorafenib acts independently of the mutational status of KRAS and BRAF (Grothey et al., 2013).

MEK

Mitogen-activated protein kinase (MAPK) cascades are key signaling pathways involved in the regulation of normal cell proliferation, survival, and differentiation. Dysregulation of MAPK cascades contribute to cancer development. Four distinct MAPK cascades have been identified. They are called extracellular signal-regulated kinase (ERK1/2), c-Jun N-terminal kinase (JNK), p38, and ERK5. Each of them is composed of three sequentially acting kinases, activating one after the other (MAPKKK/MAP3K, MAPKK/MAP2K, and MAPK) (Thompson and Lyons, 2005).

MEK proteins belong to a family of enzymes that are involved in the four MAP kinase signaling pathways (Thompson and Lyons, 2005). Seven MEK enzymes have been described each of which selective phosphorylate serine/threonine and tyrosine residues within their specific MAP kinase substrate. MEK1 and MEK2 are the most common ones and they are involved in the RAS/RAF/MEK/ERK pathway (Cargnello and Roux, 2011). This pathway is activated after ligand binding, which results in membrane-bound GTPase activation of RAS. After that, Ras recruits and activates RAF kinases (Roskoski, 2010). The activated RAF kinases interact and activate MEK1/2, which in turn catalyze the phosphorylation of particular residues in the activation of ERK1/2 (Sacks, 2006). Unlike RAF and MEK1/2 kinases, ERK1 and ERK2 have a wide variety of cytosolic and nuclear substrates (Cargnello and Roux, 2011). Activated ERKs function in a diverse cellular process like proliferation, survival, differentiation, motility, and angiogenesis. The RAS/RAF/MEK/ERK pathway is activated in human cancers via several different mechanisms. Increased ERK1/2 signaling is often due to direct mutational activation of the RAS and B-RAF genes. This results in ERK activation and ERK-dependent growth transformation (Roskoski, 2010). Furthermore, the only known substrates of RAF are MEK1 and MEK2, and no substrates for MEK have been identified other than ERK1 and ERK2, which means MEK inhibitors would be a potent inhibitor of RAS- and RAF-mediated activation of ERK. This hypothesis promotes treatment by MEK inhibitors in colorectal cancer because KRAS and BRAF mutation–positive colorectal cancer tumor cells are expected to exhibit elevated ERK activation. Therefore ERK1/2 levels could be used as a biomarker of response to MEK inhibition. In the study of Yeh et al. (2009), they found that the majority of colorectal cancer cell lines show growth inhibition using MEK inhibitors, specifically those that are BRAF or KRAS mutation positive; however, ERK activation did not correlate reliably with BRAF and KRAS mutation status. It was noted that ERK is not differentially activated in tumor tissue. These results show complexities of MEK inhibitors as anti-Ras therapy.

Currently 13 MEK inhibitors have been clinically tested, but only trametinib as a selective inhibitor of MEK 1 and 2 has emerged as the first MEK inhibitor to show favorable clinical efficacy (Akinleye et al., 2013). Initial clinical results of MEK inhibitors have yielded limited single-agent activity in colorectal cancer (Adjei et al., 2008; Balmanno et al., 2009; Bennouna et al., 2011). Recently Spreafico et al. (2013) designed a study to test the combination of a MEK Inhibitor, selumetinib, and the Wnt/calcium pathway modulator cyclosporin A in preclinical models of CRC to overcome resistance to MEK inhibition. It was suggested that the combination of MEK blockade and Wnt pathway modulation has shown synergistic antiproliferative effects in preclinical colorectal cancer models (Spreafico et al., 2013).

MEK inhibitors show promise in colorectal cancer. Studies that promote the combination of BRAF/MEK/EGFR inhibitors are still recruiting patients and results are awaiting (ClinicalTrials.gov Identifier:NCT01750918).

MTOR

Mammalian target of rapamycin (mTOR), a serine/threonine tyrosine kinase, regulates cell division and growth, largely by promoting key anabolic processes. Upon activation, mTOR relays the cellular signal to downstream effectors to stimulate cell growth, proliferation, and angiogenesis (Laplante and Sabatini, 2012). The PI3K/Akt/mTOR pathway is activated by PI3K gene mutation and amplification, AKT mutation and amplification, or loss of the PTEN tumor suppressor. Dysregulation of this pathway is seen in 40%–60% of patients with colon cancer (Laplante and Sabatini, 2012).

mTOR forms two distinct complexes, namely, mTORC1 and mTORC2 (Kim et al., 2002). The complexes are constituted by different proteins and have distinct functions in the cell cycle. mTORC1 activates S6K1 and 4EBP1, which are involved in mRNA translation. mTORC2 activates PKC-α and AKT, and regulates the actin cytoskeleton (Populo et al., 2012). mTORC1 signaling occurs as an early event in the process of tumorigenesis. It has been described in a mouse model of adenomatous polyposis (FAP) and as an intestinal polyp formation. Therefore mTORC1 could be a target for drug development of colon polyps and cancer (Crunkhorn, 2015).

Rapamycin, a macrolide antibiotic, was the first mTOR inhibitor discovered. Several derivatives of rapamycin (e.g., sirolimus) or with more favorable pharmacokinetic and solubility properties (e.g., temsirolumus and everolimus) have been synthesized. The mTORC1 complex is sensitive to rapamycin; however, mTORC2 is considered resistant to rapamycin (Klümpen et al., 2010).

Combinational therapies consisting of mTOR inhibitors and other agents such as antiangiogenic molecular or anti-EGFR agents have shown promising results. One example of this combination could be the trial of Gulhati et al. (2012). In this trial, sorafenib, a multikinase inhibitor RAF, VEGFR, and PDGFR, was used in combination with rapamycin in a preclinical setting. A combination of rapamycin with sorafenib synergistically inhibits proliferation of CRC cells. CRCs with KRAS and PIK3CA mutations are partially sensitive to either rapamycin or sorafenib monotherapy, but highly sensitive to combination treatment. It was concluded that the combination with sorafenib enhances therapeutic efficacy of rapamycin on induction of apoptosis and inhibition of cell-cycle progression, migration, and invasion of CRCs (Gulhati et al., 2012). The phase I study evaluating efficacy and tolerability of the antiangiogenic agent tivozanib (an oral VEGF receptor-1, -2, -3 inhibitor) plus everolimus in metastatic CRC indicated that the oral combination of tivozanib and everolimus was well tolerated, with stable disease achieved in 50% of patients with refractory, metastatic colorectal cancer (Wolpin et al., 2013). mTOR inhibitors have been evaluated in combination with antiEGFR mAbs such as cetuximab and panitumumab (Cuinci et al., 2014; McRee, 2014). In a phase 1 trial of everolimus a combination with cetuximab was safely administered in patients with refractory CRC (Cuinci et al., 2014). Another phase I trial of a regimen of everolimus in addition to 5-FU/LV and mFOLFOX6 appears safe and tolerable, but the further addition of panitumumab resulted in an unacceptable level of toxicity (Cuinci et al., 2014; McRee, 2014).

In conclusion, the mTOR pathway is another potential target of therapy in metastatic colorectal cancer patients and the combinational approach may further increase the efficacy of treatment.

FROM GENOME-WIDE ASSOCIATION STUDY (GWAS) TO GENETIC RISK SCORE (GRS) FOR CRC

Since the completion of both the Human Genome Project (HGP) (Venter et al., 2001) and the International HapMap project (Frazer et al., 2007), recent progress in genotyping technology has facilitated the use of GWAS as a "gene-hunting study" for linking specific genetic variants with human disease by many thousand loci simultaneously (Hindorff et al., 2009; Visscher et al., 2017).

One of the essential ideas underpinning precision medicine is to determine the information about the genetic risk factors for different diseases in order to inform patients and to introduce proactive changes in behavior such as environmental/lifestyle factors. Thus, understanding gene–environment interactions (G × E) relevant to genetic variations for common and complex human diseases is an important challenge for precision medicine (Simon et al., 2016). Accordingly, starting with the determination of phenotype susceptibilities, disease conditions, and drug efficacy or side effects, GWAS, which allow interrogation of more than a million SNPs in the genome, aims to discover genomic level variance among individuals or different case-control groups within a population in a holistic and agnostic manner (Hindorff et al., 2009; Stranger et al., 2011). That is to say, GWAS are a type of case-control study employing whole genome comparisons of allele frequencies in which individuals with the disease condition being studied are compared to similar individuals without the disease condition under study (i.e., to find correlations between specific SNPs and a phenotype). Furthermore, GWAS is a very useful preeminent tool for identifying common SNPs with a minor allele frequency (MAF) >5% as a risk factor associated with multifactorial complex disorder (Panagiotou et al., 2010; De La Vega, 2011; Lowe and Reddy, 2015), which accelerated the identification

of predictive and prognostic markers for disease susceptibility and PGx (Daly, 2010). In order to be significantly correlated with complex disease traits, in GWAS, the requisite is that SNPs should meet the stringent genome-wide threshold that is usually set to $P < 5 \times 10^{-8}$ (Zeng et al., 2015).

Although common genetic variants are largely responsible for complex diseases such as CRC, several strong lines of evidence indicate that rare variants, which are MAF frequencies <1%–5%, play a crucial role in complex disease etiology with possible larger genetic effects than common variants (Lettre, 2014; Bomba et al., 2017). Even if GWAS have been successful in identifying common genetic variations with well-established "risk" loci associated to various complex diseases, because of the small effect of these individual SNPs on complex common disease, the clinical utility of these genetic variations is skeptical in personalized risk prediction related to lifestyle, demographic, and clinical factors for the disease in question (Visscher et al., 2017). Moreover, detecting rare variants from GWAS data is difficult and often explains only a small proportion of trait heritability, in explaining the "missing heritability" of complex human trait (Blanco-Gómez et al., 2016). The advance of GWAS has identified multiple common genetic variants influencing susceptibility to different types of cancers, including CRC, which has a strong heritable background with an increasing number of susceptibility loci. To date, GWAS studies on CRC have identified common SNP variants conferring CRC susceptibility risk at more than 40 loci (Whiffin et al., 2014; Abulí et al., 2016; Frampton and Houlston, 2017). However, most individual genetic risk variants at these loci confer a modest effect in explaining only a small fraction of the CRC risk (relative risk-RR) as well as small fraction of the heritable component leading to limited ability of genetic testing for CRC prediction. Hence, since it is intimidating to consider any single SNP as a clinical test for CRC risk susceptibility, there is no clinical utility for public health. Likewise, the value of these SNPs as prognosticators has been explored only in a few studies (Dai et al., 2012; Noci et al., 2016).

Like most of the complex diseases, genetic roots of CRC are multifactorial, nondeterministic, and genetic variants scattered across the genome, contributing small risks for CRC. That is to say, information from multiple SNPs is needed to characterize genetic susceptibility to CRC. Respectively, in the context of precision medicine, since the risk estimation, risk stratification, and appropriate treatment for CRC events is crucial for personalized prevention counseling and therapy, it has been suggested that combining multiple single genetic variants with minor effects into polygenic risk scores (PRS) (also called genetic risk score, or genome-wide score) from GWAS along with conventional nongenetic factors might reduce bias and might improve power disease risk prediction for complex disorders including CRC (Garcia-Closas et al., 2014; Cooke and Igo, 2016; Chen et al., 2016; Frampton et al., 2016; Läll et al., 2017). PRS/GRS, which is a numeric summary measure of genetic risk, does not depend on single genetic variants (a single common SNP), but is a number based on variation in multiple sets of SNPs estimated from GWAS to serve as the best prediction for the trait (Dudbridge, 2013; Krapohl et al., 2017). Accordingly, since the PRS/GRS aggregates the effects of thousands of SNPs from GWAS, it appears to be a more realistic tool and so there is growing interest in constructing whether the PRS/GRS is associated with the CRC risk. Personalized screening using PRS/GRS has potential implication in optimizing population screening by stratifying for CRC patients according to their PRS/GRS as to low, medium, or high risk leading to an earlier screening for high-risk individuals as well as a decrease in the number needed to treat (NNT) for chemoprevention (Pharoah et al., 2008; Shaik et al., 2015; Frampton et al., 2016). Additionally, an individual with a high gene risk score for CRC but who has not yet developed the disease, could be advised to change his or her lifestyle factors (Maher, 2015).

Taken together, since aggregate multiple genetic loci identified by GWAS will influence the CRC risk, the combination of genetic factors and conventional nongenetic factors will lead to more precise and personalized risk prevention and a predictive model for colorectal cancer (Iwasaki et al., 2017; Frampton and Houlston, 2017).

PRECISION/PERSONALIZED MEDICINE APPROACHES FOR COLON CANCER DRIVEN BY SYSTEMS BIOLOGY

Microbiome-driven carcinogenesis in colorectal cancer (metagenomics)

The terms *microbiome* and *microbiota* are used often interchangeably; however, whereas the human microbiota that could be considered an environmental factor modulating the host metabolism comprehends the populations of more than 400 microbial species (the commensal bacteria as symbiotic and pathogenic, viruses and fungi, single-celled animals, such as protists, and archaea) with more than 100 trillion cells (approximately 10 times the total number of cells in the human body) encoding 100-fold more unique genes than our own genome (Ley et al., 2006) living in or on the human body as important determinants of health and disease, the genetic constitutes of microbial cells are called microbiome ("second genome"). It is also important to note that the human microbiome contains 150 times more genes than the human genome (Qin et al., 2010). The composition of the gut microbiome was originally coined by Joshua Lederberg as "the ecological community of symbiotic, and pathogenic microorganisms sharing our body space" and describing the collective genome of our indigenous microbes (microflora), and so it is equivalent to the gut metagenome (Lederberg and McCray, 2001). Despite the importance of genetic polymorphisms in human disease risk management and in therapeutic outcomes, it is well recognized the role of the microbiome which has been, more often than not, overlooked (Grice and Segre, 2012). Therefore, manipulation of the gut microbiota may be an important therapeutic strategy in order to dissect and characterize the association of the modulation and perturbation of the human microbiome with disease as well as to characterize a strategy regulating energy balance (harvested from food) in the body (Kim B-S et al., 2013; Manzat-Saplacan et al., 2015). Metagenomics, which is one of the fastest burgeoning fields of the important "-omics" languages (Maccaferri et al., 2011; Banerjee et al., 2015; Proctor, 2016; Hizel, 2017), has since its advent gained importance in bringing out the crucial link between sporadic CRC (Banerjee et al., 2015; Yu et al., 2017) as well as metabolic disorders, such as obesity and type 2 diabetes (Baothman et al., 2016). Metagenomics analyzes the structure and functional potential of microbial communities in their native habitats through the sequencing of PCR amplicons from the ribosomal 16S (NGS-based 16S rRNA sequencing) and whole metagenome shotgun (WMS) sequencing (Thomas et al., 2012; Jovel et al., 2016). Respectively, accumulated metagenomic studies point out the important crucial role of gut microbial dysbiosis (i.e., microbial imbalance on or inside the body), which fosters the discharge of bacterial genotoxins, metabolites leading to chronic inflammation and oxidative DNA damage in the etiology, and progress of sporadic CRC and also particular enriched bacterial species identified in fecal microbiomes from CRC compared to healthy fecal microbiomes (Sobhani et al., 2013; Gao et al., 2015; Ray and Kidane, 2016). Accordingly, the presence of several oral pathogens, particularly enriched abundance of *Fusobacterium*, positively correlated with lymph node metastasis, have been reported underlying the importance of its investigation as a marker for colorectal cancer presence, risk, or prognosis (Castellarin et al., 2012; Flanagan et al., 2014; Purcell et al., 2017).

Diet is a major external force shaping gut communities and the clear association related to contribution of diet (meat, fruits, vegetables, and fiber) to microbiota (Read and Holmes, 2017; Akin and Tözün, 2014) and to pathogenesis of CRC (Ryan-Harshman and Aldoori, 2007) is increasingly well acknowledged. Respectively, since microbiota has a crucial role in maintaining gut homeostasis and host health, diet–microbiota–microbiome interactions cannot be defined without reflection on host–diet interactions. A recent large cohort GWAS identified a significant interaction between rs4143094 SNP on chromosome 10p14 near the gene *GATA3* (GATA binding protein 3), though interaction of dietary factors with genetic variants to modify risk of colorectal cancer is still under challenging consideration (Figueiredo et al., 2014).

There is evidence to suggest an inverse relationship between the level of dietary fibers, which is a source of short-chain fatty acid (SCFA), and the incidence of human CRC (Park et al., 2005; Donohoe et al., 2011; Vital et al., 2014). As the main energy sources for the host, SCFAs can be used for de novo synthesis of lipids and glucose (den Besten et al., 2013). Distinct composition of gut microbiota produces different SCFAs, and butyrate is one of the SCFA end products of colonic fermentation known to modulate several cellular processes such as cell differentiation and inhibition of cell proliferation in different tumor cell lines (Hague et al., 1996; Hizel et al., 1999; Comalada et al., 2006).

Being the main energy source for colonocytes, production of butyrate plays a critical role in health and disease (Vital et al., 2014). Furthermore, the effect of butyrate on epigenetic gene regulation is also demonstrated. Hence, different compositions of gut microbiota in cancer as well as in obesity and type 2 diabetes could affect the epigenetic regulation of genes, such as genes encoding SCFA receptors FFAR2 and FFAR3 causing changes in gene expression and signaling of FFARs by histone deacetylases (HDACs) inhibition and hyperacetylation (Davie, 2003).

Knowledge of the composition of the gut microbiome ecosystem and gut microbial marker, such as the ratio of SCFA, are future important key elements of precision medicine, though much more research is needed to achieve more precise diagnosis and therapeutic intervention and might eventually aid in lifestyle interventions for disease prevention and/or modulation (Kasubuchi et al., 2015). Recent metagenomic analysis underscores the acetyl-CoA pathway as the main pathway for butyrate production in healthy individuals. Accordingly, dietary interventions such as high dietary fiber intake, which are the important source of SCFAs influencing microbial composition, could be considered as an option in personal nutrition in order to increase, for example, butyrate concentrations and reduce insulin resistance for the engagement against metabolic syndrome such as diabetes and also it could be an important avenue in drug development (example NaB might be a promising molecule) for the prevention and treatment of CRC (Davie, 2003; Alenghat, 2015; Kasubuchi et al., 2015).

Taken together, since every individual is different and because of the potentially modifiable nature of gut microbiota identifying the role of particular bacterial species in CRC development, the increased understanding of particular bacterial species in sporadic CRC pathogenesis through metagenomic systems biology approach will certainly help to improve CRC predictions and prevention (Eklöf et al., 2017; Purcell et al., 2017). Hence, besides genomic information, better comprehension of the impact of gut microbial dysbiosis together with novel gut microbiota genetic biomarkers could provide encouraging direction in early diagnosis, prognosis, prevention, and treatment of CRC (Sobhani et al., 2013; Cho et al., 2014; Gao et al., 2015; Manzat-Saplacan et al., 2015; Purcell et al., 2017; Yu et al., 2017).

Metabolomics of colon cancer tissues reflects "metabolic codes" of patients

CRC is linked to mutations in DNA, amino acids, pentose-phosphate pathway carbohydrates, and glycolytic, gluconeogenic, and tricarboxylic acid intermediates (Brown et al., 2016). In the year 2016, Brown et al. (2016) utilized a nontargeted global metabolome perspective to investigate human CRC, adjacent mucosa, and stool. In this research, they identified metabolite profile variations between CRC and adjacent mucosa from patients who had parts of their colon removed. Furthermore, the analyses of the metabolic pathways unveiled connections among complex networks of metabolites (Halama et al., 2015; Brown et al., 2016; Farshidfar et al., 2016). The extensive and thorough characterization of tumor phenotypes facilitated the therapeutic approaches (Halama et al., 2015). Lately, omics strategies have shed light upon tumor biology. Such applications have been to a large degree executed to avail biosignatures to examine the disease and ameliorate therapeutic results. The orchestration of metabolomics for studying tumorigenesis is particularly instrumental, since it demonstrates the biochemical outcome of a number of tumor-specific functional changes associated with the disease (Halama et al., 2015; Farshidfar et al., 2016). In 2015 Halama et al. experimented utilizing nontargeted metabolomics-based mass spectroscopy together with ultra-high-performance liquid chromatography and gas chromatography in order to perform metabolic phenotyping of four cancer cell lines: two colon cancer (HCT15, HCT116) and two ovarian cancer (OVCAR3, SKOV3). They then applied the MetaP server to statistically analyze their data (Halama et al., 2015). According to Halama et al.'s study, where a total of 225 metabolites were detected in all colon cancer cell lines mentioned herein, 67 of these molecules exceptionally discriminated colon cancer cells from ovarian cancer cells. Metabolic biomarkers identified in this study suggested elevation of β-oxidation and urea cycle metabolism in colon cancer cell lines. As a conclusion Halama et al.'s study provided a panel of specific metabolic identifiers between colon and ovarian cancer cell lines. The novel findings can be considered as potential drug targets. Furthermore, these potential hits can be assessed further in primary cells, biofluids, and tissue samples as biosignatures for CRC (Halama et al., 2015).

Glycomics, as part of an array of metastatic codes

Although carbohydrates are considered to be the most abundant macromolecules in nature, they have been underestimated and considered less functionally significant than nucleic acids and proteins. Due to their large and complex heterogeneity, carbohydrates and glycoconjugates have been difficult to isolate from their natural sources owing to the lack of efficient technologies. Consequently, this diminished the recognition of their significance in the most basic biological practices, leading to the lack of exploration of the carbohydrates in the biological arena. Nevertheless, in the past few decades complex carbohydrate expression has been acknowledged to be crucial in the establishment of living networks, including cellular signaling, cellular differentiation, and the immune system (Adamczyk et al., 2012; Drake, 2015; Siobhan et al., 2015). The recognition of this structure and functional connection of carbohydrates facilitated chemical and biological protocols to unveil new areas in molecular biology, coining the term *glycobiology* in the early 1980s. State-of-the-art protocols have thrived ever since, leading to the examination of "glycomaterials" for their glycan function (Adamczyk et al., 2012; Drake, 2015; Siobhan et al., 2015). The normal activity of glycosylation is interfered during cancer cell formation (Kim and Varki, 1997; Kannagi et al., 2004). These alterations in the tumor cells lead to the restructuring of the cell surface glycans of the transformed tumors regulating metastatic potential of the cancer cells (Shriver et al., 2004). Additionally, parallel to the alterations in the glycosylation status of the cancer cell surfaces, gene expression profiles of the carbohydrate binding proteins get modified during this neoplastic transformation causing a comprehensive alteration between the glycans and their associated receptors (Kannagi et al., 2004; Shriver et al., 2004). The mechanism via which glycosylation modification triggers cancer cell metastasis and invasion are yet to be determined, but the functions of the particular cell surface bound glycoproteins and their carbohydrate motifs have been recently identified (Dall'Olio et al., 2012; Geng et al., 2012). Due to the recent developments in the glycoanalytical methodologies, a better comprehension of the carbohydrate connection to the cell surface lipids and proteins has emerged (Siobhan et al., 2015). Learning how the glycan determinants work on cancer-associated proteins helped to unveil a new level of complexity, facilitating better comprehension of how neoplasia changes the normal process in cells. Additionally, the orchestration of glycan determinants on cancer can have connections with cancer cell metastasis, proliferation, survival, and immune escape. This extensive knowledge about the cancer glycome has implications to be very effective in the cancer field (Siobhan et al., 2015). However, much still remains to be unveiled pertaining to the functional interactions of cell surface glycans with the regulation of the metastatic potential of the cancer cells. Thorough knowledge of carbohydrate remodeling in cancer will require complete profiling of glycosylation patterns in the tumor microenvironment (Siobhan et al., 2015).

Interactomics in colon cancer tissues: Where are we?

Chromatin looping can make it possible for distant regulatory elements to affect the gene expression of their target genes located far away upstream or downstream from where these cis- and transregulatory elements are located. Chromosome conformation capture (3C2)–based methods can be utilized to determine physical interactions between enhancers and promoters (Spilianakis et al., 2005; Simonis et al., 2006; Jäger et al., 2015). GWAS helped determine SNPs that are linked to complex diseases. It has been determined that there may exist many SNPs that are located within regulatory elements and influence through long-range regulation of gene expression (Pomerantz et al., 2009; Ahmadiyeh et al., 2010; Zhang et al., 2012). Consequently, protein–DNA interaction can be facilitated upon dynamic modification of chromatin architecture.

Functionality of proteins in a cell depends on many parameters. Protein–protein interactions and the interactions of the proteins with drugs and other macromolecules in the cell contribute to the formation of cellular phenotypes. Thus far it is not very clear as to how these interactions are regulated. Nevertheless, there is a wide range of evidence that neoplasia is caused by impairments of networks operated by dysfunctional proteins. Computer

databases are being utilized to elucidate and build libraries between proteins and other macromolecules. These efforts help to reveal large network graphs of interactoms exhibiting the complexity of association of differentially expressed proteins or genes in a disease status (Sjöblom et al., 2006). According to the Nibbe et al. (2009) study investigating the interactom graph, which portrays 70 different proteins known to interact with the so-called gatekeeper gene APC, reveals mutations in over 90% of CRC tumors (Sjöblom et al., 2006). Conversely, in another well-studied WNT-signaling pathway, also known to be dysregulated in CRC, there were relatively few interactions involving APC (Sjöblom et al., 2006). Evidently, the variations in the activity of these interactions may explain the heterogeneity of tumors in patients at large. The dissimilarity in the aggressiveness of the patient tumors may be the origin of the differential outcome of treatments repeatedly observed in the clinic (Yu et al., 2005). Improved biosignatures of CRC will be essential to elucidate the coordinated, differential expression of numerous proteins and/or genes that synergistically expedite or delay the orchestration of networks responsible for CRC (De Las Rivas and de Luis, 2004; Mathivanan et al., 2006; McMurray et al., 2008). Previously, Mathivanan et al. (2006) evaluated this in the human protein–protein interaction (PPI) data set in the public domain. Human PPIs mark a new chapter in biomarker discovery in neoplasia as they present an operative setting in which the PPIs examine the mechanistic role of proteins and genes found to be notably differentially expressed by traditional means (Mathivanan et al., 2006). Interactomics provide a database of protein–protein networks, which may be examined for differentially active interactions between tumors and control samples. This in turn provides a strong basis for integrative strategy, facilitating enhanced and functional biosignatures of cancer. A proteomics-first approach furnishes an unconventional method to explore the functional subsets of proteins discriminative of CRC. Consequently, to put all the perspectives mentioned herein in a nutshell, systems biology–based strategies combine all the details about the cell, and create a potential to convey more powerful and precise classifiers of diseases, in contrast with traditional protocols (Mathivanan et al., 2006; Jonsson and Bates, 2007; Nibbe et al., 2009; Hudler et al., 2014; Tanase et al., 2016; Brandi et al., 2017).

HIGH-THROUGHPUT GENOMIC TECHNOLOGIES FOR DETECTING POTENTIAL "DRIVER" GENOMIC ALTERATIONS AS A RISK ASSESSMENT TOOL AND FOR TREATMENT SELECTION IN CRC

The application of personalized cancer medicine (PCM) relies upon the following arguments:

1. Human diseases involve numerous genetic anomalies.
2. Oncogenes and tumor suppressor genes are responsible for a certain subset of these anomalies.
3. Individualized anticancer agents are available that effectively regulate these targets (Tran et al., 2012).

This section focuses on the technologies underlying high-throughput genomics for detecting potential genomic alterations as risk evaluation for CRC. Additionally, it inspects the preliminary results of next-generation genome sequencing analysis of CRC genomes for personalized medicine and microarrays for diagnostic and treatment selection in CRC, and reviews the challenges met in the discovery phase of new genetic aberrations. Since the completion of the Human Genome Project, the appearance of the scientific age of "omics" has transformed the cancer research. The International Cancer Genome Consortium (ICGC) is synchronizing scientific research targeted at discovering genomic modifications that are linked to cancer (Tran et al., 2012; Sikdar et al., 2016; ICGC International Data Access Committee, 2016). Discoveries along these lines greatly depend on the development of novel tools in molecular diagnostics, especially in the area of whole genome sequencing. The enhanced timelines and cost-effective technologies have expedited discovery in cancer genomics as well as in translational medicine (Tran et al., 2012). Nevertheless, challenges still exist prior to PCM becoming available for the population at large on an everyday bases.

Next-generation analysis of CRC genomes for precision medicine

The advancement in the new DNA sequencing technology recognized as next-generation sequencing (NGS) has transformed the speed and throughput of classifying cancer-related genomic aberrations (Tran et al., 2012; Kim TM et al., 2013; Pellicer et al., 2017; Wang S et al., 2017). The challenge is how to take advantage of this state-of-the-art technology to better comprehend the fundamental molecular mechanism of colorectal cancer and to better recognize clinically applicable genetic/genomic biosignatures for diagnosis and precision medicine (Kim TM et al., 2013). Lately, the development of NGS has dramatically altered the speed and throughput of DNA sequencing (Ajay et al., 2011). Conventionally, genome sequencing has been expensive and laborious, but novel methodologies have reduced both of these restraints (Metzker, 2009; Shendure, 2011). The growing number of tailored therapeutics has generated an increased demand for genetic/genomic anomalies in clinical samples in a more convenient (cheaper and timely) manner (Tran et al., 2012; Armengol et al., 2016). The collection of repetitive mutations is on the rise, and these mutations have been attractive in testing to assess their part in predicting mutations for a number of molecularly targeted factors in development. This demands a necessity for higher-throughput genotyping technologies (D'Haene et al., 2015; Armengol et al., 2016).

High-throughput genotyping technologies, composed of microarrays and multiplexed tests, have been efficiently utilized for genotyping clinical CRC specimens (Dias-Santagata et al., 2010; D'Haene et al., 2015). CRC can be regarded as a complex disease considering the common variant hypothesis with a polygenic model of inheritance. The genetic constituents of common complex diseases correlate with the variants of moderate outcome (Esteban-Jurado et al., 2014). Thus far, 30 common, low-penetrance susceptibility variants have been discovered for CRC (Esteban-Jurado et al., 2014). Lately, novel sequencing methodologies including whole genome and exome sequencing enable a new approach to expedite the discovery of previously undescribed susceptibility genes responsible for human diseases. Through utilizing whole genome sequencing, germ-line mutations in the POLE and POLD1 genes have been identified to be accountable for a novel form of CRC genetic susceptibility named polymerase proofreading-associated polyposis (Esteban-Jurado et al., 2014). Consequently, simultaneous sequencing of gene groups via NGS will help analyze all applicable driver mutations of CRC in parallel, which will be required in routine clinical applications in the coming years (Huth et al., 2014).

Microarrays in diagnostics and treatment selection in CRC

Administration of new molecular diagnostic assays may facilitate a better survival rate for CRC patients (Huth et al., 2014; Mahasneh et al., 2017). Protocols for molecular evaluation of single genes (i.e., TP53 and/or KRAS) in addition to microarray-based methodologies are inevitable. Innovative screening applications are needed to be able to detect precancerous state and early-stage malignancies in healthy populations to facilitate corrective therapeutic interventions (Huth et al., 2014; Mahasneh et al., 2017). A recently developed noninvasive multitarget stool DNA kit (Cologuard™) for CRC screening involves KRAS mutation analysis, as well as testing of the methylation condition of the NDRG4 and BMP3 genes (Huth et al., 2014; Mahasneh et al., 2017). Additionally, besides the fecal occult blood testing (FOBT) and colonoscopy, other novel assays also offer approaches for early detection of CRC such as the blood-based Septin 9 DNA methylation test (Huth et al., 2014).

Liquid biopsy analysis also presents diagnostic potential for screening CRC developing resistance to treatment. Determination of additional novel molecular biomarkers and diagnostic methods in CRC will facilitate early detection and targeted therapy of colorectal cancer (Huth et al., 2014; Mahasneh et al., 2017).

The MS status of CRC could also be predicted depending on miRNA expression profiles (Schepeler et al., 2008; Carvalho et al., 2017; Liu D et al., 2017; Strubberg and Madison, 2017). Spotted locked nucleic acid (LNA)–based oligonucleotide microarrays can be utilized to screen the expression profiles of miRNAs. Hence, miRNAs can also be presented as promising tools to classify colon cancers as either MSI or MSS (T Schepeler et al., 2008; Carvalho et al., 2017; Liu J et al., 2017; Strubberg and Madison, 2017). Mutations within the TP53 gene are the most common genetic/genomic aberrations in human cancers such as CRC. The GeneChip p53 assay is based on the recently designed oligonucleotide microarray technology (Takahashi et al., 2003).

Considering the presence of epigenetic changes and posttranslational modifications in CRC,

development of diagnostic technologies will also facilitate screening protocols of CRC in a multistep process that results from genetic or epigenetic alterations. Cancers with high degrees of methylation (the CpG island methylator phenotype, CIMP) display a clinically discrete group that is classified by epigenetic variability (Issa, 2004, 2008). Thus, a methylome signature in CRC should be described employing a methylation microarray analysis (i.e., Illumina HumanMethylation27 array) (Issa, 2004). This will lead to identification of a defined list for methylome specific genes that will eventually provide better clinical management of CRC patients in the future (Issa, 2004; Ashktorab et al., 2013; Kim et al., 2014). Nevertheless, additional studies are necessary to improve specificity and sensitivity prior to initiating clinical trials.

CRC is a multifactorial condition that emerges due to genomics, genetics, and epigenetic changes in a lot of tumor suppressor genes, oncogenes, mismatch repair genes, and cell cycle regulatory elements. These molecular aberrations are considered as candidate CRC biosignatures. They can present the physicians with diagnostic, prognostic, and treatment response details. The overall aim is to determine applicable, affordable, and appropriate biosignatures, which will be instrumental in patient care decisions, resulting in direct benefits to patients (Mahasneh et al., 2017).

It is essential that the scientific methods continue to be tested prior to administering the novel omics-based technologies toward the management of patient care. Nevertheless, these novel applications provide substantial capacity for improved care of cancer patients notably by enabling novel trial designs, which will provide us with the answers to rapidly translate these findings from bench to bedside.

CONCLUSION AND PERSPECTIVE

In this postgenomic era, fast-tracked development of omics languages are on the verge of transforming health care leading to better understanding of the molecular underpinnings of different polygenic multifactorial chronic complex diseases including cancer, which represents the preponderance of health care. Hence, together with environmental factors, the deployment of precision medicine relies on the ability to identify accurately relevant patient subgroups from extensive clinical and precise genomic data sets in order to create a "bespoke" health care approach to the detection, prevention, and treatment of diseases (Hizel, 2017).

Taking into consideration the genetic environment of diseases in individual patients together with the nongenetic exposures, personalized genetic screening could answer some important pressing questions related to the existence of disease, prediction of disease occurrence, treatment choice through PGx testing (Hizel, 2017), and stratification of subjects according to their polygenic risk score (PRS) (low, medium or high) leading to a decrease in the number needed to treat (NNT) for chemoprevention (Frampton et al., 2016). However, in spite of the quickening pace of omics technologies and a host of available literature on biomarkers in CRC, currently only a few biomarkers are used in daily clinical practice such as KRAS, BRAF, and MSI (Ballman, 2015; Carethers and Jung, 2015; Gatalica et al., 2015; Mahasneh et al., 2017). Even if UGT1A1*28 PGx testing is approved by the FDA, it is not routinely used for prediction of toxicity of irinotecan in clinical practice resulting in the dearth of alternative treatments for risk genotype carrier patients (Innocenti and Ratain, 2006; Vivot et al., 2015). To this end, considering the high incidence of CRC with high significant cost for society, there is a growing need for validation of efficient prognostic or predictive biomarkers in large prospective cohorts (Duffy and Crown, 2008; Danese and Montagnana, 2017; Kamel and Al-Amodi, 2017; Lemery et al., 2017). Moreover, since science is moving toward more interdisciplinary involvement of multiple disciplines and because of the multidisciplinary nature of clinical cancer genomics, an "integrated knowledge ecosystem" with contributions from all sectors of science and society is needed for successful application of precision genomic medicine (Lemay et al., 2017).

Last, precision medicine could be desirable for the society inasmuch as it will be consistent with social, ethical, local technological, and local disease burden realities of the society in countries with different income levels (Hizel, 2017).

REFERENCES

Ab Mutalib NS, Md Yusof NF, Abdul SN, and Jamal R. Pharmacogenomics DNA biomarkers in colorectal cancer: Current update. *Front Pharmacol* 8;2017:736.

Abdulla MH, Valli-Mohammed MA, Al-Khayal K et al. Cathepsin B expression in colorectal cancer in a Middle East population: Potential value as a tumor biomarker for late disease stages. *Oncol Rep* 37(6);2017:3175–80.

Abulí A, Castells A, Bujanda L et al. Genetic variants associated with colorectal adenoma susceptibility. *PLOS ONE* 11(4);2016:e0153084.

Adamczyk B, Tharmalingam T, and Rudd PM. Glycans as cancer biomarkers. *Biochim Biophys Acta* 1820(9);2012:1347–53.

Adams JM, and Cory S. The Bcl-2 apoptotic switch in cancer development and therapy. *Oncogene* 26(9);2007:1324–37.

Adjei AA, Cohen RB, Franklin W et al. Phase I pharmacokinetic and pharmacodynamic study of the oral, small-molecule mitogen-activated protein kinase kinase 1/2 inhibitor AZD6244 (ARRY-142886) in patients with advanced cancers. *J Clin Oncol* 26;2008:2139–46.

Afzal S, Gusella M, Vainer B et al. Combinations of polymorphisms in genes involved in the 5-Fluorouracil metabolism pathway are associated with gastrointestinal toxicity in chemotherapy treated colorectal cancer patients. *Clin Cancer Res* 17(11);2011:3822–9.

Aguirre-Portolés C, Fernández LP, and Ramírez de Molina A. Precision nutrition for targeting lipid metabolism in colorectal cancer. *Nutrients* 9(10);2017, pii: E1076.

Ahlquist T, Bottillo I, Danielsen SA et al. RAS signaling in colorectal carcinomas through alteration of RAS, RAF, NF1, and/or RASSF1A. *Neoplasia* 10(7);2008:680–6.

Ahmadiyeh N, Pomerantz MM, Grisanzio C et al. 8q24 prostate, breast, and colon cancer risk loci show tissue-specific long range interaction with MYC. *Proc Natl Acad Sci USA* 107(21);2010:9742–6.

Ajay SS, Parker SC, Abaan HO, Fajardo KV, and Margulies EH. Accurate and comprehensive sequencing of personal genomes. *Genome Res* 21(9);2011:1498–5.

Akin H, and Tözün N. Diet, microbiota, and colorectal cancer. *J Clin Gastroenterol* 48(Suppl. 1);2014:S67–9.

Akinleye A, Furqan M, Mukhi N, Ravella P, and Liu D. MEK and the inhibitors: From bench to bedside. *J Hematol Oncol* 6;2013:27.

Akshatha C, Mysorekar V, Arundhathi S, Arul P, Raj A, and Shetty S. Correlation of p53 Overexpression with the clinicopathological prognostic factors in colorectal adenocarcinoma. *J Clin Diagn Res* 10(12);2016:EC05–8.

Aleksandrova K, Pischon T, Jenab M et al. Combined impact of healthy lifestyle factors on colorectal cancer: A large European cohort study. *BMC Med* 12;2014:168.

Alenghat T. Epigenomics and the microbiota. *Toxicol Pathol* 43(1);2015:101–6.

Alexandrova EM, Mirza SA, Xu S, Schulz-Heddergott R, Marchenko ND, and Moll UM. p53 loss-of-heterozygosity is a necessary prerequisite for mutant p53 stabilization and gain-of-function in vivo. *Cell Death Dis* 8(3);2017:e2661.

Amado RG, Wolf M, Peeters M et al. Wild-type KRAS is required for panitumumab efficacy in patients with metastatic colorectal cancer. *ICE* 26(10);2008:1626–34.

Amelio I, Knight RA, Lisitsa A, Melino G, and Antonov AV. p53MutaGene: An online tool to estimate the effect of p53 mutational status on gene regulation in cancer. *Cell Death Dis* 7;2016:e2148.

Ando Y, Chida M, Nakayama K, Saka H, and Kamataki T. The UGT1A1*28 allele is relatively rare in a Japanese population. *Pharmacogenetics* 8(4);1998:357–60.

Andreyev HJ, Norman AR, Cunningham D, Oates JR, and Clarke PA. Kirsten ras mutations in patients with colorectal cancer: The multicenter "RASCAL" study. *Natl Cancer Inst* 90(9);1998:675–84.

Andreyev HJ, Norman AR, Cunningham D et al. Kirsten ras mutations in patients with colorectal cancer: The "RASCAL II" study. *Br J Cancer* 85;2001:692–6.

Antelo M, Balaguer F, Shia J et al. A high degree of LINE-1 hypomethylation is a unique feature of early-onset colorectal cancer. *PLOS ONE* 7(9);2012:e45357.

Aparicio J, Fernandez-Martos C, Vincent JM et al. FOLFOX alternated with FOLFIRI as first-line chemotherapy for metastatic colorectal cancer. *Clin Colorectal Cancer* 5(4);2005:263–7.

Armaghany T, Wilson JD, Chu Q, and Mills G. Genetic alterations in colorectal cancer. *Gastrointest Cancer Res* 5(1);2012:19–27.

Armanios M, and Blackburn EH. The telomere syndromes. *Nat Rev Genet* 13;2012:693–704.

Armengol G, Sarhadi VK, Ghanbari R et al. Driver gene mutations in stools of colorectal carcinoma patients detected by targeted next-generation sequencing. *J Mol Diagn* 18(4);2016:471–9.

Arnold M, Sierra MS, Laversanne M, Soerjomataram I, Jemal A, and Bray F. Global patterns and trends in colorectal cancer incidence and mortality. *Gut* 66(4);2017:683–691.

Aschacher T, Wolf B, Enzmann F et al. LINE-1 induces hTERT and ensures telomere maintenance in tumour cell lines. *Oncogene* 35(1);2016:94–104.

Ashkenazi A. Directing cancer cells to self-destruct with pro-apoptotic receptor agonists. *Nat Rev Drug Discov* 7(12);2008:1001–12.

Ashktorab H, Belgrave K, Hosseinkhah F et al. Global histone H4 acetylation and HDAC2 expression in colon adenoma and carcinoma. *Dig Dis Sci* 54(10);2009:2109–17.

Ashktorab H, Rahi H, Wansley D et al. Toward a comprehensive and systematic methylome signature in colorectal cancers. *Epigenetics* 8;2013:807–15.

Atreya CE, Cutsem V, Bendell J et al. Updated efficacy of the MEK inhibitor trametinib (T), BRAF inhibitor dabrafenib (D), and anti-EGFR antibody panitumumab (P) in patients (pts) with BRAF V600E mutated (BRAFm) metastatic colorectal cancer (mCRC). *J Clin Oncol* 33;2015 (suppl; abstr 103).

Augustine T, Maitra R, and Goel S. Telomere length regulation through epidermal growth factor receptor signaling in cancer. *Genes Cancer* 8(5–6);2017:550–8.

Augustine TA, Baig M, Sood A et al. Telomere length is a novel predictive biomarker of sensitivity to anti-EGFR therapy in metastatic colorectal cancer. *Br J Cancer* 112(2);2015:313–8.

Aydin Son Y, Tüzmen Ş, Hızel C. Designing and implementing pharmacogenomics study. In: *Omics for Personalized Medicine.* Barh D, Dahawan D, Ganguly NK (Eds), New Delhi, Springer, 2013, pp. 97–117.

Baba Y, Nosho K, Shima K et al. Relationship of CDX2 loss with molecular features and prognosis in colorectal cancer. *Clin Cancer Res* 15(4);2009:4665–73.

Babaei H, Zeinalian M, Emami MH, Hashemzadeh M, Farahani N, and Salehi R. Simplified microsatellite instability detection protocol provides equivalent sensitivity to robust detection strategies in Lynch syndrome patients. *Cancer Biol Med* 14(2);2017:142–50.

Bai Y, Wu HW, Ma X, Liu Y, and Zhang YH. Relationship between UGT1A1*6/*28 gene polymorphisms and the efficacy and toxicity of irinotecan-based chemotherapy. *Onco Targets Ther* 10;2017:3071–81.

Balc'h EL, Grandin N, Demattei MV et al. Measurement of telomere length in colorectal cancers for improved molecular diagnosis. *Int J Mol Sci* 18(9);2017, pii: E1871.

Ballester V, Rashtak S, and Boardman L. Clinical and molecular features of young-onset colorectal cancer. *World J Gastroenterol* 22(5);2016:1736–44.

Ballman KV. Biomarker: Predictive or prognostic? *J Clin Oncol* 33(33);2015:3968–71.

Balmanno K, Chell SD, Gillings AS, Hayat S, and Cook SJ. Intrinsic resistance to the MEK1/2 inhibitor AZD6244 (ARRY-142886) is associated with weak ERK1/2 signalling and/or strong PI3K signalling in colorectal cancer cell lines. *Int J Cancer* 125;2009:2332–41.

Bandrés E, Cubedo E, Agirre X et al. Identification by Real-time PCR of 13 mature microRNAs differentially expressed in colorectal cancer and non-tumoral tissues. *Mol Cancer* 5;2006:29.

Banerjee J, Mishra N, and Dhas Y. Metagenomics: A new horizon in cancer research. *Meta Gene* 5;2015:84–9.

Baothman OA, Zamzami MA, Taher I, Abubaker J, and Abu-Farha M. The role of gut microbiota in the development of obesity and diabetes. *Lipids Health Dis* 15;2016:108.

Barault L, Charon-Barra C, Jooste V et al. Hypermethylator phenotype in sporadic colon cancer: Study on a population-based series of 582 cases. *Cancer Res* 68(20);2008a:8541–6.

Barault L, Veyries N, Jooste V et al. Mutations in the RAS-MAPK, PI(3)K (phosphatidylinositol-3-OH kinase) signaling network correlate with poor survival in a population-based series of colon cancers. *Int J Cancer* 122(10);2008b:2255–9.

Barbarino JM, Haidar CE, Klein TE, and Altman RB. PharmGKB summary: Very important pharmacogene information for UGT1A1. *Pharmacogenet Genomics* 24(3);2014:177–83.

Bardhan K, and Liu K. Epigenetics and colorectal cancer pathogenesis. *Cancers (Basel)* 5(2);2013:676–713.

Barh D, Azevedo V. *Omics Technologies and Bio-Engineering, Volume 1: Towards Improving Quality of Life.* London, Elsevier, 2017.

Barski A, Cuddapah S, Cui K et al. High-resolution profiling of histone methylations in the human genome. *Cell* 129(4);2007:823–37.

Bashir B, and Snook AE. Immunotherapy regimens for metastatic colorectal carcinomas. *Hum Vaccin Immunother* 2017:1–5.

Bazan V, Migliavacca M, Zanna I et al. Specific codon 13 K-ras mutations are predictive of clinical outcome in colorectal cancer patients, whereas codon 12 K-ras mutations are associated with mucinous histotype. *Ann Oncol* 13;2002:1438–46.

Becquemont L. Pharmacogenomics of adverse drug reactions: Practical applications and perspectives. *Pharmacogenomics* 10(6);2009:96.

Bélanger A, Tojcic J, Harvey M, and Guillemette C. Regulation of UGT1A1 and HNF1 transcription factor gene expression by DNA methylation in colon cancer cells. *BMC Mol Biol* 11;2010:9.

Bennouna J, Lang I, Valladares-Ayerbes M et al. A phase II, open-label, randomised study to assess the efficacy and safety of the MEK1/2 inhibitor AZD6244 (ARRY-142886) versus capecitabine monotherapy in patients with colorectal cancer who have failed one or two prior chemotherapeutic regimens. *Invest New Drugs* 29;2011:1021–8.

Berger FG, and Berger SH. Thymidylate synthase as a chemotherapeutic drug target: Where are we after fifty years? *Cancer Biol Ther* 5(9);2006:1238–41.

Berquin IM, Cao L, Fong D, and Sloane BF. Identification of two new exons and multiple transcription start points in the 5′-untranslated region of the human cathepsin-B-encoding gene. *Gene* 159(2);1995:143–9.

Bertholee D, Maring JG, and van Kuilenburg AB. Genotypes affecting the pharmacokinetics of anticancer drugs. *Clin Pharmacokinet* 56(4);2017:317–37.

Bertino JR, and Banerjee D. Thymidylate synthase as an oncogene? *Cancer Cell* 5(4);2004:301–2.

Bertorelle R, Briarava M, Rampazzo E et al. Telomerase is an independent prognostic marker of overall survival in patients with colorectal cancer. *Br J Cancer* 108(2);2013: 278–84.

Beutler E, Gelbart T, and Demina A. Racial variability in the UDP-glucuronosyltransferase 1 (UGT1A1) promoter: A balanced polymorphism for regulation of bilirubin metabolism? *Proc Natl Acad Sci USA* 95(14);1998:8170–4.

Bhandari A, Woodhouse M, and Gupta S. Colorectal cancer is a leading cause of cancer incidence and mortality among adults younger than 50 years in the USA: A SEER-based analysis with comparison to other young-onset cancers. *J Investig Med* 65(2);2017:311–5.

Bhati A, Garg H, Gupta A, Chhabra H, Kumari A, and Patel T. Omics of cancer. *Asian Pac J Cancer Prev* 13(9);2012:4229–33.

Bhurgri H, and Samiullah S. Colon cancer screening—Is it time yet? *J Coll Physicians Surg Pak* 27(6);2017:327–8.

Bian B, Mongrain S, Cagnol S et al. Cathepsin B promotes colorectal tumorigenesis, cell invasion, and metastasis. *Mol Carcinog* 55(5);2016:671–87.

Biason P, Masier S, and Toffoli G. UGT1A1*28 and other UGT1A polymorphisms as determinants of irinotecan toxicity. *J Chemother* 20(2);2008:158–65.

Bilgin B, Sendur MA, Bülent Akıncı M, Şener Dede D, and Yalçın B. Targeting the PD-1 pathway: A new hope for gastrointestinal cancers. *Curr Med Res Opin* 33(4);2017:749–59.

Blanco-Gómez A, Castillo-Lluva S, Del Mar Sáez-Freire M, Hontecillas-Prieto L, Mao JH, Castellanos-Martín A, and Pérez-Losada J. Missing heritability of complex diseases: Enlightenment by genetic variants from intermediate phenotypes. *BioEssays* 38(7);2016:664–73.

Blons H, Emile JF, Le Malicot K et al. Prognostic value of KRAS mutations in stage III colon cancer: Post-hoc analysis of the PETACC8 phase III trial dataset. *Ann Oncol* 25(12);2014.

Boisdron-Celle M, Remaud G, Traore S, Poirier AL, Gamelin L, Morel A, and Gamelin E. 5-Fluorouracil-related severe toxicity: A comparison of different methods for the pretherapeutic detection of dihydropyrimidine dehydrogenase deficiency. *Cancer Lett* 249;2007:271–82.

Bokemeyer C, Bondarenko I, Hartmann JT et al. Efficacy according to biomarker status of cetuximab plus FOLFOX-4 as first-line treatment for metastatic colorectal cancer: The OPUS study. *Ann Oncol* 22(7);2011:1535–46.

Bokemeyer C, Bondarenko I, Makhson A et al. Fluorouracil, leucovorin, and oxaliplatin with and without cetuximab in the first-line treatment of metastatic colorectal cancer. *J Clin Oncol* 27(5);2009:663–71.

Bokemeyer C, Van Cutsem E, Rougier P et al. Addition of cetuximab to chemotherapy as first-line treatment for KRAS wild-type metastatic colorectal cancer: Pooled analysis of the CRYSTAL and OPUS randomised clinical trials. *Eur J Cancer* 48(10);2012:1466–75.

Boland CR, and Goel A. Microsatellite instability in colorectal cancer. *Gastroenterology* 138(6);2010:2073–2087.e3.

Boland CR, Thibodeau SN, Hamilton SR et al. A National Cancer Institute Workshop on Microsatellite Instability for cancer detection and familial predisposition: Development of international criteria for the determination of microsatellite instability in colorectal cancer. *Cancer Res* 58(22);1998:5248–57.

Boland PM, and Ma WW. Immunotherapy for colorectal cancer. *Cancers (Basel)* 9(5);2017, pii: E50.

Bolden JE, Peart MJ, and Johnstone RW. Anticancer activities of histone deacetylase inhibitors. *Nat Rev Drug Discov* 5(9);2006:769–84.

Bomba L, Walter K, and Soranzo N. The impact of rare and low-frequency genetic variants in common disease. *Genome Biol* 18(1);2017:77.

Bond CE, Nancarrow DJ, Wockner LF, Wallace L, Montgomery GW, Leggett BA, and Whitehall VL. Microsatellite stable colorectal cancers stratified by the BRAF V600E mutation show distinct patterns of chromosomal instability. *PLOS ONE* 9(3); 2014:e91739.

Bosma PJ, Chowdhury JR, Bakker C et al. The genetic basis of the reduced expression of bilirubin UDP-glucuronosyltransferase 1 in Gilbert's syndrome. *N Engl J Med* 333(18);1995:1171–5.

Bourlioux P, Koletzko B, Guarner F, and Braesco V. The intestine and its microflora are partners for the protection of the host: Report on the Danone Symposium "The Intelligent Intestine," held in Paris, June 14, 2002. *Am J Clin Nutr* 78(4);2003:675–83.

Brandi J, Manfredi M, Speziali G, Gosetti F, Marengo E, and Cecconi D. Proteomic approaches to decipher cancer cell secretome. *Semin Cell Dev Biol* 2017, pii: S1084-9521(17)30301-30304. doi: 10.1016/j.semcdb.2017.06.030.

Braun MS, Richman SD, Thompson L et al. Association of molecular markers with toxicity outcomes in a randomized trial of chemotherapy for advanced colorectal cancer: The FOCUS trial. *J Clin Oncol* 27(33);2009:5519–28.

Braun MS, and Seymour MT. Balancing the efficacy and toxicity of chemotherapy in colorectal cancer. *Ther Adv Med Oncol* 3(1);2011:43–52.

Brosens RP, Belt EJ, Haan JC et al. Deletion of chromosome 4q predicts outcome in stage II colon cancer patients. *Anal Cell Pathol (Amst)* 33(2);2010:95–104.

Brown DG, Rao S, Weir TL, O'Malia J, Bazan M, Brown RJ, and Ryan EP. Metabolomics and metabolic pathway networks from human colorectal cancers, adjacent mucosa, and stool. *Cancer Metab* 4;2016:11.

Brule SY, Jonker DJ, Karapetis CS et al. Location of colon cancer (right-sided versus left-sided) as a prognostic factor and a predictor of benefit from cetuximab in NCIC CO.17. *Eur J Cancer* 51(11);2015:1405–14.

Buecher B, Cacheux W, Rouleau E, Dieumegard B, Mitry E, and Lièvre A. Role of microsatellite instability in the management of colorectal cancers. *Dig Liver Dis* 45(6);2013:441–9.

Cacev T, Jokić M, Spaventi R, Pavelić K, and Kapitanović S. Loss of heterozygosity testing using real-time PCR analysis of single nucleotide polymorphisms. *J Cancer Res Clin Oncol* 132(3);2006:200–4.

Calvo E, Walko C, Dees EC, and Valenzuela B. Pharmacogenomics, pharmacokinetics, and pharmacodynamics in the era of targeted therapies. *Am Soc Clin Oncol Educ Book* 35;2016:e175–84.

Campbell JM, Stephenson MD, Bateman E, Peters MD, Keefe DM, and Bowen JM. Irinotecan-induced toxicity pharmacogenetics: An umbrella review of systematic reviews and meta-analyses. *Pharmacogenomics J* 17(1);2017:21–8.

Campos FG, Figueiredo MN, and Martinez CA. Colorectal cancer risk in hamartomatous polyposis syndromes. *World J Gastrointest Surg* 7(3);2015:25–32.

Capella G, Cronauer-Mitra S, Pienado MA, and Perucho M. Frequency and spectrum of mutations at codons 12 and 13 of the c-K-ras gene in human tumors. *Environ Health Perspect* 93;1991:125–31.

Capello M, Lee M, Wang H et al. Carboxylesterase 2 as a determinant of response to irinotecan and neoadjuvant FOLFIRINOX therapy in pancreatic ductal adenocarcinoma. *J Natl Cancer Inst* 107(8);2015, pii: djv132.

Capitain O, Boisdron-Celle M, Poirier AL, Abadie-Lacourtoisie S, Morel A, and Gamelin E. The influence of fluorouracil outcome parameters on tolerance and efficacy in patients with advanced colorectal cancer. *Pharmacogenomics J* 8(4);2008:256–67.

Carethers JM, and Jung BH. Genetics and genetic biomarkers in sporadic colorectal cancer. *Gastroenterology* 149(5);2015:1177–90.

Cargnello M, and Roux PP. Activation and function of the MAPKs and their substrates, the MAPK-activated protein kinases. *Microbiol Mol Biol Rev* 75(1);2011:50–83.

Carlini LE, Meropol NJ, Bever J et al. UGT1A7 and UGT1A9 polymorphisms predict response and toxicity in colorectal cancer patients treated with capecitabine/irinotecan. *Clin Cancer Res* 11;2005:1226–36.

Carreras CW, and Santi DV. The catalytic mechanism and structure of thymidylate synthase. *Annu Rev Biochem* 64;1995:721–62.

Carvalho TI, Novais PC, Lizarte FSN et al. Analysis of gene expression EGFR and KRAS, microRNA-21 and microRNA-203 in patients with colon and rectal cancer and correlation with clinical outcome and prognostic factors. *Acta Cir Bras* 32(3);2017:243–50.

Castellarin M, Warren RL, Freeman JD et al. Fusobacterium nucleatum infection is prevalent in human colorectal carcinoma. *Genome Res* 22(2);2012:299–306.

Catalano V, Loupakis F, Graziano F et al. Mucinous histology predicts for poor response rate and overall survival of patients with colorectal cancer and treated with first-line oxaliplatin- and/or irinotecan-based chemotherapy. *Br J Cancer* 100(6);2009:881–7.

Cathomas G. PIK3CA in colorectal cancer. *Front Oncol* 4;2014:35.

Caudle KE, Thorn CF, Klein TE, Swen JJ, McLeod HL, Diasio RB, and Schwab M. Clinical Pharmacogenetics Implementation Consortium guidelines for dihydropyrimidine dehydrogenase genotype and fluoropyrimidine dosing. *Clin Pharmacol Ther* 94(6);2013:640–5.

Cecchin E, Corona G, Masier S et al. Carboxylesterase isoform 2 mRNA expression in peripheral blood mononuclear cells is a predictive marker of the irinotecan to SN38 activation step in colorectal cancer patients. *Clin Cancer Res* 11(19 Pt 1);2005:6901–7.

Cecchin E, Innocenti F, D'Andrea M et al. Predictive role of the UGT1A1, UGT1A7, and UGT1A9 genetic variants and their haplotypes on the outcome of metastatic colorectal cancer patients treated with fluorouracil, leucovorin, and irinotecan. *J Clin Oncol* 27(15);2009:2457–65.

Cecchin E, Roncato R, Guchelaar HJ, Toffoli G, for the Ubiquitous Pharmacogenomics Consortium. Ubiquitous pharmacogenomics (U-PGx): The time for implementation is now. An Horizon2020 program to drive pharmacogenomics into clinical practice. *Curr Pharm Biotechnol* 18(3);2017:204–9.

Chalhoub N, and Baker SJ. PTEN and the PI3-kinase pathway in cancer. *Annu Rev Pathol* 4;2009:127–50.

Chang L, Chang M, Chang HM, and Chang F. Microsatellite instability: A predictive biomarker for cancer immunotherapy. *Appl Immunohistochem Mol Morphol* 26(2);2017 Sep 4:e15–21.

Chang SC, Lin JK, Lin TC, and Liang WY. Loss of heterozygosity: An independent prognostic factor of colorectal cancer. *World J Gastroenterol* 11(6);2005:778–84.

Charasson V, Haaz MC, and Robert J. Determination of drug interactions occurring with the metabolic pathways of irinotecan. *Drug Metab Dispos* 30(6);2002:731–3.

Chen H, Liu X, Brendler CB et al. Adding genetic risk score to family history identifies twice as many high-risk men for prostate cancer: Results from the prostate cancer prevention trial. *Prostate* 76(12);2016:1120–9.

Chen S, Villeneuve L, Jonker D et al. ABCC5 and ABCG1 polymorphisms predict irinotecan-induced severe toxicity in metastatic colorectal cancer patients. *Pharmacogenet Genomics* 25(12);2015:573–83.

Chen S, Yueh MF, Bigo C et al. Intestinal glucuronidation protects against chemotherapy-induced toxicity by irinotecan (CPT-11). *Proc Natl Acad Sci USA* 110(47);2013:19143–8.

Cheng H, Zhang L, Cogdell DE et al. Circulating plasma MiR-141 is a novel biomarker for metastatic colon cancer and predicts poor prognosis. *PLOS ONE* 6(3); 2011:e17745.

Cheng L, Li M, Hu J et al. UGT1A1*6 polymorphisms are correlated with irinotecan-induced toxicity: A system review and meta-analysis in Asians. *Cancer Chemother Pharmacol* 73(3);2014:551–60.

Chin LJ, Ratner E, Leng S et al. A SNP in a let-7 microRNA complementary site in the KRAS 3′ untranslated region increases non-small cell lung cancer risk. *Cancer Res* 68(20);2008:8535–40.

Cho M, Carter J, Harari S, and Pei Z. The interrelationships of the gut microbiome and inflammation in colorectal carcinogenesis. *Clin Lab Med* 34(4);2014:699–710.

Chong H, Vikis HG, and Guan KL. Mechanisms of regulating the Raf kinase family. *Cell Signal* 15;2003:463–9.

Chua TC, and Morris DL. Resectable colorectal liver metastases: Optimal sequencing of chemotherapy. *J Gastrointest Cancer* 43(3);2012:496–8.

Coffman BL, Rios GR, King CD, and Tephly TR. Human UGT2B7 catalyzes morphine glucuronidation. *Drug Metab Dispos* 25;1997:1–4.

Collins K, and Mitchell JR. Telomerase in the human organism. *Oncogene* 21(4);2002:564–79.

Comalada M, Bailon E, de Haro O et al. The effects of short-chain fatty acids on colon epithelial proliferation and survival depend on the cellular phenotype. *J Cancer Res Clin Oncol* 132(8);2006:487–97.

Connell LC, Mota JM, Braghiroli MI, and Hoff PM. The rising incidence of younger patients with colorectal cancer: Questions about screening, biology, and treatment. *Curr Treat Options Oncol* 18(4);2017:23.

Connolly K, Brungs D, Szeto E, and Epstein RJ. Anticancer activity of combination targeted therapy using cetuximab plus vemurafenib for refractory BRAF (V600E)-mutant meta-static colorectal carcinoma. *Curr Oncol* 21(1);2014:e151–1.

Cooke Bailey JN, and Igo RP Jr. Genetic risk scores. *Curr Protoc Hum Genet* 91;2016:1.29.1–9.

Corcoran RB, Atreya CE, Falchook GS et al. Combined BRAF and MEK inhibition with dabrafenib and trametinib in BRAF V600–mutant colorectal cancer. *J Clin Oncol* 33(34);2015:4023–31.

Cortes-Ciriano I, Lee S, Park WY, Kim TM, and Park PJ. A molecular portrait of microsatel-lite instability across multiple cancers. *Nat Commun* 8;2017:15180.

Corté H, Manceau G, Blons H, and Laurent-Puig P. MicroRNA and colorectal cancer. *Dig Liver Dis* 44(3);2012:195–200.

Costa E. Hematologically important mutations: Bilirubin UDPglucuronosyltransferase gene mutations in Gilbert and Crigler-Najjar syn-dromes. *Blood Cells Mol Dis* 36;2006:77–80.

Court MH, and Greenblatt DJ. Molecular genetic basis for deficient acetaminophen glucuroni-dation by cats: UGT1A6 is a pseudogene, and evidence for reduced diversity of expressed hepatic UGT1A isoforms. *Pharmacogenetics* 10(4);2000:355–69.

Crona DJ, Ramirez J, Qiao W et al. Clinical validity of new genetic biomarkers of irinotecan neutropenia: An independent replication study. *Pharmacogenomics J* 16(1);2016:54–9.

Crunkhorn S. mTOR inhibition curbs colorectal cancer. *Nat Rev Drug Discov* 14;2015:14–5.

Cui C, Shu C, Cao D et al. UGT1A1*6, UGT1A7*3 and UGT1A9*1b polymorphisms are predic-tive markers for severe toxicity in patients with metastatic gastrointestinal cancer treated with irinotecan-based regimens. *Oncol Lett* 12(5);2016:4231–7.

Cuinci C, Perini RF, Avadhani AN et al. Phase 1 and pharmacodynamic trial of everolimus in combination with cetuximab in patients with advanced cancer. *Cancer* 120(1);2014:77–85.

Cunningham D, Atkin W, Lenz HJ, Lynch HT, Minsky B, Nordlinger B, and Starling N. Colorectal cancer. *Lancet* 375;2010:1030–47.

Côté JF, Kirzin S, Kramar A et al. UGT1A1 poly-morphism can predict hematologic toxicity in patients treated with irinotecan. *Clin Cancer Res* 13(11);2007:3269–75.

Dai J, Gu J, Huang M, Eng C, Kopetz ES, Ellis LM, Hawk E, and Wu X. GWAS-identified colorectal cancer susceptibility loci associ-ated with clinical outcomes. *Carcinogenesis* 33(7);2012:1327–31.

Dall'Olio F, Malagolini N, Trinchera M, and Chiricolo M. Mechanisms of cancer-asso-ciated glycosylation changes. *Front Biosci* 17;2012:670–99.

Daly AK. Genome-wide association stud-ies in pharmacogenomics. *Nat Rev Genet* 11(4);2010:241–6.

Danese E, and Montagnana M. Epigenetics of colorectal cancer: Emerging circulating diag-nostic and prognostic biomarkers. *Ann Transl Med* 5(13);2017:279.

Das C, and Tyler JK. Histone exchange and histone modifications during transcrip-tion and aging. *Biochim Biophys Acta* 1819(3–4);2013:332–42.

Davie JR. Inhibition of histone deacety-lase activity by butyrate. *J Nutr* 133(7 Suppl.);2003:2485S–93S.

Day FL, Jorissen RN, Lipton L et al. PIK3CA and PTEN gene and exon mutation-specific clinicopathologic and molecular associa-tions in colorectal cancer. *Clin Cancer Res* 19(12);2013:3285–96.

de Jong FA, van der Bol JM, Mathijssen RH, van Gelder T, Wiemer EA, Sparreboom A, and Verweij J. Renal function as a predictor of iri-notecan-induced neutropenia. *Clin Pharmacol Ther* 84;2008a:254–62.

de Jong FA, Sparreboom A, Verweij J, and Mathijssen RH. Lifestyle habits as a contribu-tor to anti-cancer treatment failure. *Eur J Cancer* 44;2008b:374–82.

De La Vega FM, Bustamante CD, and Leal SM. Genome-wide association mapping and rare alleles: From population genomics to personalized medicine—Session introduction. *Pac Symp Biocomput* 2011:74–5.

De Lange T. How telomeres solve the end-protection problem. *Science* 326;2009:948–52.

De Las Rivas J, and de Luis A. Interactome data and databases: Different types of protein interaction. *Comp Funct Genomics* 5;2004:173–8.

De Mattia E, and Toffoli G. C677 T and A1298C MTHFR polymorphisms, a challenge for antifolate and fluoropyrimidine-based therapy personalisation. *Eur J Cancer* 45(8);2009:1333–51.

den Besten G, van Eunen K, Groen AK, Venema K, Reijingoud DJ, and Bakker BM. The role of short-chain fatty acids in the interplay between diet, gut microbiota, and host energy metabolism. *J Lipid Res* 54;2013:2325–40.

De Palma M, and Hanahan D. The biology of personalized cancer medicine: Facing individual complexities underlying hallmark capabilities. *Mol Oncol* 6(2);2012:111–27.

De Rook W, Claes B, Bernasconi D et al. Effects of KRAS, BRAF, NRAS, and PIK3CA mutations on the efficacy of cetuximab plus chemotherapy in chemotherapy-refractory metastatic colorectal cancer: A retrospective consortium analysis. *Lancet Oncol* 11(8);2010:753–62.

Deenen MJ, Cats A, Beijnen JH, and Schellens JH. Part 4: Pharmacogenetic variability in anticancer pharmacodynamic drug effects. *Oncologist* 16(7);2011a:1006–20.

Deenen MJ, Meulendijks D, Cats A et al. Upfront genotyping of DPYD/2A to individualize fluoropyrimidine therapy: A safety and cost analysis. *J Clin Oncol* 34(3);2016:227–3.

Deenen MJ, Tol J, Burylo AM et al. Relationship between single nucleotide polymorphisms and haplotypes in DPYD and toxicity and efficacy of capecitabine in advanced colorectal cancer. *Clin Cancer Res* 17(10);2011b:3455–68.

Deininger P. Genetic instability in cancer: Caretaker and gatekeeper genes. *Ochsner J* 1(4);1999:206–9.

Del Re M, Di Paolo A, van Schaik RH, Bocci G, Simi P, Falcone A, and Danesi R. Dihydropyrimidine dehydrogenase polymorphisms and fluoropyrimidine toxicity: Ready for routine clinical application within personalized medicine? *EPMA J* 1(3);2010:495–502.

Deng G, Kakar S, Tanaka H, Matsuzaki K, Miura S, Sleisenger MH, and Kim YS. Proximal and distal colorectal cancers show distinct gene-specific methylation profiles and clinical and molecular characteristics. *Eur J Cancer* 44;2008:1290–301.

Derks S, and Diosdado B. Personalized cancer medicine: Next steps in the genomic era. *Cell Oncol (Dordr)* 38(1);2015:1–2.

Derwinger K, Wettergren Y, Odin E, Carlsson G, and Gustavsson B. A study of the MTHFR gene polymorphism C677T in colorectal cancer. *Clin Colorectal Cancer.* 8(1);2009:43–8.

Des Guetz G, Schischmanoff O, Nicolas P, Perret GY, Morere JF, and Uzzan B. Does microsatellite instability predict the efficacy of adjuvant chemotherapy in colorectal cancer? A systematic review with meta-analysis. *Eur J Cancer* 45(10);2009:1890–6.

Devaud N, and Gallinger S. Chemotherapy of MMR-deficient colorectal cancer. *Fam Cancer* 12(2);2013:301–6.

Devilee P, Cleton-Jansen AM, and Cornelisse CJ. Ever since Knudson. *Trends Genet* 17(10);2001:569–73.

de Wildt SN, Kearns GL, Leeder JS, and van den Anker JN. Glucuronidation in humans. Pharmacogenetic and developmental aspects. *Clin Pharmacokinet* 36;1999:439–52.

D'Haene N, Le Mercier M, De N et al. Clinical validation of targeted next generation sequencing for colon and lung cancers. *PLOS ONE* 10(9);2015:e0138245.

Di Martino MT, Arbitrio M, Leone E et al. Single nucleotide polymorphisms of ABCC5 and ABCG1 transporter genes correlate to irinotecan-associated gastrointestinal toxicity in colorectal cancer patients: A DMET microarray profiling study. *Cancer Biol Ther* 12(9);2011:780–7.

Di Nicolantonio F, Martini M, Molinari F et al. Wild-type BRAF is required for response to panitumumab or cetuximab in metastatic colorectal cancer. *J Clin Oncol* 26(35);2008:5705–12.

Dias-Santagata D, Akhavanfard S, David SS et al. Rapid targeted mutational analysis of human tumours: A clinical platform to guide personalized cancer medicine. *EMBO Mol Med* 2;2010:146–58.

Dienstmann R, Vermeulen L, Guinney J, Kopetz S, Tejpar S, and Tabernero J. Consensus molecular subtypes and the evolution of precision medicine in colorectal cancer. *Nat Rev Cancer* 17(2);2017:79–92.

Dix BR, Robbins P, Soong R, Jenner D, House AK, and Iacopetta BJ. The common molecular genetic alterations in Dukes' B and C colorectal carcinomas are not short-term prognostic indicators of survival. *Int J Cancer* 59(6);1994:747–51.

Domingo E, Niessen RC, Oliveira C et al. BRAF-V600E is not involved in the colorectal tumorigenesis of HNPCC in patients with functional MLH1 and MSH2 genes. *Oncogene* 24(24);2005:3995–8.

Donohoe DR, Garge N, Zhang X, Sun W, O'Connell TM, Bunger MK, and Bultman SJ. The microbiome and butyrate regulate energy metabolism and autophagy in the mammalian colon. *Cell Metab* 13(15);2011:517–26.

Doullard JY, Oliner KS, Siena S et al. Panitumumab-FOLFOX4 treatment and RAS mutations in colorectal cancer. *NEJM* 369(11);2013:1023–34.

Drake RR. Glycosylation and cancer: Moving glycomics to the forefront. *Adv Cancer Res* 126;2015:1–10.

Drew DA, Nishihara R, Lochhead P et al. A prospective study of smoking and risk of synchronous colorectal cancers. *Am J Gastroenterol* 112(3);2017:493–501.

Du L, Kim JJ, Shen J, Chen B, and Dai N. KRAS and TP53 mutations in inflammatory bowel disease-associated colorectal cancer: A meta-analysis. *Oncotarget* 8(13);2017:22175–86.

Dudbridge F. Power and predictive accuracy of polygenic risk scores. *PLoS Genet* 9(3);2013:e1003348.

Duffy MJ, and Crown J. A personalized approach to cancer treatment: How biomarkers can help. *Clin Chem* 54(11);2008:1770–9.

Dukes CE. The classification of cancer of the rectum. *J Pathol Bactériol* 35;1932:323–32.

Dumez H, Guetens G, De Boeck G, Highley MS, de Bruijn EA, van Oosterom AT, and Maes RA. In vitro partition of irinotecan (CPT-11) in human volunteer blood: The influence of concentration, gender and smoking. *Anticancer Drugs* 16;2005:893–5.

Ehrlich M. DNA hypomethylation in cancer cells. *Epigenomics* 1(2);2009:239–59.

Eklöf V, Löfgren-Burström A, Zingmark C et al. Cancer-associated fecal microbial markers in colorectal cancer detection. *Int J Cancer* 141(12);2017:2528–36.

Emami AH, Sadighi S, Shirkoohi R, and Mohagheghi MA. Prediction of response to irinotecan and drug toxicity based on pharmacogenomics test: A prospective case study in advanced colorectal cancer. *Asian Pac J Cancer Prev* 18(10);2017:2803–7.

Engelhardt M, Drullinsky P, Guillem J, and Moore MA. Telomerase and telomere length in the development and progression of premalignant lesions to colorectal cancer. *Clin Cancer Res* 3(11);1997:1931–41.

Eshghifar N, Farrokhi N, Naji T, and Zali M. Tumor suppressor genes in familial adenomatous polyposis. *Gastroenterol Hepatol Bed Bench* 10(1);2017:3–13.

Esteban-Jurado C, Garre P, Vila M et al. New genes emerging for colorectal cancer predisposition. *World J Gastroenterol* 20(8);2014:1961–71.

Etienne-Grimaldi MC, Boyer JC, Beroud C et al. New advances in DPYD genotype and risk of severe toxicity under capecitabine. *PLOS ONE* 12(5);2017:e0175998.

Etienne-Grimaldi MC, Boyer JC, Thomas F et al. UGT1A1 genotype and irinotecan therapy: General review and implementation in routine practice. *Fundam Clin Pharmacol* 29(3);2015:219–37.

Etienne-Grimaldi MC, Francoual M, Formento JL, and Milano G. Methylenetetrahydrofolate reductase (MTHFR) variants and fluorouracil-based treatments in colorectal cancer. *Pharmacogenomics* 8(11);2007:1561–6.

Etienne-Grimaldi MC, Milano G, Maindrault-Goebel F et al. Methylenetetrahydrofolate reductase (MTHFR) gene polymorphisms and FOLFOX response in colorectal cancer patients. *Br J Clin Pharmacol* 69;2010:58–66.

Evaluation of Genomic Applications in Practice and Prevention (EGAPP) Working Group. Recommendations from the EGAPP Working Group: Can UGT1A1 genotyping reduce morbidity and mortality in patients with metastatic colorectal cancer treated with irinotecan? *Genet Med* 11(1);2009:15–20.

Fallik D, Borrini F, Boige V et al. Microsatellite instability is a predictive factor of the tumor response to irinotecan in patients with advanced colorectal cancer. *Cancer Res* 63(18);2003:5738–44.

Fariña-Sarasqueta A. The BRAF V600E mutation is an independent prognostic factor for survival in stage II and stage III colon cancer patients. *Ann Oncol* 21;2010:2396–402.

Farshidfar F, Weljie AM, Kopciuk KA et al. A validated metabolomic signature for colorectal cancer: Exploration of the clinical value of metabolomics. *Br J Cancer* 115(7);2016:848–57.

Fearon ER, and Vogelstein B. A genetic model for colorectal tumorigenesis. *Cell* 61;1990:759–67.

Fedewa SA, Flanders WD, Ward KC, Lin CC, Jemal A, Goding Sauer A, Doubeni CA, and Goodman M. Racial and ethnic disparities in interval colorectal cancer incidence: A population-based cohort study. *Ann Intern* 166;2017:857–66.

Feinberg AP, and Vogelstein B. Hypomethylation distinguishes genes of some human cancers from their normal counterparts. *Nature* 301(5895);1983:89–92.

Fennelly C, and Amaravadi RK. Lysosomal biology in cancer. *Methods Mol Biol* 1594;2017:293–308.

Ferlay J, Soerjomataram I, Dikshit R et al. Cancer incidence and mortality worldwide: Sources, methods and major patterns in GLOBOCAN 2012. *Int J Cancer* 136(5);2015:E359–86.

Fernández-Marcelo T, Sánchez-Pernaute A, Pascua I, De Juan C, Head J, Torres-García AJ, and Iniesta P. Clinical relevance of telomere status and telomerase activity in colorectal cancer. *PLOS ONE* 11(2);2016:e0149626.

Ferraldeschi R. Pharmacogenetics in colorectal cancer. In: *Pharmacogenetics: Making Cancer Treatment Safer and More Effective.* Neman WG (Ed.), Springer, New York, 2010, pp. 61–86.

Ferraldeschi R, and Newman WG. Pharmacogenetics and pharmacogenomics: A clinical reality. *Ann Clin Biochem* 48(Pt 5);2011:410–7.

Figueiredo JC, Hsu L, Hutter CM et al. Genome-wide diet-gene interaction analyses for risk of colorectal cancer. *PLoS Genet* 10(4);2014:e1004228.

Flanagan L, Schmid J, Ebert M et al. Fusobacterium nucleatum associates with stages of colorectal neoplasia development, colorectal cancer and disease outcome. *Eur J Clin Microbiol Infect Dis* 33(8);2014:1381–90.

Frampton M, and Houlston RS. Modeling the prevention of colorectal cancer from the combined impact of host and behavioral risk factors. *Genet Med* 19(3);2017:314–21.

Frampton MJ, Law P, Litchfield K et al. Implications of polygenic risk for personalised colorectal cancer screening. *Ann Oncol* 27(3);2016:429–34.

Francisco LM, Salinas VH, Brown KE, Vanguri VK, Freeman GJ, Kuchroo VK, and Sharpe AH. PD-L1 regulates the development, maintenance, and function of induced regulatory T cells. *J Exp Med* 206(13);2009:3015–29.

Fransen K, Klintenas M, Osterstrom A, Dimberg J, Monstein HJ, and Söderkvist P. Mutation analysis of the BRAF, ARAF and RAF-1 genes in human colorectal adenocarcinomas. *Carcinogenesis* 25(4);2004:527–33.

Frazer KA, Ballinger DG, Cox DR et al. A second generation human haplotype map of over 3.1 million SNPs. *Nature* 449(7164);2007:851–61.

Frosst P, Blom HJ, Milos R et al. A candidate genetic risk factor for vascular disease: A common mutation in methylenetetrahydrofolate reductase. *Nat Genet* 10;1995:111–3.

Frueh FW, and Gurwitz D. From pharmacogenetics to personalized medicine: A vital need for educating health professionals and the community. *Pharmacogenomics* 5(5);2004:571–9.

Fujita K, Kubota Y, Ishida H, and Sasaki Y. Irinotecan, a key chemotherapeutic drug for metastatic colorectal cancer. *World J Gastroenterol* 21(43);2015:12234–48.

Fujiyoshi K, Yamaguchi T, Kakuta M et al. Predictive model for high-frequency microsatellite instability in colorectal cancer patients over 50 years of age. *Cancer Med* 6(6);2017:1255–63.

Funkhouser WK Jr, Lubin IM, Monzon FA, Zehnbauer BA, Evans JP, Ogino S, and Nowak JA. Relevance, pathogenesis, and testing algorithm for mismatch repair-defective colorectal carcinomas: A report of the association for molecular pathology. J Mol Diagn 14(2);2012:91–103.

Gagne JF, Montminy V, Belanger P, Journault K, Gaucher G, and Guillemette C. Common human UGT1A polymorphisms and the altered metabolism of irinotecan active metabolite 7-ethyl-10-hydroxycamptothecin (SN-38). Mol Pharmacol 62;2002:608–17.

Gagnon J, Bernard O, Villeneuve L, Têtu B, and Guillemette C. Irinotecan inactivation is modulated by epigenetic silencing of UGT1A1 in colon cancer. Clin Cancer Res 12;2006:1850–8.

Gallagher DJ, and Kemeny N. Metastatic colorectal cancer: From improved survival to potential cure. Oncology 78;2010:237–48.

Galon J, Costes A, Sanchez-Cabo F et al. Type, density, and location of immune cells within human colorectal tumors predict clinical outcome. Science 313(5795);2006:1960–4.

Gao J, Zhou J, Li Y, Lu M, Jia R, and Shen L. UGT1A1 6/28 polymorphisms could predict irinotecan-induced severe neutropenia not diarrhea in Chinese colorectal cancer patients. Med Oncol 30(3);2013:604.

Gao Z, Guo B, Gao R, Zhu Q, and Qin H. Microbiota disbiosis is associated with colorectal cancer. Front Microbiol 6;2015:20.

Garcia-Aranda C, de Juan C, Diaz-Lopez A et al. Correlations of telomere length, telomerase activity, and telomeric-repeat binding factor 1 expression in colorectal carcinoma. Cancer 106(3);2006:541–51.

Garcia-Closas M, Gunsoy NB, and Chatterjee N. Combined associations of genetic and environmental risk factors: Implications for prevention of breast cancer. J Natl Cancer Inst 106(11);2014.

Gatalica Z. Thymidylate synthase over-expression underlies the observed lack of 5-FU therapy benefit for MSI-H colorectal cancers. Ann Oncol 25(Suppl. 4);2014:iv167–209.

Gatalica Z, Vijayvergia N, Vranic S, Xiu J, Reddy S, Lynch HT, and El-Deiry WS. Therapeutic biomarker differences between MSI-H and MSS colorectal cancers. J Clin Oncol 33(Suppl; abstr 3597);2015:3597.

Gatalica Z, Vranic S, Xiu J, Swensen J, and Reddy S. High microsatellite instability (MSI-H) colorectal carcinoma: A brief review of predictive biomarkers in the era of personalized medicine. Fam Cancer 15(3);2016:405–12.

Geng Y, Marshall JR, and King MR. Glycomechanics of the metastatic cascade: Tumor cell-endothelial cell interactions in the circulation. Ann Biomed Eng 40;2012:790–805.

Gertler R, Rosenberg R, Stricker D, Friederichs J, Hoos A, Werner M, Ulm K, Holzmann B, Nekarda H, and Siewert J-R. Telomere length and human telomerase reverse transcriptase expression as markers for progression and prognosis of colorectal carcinoma. J Clin Oncol 22(10);2004:1807–14.

Glimelius B, Garmo H, Berglund A, Fredriksson LA, Berglund M, Kohnke H, Byström P, Sørbye H, and Wadelius M. Prediction of irinotecan and 5-fluorouracil toxicity and response in patients with advanced colorectal cancer. Pharmacogenomics J 11;2011:61–71.

Goel A, and Boland CR. Epigenetics of colorectal cancer. Gastroenterology 143(6);2012:1442–60.

Goey AK, Mooiman KD, Beijnen JH, Schellens JH, and Meijerman I. Relevance of in vitro and clinical data for predicting CYP3A4-mediated herb-drug interactions in cancer patients. Cancer Treat Rev 39(7);2013 8:773–83.

Gong J, Wang C, Lee PP, Chu P, and Fakih M. Response to PD-1 blockade in microsatellite stable metastatic colorectal cancer harboring a POLE mutation. J Natl Compr Canc Netw 15(2);2017:142–7.

Gong Q, Chan SJ, Bajkowski AS, Steiner DF, and Frankfater A. Characterization of the cathepsin B gene and multiple mRNAs in human tissues: Evidence for alternative splicing of cathepsin B pre-mRNA. DNA Cell Biol 12;1993:299–309.

Gonsalves WI, Mahoney MR, Sargent DJ et al. Patient and tumor characteristics and BRAF and KRAS mutations in colon cancer, NCCTG/alliance N0147. J Natl Cancer Inst 106(7);2014, pii: dju106.

González N, Prieto I, Del Puerto-Nevado L et al. 2017 update on the relationship between diabetes and colorectal cancer: Epidemiology, potential molecular mechanisms and therapeutic implications. *Oncotarget* 8(11);2017:18456–85.

Gosens MJ, Moerland E, Lemmens VP, Rutten HT, Tan-Go I, and van den Brule AJ. Thymidylate synthase genotyping is more predictive for therapy response than immunohistochemistry in patients with colon cancer. *Int J Cancer* 123(8);2008:1941–9.

Goyette P, Sumner JS, Milos R, Duncan AM, Rosenblatt DS, Matthews RG, and Rozen R. Human methylenetetrahydrofolate reductase: Isolation of cDNA, mapping and mutation identification. *Nat Genet* 7(2);1994:195–200.

Grady WM, and Carethers JM. Genomic and epigenetic instability in colorectal cancer pathogenesis. *Gastroenterology* 135(4);2008:1079–99.

Grice EA, and Segre JA. The human microbiome: Our second genome. *Annu Rev Genomics Hum Genet* 13;2012:151–70.

Grothey A, Van Cutsem E, Sobrero A, Siena S, Falcone A, and Ychou M. Regorafenib monotherapy for previously treated metastatic colorectal cancer (CORRECT): An international, multicentre, randomised, placebo-controlled, phase 3 trial. *Lancet* 381(9863);2013:303–12.

Gu M, Nishihara R, Chen Y et al. Aspirin exerts high anti-cancer activity in PIK3CA-mutant colon cancer cells. *Oncotarget* 8(50);2017:87379–89.

Guerrero S, Casanova I, Farré L, Mazo A, Capellà G, and Mangues R. K-ras codon 12 mutation induces higher level of resistance to apoptosis and predisposition to anchorage-independent growth than codon 13 mutation or proto-oncogene overexpression. *Cancer Res* 60(23);2000:6750–6.

Guillemette C. Pharmacogenomics of human UDP glucuronosyltransferase enzymes. *Pharmacogenomics J* 3(3);2003:136–58.

Guillemette C, Lévesque É, and Rouleau M. Pharmacogenomics of human uridine diphospho-glucuronosyltransferases and clinical implications. *Clin Pharmacol Ther* 96(3);2014:324–39.

Gulhati P, Zaytseva YY, Valentino JD et al. Sorafenib enhances the therapeutic efficacy of rapamycin in colorectal cancers harboring oncogenic KRAS and PIK3CA. *Carcinogenesis* 33;2012:1782–90.

Gupta E, Lestingi TM, Mick R, Ramirez J, Vokes EE, and Ratain MJ. Metabolic fate of irinotecan in humans: Correlation of glucuronidation with diarrhea. *Cancer Res* 54(14);1994:3723–5.

Gusella M, and Padrini R. G > C SNP of thymidylate synthase with respect to colorectal cancer. *Pharmacogenomics* 8(8);2007:985–96.

Guzińska-Ustymowicz K. MMP-9 and cathepsin B expression in tumor budding as an indicator of a more aggressive phenotype of colorectal cancer (CRC). *Anticancer Res* 26;2006:1589–94.

Haaz MC, Rivory L, Riché C, Vernillet L, and Robert J. Metabolism of irinotecan (CPT-11) by human hepatic microsomes: Participation of cytochrome P-450 3A and drug interactions. *Cancer Res* 58(3);1998:468–72.

Haddad J, Slika S, and Mahfouz R. Epidermal growth factor receptor (EGFR) in the era of precision medicine: The tale of a perfect example of targeted therapy. A review. *Meta Gene (Elsevier)* 11;2017:157–63.

Hague A, Butt AJ, and Paraskeva C. The role of butyrate in human colonic epithelial cells: An energy source or inducer of differentiation and apoptosis? *Proc Nutr Soc* 55;1996:937–43.

Hahn MM, de Voer RM, Hoogerbrugge N, Ligtenberg MJ, Kuiper RP, and van Kessel AG. The genetic heterogeneity of colorectal cancer predisposition—Guidelines for gene discovery. *Cell Oncol (Dordr)* 39(6);2016:491–510.

Halama A, Guerrouahen BS, Pasquier J, Diboun I, Karoly ED, Suhre K, and Rafii A. Metabolic signatures differentiate ovarian from colon cancer cell lines. *J Transl Med* 13;2015:223.

Half E, Bercovich D, and Rozen P. Familial adenomatous polyposis. *Orphanet J Rare Dis* 4;2009:22.

Hamet P. Future needs in exploration of gene-environment interactions. *J Hypertens* 30;2012:1915–6.

Hammond WA, Swaika A, and Mody K. Pharmacologic resistance in colorectal cancer: A review. *Ther Adv Med Oncol* 8(1);2016:57–84.

Han FF, Guo CL, Yu D, Zhu J, Gong LL, Li GR, and Lv YL. Associations between UGT1A1*6 or UGT1A1*6/*28 polymorphisms and irinotecan-induced neutropenia in Asian cancer patients. *Cancer Chemother Pharmacol* 73(4);2014:779–88.

Hanahan D, and Weinberg RA. Hallmarks of cancer: The next generation. *Cell* 144(5);2011:646–74.

Hart IR, and Saini A. Biology of tumour metastasis. *Lancet* 339;1992:1453–7.

Hazama S, Mishima H, Tsunedomi R et al. UGT1A1*6, 1A7*3, and 1A9*22 genotypes predict severe neutropenia in FOLFIRI-treated metastatic colorectal cancer in two prospective studies in Japan. *Cancer Sci* 104(12);2013:1662–9.

Health Quality Ontario. Fecal occult blood test for colorectal cancer screening: An evidence-based analysis. *Onta Health Technol Assess Ser* 9(10);2009:1–40.

Heidelberger C, Chaudhuri N, Danneberg P et al. Fluorinated pyrimidines, a new class of tumour-inhibitory compounds. *Nature* 179;1957:663–6.

Henricks LM, Lunenburg CA, Meulendijks D et al. Translating DPYD genotype into DPD phenotype: Using the DPYD gene activity score. *Pharmacogenomics* 16(11);2015:1277–86.

Heo M, Allison DB, and Fontaine KR. Overweight, obesity, and colorectal cancer screening: Disparity between men and women. *BMC Public Health* 4;2004:53.

Herszènyi L, Plebani M, Carraro P et al. The role of cysteine and serine proteases in colorectal carcinoma. *Cancer* 86(7);1999:1135–42.

Herszényi L, Barabás L, Hritz I, István G, and Tulassay Z. Impact of proteolytic enzymes in colorectal cancer development and progression. *World J Gastroenterol* 20(37);2014:13246–57.

Hindorff LA, Sethupathy P, Junkins HA, Ramos EM, Mehta JP, Collins FS, and Manolio TA. Potential etiologic and functional implications of genome-wide association loci for human diseases and traits. *Proc Natl Acad Sci USA* 106(23);2009:9362–7.

Hitchins MP, Wong JJ, Suthers G, Suter CM, Martin DI, Hawkins NJ, and Ward RL. Inheritance of a cancer-associated MLH1 germ-line epimutation. *N Engl J Med* 356(7);2007:697–705.

Hizel C. Omics' language metagenomics is in wait for precision medicine as a new clinical frontier. *Focus on Colorectal Cancer Precis Med J* 2(1);2017:1–5.

Hizel C, Ferrara M, Cure H et al. Evaluation of the 5′ spliced form of human cathepsin B mRNA in colorectal mucosa and tumors. *Oncol Rep* 5(1);1998:31–4.

Hizel C, Maurizis JC, Rio P, Communal Y, Chassagne J, Favy D, Biqnon YJ, and Bernard-Gallon DJ. Isolation, purification and quantification of BRCA1 protein from tumour cells by affinity perfusion chromatography. *J Chromatogr B Biomed Sci Appl* 721(2);1999:163–70.

Hizel C, Tremblay J, Bartlett G, Hamet P. Introduction: Every individual is different and precision medicine offers options for disease control and treatment. In: *Progress and Challenges in Precision Medicine*. Verma M, and Barh D (Eds), Elsevier Press, San Diego, 2017, pp. 1–33.

Holderfield M, Deuker MM, McCormick F, and McMahon M. Targeting RAF kinases for cancer therapy: BRAF-mutated melanoma and beyond. *Nat Rev Cancer* 14(7);2014:455–67.

Holmes EC. Immunology of tumor infiltrating lymphocytes. *Ann Surg* 201(2);1985:158–63.

Hong DS, Morris VK, Fu S et al. Phase 1B study of vemurafenib in combination with irinotecan and cetuximab in patients with BRAF-mutated advanced cancers and metastatic colorectal cancer. *J Clin Oncol* 32;2014: 3516, 5s, (suppl; abstr 3516).

Horie N, Aiba H, Oguro K, Hojo H, and Takeishi K. Functional analysis and DNA polymorphism of the tandemly repeated sequences in the 5′-terminal regulatory region of the human gene for thymidylate synthase. *Cell Struct Funct* 20(3);1995:191–7.

Hoskins JM, Goldberg RM, Qu P, Ibrahim JG, and McLeod HL. UGT1A1*28 genotype and irinotecan-induced neutropenia: Dose matters. *J Natl Cancer Inst* 99(17);2007:1290–5.

Hu ZY, Yu Q, Pei Q, and Guo C. Dose-dependent association between UGT1A1*28 genotype and irinotecan-induced neutropenia: Low doses also increase risk. *Clin Cancer Res* 16(15);2010:3832–42.

Huang Y, Shen XJ, Zou Q, Wang SP, Tang SM, and Zhang GZ. Biological functions of microRNAs: A review. *J Physiol Biochem* 67(1);2011:129–39.

Huang Z, Huang D, Ni S, Peng Z, Sheng W, and Du X. Plasma microRNAs are promising novel biomarkers for early detection of colorectal cancer. *Int J Cancer* 127(1);2010:118–26.

Hudler P, Kocevar N, and Komel R. Proteomics approaches in biomarker discovery: New perspectives in cancer diagnostics. *Sci World J* 2014;2014:260348.

Hughes LA, Melotte V, de Schrijver J et al. The CpG island methylator phenotype: What's in a name? *Cancer Res* 73(19);2013:5858–68.

Huth L, Jäkel J, and Dahl E. Molecular diagnostic applications in colorectal cancer. *Microarrays* 3(3);2014:168–79.

Iacopetta B. Are there two sides to colorectal cancer? *Int J Cancer* 101(5);2002:403–8.

Iacopetta B. TP53 mutation in colorectal cancer. *Hum Mutat* 21(3);2003:271–6.

Iacopetta B, Russo A, Bazan V et al. Functional categories of TP53 mutation in colorectal cancer: Results of an international collaborative study. *Ann Oncol* 17(5);2006:842–7.

ICGC Data Access Compliance Office, and ICGC International Data Access Committee. Analysis of five years of controlled access and data sharing compliance at the International Cancer Genome Consortium. *Nat Genet* 48(3);2016:224–55.

Ide T, Kitajima Y, Ohtaka K, Mitsuno M, Nakafusa Y, and Miyazaki K. Expression of the hMLH1 gene is a possible predictor for the clinical response to 5-fluorouracil after a surgical resection in colorectal cancer. *Oncol Rep* 19;2008:1571–6.

Ikenoue T, Hikiba Y, Kanai F et al. Functional analysis of mutations within the kinase activation segment of B-Raf in human colorectal tumors. *Cancer Res* 63(23);2003:8132–7.

Inamura K. Colorectal cancers: An update on their molecular pathology. *Cancers (Basel)* 10(1);2018, pii: E26.

Ingelman-Sundberg M. Pharmacogenetics: An opportunity for a safer and more efficient pharmacotherapy. *J Intern Med* 250(3);2001:186–200.

Innocenti F, Grimsley C, Das S et al. Haplotype structure of the UDP-glucuronosyltransferase 1A1 promoter in different ethnic groups. *Pharmacogenetics* 12;2002:725–33.

Innocenti F, Kroetz DL, Schuetz E et al. Comprehensive pharmacogenetic analysis of irinotecan neutropenia and pharmacokinetics. *J Clin Oncol* 27(16);2009:2604–14.

Innocenti F, and Ratain MJ. Irinotecan treatment in cancer patients with UGT1A1 polymorphisms. *Oncology (Williston Park)* 17(5 Suppl. 5);2003:52–5.

Innocenti F, and Ratain MJ. "Irinogenetics" and UGT1A: From genotypes to haplotypes. *Clin Pharmacol Ther* 75(6);2004:495–500.

Innocenti F, and Ratain MJ. Pharmacogenetics of irinotecan: Clinical perspectives on the utility of genotyping. *Pharmacogenomics* 7(8);2006:1211–21.

Innocenti F, Undevia SD, Iyer L et al. Genetic variants in the UDP-glucuronosyltransferase 1A1 gene predict the risk ofsevere neutropenia of irinotecan. *J Clin Oncol* 22(8);2004:1382–8.

Ishihama K, Yamakawa M, Semba S, Takeda H, Kawata S, Kimura S, and Kimura W. Expression of HDAC1 and CBP/p300 in human colorectal carcinomas. *J Clin Pathol* 60(11);2007:1205–10.

Isobe M, Emanuel BS, Givol D, Oren M, and Croce CM. Localization of gene for human p53 tumour antigen to band 17p13. *Nature* 320(6057);1986:84–5.

Issa IA, and Noureddine M. Colorectal cancer screening: An updated review of the available options. *World J Gastroenterol* 23(28);2017:5086–96.

Issa JP. CpG island methylator phenotype in cancer. *Nat Rev Cancer* 4(12);2004:988–93.

Issa JP. Colon cancer: It's CIN or CIMP. *Clin Cancer Res* 14;2008:5939–40.

Ivancich M, Schrank Z, Wojdyla L, Leviskas B, Kuckovic A, Sanjali A, and Puri N. Treating cancer by targeting telomeres and telomerase. *Antioxidants (Basel)* 6(1);2017, pii: E15.

Iwasaki M, Tanaka-Mizuno S, Kuchiba A et al. Inclusion of a genetic risk score into a validated risk prediction model for colorectal cancer in Japanese men improves performance. *Cancer Prev Res (Phila)* 10(9);2017 Jul 20:535–41, doi: 10.1158/1940-6207.CAPR-17-0141 [Epub ahead of print].

Iyer L, Das S, Janisch L et al. UGT1A1*28 polymorphism as a determinant of irinotecan disposition and toxicity. *Pharmacogenomics J* 2(1);2002:43–7.

Jackson CS, Oman M, Patel AM, and Vega KJ. Health disparities in colorectal cancer among racial and ethnic minorities in the United States. *J Gastrointest Oncol* 7(Suppl. 1);2016:S32–43.

Jäger R, Migliorini G, Henrion M et al. Capture Hi-C identifies the chromatin interactome of colorectal cancer risk loci. *Nat Commun* 6;2015:6178.

Jaiswal AS, Balusu R, and Narayan S. Involvement of adenomatous polyposis coli in colorectal tumorigenesis. *Front Biosci* 10;2005:1118–34.

Jakobsen A, Nielsen NJ, Gyldenkerne N, and Lindeberg J. Thymidylate synthase and methylenetetrahydrofolate reductase gene polymorphism in normal tissue as predictors of fluorouracil sensitivity. *J Clin Oncol* 23;2005:1365–9.

Jang E, and Chung DC. Hereditary colon cancer: Lynch syndrome. *Gut Liver* 4(2);2010:151–60.

Jelsig AM. Hamartomatous polyps—A clinical and molecular genetic study. *Dan Med J* 63(8);2016, pii: B5280.

Jen J, Kim H, Piantadosi S et al. Allelic loss of chromosome 18q and prognosis in colorectal cancer. *N Engl J Med* 331(4);1994:213–21.

Jensen SA, Vainer B, Kruhoffer M, and Sorensen JB. Microsatellite instability in colorectal cancer and association with thymidylate synthase and dihydropyrimidine dehydrogenase expression. *BMC Cancer* 9;2009:25.

Jia X, Shanmugam C, Paluri RK et al. Prognostic value of loss of heterozygosity and subcellular localization of SMAD4 varies with tumor stage in colorectal cancer. *Oncotarget* 8(12);2017:20198–212.

Jin B, Li Y, and Robertson KD. DNA methylation: Superior or subordinate in the epigenetic hierarchy? *Genes Cancer* 2(6);2011:607–17.

Johnson SM, Grosshans H, Shingara J et al. RAS is regulated by the let-7 microRNA family. *Cell* 120(5);2005:635–47.

Jonsson PF, and Bates PA. Global topological features of cancer proteins in the human interactome. *Bioinformatics* 22(18);2007: 2291–7.

Jovel J, Patterson J, Wang W et al. Characterization of the gut microbiome using 16S or shotgun metagenomics. *Front Microbiol* 7;2016:459.

Kajihara M, Takakura K, Kanai T et al. Dendritic cell-based cancer immunotherapy for colorectal cancer. *World J Gastroenterol* 22(17);2016:4275–86.

Kalow W. *Pharmacogenetics: Heredity and the Response to Drugs.* W.B. Saunders Co., Philadelphia, 1962.

Kamel HFM, and Al-Amodi HSAB. Exploitation of gene expression and cancer biomarkers in paving the path to era of personalized medicine. *Genomics Proteomics Bioinformatics* 15(4);2017:220–35.

Kamoshida S, Matsuoka H, Ishikawa T, Maeda K, Shimomura R, Inada K, and Tsutsumi Y. Immunohistochemical evaluation of thymidylate synthase (TS) and p16INK4a in advanced colorectal cancer: Implication of TS expression in 5-FU-based adjuvant chemotherapy. *Jpn J Clin Oncol* 34(10);2004:594–601.

Kannagi R, Izawa M, Koike T, Miyazaki K, and Kimura N. Carbohydrate-mediated cell adhesion in cancer metastasis and angiogenesis. *Cancer Sci* 95;2004:377–84.

Kasahara M, Takahashi Y, Nagata T et al. Thymidylate synthase expression correlates closely with E2F1 expression in colon cancer. *Clin Cancer Res* 6(7);2000:2707–11.

Kasubuchi M, Hasegawa S, Hiramatsu T, Ichimura A, and Kimura I. Dietary gut microbial metabolites, short-chain fatty acids, and host metabolic regulation. *Nutrients* 7(4);2015:2839–49.

Kawakami H, Zaanan A, and Sinicrope FA. Microsatellite instability testing and its role in the management of colorectal cancer. *Curr Treat Options Oncol* 16(7);2015:30.

Kawakami K, and Watanabe G. Identification and functional analysis of single nucleotide polymorphism in the tandem repeat sequence of thymidylate synthase gene. *Cancer Res* 63(18);2003:6004–7.

Kawato Y, Aonuma M, Hirota Y, Kuga H, and Sato K. Intracellular roles of SN-38, a metabolite of the camptothecin derivative CPT-11, in the antitumor effect of CPT-11. *Cancer Res* 51;1991:4187–91.

Kefeli U, Ucuncu Kefeli A, Cabuk D et al. Netrin-1 in cancer: Potential biomarker and therapeutic target? *Tumour Biol* 39(4);2017:1010428317698388.

Kefford R, Arkenau H, Brown MP et al. Phase I/II study of GSK2118436, a selective inhibitor of oncogenic mutant BRAF kinase, in patients with metastatic melanoma and other solid tumours. *J Clin Oncol* 28(Suppl.);2010, abstract 10523.

Kehrer DF, Mathijssen RH, Verweij J, de Bruijn P, and Sparreboom A. Modulation of irinotecan metabolism by ketoconazole. *J Clin Oncol* 20;2002:3122–9.

Khan A, Krishna M, Baker S, and Banner B. Cathepsin B and tumor-associated laminin expression in the progression of colorectal adenoma to carcinoma. *Mod Pathol* 11;2004:704–8.

Khanna R, Morton CL, Danks MK, and Potter PM. Proficient metabolism of irinotecan by a human intestinal carboxylesterase. *Cancer Res* 60;2000:4725–8.

Kiang TK, Ensom MH, and Chang TK. UDP-glucuronosyl-transferases and clinical drug-drug interactions. *Pharmacol Ther* 106;2005:97–132.

Kim B, Kang S, Jeong G, Park SB, and Kim SJ. Identification and comparison of aberrant key regulatory networks in breast, colon, liver, lung, and stomach cancers through methylome database analysis. *PLOS ONE* 9(5);2014:e97818.

Kim B-S, Jeon Y-S, and Chun J. Current status and future promise of the human microbiome. *Pediatr Gastroenterol Hepatol Nutr* 16;2013:71–7.

Kim D, Kim SY, Lee JS et al. Primary tumor location predicts poor clinical outcome with cetuximab in RAS wild-type metastatic colorectal cancer. *BMC Gastroenterol* 17(1);2017:121.

Kim DH, Sarbassov DD, Ali SM et al. mTOR interacts with raptor to form a nutrient sensitive complex that signals to cell growth machinery. *Cell* 110;2002:163–75.

Kim J, Cho YA, Kim DH et al. Dietary intake of folate and alcohol, MTHFR C677 T polymorphism, and colorectal cancer risk in Korea. *Am J Clin Nutr* 95(2);2012:405–12.

Kim TM, Lee SH, and Chung YJ. Clinical applications of next-generation sequencing in colorectal cancers. *World J Gastroenterol* 19(40);2013:6784–93.

Kim YJ, and Varki A. Perspectives on the significance of altered glycosylation of glyco-proteins in cancer. *Glycoconj J* 14;1997:569–76.

Kittles R. Genes and environments: Moving toward personalized medicine in the context of health disparities. *Ethn Dis* 22;2012:S1-43–6.

Klose RJ, and Zhang Y. Regulation of histone methylation by demethylimination and demethylation. *Nat Rev Mol Cell Biol* 8(4);2007:307–18.

Klümpen HJ, Beijnen JH, Gurney H, and Schellens JH. Inhibitors of mTOR. *Oncologist* 15(12);2010:1262–9.

Knudson AG. Mutation and cancer: Statistical study of retinoblastoma. *Proc Natl Acad Sci USA* 68(4);1971:820–3.

Ko LJ, and Prives C. p53: Puzzle and paradigm. *Genes Dev* 10(9);1996:1054–72.

Koga Y, Yasunaga M, Takahashi A et al. MicroRNA expression profiling of exfoliated colonocytes isolated from feces for colorectal cancer screening. *Cancer Prev Res (Phila)* 3(11);2010:1435–42.

Koido S, Homma S, Takahara A et al. Immunotherapy synergizes with chemotherapy targeting pancreatic cancer. *Immunotherapy* 4(1);2012:5–7.

Kopetz S, Desai J, Chan E et al. Phase II pilot study of vemurafenib in patients with metastatic BRAF-mutated colorectal cancer. *J Clin Oncol* 33(34);2015:4032–8.

Krapohl E, Patel H, Newhouse S et al. Multipolygenic score approach to trait prediction. *Mol Psychiatry* 2017 Aug 8, doi: 10.1038/mp.2017.163 [Epub ahead of print].

Kruszewski WJ, Rzepko R, Wojtacki J, Skokowski J, Kopacz A, Jaśkiewicz K, and Drucis K. Overexpression of cathepsin B correlates with angiogenesis in colon adenocarcinoma. *Neoplasma* 51;2004:38–43.

Kummar S, Gutierrez M, Doroshow JH, and Murgo AJ. Drug development in oncology: Classical cytotoxics and molecularly targeted agents. *Br J Clin Pharmacol* 62(1);2006:15–26.

Kweekel DM, Gelderblom H, Van der Straaten T, Antonini NF, Punt CJ, Guchelaar HJ, Dutch Colorectal Cancer Group study. UGT1A1*28 genotype and irinotecan dosage in patients with metastatic colorectal cancer: A Dutch Colorectal Cancer Group study. *Br J Cancer* 99;2008:275–82.

LaFave MC, and Sekelsky J. Mitotic recombination: Why? When? How? Where? *PLoS Genet* 5(3);2009:e1000411.

Läll K, Mägi R, Morris A, Metspalu A, and Fischer K. Personalized risk prediction for type 2 diabetes: The potential of genetic risk scores. *Genet Med* 201719(3);2017:322–9.

Lam K, Pan K, Linnekamp JF, Medema JP, and Kandimalla R. DNA methylation based biomarkers in colorectal cancer: A systematic review. *Biochim Biophys Acta* 1866(1);2016:106–20.

Lane DP. p53, guardian of the genome. *Nature* 358;1992:15–6.

Lankisch TO, Vogel A, Eilermann S et al. Identification and characterization of a functional TATA box polymorphism of the UDP glucuronosyltransferase 1A7 gene. *Mol Pharmacol* 67(5);2005:1732–9. Epub 2005 Feb 16.

Laplante M, and Sabatini DM. mTor signaling in growth control and disease. *Cell* 149(2);2012:274–93.

Laurent-Puig P, Cayre A, Manceau G et al. Analysis of PTEN, BRAF, and EGFR status in determining benefit from cetuximab therapy in wild-type KRAS metastatic colon cancer. *J Clin Oncol* 27(35);2009:5924–30.

Lederberg J, and McCray AT. "Ome Sweet" Omics—A genealogical treasury of words. *Scientist* 15;2001:8.

Lee JK, and Chan AT. Molecular prognostic and predictive markers in colorectal cancer: Current status. *Curr Colorectal Cancer Rep* 7(2);2011:136–44.

Lee SY, Miyai K, Han HS et al. Microsatellite instability, EMAST, and morphology associations with T cell infiltration in colorectal neoplasia. *Dig Dis Sci* 57(1);2012:72–8.

Leggett B, and Whitehall V. Role of the serrated pathway in colorectal cancer pathogenesis. *Gastroenterology* 138(6);2010:2088–100.

Lemay V, Hamet P, Hizel C, Lemaire E, Tremblay Y. Personalized medicine: Interdisciplinary perspective, world tidal wave and potential growth for the emerging countries. In: *Progress and Challenges in Precision Medicine*. Verma M, and Barh D (Eds), Elsevier Press, San Diego, 2017, pp. 301–14.

Lemery S, Keegan P, and Pazdur R. First FDA approval agnostic of cancer site—When a biomarker defines the indication. *N Engl J Med* 377(15);2017:1409–12.

Lengauer C, Kinzler KW, and Vogelstein B. Genetic instability in colorectal cancers. *Nature* 386(6625);1997:623–7.

Leroy B, Anderson M, and Soussi T. TP53 mutations in human cancer: Database reassessment and prospects for the next decade. *Hum Mutat* 35(6);2014:672–88.

Lettre G. Rare and low-frequency variants in human common diseases and other complex traits. *J Med Genet* 51(11);2014:705–14.

Levine AJ. p53, the cellular gatekeeper for growth and division. *Cell* 88;1997:323–31.

Ley RE, Peterson DA, and Gordon JI. Ecological and evolutionary forces shaping microbial diversity in the human intestine. *Cell* 124;2006:837–48.

Li J, Qin S, Xu R et al. Regorafenib plus best supportive care versus placebo plus best supportive care in Asian patients with previously treated metastatic colorectal cancer (CONCUR): A randomised, double-blind, placebo-controlled, phase 3 trial. *The Lancet Oncol* 16(6);2015:619–29.

Li W, Zhang H, Assaraf YG et al. Overcoming ABC transporter-mediated multidrug resistance: Molecular mechanisms and novel therapeutic drug strategies. *Drug Resist Updat* 27;2016:14–29.

Li XL, Zhou J, Chen ZR, and Chng WJ. P53 mutations in colorectal cancer—Molecular pathogenesis and pharmacological reactivation. *World J Gastroenterol* 21(1);2015:84–93.

Liao X, Lochhead P, Nishihara R et al. Aspirin use, tumor PIK3CA mutation, and colorectal-cancer survival. *N Engl J Med* 367(17);2012:1596–606.

Lievre A, Bachet J, Le Corre D et al. KRAS mutation status is predictive of response to cetuximab therapy in colorectal cancer. *Cancer Res* 66(8);2006:3992–5.

Lima A, Azevedo R, Sousa H, Seabra V, and Medeiros R. Current approaches for TYMS polymorphisms and their importance in molecular epidemiology and pharmacogenetics. *Pharmacogenomics* 14(11);2013:1337–51.

Lindpaintner K. The impact of pharmacogenetics and pharmacogenomics on drug discovery. *Nat Rev Drug Discov* 1;2002:463–9.

Link A, Balaguer F, Shen Y, Nagasaka T, Lozano JJ, Boland CR, and Goel A. Fecal MicroRNAs as novel biomarkers for colon cancer screening. *Cancer Epidemiol Biomarkers Prev* 19(7);2010:1766–74.

Liotta LA, Nageswara RC, and Wewer UM. Biochemical interactions of tumor cells with the basement membrane. *Ann Rev Biochem* 55;1986:1037–57.

Liu D, Li J, Gao J, Li Y, Yang R, and Shen L. Examination of multiple UGT1A and DPYD polymorphisms has limited ability to predict the toxicity and efficacy of metastatic colorectal cancer treated with irinotecan-based chemotherapy: A retrospective analysis. *BMC Cancer* 17(1);2017:437.

Liu DP, Song H, and Xu Y. A common gain of function of p53 cancer mutants in inducing genetic instability. *Oncogene* 29(7);2010:949–56.

Liu J, Liu F, Li X, Song X, Zhou L, and Jie J. Screening key genes and miR-NAs in early-stage colon adenocarcinoma by RNA-sequencing. *Tumour Biol* 39(7);2017:1010428317714899.

Liu J, Zhang C, and Feng Z. Tumor suppressor p53 and its gain-of-function mutants in cancer. *Acta Biochim Biophys Sin (Shanghai)* 46(3);2014:170–9.

Liu XH, Lu J, Duan W et al. Predictive value of UGT1A1*28 polymorphism in irinotecan-based chemotherapy. *J Cancer* 8(4);2017:691–703.

Lochhead P, Kuchiba A, Imamura Y et al. Microsatellite instability and BRAF mutation testing in colorectal cancer prognostication. *J Natl Cancer Inst* 105(15);2013:1151–6.

Longley DB, Harkin DP, and Johnston PG. 5-fluorouracil: Mechanisms of action and clinical strategies. *Nat Rev Cancer* 3(5);2003:330–8.

Loree JM, and Kopetz S. Recent developments in the treatment of metastatic colorectal cancer. *Ther Adv Med Oncol* 9(8);2017:551–64.

Loupakis F, Ruzzo A, Cremolini C et al. KRAS codon 61, 146 and BRAF mutations predict resistance to cetuximab plus irinotecan in KRAS codon 12 and 13 wild-type metastatic colorectal cancer. *Br J Cancer* 101(4);2009:715–21.

Low SK, Takahashi A, Mushiroda T, and Kubo M. Genome-wide association study: A useful tool to identify common genetic variants associated with drug toxicity and efficacy in cancer pharmacogenomics. *Clin Cancer Res* 20(10);2014:2541–52.

Lowe WL Jr, and Reddy TE. Genomic approaches for understanding the genetics of complex disease. *Genome Res.* 25;2015:1432–41.

Lungulescu CV, Răileanu S, Afrem G et al. Histochemical and immunohistochemical study of mucinous rectal carcinoma. *J Med Life* 10(2);2017:139–43.

Lynch D, and Murphy A. The emerging role of immunotherapy in colorectal cancer. *Ann Transl Med* 4(16);2016:305.

Lynch HT, and Shaw TG. Practical genetics of colorectal cancer. *Chin Clin Oncol* 2(2);2013:12.

Lévesque E, Bélanger AS, Harvey M, Couture F, Jonker D, and Innocenti F. Refining the UGT1A haplotype associated with irinotecan-induced hematological toxicity in metastatic colorectal cancer patients treated with 5 fluorouracil/irinotecan-based regimens. *J Pharmacol Exp Ther* 345(1);2013:95–101.

Maccaferri S, Biagi E, and Brigidi P. Metagenomics: Key to human gut microbiota. *Dig Dis* 29;2011:525–30.

Mahasneh A, Al-Shaheri F, and Jamal E. Molecular biomarkers for an early diagnosis, effective treatment and prognosis of colorectal cancer. Current updates. *Exp Mol Pathol* 102(3);2017:475–83.

Maher BS. Polygenic scores in epidemiology: Risk prediction, etiology, and clinical utility. *Curr Epidemiol Rep* 2(4);2015:239–44.

Makondi PT, Chu CM, Wei PL, and Chang YJ. Prediction of novel target genes and pathways involved in irinotecan-resistant colorectal cancer. *PLOS ONE* 12(7);2017:e0180616.

Malkin D. Li-fraumeni syndrome. *Genes Cancer* 2(4);2011:475–84.

Malumbres M, and Barbacid M. RAS oncogenes: The first 30 years. *Nat Rev Cancer* 3(6);2003:459–65.

Mandola MV, Stoehlmacher J, Muller-Weeks S, Cesarone G, Yu MC, Lenz HJ, and Ladner RD. A novel single nucleotide polymorphism within the 5′ tandem repeat polymorphism of the thymidylate synthase gene abolishes USF-1 binding and alters transcriptional activity. *Cancer Res* 63;2003:2898–904.

Manning BD, and Cantley LC. AKT/PKB signaling: Navigating downstream. *Cell* 129(7);2007: 1261–74.

Manson LE, van der Wouden CH, Swen JJ, and Guchelaar HJ. The Ubiquitous Pharmacogenomics consortium: Making effective treatment optimization accessible to every European citizen. *Pharmacogenomics* 18(11);2017:1041–5.

Manzat-Saplacan RM, Mircea PA, Balacescu L, Chira RI, Berindan-Neagoe I, and Balacescu O. Can we change our microbiome to prevent colorectal cancer development? *Acta Oncol* 54(8);2015:1085–95.

Mao C, Yang ZY, Hu XF, Chen Q, and Tang JL. PIK3CA exon 20 mutations as a potential biomarker for resistance to anti-EGFR monoclonal antibodies in KRAS wild-type metastatic colorectal cancer: A systematic review and meta-analysis. *Ann Oncol* 23(6);2012:1518–25.

Mao M, Tian F, Mariadason JM et al. Resistance to BRAF inhibition in BRAF-mutant colon cancer can be overcome with PI3K inhibition or demethylating agents. *Clin Cancer Res* 19(3);2012:657–67.

Marcuello E, Altés A, Menoyo A, Del Rio E, Gómez-Pardo M, and Baiget M. UGT1A1 gene variations and irinotecan treatment in patients with metastatic colorectal cancer. *Br J Cancer* 91(4);2004:678–82.

Marcuello E, Páez D, Paré L, Salazar J, Sebio A, del Rio E, and Baiget M. A genotype-directed phase I-IV dose-finding study of irinotecan in combination with fluorouracil/leucovorin as first-line treatment in advanced colorectal cancer. *Br J Cancer* 105(1);2011 Jun 28:53–7.

Marques SC, and Ikediobi ON. The clinical application of UGT1A1 pharmacogenetic testing: Gene-environment interactions. *Hum Genomics* 4(4);2010:238–49.

Marsh S, Ameyaw MM, Githang'a J, Indalo A, Ofori-Adjei D, and McLeod HL. Novel thymidylate synthase enhancer region alleles in African populations. *Hum Mutat* 16;2000:528.

Marsh S, Collie-Duguid ES, Li T, Liu X, and McLeod HL. Ethnic variation in the thymidylate synthase enhancer region polymorphism among Caucasian and Asian populations. *Genomics* 58;1999:310–2.

Marsh S, McKay JA, Cassidy J, and McLeod HL. Polymorphism in the thymidylate synthase promoter enhancer region in colorectal cancer. *Int J Oncol* 19(2);2001:383–6.

Marsh S, and McLeod HL. Thymidylate synthase pharmacogenetics in colorectal cancer. *Clin Colorectal Cancer* 1(3);2001:175–8.

Massacesi C, Terrazzino S, Marcucci F et al. Uridine diphosphate glucuronosyl transferase 1A1 promoter polymorphism predicts the risk of gastrointestinal toxicity and fatigue induced by irinotecan-based chemotherapy. *Cancer* 106(5);2006:1007–16.

Mathijssen RH, van Alphen RJ, Verweij J, Loos WJ, Nooter K, Stoter G, and Sparreboom A. Clinical pharmacokinetics and metabolism of irinotecan (CPT-11). *Clin Cancer Res* 7(8);2001:2182–94.

Mathivanan S, Periaswamy B, Gandhi TK et al. An evaluation of human protein-protein interaction data in the public domain. *BMC Bioinformatics* 7(Suppl 5);2006:S19.

Maughan TS, Adams RA, Smith CG et al. Addition of cetuximab to oxaliplatin-based first-line combination chemotherapy for treatment of advanced colorectal cancer: Results of the randomised phase 3 MRC COIN trial. *Lancet* 377(9783);2011:2103–14.

Mayrhofer M, Kultima HG, Birgisson H et al. 1p36 deletion is a marker for tumour dissemination in microsatellite stable stage II-III colon cancer. *BMC Cancer* 14;2014:872.

Mazelin L, Bernet A, Bonod-Bidaud C et al. Netrin-1 controls colorectal tumorigenesis by regulating apoptosis. *Nature* 431(7004);2004:80–4.

McBride OW, Merry D, and Givol D. The gene for human p53 cellular tumor antigen is located on chromosome 17 short arm (17p13). *Proc Natl Acad Sci USA* 83(1);1986:130–4.

McDermott DF, and Atkins MB. PD-1 as a potential target in cancer therapy. *Cancer Med* 2(5);2013:662–73.

McMurray HR, Sampson ER, Compitello G et al. Synergistic response to oncogenic mutations defines gene class critical to cancer phenotype. *Nature* 453(7198);2008:1112.

McRee AJ, Davies JM, Sanoff HG et al. A phase I trial of everolimus in combination with 5-FU/LV, mFOLFOX6 and mFOLFOX6 plus panitumumab in patients with refractory solid tumors. *Cancer Chemother Pharmacol* 74(1);2014:117–23.

Meguid RA, Slidell MB, Wolfgang CL, Chang DC, and Ahuja N. Is there a difference in survival between right- versus left-sided colon cancers? *Ann Surg Oncol* 15(9);2008:2388–94.

Mei ZB, Duan CY, Li CB, Cui L, and Ogino S. Prognostic role of tumor PIK3CA mutation in colorectal cancer: A systematic review and meta-analysis. *Ann Oncol* 27(10);2016:1836–48.

Merino D, and Malkin D. p53 and hereditary cancer. *Subcell Biochem* 85;2014:1–16.

Metzker ML. Sequencing technologies: The next generation. *Nat Rev Genet* 11(1);2009:31–46.

Meulendijks D, Cats A, Beijnen JH, and Schellens JH. Improving safety of fluoropyrimidine chemotherapy by individualizing treatment based on dihydropyrimidine dehydrogenase activity—Ready for clinical practice? *Cancer Treat Rev* 50;2016a:23–34.

Meulendijks D, Henricks LM, van Kuilenburg AB et al. Patients homozygous for DPYD c.1129-5923C > G/haplotype B3 have partial DPD deficiency and require a dose reduction when treated with fluoropyrimidines. *Cancer Chemother Pharmacol* 78(4);2016b:875–80.

Michael MZ, O'Connor SM, van Holst Pellekaan NG, Young GP, and James RJ. Reduced accumulation of specific microRNAs in colorectal neoplasia. *Mol Cancer Res* 1(12);2003:882–91.

Michael-Robinson JM, Biemer-Hüttmann A, Purdie DM et al. Tumour infiltrating lymphocytes and apoptosis are independent features in colorectal cancer stratified according to microsatellite instability status. *Gut* 48(3);2001:360–6.

Mikolasevic I, Orlic L, Stimac D, Hrstic I, Jakopcic I, and Milic S. Non-alcoholic fatty liver disease and colorectal cancer. *Postgrad Med J* 93(1097);2017:153–8.

Milos PM, and Seymour AB. Emerging strategies and applications of pharmacogenomics. *Hum Genomics* 1;2004:444–55.

Mishra N, and Hall J. Identification of patients at risk for hereditary colorectal cancer. *Clin Colon Rectal Surg* 25(2);2012:67–82.

Moerkerk P, Arends JW, Van Driel M, de Bruïne A, de Goeij A, and ten Kate J. Type and number of Ki-ras point mutations relate to stage of human colorectal cancer. *Cancer Res* 54(13);1994:3376–8.

Moertel C, Fleming T, Macdonald J et al. Levamisole and fluorouracil for adjuvant therapy of resected colon carcinoma. *N Engl J Med* 322;1990:352–8.

Moraes F, and Góes A. A decade of human genome project conclusion: Scientific diffusion about our genome knowledge. *Biochem Mol Biol Educ* 44(3);2016:215–23.

Mori R, Futamura M, Tanahashi T, Tanaka Y, Matsuhashi N, Yamaguchi K, and Yoshida K. 5FU resistance caused by reduced fluorodeoxyuridine monophosphate and its reversal using deoxyuridine. *Oncol Lett* 14(3);2017:3162–8.

Moyzis RK, Buckingham JM, Cram LS, Dani M, Deaven LL, Jones MD, Meyne J, Ratliff RL, and Wu JR. A highly conserved repetitive DNA sequence, (TTAGGG)n, present at the telomeres of human chromosomes. *Proc Natl Acad Sci USA* 85(18);1988:6622–6.

Muller PA, and Vousden KH. p53 mutations in cancer. *Nat Cell Biol* 15(1);2013:2–8.

Murnane MJ, Sheahan K, Ozdemirli M, and Shuja S. Stage-specific increases in cathepsin B messenger RNA content in human colorectal carcinoma. *Cancer Res* 51;1991:1137–42.

Mzahma R, Kharrat M, Fetiriche F et al. The relationship between telomere length and clinicopathologic characteristics in colorectal cancers among Tunisian patients. *Tumour Biol* 36(11);2015:8703–13.

Nakajima G, Hayashi K, Xi Y et al. Non-coding MicroRNAs hsa-let-7 g and hsa-miR-181b are associated with chemoresponse to S-1 in colon cancer. *Cancer Genomics Proteomics* 3(5);2006:317–24.

Namasivayam V, and Lim S. Recent advances in the link between physical activity, sedentary behavior, physical fitness, and colorectal cancer. *F1000Res* 6;2017:199.

Nash GM, Gimbel M, Shia J et al. KRAS mutation correlates with accelerated metastatic progression in patients with colorectal liver metastases. *Ann Surg Oncol* 17;2010:572–8.

Nazemalhosseini Mojarad E, Kuppen PJ, Aghdaei HA, and Zali MR. The CpG island methylator phenotype (CIMP) in colorectal cancer. *Gastroenterol Hepatol Bed Bench* 6(3);2013:120–8.

Ng EK, Chong WW, Jin H et al. Differential expression of microRNAs in plasma of patients with colorectal cancer: A potential marker for colorectal cancer screening. *Gut* 58(10);2009:1375–81.

Nguyen A, Desta Z, and Flockhart DA. Enhancing race-based prescribing precision with pharmacogenomics. *Clin Pharmacol Ther* 81(3);2007:323–5.

Nibbe RK, Markowitz S, Myeroff L, Ewing R, and Chance MR. Discovery and scoring of protein interaction subnetworks discriminative of late stage human colon cancer. *Mol Cell Proteomics* 8(4);2009:827–45.

Nicolaides NC, Papadopoulos N, Liu B et al. Mutations of two PMS homologues in hereditary nonpolyposis colon cancer. *Nature* 371(6492);1994:75–80.

Nielsen DL, Palshof JA, Brünner N, Stenvang J, and Viuff BM. Implications of ABCG2 expression on irinotecan treatment of colorectal cancer patients: A review. *Int J Mol Sci* 18(9);2017, pii: E1926.

Nishihara R, Morikawa T, Kuchiba A et al. A prospective study of duration of smoking cessation and colorectal cancer risk by epigenetics-related tumor classification. *Am J Epidemiol* 178(1);2013:84–100.

Nishisho I, Nakamura Y, Miyoshi Y, Miki Y, Ando H, Horii A, Koyama K, Utsunomiya J, Baba S, and Hedge P. Mutations of chromosome 5q21 genes in FAP and colorectal cancer patients. *Science* 253(5020);1991:665–9.

Noci S, Dugo M, Bertola F, Melotti F, Vannelli A, Dragani TA, and Galvan A. A subset of genetic susceptibility variants for colorectal cancer also has prognostic value. *Pharmacogenomics J* 16(2);2016:173–9.

Nordlinger B, Van Cutsem E, Gruenberger T et al. Combination of surgery and chemotherapy and the role of targeted agents in the treatment of patients with colorectal liver metastases: Recommendations from an expert panel. *Ann Oncol* 20;2009:985–92.

Nosho K, Irahara N, Shima K et al. Comprehensive biostatistical analysis of CpG island methylator phenotype in colorectal cancer using a large population-based sample. *PLOS ONE* 3(11);2008:e3698.

Ogino S, Nosho K, Kirkner GJ et al. CpG island methylator phenotype, microsatellite instability, BRAF mutation and clinical outcome in colon cancer. *Gut* 58(1);2009b:90–6.

Ogino S, Nosho K, Kirkner GJ et al. PIK3CA mutation is associated with poor prognosis among patients with curatively resected colon cancer. *J Clin Oncol* 27(9);2009a:1477–84.

Oldenhuis CN, Oosting SF, Gietema JA, and de Vries EG. Prognostic versus predictive value of biomarkers in oncology. *Eur J Cancer* 44(7);2008:946–53.

O'Sullivan RJ, and Karlseder J. Telomeres: Protecting chromosomes against genome instability. *Nat Rev Mol Cell Biol* 11;2010:171–81.

Oudejans JJ, Slebos RJC, Zoetmulder FAN, Mooi WJ, and Rodenhuis S. Differential activation of ras genes by point mutation in human colon cancer with metastases to either lung or liver. *Int J Cancer* 49(6);1991:875–9.

Ozaslan M, and Aytekin T. Loss of heterozygosity in colorectal cancer. *Afr J Biotechnol* 8;2009:7308–12.

Pagès F, Galon J, Dieu-Nosjean MC, Tartour E, Sautès-Fridman C, and Fridman WH. Immune infiltration in human tumors: A prognostic factor that should not be ignored. *Oncogene* 29(8);2010:1093–102.

Paige AJ. Redefining tumour suppressor genes: Exceptions to the two-hit hypothesis. *Cell Mol Life Sci Oct* 60(10);2003:2147–63.

Paleari L, Puntoni M, Clavarezza M, DeCensi M, Cuzick J, and DeCensi A. PIK3CA mutation, aspirin use after diagnosis and survival of colorectal cancer. A systematic review and meta-analysis of epidemiological studies. *Clin Oncol (R Coll Radiol)* 28(5);2016:317–26.

Palomaki GE, Bradley LA, Douglas MP, Kolor K, and Dotson WD. Can UGT1A1 genotyping reduce morbidity and mortality in patients with metastatic colorectal cancer treated with irinotecan? An evidence-based review. *Genet Med* 11(1);2009:21–34.

Panagiotou OA, Evangelou E, and Ioannidis JP. Genome-wide significant associations for variants with minor allele frequency of 5% or less—An overview: A HuGE review. *Am J Epidemiol* 172(8);2010:869–89.

Pancione M, Remo A, and Colantuoni V. Genetic and epigenetic events generate multiple pathways in colorectal cancer progression. *Patholog Res Int* 2012;2012:509348.

Panczyk M. Pharmacogenetics research on chemotherapy resistance in colorectal cancer over the last 20 years. *World J Gastroenterol* 20(29);2014:9775–827.

Papadatos-Pastos D, De Miguel Luken MJ, and Yap TA. Combining targeted therapeutics in the era of precision medicine. *Br J Cancer* 112(1);2015:1–3.

Pardoll DM. The blockade of immune checkpoints in cancer immunotherapy. *Nat Rev Cancer* 12(4);2012:252–64.

Park Y, Hunter DJ, Spiegelman D et al. Dietary fiber intake and risk of colorectal cancer: A pooled analysis of prospective cohort studies. *JAMA* 294(22);2005:2849–57.

Parodi L, Pickering E, Cisar LA, Lee D, and Soufi-Mahjoubi R. Utility of pretreatment bilirubin level and UGT1A1 polymorphisms in multivariate predictive models of neutropenia associated with irinotecan treatment in previously untreated patients with colorectal cancer. *Arch Drug Inf* 1(3);2008:97–106.

Parsons MT, Buchanan DD, Thompson B, Young JP, and Spurdle AB. Correlation of tumour BRAF mutations and MLH1 methylation with germline mismatch repair (MMR) gene mutation status: A literature review assessing utility of tumour features for MMR variant classification. *J Med Genet* 49(3);2012:151–7.

Pavicic W, Joensuu EI, Nieminen T, and Peltomäki P. LINE-1 hypomethylation in familial and sporadic cancer. *J Mol Med (Berl)* 90(7);2012:827–35.

Pawlik TM, Raut CP, and Rodriguez-Bigas MA. Colorectal carcinogenesis: MSI-H versus MSI-L. *Dis Markers* 20(4–5);2004:199–206.

Payne SR, and Kemp CJ. Tumor suppressor genetics. *Carcinogenesis* 26(12);2005:2031–45.

Pearson G, Robinson F, Beers Gibson T, Xu BE, Karandikar M, Berman K, and Cobb MH. Mitogen-activated protein (MAP) kinase pathways: Regulation and physiological functions. *Endocr Rev* 22(2);2001:153–83.

Pech MF, Garbuzov A, Hasegawa K et al. High telomerase is a hallmark of undifferentiated spermatogonia and is required for maintenance of male germline stem cells. *Genes Dev* 29(23);2015:2420–34.

Peeters M, Douillard JY, Van Cutsem E, Siena S, Zhang K, Williams R, and Wiezorek J. Mutant KRAS codon 12 and 13 alleles in patients with metastatic colorectal cancer: Assessment as prognostic and predictive biomarkers of response to panitumumab. *J Clin Oncol* 31(6);2013:759–65.

Peeters M, Oliner KS, Price TJ et al. Analysis of KRAS/NRAS mutations in a phase III study of panitumumab with FOLFIRI compared with FOLFIRI alone as second-line treatment for metastatic colorectal cancer. *Clin Cancer Res* 21(24);2015:5469–79.

Peeters M, Price TJ, Cervantes A et al. Final results from a randomized phase 3 study of FOLFIRI {+/−} panitumumab for second-line treatment of metastatic colorectal cancer. *Ann Oncol* 25(1);2014:107–16.

Pellicer M, García-González X, García MI et al. Use of exome sequencing to determine the full profile of genetic variants in the fluoropyrimidine pathway in colorectal cancer patients affected by severe toxicity. *Pharmacogenomics* 18(13);2017:1215–23.

Peláez IM, Kalogeropoulou M, Ferraro A, Voulgari A, Pankotai T, Boros I, and Pintzas A. Oncogenic RAS alters the global and gene-specific histone modification pattern during epithelial-mesenchymal transition in colorectal carcinoma cells. *Int J Biochem Cell Biol* 42(6);2010:911–20.

Peng H, Duan Z, Pan D, Wen J, and Wei X. UGT1A1 gene polymorphism predicts irinotecan-induced severe neutropenia and diarrhea in Chinese cancer patients. *Clin Lab* 63(9);2017:1339–46.

Perera MA, Innocenti F, and Ratain MJ. Pharmacogenetic testing for uridine diphosphate glucuronosyltransferase 1A1 polymorphisms: Are we there yet? *Pharmacotherapy* 28;2008:755–68.

Pesenti C, Gusella M, Sirchia SM, and Miozzo M. Germline oncopharmacogenetics, a promising field in cancer therapy. *Cell Oncol (Dordr)* 38(1);2015:65–89.

Peyssonnoux C, and Eychene A. The Raf/MEK/ERK pathway: New concepts of activation. *Biol Cell* 93(1–2);2001:53–62.

Pharoah PD, Antoniou AC, Easton DF, and Ponder BA. Polygenes, risk prediction, and targeted prevention of breast cancer. *N Engl J Med* 358(26);2008:2796–803.

Philippi C, Loretz B, Schaefer UF, and Lehr CM. Telomerase as an emerging target to fight cancer—Opportunities and challenges for nanomedicine. *J Control Release* 146(2);2010:228–40.

Phillips SM, Banerjea A, Feakins R et al. Tumour-infiltrating lymphocytes in colorectal cancer with microsatellite instability are activated and cytotoxic. *Br J Surg* 91(4);2004:469–75.

Pierotti MA. The molecular understanding of cancer: From the unspeakable illness to a curable disease. *Ecancermedicalscience* 11;2017:747.

Pietrantonio F, Petrelli F, Coinu A et al. Predictive role of BRAF mutations in patients with advanced colorectal cancer receiving cetuximab and panitumumab: A meta-analysis. *Eur J Cancer* 51(5);2015:587–94.

Pino MS, and Chung DC. Microsatellite instability in the management of colorectal cancer. *Expert Rev Gastroenterol Hepatol* 5(3);2011:385–99.

Pirmohamed M. Pharmacogenetics and pharmacogenomics. *Br J Clin Pharmacol* 52(4);2001:345–7.

Pizzolato JF, and Saltz LB. The camptothecins. *Lancet* 361(9376);2003:2235–42.

Piñol-Felis C, Fernández-Marcelo T, Viñas-Salas J, and Valls-Bautista C. Telomeres and telomerase in the clinical management of colorectal cancer. *Clin Transl Oncol* 19(4);2017:399–408.

Pomerantz MM, Ahmadiyeh N, Jia L et al. The 8q24 cancer risk variant rs6983267 shows long-range interaction with MYC in colorectal cancer. *Nat Genet* 41(8);2009:882–4.

Popat S, Chen Z, Zhao D et al. A prospective, blinded analysis of thymidylate synthase and p53 expression as prognostic markers in the adjuvant treatment of colorectal cancer. *Ann Oncol* 17(12);2006:1810–7.

Popat S, Hubner R, and Houlston RS. Systematic review of microsatellite instability and colorectal cancer prognosis. *J Clin Oncol* 23(3);2005:609–18.

Popovici V, Budinska E, Bosman F, Tejpar S, Roth A, and Delorenzi M. Context dependent interpretation of the prognostic value of BRAF and KRAS mutations in colorectal cancer. *BMC Cancer* 13(1);2013:439.

Populo H et al. mTOR signaling pathway in human cancer. *Int. J Mol Sci* 13;2012:1886–191.

Power DG, and Kemeny NE. Role of adjuvant therapy after resection of colorectal cancer liver metastases. *J Clin Oncol* 28(13);2010:2300–9.

Pox CP. Controversies in colorectal cancer screening. *Digestion* 89(4);2014:274–81.

Pozzo C, Barone C, and Kemeny NE. Advances in neoadjuvant therapy for colorectal cancer with liver metastases. *Cancer Treat Rev* 34(4);2008:293–301.

Prahallad A, Sun C, Huang S et al. Unresponsiveness of colon cancer to BRAF (V600E) inhibition through feedback activation of EGFR. *Nature* 483(7387); 2012:100–3.

Premawardhena A, Fisher CA, Liu YT et al. The global distribution of length polymorphisms of the promoters of the glucuronosyltransferase 1 gene (UGT1A1): Hematologic and evolutionary implications. *Blood Cells Mol Dis* 31(1);2003:98–101.

Proctor LM. The National Institutes of Health Human Microbiome Project. *Semin Fetal Neonatal Med* 2016, pii: S1744-165X(16)30016-6.

Pu XX, Huang GL, Guo HQ et al. Circulating miR-221 directly amplified from plasma is a potential diagnostic and prognostic marker of colorectal cancer and is correlated with p53 expression. *J Gastroenterol Hepatol* 25(10);2010:1674–80.

Puccini A, Berger MD, Naseem M et al. Colorectal cancer: Epigenetic alterations and their clinical implications. *Biochim Biophys Acta* 1868(2);2017:439–48.

Pullarkat ST, Stoehlmacher J, Ghaderi V et al. Thymidylate synthase gene polymorphism determines response and toxicity of 5-FU chemotherapy. *Pharmacogenomics J* 1;2001:65–70.

Purcell RV, Visnovska M, Biggs PJ, Schmeier S, and Frizelle FA. Distinct gut microbiome patterns associate with consensus molecular subtypes of colorectal cancer. *Sci Rep* 7(1);2017:11590.

Qin J, Li R, Raes J et al. A human gut microbial gene catalogue established by metagenomics sequencing. *Nature* 464(7285);2010:59–65.

Rahman L, Voeller D, Rahman M et al. Thymidylate synthase as an oncogene: A novel role for an essential DNA synthesis enzyme. *Cancer Cell* 5(4);2004:341–51.

Rajagopalan H, Bardelli A, Lengauer C, Kinzler KW, Vogelstein B, and Velculescu VE. Tumorigenesis: RAF/RAS oncogenes and mismatch-repair status. *Nature* 418(6901);2002:934.

Ramchandani RP, Wang Y, Booth BP, Ibrahim A, Johnson JR, Rahman A, Mehta M, Innocenti F, Ratain MJ, and Gobburu JV. The role of SN-38 exposure, UGT1A1*28 polymorphism, and baseline bilirubinlevel in predicting severe irinotecan toxicity. *J Clin Pharmacol* 47(1);2007:78–86.

Raskov H, Pommergaard HC, Burcharth J, and Rosenberg J. Colorectal carcinogenesis—Update and perspectives. *World J Gastroenterol* 20(48);2014:18151–64.

Rattray NJW, Charkoftaki G, Rattray Z, Hansen JE, Vasiliou V, and Johnson CH. Environmental influences in the etiology of colorectal cancer: The premise of metabolomics. *Curr Pharmacol Rep* 3(3);2017:114–25.

Ray D, and Kidane D. Gut microbiota imbalance and base excision repair dynamics in colon cancer. *J Cancer* 7(11);2016:1421–30.

Read MN, and Holmes AJ. Towards an integrative understanding of diet-host-gut microbiome interactions. *Front Immunol* 8;2017:538.

Remvikos Y, Laurent-Puig P, Salmon RJ, Frelat G, Dutrillaux B, and Thomas G. Simultaneous monitoring of p53 protein and DNA content of colorectal adenocarcinomas by flow cytometry. *Int J Cancer* 45;1990:450–6.

Ribic CM, Sargent DJ, Moore MJ et al. Tumor microsatellite-instability status as a predictor of benefit from fluorouracil-based adjuvant chemotherapy for colon cancer. *N Engl J Med* 349(3);2003:247–57.

Rijcken FE, Hollema H, and Kleibeuker JH. Proximal adenomas in hereditary non-polyposis colorectal cancer are prone to rapid malignant transformation. *Gut* 50(3);2002:382–6.

Ritchie MD. The success of pharmacogenomics in moving genetic association studies from bench to bedside: Study design and implementation of precision medicine in the post-GWAS era. *Hum Genet* 131(10);2012:1615–26.

Robles AI, and Harris CC. Clinical outcomes and correlates of TP53 mutations and cancer. *Cold Spring Harb Perspect Biol* 2(3);2010: a001016.

Rodríguez-Antona C, and Taron M. Pharmacogenomic biomarkers for personalized cancer treatment. *J Intern Med* 277(2);2015:201–17.

Roper J and Hung KE. Molecular mechanisms of colorectal carcinogenesis. In: *Molecular Pathogenesis of Colorectal Cancer*. Haigis KM (Ed), Springer Science+Business Media, New York, 2013, pp. 25–65.

Roskoski R Jr. RAF protein-serine/threonine kinases: Structure and regulation. *Biochem Biophys Res Commun* 399(3);2010:313–7.

Rosmarin D, Palles C, Church D et al. Genetic markers of toxicity from capecitabine and other fluorouracil-based regimens: Investigation in the QUASAR2 study, systematic review, and meta-analysis. *J Clin Oncol* 32(10);2014:1031–9.

Rosner GL, Panetta JC, Innocenti F, and Ratain MJ. Pharmacogenetic pathway analysis of irinotecan. *Clin Pharmacol Ther* 84(3);2008:393–402.

Rosty C, Young JP, Walsh MD et al. PIK3CA activating mutation in colorectal carcinoma: Associations with molecular features and survival. *PLOS ONE* 8(6);2013:e65479.

Roth AD. Prognostic role of KRAS and BRAF in stage II and III resected colon cancer: Results of the translational study on the PETACC-3, EORTC 40993, SAKK 60-00 trial. *J Clin Oncol* 28(3);2010:466–74.

Rovcanin B, Ivanovski I, Djuric O, Nikolic D, Petrovic J, and Ivanovski P. Mitotic crossover—An evolutionary rudiment which promotes carcinogenesis of colorectal carcinoma. *World J Gastroenterol* 20(35);2014:12522–5.

Rubbi CP, and Milner J. P53: Gatekeeper, caretaker or both?. In: *25 Years of p53 Research*. Hainaut P, Wiman KG (Eds). Springer, Dordrecht, 2007, pp. 233–253.

Ruden M, and Puri N. Novel anticancer therapeutics targeting telomerase. *Cancer Treat Rev* 39;2013:444–56.

Russo A, Bazan V, Iacopetta B, Kerr D, Soussi T, Gebbia N; TP53-CRC Collaborative Study Group. The TP53 colorectal cancer international collaborative study on the prognostic and predictive significance of p53 mutation: Influence of tumor site, type of mutation, and adjuvant treatment. *J Clin Oncol* 23(30);2005:7518–28.

Rustum YM. Thymidylate synthase: A critical target in cancer therapy? *Front Biosci* 9;2004:2467–73.

Ruzzo A, Graziano F, Loupakis F et al. Pharmacogenetic profiling in patients with advanced colorectal cancer treated with first-line FOLFIRI chemotherapy. *Pharmacogenomics J* 8;2008:278–88.

Ryan-Harshman M, and Aldoori W. Diet and colorectal cancer: Review of the evidence. *Can Fam Physician* 53(11);2007:1913–20.

Ryland GL, Doyle MA, Goode D, Boyle SE, Choong DY, and Rowley SM. Loss of heterozygosity: What is it good for? *BMC Med Genomics* 8;2015:45.

Sachan M, and Kaur M. Epigenetic modifications. Therapeutic potential in cancer. *Braz Arch Biol Technol* 58(4);2015:526–39.

Sacks DB. The role of scaffold proteins in MEK/ERK signalling. *Biochem Soc Trans* 34(Pt 5); 2006:833–6.

Sai K, Saito Y, Fukushima-Uesaka H et al. Impact of CYP3A4 haplotypes on irinotecan pharmacokinetics in Japanese cancer patients. *Cancer Chemother Pharmacol* 62(3);2008:529–37.

Saito Y, Sai K, Maekawa K et al. Close association of UGT1A9 IVS1 + 399C > T with UGT1A1*28, *6, or *60 haplotype and its apparent influence on 7-ethyl-10-hydroxycamptothecin (SN-38) glucuronidation in Japanese. *Drug Metab Dispos* 37(2);2009:272–6.

Sakakibara T, Hibi K, Koike M, Fujiwara M, Kodera Y, Ito K, and Nakao A. Plasminogen activator inhibitor-1 as a potential marker for the malignancy of colorectal cancer. *Br J Cancer* 93(7);2005:799–803.

Saltz LB, Cox JV, Blanke C et al. Irinotecan plus fluorouracil and leucovorin for metastatic colorectal cancer. Irinotecan Study Group. *N Engl J Med* 343(13);2000:905–14.

Samalin E, Bouché O, Thézenas S et al. Sorafenib and irinotecan (NEXIRI) as second- or later-line treatment for patients with metastatic colorectal cancer and KRAS-mutated tumours: A multicentre Phase I/II trial. *British J Cancer* 110(5);2014:1148–54.

Samuels Y, Diaz LA Jr, Schmidt-Kittler O et al. Mutant PIK3CA promotes cell growth and invasion of human cancer cells. *Cancer Cell* 7(6);2005:561–73.

Samuels Y, Wang Z, Bardelli A et al. High frequency of mutations of the PIK3CA gene in human cancers. *Science* 304(5670);2004:554.

Sanchez-Castañón M, Er TK, Bujanda L, and Herreros-Villanueva M. Immunotherapy in colorectal cancer: What have we learned so far? *Clin Chim Acta* 460;2016:78–87.

Sandel MH, Speetjens FM, Menon AG et al. Natural killer cells infiltrating colorectal cancer and MHC class I expression. *Mol Immunol* 42(4);2005:541–6.

Sankila R, Aaltonen LA, Jarvinen HJ, and Mecklin JP. Better survival rates in patients with MLH1-associated hereditary colorectal cancer. *Gastroenterology* 110(3);1996:682–7.

Santos A, Zanetta S, Cresteil T et al. Metabolism of irinotecan (CPT-11) by CYP3A4 and CYP3A5 in humans. *Clin Cancer Res* 6(5);2000:2012–20.

Sasaki T, Fujita K, Sunakawa Y et al. Concomitant polypharmacy is associated with irinotecan related adverse drug reactions in patients with cancer. *Int J Clin Oncol* 18(4);2013:735–42.

Saxonov S, Berg P, and Brutlag DL. A genome-wide analysis of CpG dinucleotides in the human genome distinguishes two distinct classes of promoters. *Proc Natl Acad Sci USA* 103(5);2006:1412–7.

Schepeler T, Reinert JT, Ostenfeld MS et al. Diagnostic and prognostic microRNAs in stage II colon cancer. *Cancer Res* 68(15);2008:6416–24.

Schmoll HJ, Van Cutsem E, Stein A et al. ESMO consensus guidelines for management of patients with colon and rectal cancer. A personalized approach to clinical decision making. *Ann Oncol* 23(10);2012:2479–516.

Schork N, Murray SS, Frazer KA, and Topol EJ. Common vs. rare allele hypotheses for complex diseases. *Curr Opin Genet Dev* 19;2009:212–9.

Schwarzenbach H. Predictive diagnostics in colorectal cancer: Impact of genetic polymorphisms on individual outcomes and treatment with fluoropyrimidine-based chemotherapy. *EPMA J* 1(3);2010:485–94.

Senter L, Clendenning M, Sotamaa K et al. The clinical phenotype of Lynch syndrome due to germ-line PMS2 mutations. *Gastroenterology* 135(2);2008:419–28.

Sevinc A, Kalender ME, Altinbas M, Ozkan M, Dikilitas M, Camci C; Anatolian Society of Medical Oncology (ASMO). Irinotecan as a second-line monotherapy for small cell lung cancer. *Asian Pac J Cancer Prev* 2(4);2011:1055–9.

Shaik AP, Shaik AS, and Al-Sheikh YA. Colorectal cancer: A review of the genome-wide association studies in the kingdom of Saudi Arabia. *Saudi J Gastroenterol* 21(3);2015:123–8.

Shammas MA. Telomeres, lifestyle, cancer, and aging. *Curr Opin Clin Nutr Metab Care* 14(1);2011:28–34.

Shaw RJ, and Cantley LC. Ras, PI(3)K and mTOR signalling controls tumour cell growth. *Nature* 441(7092);2006:424–30.

Shay JW, and Wright WE. Telomeres and telomerase in normal and cancer stem cells. *FEBS Lett* 584(17);2010:3819–25.

Shay JW, and Wright WE. Role of telomeres and telomerase in cancer. *Semin Cancer Biol* 21(6);2011:349–53.

Sheffield LJ, and Phillimore HE. Clinical use of pharmacogenomic tests. *Clin Biochem Rev* 30;2009:55–65.

Shendure J. Next-generation human genetics. *Genome Biol* 12(9);2011:408.

Shima H, Hiyama T, Tanaka S et al. Loss of heterozygosity on chromosome 10p14-p15 in colorectal carcinoma. *Pathobiology* 72(4);2005:220–4.

Shimoyama S. Pharmacogenetics of irinotecan: An ethnicity-based prediction of irinotecan adverse events. *World J Gastrointest Surg* 2(1);2010:14–21.

Shin JS, Hong A, Solomon MJ, and Lee CS. The role of telomeres and telomerase in the pathology of human cancer and aging. *Pathology* 38;2006:103–13.

Shriver Z, Raguram S, and Sasisekharan R. Glycomics: A pathway to a class of new and improved therapeutics. *Nat Rev Drug Discov* 3;2004:863–73.

Siddiqui AD, and Piperdi B. KRAS mutation in colon cancer: A marker of resistance to EGFR-I Therapy. *Ann Surg Oncol* 17(4);2010:1168–76.

Siegel RL, Miller KD, and Jemal A. Cancer statistics, 2018. *CA Cancer J Clin* 68(1);2018:7–30.

Sikdar S, Datta S, and Datta S. Exploring the importance of cancer pathway by meta-analysis of differential protein expression networks in three different cancers. *Biol Direct* 11(1);2016:65.

Simon PH, Sylvestre MP, Tremblay J, and Hamet P. Key considerations and methods in the study of gene-environment interactions. *Am J Hypertens* 29(8);2016 Apr 1:891–9, pii: hpw021.

Simonis M, Klous P, Splinter E et al. Nuclear organization of active and inactive chromatin domains uncovered by chromosome conformation capture-on-chip (4C). *Nat Genet* 38(11);2006:1348–54.

Sinicrope FA, and Sargent DJ. Molecular pathways: Microsatellite instability in colorectal cancer: Prognostic, predictive, and therapeutic implications. *Clin Cancer Res* 18;2012:1506–12.

Siobhan V, Glavey SV, Huynh D, Reagan MR et al. The cancer glycome: Carbohydrates as mediators of metastasis. *Blood Rev* 29(4);2015:269–79.

Sjöblom T, Jones S, Wood LD et al. The consensus coding sequences of human breast and colorectal cancers. *Science* 314(5797);2006:268–74.

Slattery ML, Wolff E, Hoffman MD, Pellatt DF, Milash B, and Wolff RK. MicroRNAs and colon and rectal cancer: Differential expression by tumor location and subtype. *Genes Chromosomes Cancer* 50(3);2011:196–206.

Sloane BF, Berquin IM. Proteases and cancer: An introduction. In: *Proteolysis and Protein*

Turnover. Bond JS and Barrett AJ (Eds), Portland Press (Ed.), London, 1993, pp. 225–31.

Smith G, Carey FA, Beattie J et al. Mutations in APC, Kirsten-ras, and p53—Alternative genetic pathways to colorectal cancer. *Proc Natl Acad Sci USA* 99;2002:9433–8.

Smith NF, Figg WD, and Sparreboom A. Pharmacogenetics of irinotecan metabolism and transport: An update. *Toxicol In Vitro* 20(2);2006:163–75.

Sobhani I, Amiot A, Le Baleur Y, Levy M, Auriault ML, Van Nhieu J I, and Delchier JC. Microbial dysbiosis and colon carcinogenesis: Could colon cancer be considered a bacteria-related disease? *Therap Adv Gastroenterol* 6(3);2013:215–29.

Sood A, McClain D, Maitra R et al. PTEN gene expression and mutations in the PIK3CA gene as predictors of clinical benefit to anti-epidermal growth factor receptor antibody therapy in patients with KRAS wild-type metastatic colorectal cancer. *Clin Colorectal Cancer* 11(2);2012:143–50.

Sorich MJ, Wiese MD, Rowland A, Kichenadasse G, McKinnon RA, and Karapetis CS. Extended RAS mutations and anti-EGFR monoclonal antibody survival benefit in metastatic colorectal cancer: A meta-analysis of randomized controlled trials. *Ann Oncol* 26(1);2015:13–21.

Sorrell AD, Espenschied CR, Culver JO, and Weitzel JN. Tumor protein p53 (TP53) testing and Li-Fraumeni syndrome: Current status of clinical applications and future directions. *Mol Diagn Ther* 17(1);2013:31–47.

Souglakos J, Philips J, Wang R et al. Prognostic and predictive value of common mutations for treatment response and survival in patients with metastatic colorectal cancer. *Br J Cancer* 101(3);2009:465–72.

Spilianakis CG, Lalioti MD, Town T, Lee GR, and Flavell RA. Interchromosomal associations between alternatively expressed loci. *Nature* 435(7042);2005:637–45.

Spreafico A, Tentler JJ, Pitts TM et al. Rational combination of a MEK inhibitor, selumetinib, and the Wnt/calcium pathway modulator, cyclosporin A, in preclinical models of colorectal cancer. *Clin Cancer Res* 19(15);2013:4149–62.

Steck PA, Pershouse MA, Jasser SA et al. Identification of a candidate tumor suppressor gene, MMAC1, at chromosome 10q23.3 that is mutated in multiple advanced cancers. *Nat Genet* 15(4);1997:356–62.

Steelman LS, Pohnert SC, Shelton JG, Franklin RA, Bertrand FE, and McCubrey JA. JAK/STAT, Raf/MEK/ ERK, PI3K/Akt and BCR-ABL in cell cycle progression and leukemogenesis. *Leukemia* 18(2);2004:189–218.

Stewart CF, Panetta JC, O'Shaughnessy MA et al. UGT1A1 promoter genotype correlates with SN-38 pharmacokinetics, but not severe toxicity in patients receiving low-dose irinotecan. *J Clin Oncol* 25(18);2007:2594–600.

Stigliano V, Sanchez-Mete L, Martayan A, and Anti M. Early-onset colorectal cancer: A sporadic or inherited disease? *World J Gastroenterol.* 20(35);2014:12420–30.

Stoehlmacher J, Goekkurt E, Mogck U et al. Thymidylate synthase genotypes and tumour regression in stage II/III rectal cancer patients after neoadjuvant fluorouracil-based chemoradiation. *Cancer Lett* 272(2);2008:221–5.

Stracci F, Zorzi M, and Grazzini G. Colorectal cancer screening: Tests, strategies, and perspectives. *Front Public Health* 2;2014:210.

Stranger BE, Stahl EA, and Raj T. Progress and promise of genome-wide association studies for human complex trait genetics. *Genetics* 187(2);2011:367–83.

Strassburg CP. Pharmacogenetics of Gilbert's syndrome. *Pharmacogenomics* 9(6);2008:703–15.

Strassburg CP, Barut A, Obermayer-Straub P, Li Q, Nguyen N, Tukey RH, and Manns MP. Identification of cyclosporine A and tacrolimus glucuronidation in human liver and the gastrointestinal tract by a differentially expressed UDP-glucuronosyltransferase: UGT2B7J. *Hepatol* 34(6);2001:865–72.

Strassburg CP, Manns MP, and Tukey RH. Expression of the UDP-glucuronosyltransferase 1A locus in human colon. Identification a characterization of the novel extrahepatic UGT1A8. *J Biol Chem* 273;1998:8719–26.

Strassburg CP, Oldhafer K, Manns MP, and Tukey RH. Differential expression of the UGT1A locus in human liver, biliary, and gastric tissue: Identification of UGT1A7 and UGT1A10 transcripts in extrahepatic tissue. *Mol Pharmacol* 52;1997:212–20.

Strauss BS. Hypermutability in carcinogenesis. *Genetics* 148(4);1998:1619–26.

Strubberg AM, and Madison BB. MicroRNAs in the etiology of colorectal cancer: Pathways and clinical implications. *Dis Model Mech* 10(3);2017:197–214.

Stucky-Marshall L. New agents in gastrointestinal malignancies: Part 1: Irinotecan in clinical practice. *Cancer Nurs* 22(3);1999:212–9.

Sud A, Kinnersley B, and Houlston RS. Genome-wide association studies of cancer: Current insights and future perspectives. *Nat Rev Cancer* 17(11);2017:692–704.

Sudhan DR, and Siemann DW. Cathepsin L targeting in cancer treatment. *Pharmacol Ther* 155;2015:105–16.

Sugai T, Habano W, Jiao Y-F et al. Analysis of molecular alterations in left- and right-sided colorectal carcinomas reveals distinct pathways of carcinogenesis: Proposal for new molecular profile of colorectal carcinomas. *J Mol Diagn* 8(2);2006:193–201.

Sugai T, Yoshida M, Eizuka M et al. Analysis of the DNA methylation level of cancer-related genes in colorectal cancer and the surrounding normal mucosa. *Clin Epigenetics* 9;2017:55.

Sugatani J. Function, genetic polymorphism, and transcriptional regulation of human UDP-glucuronosyltransferase (UGT) 1A1. *Drug Metab Pharmacokinet* 28(2);2013:83–92. Epub 2012 Oct 23.

Sukari A, Nagasaka M, Al-Hadidi A, and Lum LG. Cancer Immunology and Immunotherapy. *Anticancer Res* 36(11);2016:5593–606.

Sun YL, Patel A, Kumar P, and Chen ZS. Role of ABC transporters in cancer chemotherapy. *Chin J Cancer* 31(2);2012:51–7.

Surget S, Khoury MP, and Bourdon JC. Uncovering the role of p53 splice variants in human malignancy: A clinical perspective. *Onco Targets Ther* 7;2013:57–68.

Svensson T, Yamaji T, Budhathoki S, Hidaka A, Iwasaki M, and Sawada N. Alcohol consumption, genetic variants in the alcohol- and folate metabolic pathways and colorectal cancer risk: The JPHC Study. *Sci Rep* 6;2016:36607.

Swen JJ, Nijenhuis M, de Boer A et al. Pharmacogenetics: From bench to byte—An update of guidelines. *Clin Pharmacol Ther* 89(5);2011:662–73.

Symonds DA, and Vickery AL. Mucinous carcinoma of the colon and rectum. *Cancer* 37(4);1976:1891–900.

Takahashi Y, Ishii Y, Nagata T, Ikarashi M, Ishikawa K, and Asai S. Clinical application of oligonucleotide probe array for full-length gene sequencing of TP53 in colon cancer. *Oncology* 64;2003:54–60.

Takahashi Y, Sugai T, Habano W et al. Molecular differences in the microsatellite stable phenotype between left-sided and right-sided colorectal cancer. *Int J Cancer* 139(1);2016:2493–501.

Takano M, and Sugiyama T. UGT1A1 polymorphisms in cancer: Impact on irinotecan treatment. *Pharmgenomics Pers Med* 10;2017:61–8.

Tampellini M, Sonetto C, and Scagliotti GV. Novel anti-angiogenic therapeutic strategies in colorectal cancer. *Expert Opin Investig Drugs* 25(5);2016:507–20.

Tan GJ, Peng ZK, Lu JP, and Tang FQ. Cathepsins mediate tumor metastasis. *World J Biol Chem* 4(4);2013:91–101.

Tanaka F, Fukuse T, Wada H, and Fukushima M. The history, mechanism and clinical use of oral 5-fluorouracil derivative chemotherapeutic agents. *Curr Pharm Biotechnol* 1(2);2000:137–64.

Tanaka M, Omura K, Watanabe Y, Oda Y, and Nakanishi I. Prognostic factors of colorectal cancer: K-ras mutation, overexpression of the P53-protein, and cell proliferative activity. *J Surg Oncol* 57(1);1994:57–64.

Tanaka T, Watanabe T, Kazama Y, Tanaka J, Kanazawa T, Kazama S, and Nagawa H. Loss of Smad4 protein expression and 18qLOH as molecular markers indicating lymph node metastasis in colorectal cancer—A study matched for tumor depth and pathology. *J Surg Oncol* 97(1);2008:69–73.

Tanase C, Albulescu R, and Neagu M. Proteomics approaches for biomarker panels in cancer. *J Immunoassay Immunochem* 37(1);2016:1–15.

Taylor WR, and Stark GR. Regulation of the G2/M transition by p53. *Oncogene* 20(15);2001:1803–15.

Teft WA, Welch S, Lenehan J et al. OATP1B1 and tumour OATP1B3 modulate exposure, toxicity, and survival after irinotecan-based chemotherapy. *Br J Cancer* 112(5);2015:857–65.

Tejpar S, Lenz H-J, Köhne C-H et al. Effect of KRAS and NRAS mutations on treatment outcomes in patients with metastatic colorectal cancer (mCRC) treated first-line with cetuximab plus FOLFOX-4: New results from OPUS study. *JCO* 32(suppl3 abstr LBA444);2014.

Terzić J, Grivennikov S, Karin E, and Karin M. Inflammation and colon cancer. *Gastroenterology* 138;2010:2101–14.e5.

Tezcan G, Tunca B, Ak S, Cecener G, and Egeli U. Molecular approach to genetic and epigenetic pathogenesis of early-onset colorectal cancer. *World J Gastrointest Oncol* 8(1);2016:83–98.

Therkildsen C, Bergmann TK, Henrichsen-Schnack T, Ladelund S, and Nilbert M. The predictive value of KRAS, NRAS, BRAF, PIK3CA and PTEN for anti-EGFR treatment in metastatic colorectal cancer: A systematic review and meta-analysis. *Acta Oncol* 53(7);2014:852–64.

Thomas T, Gilbert J, and Meyer F. Metagenomics—A guide from sampling to data analysis. *Microb Inform Exp* 2(1);2012:3.

Thompson N, and Lyons J. Recent progress in targeting the Raf/MEK/ERK pathway with inhibitors in cancer drug discovery. *Curr Opin Pharmacol* 5(4);2005:350–6.

Thélin C, and Sikka S. Epidemiology of colorectal cancer—Incidence, lifetime risk factors statistics and temporal trends. In: *Screening for Colorectal Cancer with Colonoscopy*, Ettarh R (Ed), InTech, 2015, doi: 10.5772/61945. Available from https://www.intechopen.com/books/screening-for-colorectal-cancer-with-colonoscopy/epidemiology-of-colorectal-cancer-incidence-lifetime-risk-factors-statistics-and-temporal-trends.

Toffoli G, Cecchin E, Corona G et al. The role of UGT1A1*28 polymorphism in the pharmacodynamics and pharmacokinetics of irinotecan in patients with metastatic colorectal cancer. *J Clin Oncol* 24(19);2006:3061–8.

Tran B, Dancey JE, Kamel-Reid S et al. Cancer genomics: Technology, discovery, and translation. *J Clin Oncol* 30(6);2012:647–60.

Tsutani Y, Yoshida K, Sanada Y, Wada Y, Konishi K, Fukushima M, and Okada M. Decreased orotate phosphoribosyltransferase activity produces 5-fluorouracil resistance in a human gastric cancer cell line. *Oncol Rep* 20(6);2008:1545–51.

Tukey RH, and Strassburg CP. Human UDP-glucuronosyltransferases: Metabolism, expression, and disease. *Annu Rev Pharmacol Toxicol* 40;2000:581–616.

Tukey RH, Strassburg CP, and Mackenzie PI. Pharmacogenomics of human UDP-glucuronosyltransferases and irinotecan toxicity. *Mol Pharmacol* 62(3);2002:446–50.

Tuy HD, Shiomi H, Mukaisho KI et al. ABCG2 expression in colorectal adenocarcinomas may predict resistance to irinotecan. *Oncol Lett* 12(4);2016 Oct:2752–60.

Tziotou M, Kalotychou V, Ntokou A et al. Polymorphisms of uridine glucuronosyltransferase gene and irinotecan toxicity: Low dose does not protect from toxicity. *Ecancermedicalscience* 8;2014:428.

Ulrich CM, Bigler J, Velicer CM, Greene EA, Farin FM, and Potter JD. Searching expressed sequence tag databases: Discovery and confirmation of a common polymorphism in the thymidylate synthase gene. *Cancer Epidemiol Biomarkers Prev* 9(12);2000:1381–5.

van Brummelen EMJ, de Boer A, Beijnen JH, and Schellens JHM. BRAF mutations as predictive biomarker for response to anti-EGFR monoclonal antibodies. *Oncologist* 22(7);2017:864–72.

Van Cutsem E, Köhne CH, Hitre E et al. Cetuximab and chemotherapy as initial treatment for metastatic colorectal cancer. *N Engl J Med* 360(14);2009:1408–17.

Van Cutsem E, Köhne CH, Láng I et al. Cetuximab plus irinotecan, fluorouracil, and leucovorin as first-line treatment for metastatic colorectal cancer: Updated analysis of overall survival according to tumor KRAS and BRAF mutation status. *J Clin Oncol* 29(15);2011:2011–9.

Van Cutsem E, Lenz H-J, Köhne CH et al. Fluorouracil, leucovorin, and irinotecan plus cetuximab treatment and RAS mutations in colorectal cancer. *J Clin Oncol* 33(7);2015:692–700.

Van Cutsem E, Tabernero J, Lakomy R et al. Addition of aflibercept to fluorouracil, leucovorin, and irinotecan improves survival in a phase III randomized trial in patients with metastatic colorectal cancer previously treated with an oxaliplatin-based regimen. *J Clin Oncol* 30;2012:3499–506.

van der Bol JM, Mathijssen RH, Creemers GJ et al. A CYP3A4 phenotype-based dosing algorithm for individualized treatment of irinotecan. *Clin Cancer Res* 16(2);2010:736–42.

Van der Meide PH, and Schellekens H. Cytokines and the immune response. *Biotherapy* 8(3–4);1996:243–9.

van Es HH, Bout A, Liu J et al. Assignment of the human UDP glucuronosyltransferase gene (UGT1A1) to chromosome region 2q37. *Cytogenet Cell Genet* 63(2);1993:114–6.

van Kuilenburg AB, Muller EW, Haasjes J et al. Lethal outcome of a patient with a complete dihydropyrimidine dehydrogenase (DPD) deficiency after administration of 5-fluorouracil: Frequency of the common IVS14 + 1G > A mutation causing DPD deficiency. *Clin Cancer Res* 7(5);2001:1149–53.

van Staveren MC, Guchelaar HJ, van Kuilenburg AB, Gelderblom H, and Maring JG. Evaluation of predictive tests for screening for dihydropyrimidine dehydrogenase deficiency. *Pharmacogenomics J* 13;2013:389–95.

Vandana G, Lokesh Rao Magar S, Sweth D, and Sandhya S. An expression of p53 marker in colorectal cancer with histopathological correlation. *Int Arch Integr Med* 4(12);2017:168–84.

Varley JM. Germline TP53 mutations and Li-Fraumeni syndrome. *Hum Mutat* 21(3);2003:313–20.

Velho S, Moutinho C, Cirnes L et al. BRAF, KRAS and PIK3CA mutations in colorectal serrated polyps and cancer: Primary or secondary genetic events in colorectal carcinogenesis? *BMC Cancer* 8;2008:255.

Velho S, Oliveira C, Ferreira A et al. The prevalence of PIK3CA mutations in gastric and colon cancer. *Eur J Cancer* 41(6);2005:1649–54.

Venderbosch S, van Lent-van Vliet S, de Haan AF et al. EMAST is associated with a poor prognosis in microsatellite instable metastatic colorectal cancer. *PLOS ONE* 10;2015:e0124538.

Venter JC, Adams MD, Myers EW et al. The sequence of the human genome. *Science* 291(5507);2001:1304–51.

Verdaguer H, Saurí T, and Macarulla T. Predictive and prognostic biomarkers in personalized gastrointestinal cancer treatment. *J Gastrointest Oncol* 8(3);2017:405–17.

Vignoli M, Nobili S, Napoli C et al. Thymidylate synthase expression and g Thymidylate synthase expression and genotype have no major impact on the clinical outcome of colorectal cancer patients treated with 5-fluorouracil. *Pharmacol Res* 64(3);2011:242–8.

Visscher PM, Wray NR, Zhang Q, Sklar P, McCarthy MI, Brown MA, and Yang J. 10 years of GWAS discovery: Biology, function, and translation. *Am J Hum Genet* 101(1);2017:5–22.

Vital M, Howe AC, and Tiedje JM. Revealing the bacterial butyrate synthesis pathways by analyzing (meta)genomic data. *MBio* 5(2);2014:e00889.

Vivot A, Boutron I, Ravaud P, and Porcher R. Guidance for pharmacogenomic biomarker testing in labels of FDA-approved drugs. *Genet Med* 17(9);2015:733–8.

Vogelstein B, Fearon ER, Hamilton SR et al. Genetic alterations during colorectal-tumor development. *N Engl J Med* 319(9);1988:525–32.

Vogelstein B, Fearon ER, Kern SE, Hamilton SR, Preisinger AC, Nakamura Y, and White R. Allelotype of colorectal carcinomas. *Science* 244(4901);1989:207–11.

Vogelstein B, Lane D, and Levine AJ. Surfing the p53 network. *Nature* 408(6810);2000:307–10.

Vogelstein B, Papadopoulos N, Velculescu VE, Zhou S, Diaz LA Jr, and Kinzler KW. Cancer genome landscapes. *Science* 339(6127);2013:1546–58.

Volinia S, Hiles I, Ormondroyd E, Nizetic D, Antonacci R, Rocchi M, and Waterfield MD. Molecular cloning, cDNA sequence and chromosomal localization of the human phosphatidylinositol 3-kinase p110 alpha (PIK3CA) gene. *Genomics* 24(3);1994:472–7.

Wallace K, DeToma A, Lewin DN, Sun S, Rockey D, Britten CD, Wu JD, Ba A, Alberg AJ, and Hill EG. Racial differences in stage IV colorectal cancer survival in younger and older patients. *Clin Colorectal Cancer* 16(3);2016:178–86, pii: S1533-0028(16)30260-2.

Wan J, Li H, Li Y, Zhu ML, and Zhao P. Loss of heterozygosity of Kras2 gene on 12p12-13 in Chinese colon carcinoma patients. *World J Gastroenterol* 12(7);2006:1033–7.

Wang S, Zhang C, Zhang Z, Qian W, Sun Y, and Ji B. Transcriptome analysis in primary colorectal cancer tissues from patients with and without liver metastases using next-generation sequencing. *Cancer Med* 6(8);2017:1976–87.

Wang W, Wang GQ, Sun XW et al. Prognostic values of chromosome 18q microsatellite alterations in stage II colonic carcinoma. *World J Gastroenterol* 16(47);2010:6026–34.

Wang W, Zheng L, Zhou N et al. Meta-analysis of associations between telomere length and colorectal cancer survival from observational studies. *Oncotarget* 8(37);2017:62500–7.

Watanabe T, Itabashi M, Shimada Y et al. Japanese society for cancer of the colon and rectum (JSCCR) guidelines 2014 for treatment of colorectal cancer. *Int J Clin Oncol* 20(2);2015:207–39.

Watson MM, Lea D, Rewcastle E, Hagland HR, and Søreide K. Elevated microsatellite alterations at selected tetranucleotides in early-stage-colorectal cancers with and without high-frequency microsatellite instability: Same, same but different? *Cancer Med* 5(7);2016:1580–7.

Webber EM, Kauffman TL, O'Connor E, and Goddard KA. Systematic review of the predictive effect of MSI status in colorectal cancer patients undergoing 5FU-based chemotherapy. *BMC Cancer* 15;2015:156.

Weinshilboum RM, and Wang L. Pharmacogenomics: Precision medicine and drug response. *Mayo Clin Proc* 92(11);2017:1711–22.

Weng L, Zhang L, Peng Y, and Huang RS. Pharmacogenetics and pharmacogenomics: A bridge to individualized cancer therapy. *Pharmacogenomics* 14(3);2013:315–24.

Werner RJ, Kelly AD, and Issa JJ. Epigenetics and precision oncology. *Cancer J* 23(5);2017:262–9.

Whiffin N, Hosking FJ, Farrington SM et al. Identification of susceptibility loci for colorectal cancer in a genome-wide meta-analysis. *Hum Mol Genet* 23(17);2014:4729–37.

Whitehall VL, Rickman C, Bond CE et al. Oncogenic PIK3CA mutations in colorectal cancers and polyps. *Int J Cancer* 131(4);2012:813–20.

Wiesner GL, Slavin TP, and Barnholtz-Sloan JS. Colorectal cancer. In *Genomics and Personalized Medicine*. 1st edition. Ginsbur GS, Williard HF (Eds), Academic Press, 2010, pp. 457–476.

Wilhelm SM, Carter C, Tang L, Wilkie D, McNabola A, and Rong H. BAY 43-9006 exhibits broad spectrum oral antitumor activity and targets the RAF/MEK/ERK pathway and receptor tyrosine kinases involved in tumor progression and angiogenesis. *Cancer Res* 64(19);2004:7099–109.

Wilson AJ, Byun DS, Nasser S et al. HDAC4 promotes growth of colon cancer cells via repression of p21. *Mol Biol Cell* 19(10);2008:4062–75.

Wilson AJ, Byun DS, Popova N et al. Histone deacetylase 3 (HDAC3) and other class I HDACs regulate colon cell maturation and p21 expression and are deregulated in human colon cancer. *J Biol Chem* (19);2006:13548–58.

Wilson JL, and Altman RB. Biomarkers: Delivering on the expectation of molecularly driven, quantitative health. *Exp Biol Med (Maywood)* 243(3);2017:313–22.

Wolpin B, Ng K, Zhu AX et al. Multicenter phase II study of tivozanib (AV-951) and everolimus (RAD001) for patients with refractory, metastatic colorectal cancer. *Oncologist* 18;2013:377–8.

Wong RS. Apoptosis in cancer: From pathogenesis to treatment. *J Exp Clin Cancer Res* 30;2011:87.

Wu S, Gan Y, Wang X, Liu J, Li M, and Tang Y. PIK3CA mutation is associated with poor survival among patients with metastatic colorectal cancer following anti-EGFR monoclonal antibody therapy: A meta-analysis. *J Cancer Res Clin Oncol* 139(5);2013:891–900.

Xiang X, Jada SR, Li HH et al. Pharmacogenetics of SLCO1B1 gene and the impact of *1b and *15 haplotypes on irinotecan disposition in Asian cancer patients. *Pharmacogenet Genomics* 16;2006:683–91.

Xiao Y, and Freeman GJ. The microsatellite instable subset of colorectal cancer is a particularly good candidate for checkpoint blockade immunotherapy. *Cancer Discov* 5(1);2015:16–8.

Xie FW, Peng YH, Wang WW, Chen X, Chen X, Li J, Yu ZY, and Ouyang XN. Influence of UGT1A1 gene methylation level in colorectal cancer cells on the sensitivity of the chemotherapy drug CPT-11. *Biomed Pharmacother* 68(7);2014:825–31.

Xu JM, Wang Y, Ge FJ, Lin L, Liu ZY, and Sharma MR. Severe irinotecan-induced toxicity in a patient with UGT1A1 28 and UGT1A1 6 polymorphisms. *World J Gastroenterol* 19(24);2013:3899–903.

Xu Y. Induction of genetic instability by gain-of-function p53 cancer mutants. *Oncogene* 27(25);2008:3501–7.

Xu Y, and Villalona-Calero M. Irinotecan: Mechanisms of tumor resistance and novel strategies for modulating its activity. *Ann Oncol* 13;2002:1841–51.

Yamada H, Ichikawa W, Uetake H, Shirota Y, Nihei Z, Sugihara K, and Hirayama R. Thymidylate synthase gene expression in primary colorectal cancer and metastatic sites. *Clin Colorectal Cancer* 1(3);2001:169–73.

Yamauchi M, Morikawa T, Kuchiba A et al. Assessment of colorectal cancer molecular features along bowel subsites challenges the conception of distinct dichotomy of proximal versus distal colorectum. *Gut* 61(6);2012:847–54.

Yang C, Liu Y, Xi WQ et al. Relationship between UGT1A1*6/*28 polymorphisms and severe toxicities in Chinese patients with pancreatic or biliary tract cancer treated with irinotecan-containing regimens. *Drug Des Devel Ther* 9;2015:3677–83.

Yang J, Nishihara R, Zhang X, Ogino S, and Qian ZR. Energy sensing pathways: Bridging type 2 diabetes and colorectal cancer? *J Diabetes Complications* 31(7);2017:1228–36.

Yang ZY, Wu XY, Huang YF et al. Promising biomarkers for predicting the outcomes of patients with KRAS wild-type metastatic colorectal cancer treated with anti-epidermal growth factor receptor monoclonal antibodies: A systematic review with meta-analysis. *Int J Cancer* 133(8);2013:1914–25.

Yasar U, Greenblatt DJ, Guillemette C, and Court MH. Evidence for regulation of UDP-glucuronosyltransferase (UGT)1A1 protein expression and activity via DNA methylation in healthy human livers. *J Pharm Pharmacol* 65;2013:874–83.

Yeh JJ, Routh ED, Rubinas T et al. KRAS/BRAF mutation status and ERK1/2 activation as biomarkers for MEK1/2 inhibitor therapy in colorectal cancer. *Mol Cancer Ther* 8(4);2009:834–43.

Yokota T, Ura T, Shibata N et al. BRAF mutation is a powerful prognostic factor in advanced and recurrent colorectal cancer. *Br J Cancer* 104(5);2011:856–62.

Yoon HH, Tougeron D, Shi Q et al. KRAS codon 12 and 13 mutations in relation to disease-free survival in BRAF–Wild-Type stage III colon cancers from an adjuvant chemotherapy trial (N0147 Alliance). *Clin Cancer Res* 20(11);2014:3033–43.

You JS, and Jones PA. Cancer genetics and epigenetics: Two sides of the same coin? *Cancer Cell* 22(1);2012:9–20.

Yu J, Feng Q, Wong SH et al. Metagenomic analysis of faecal microbiome as a tool towards targeted non-invasive biomarkers for colorectal cancer. *Gut* 66(1);2017:70–8.

Yu J, Shannon WD, Watson MA, and McLeod HL. Gene expression profiling of the irinotecan pathway in colorectal cancer. *Clin Cancer Res* 11(5);2005:2053–62.

Zauber P, Sabbath-Solitare M, Marotta SP, and Bishop T. Loss of heterozygosity for chromosome 18q and microsatellite instability are highly consistent across the region of the DCC and SMAD4 genes in colorectal carcinomas and adenomas. *J Appl Res* 8(1);2008:14–23.

Zeng P, Zhao Y, Qian C et al. Statistical analysis for genome-wide association study. *J Biomed Res* 29(4);2015:285–97.

Zgheib NK, Arawi T, Mahfouz RA, and Sabra R. Attitudes of health care professionals toward pharmacogenetic testing. *Mol Diagn Ther* 15;2011:115–22.

Zhang C, Chen X, Li L, Zhou Y, Wang C, and Hou S. The association between telomere length and cancer prognosis: Evidence from a meta-analysis. *PLOS ONE* 10(7);2015:e0133174.

Zhang CM, Lv JF, Gong L, Yu LY, Chen XP, Zhou HH, and Fan L. Role of deficient mismatch repair in the personalized management of colorectal cancer. *Int J Environ Res Public Health* 13(9);2016, pii: E892.

Zhang X, Cowper-Sal lari R, Bailey SD, Moore JH, and Lupien M. Integrative functional genomics identifies an enhancer looping to the SOX9 gene disrupted by the 17q24.3 prostate cancer risk locus. *Genome Res* 22(8);2012:1437–46.

Zhang X, Kelaria S, Kerstetter J, and Wang J. The functional and prognostic implications of regulatory T cells in colorectal carcinoma. *J Gastrointest Oncol* 6(3);2015:307–13.

Zhang X, Yin JF, Zhang J, Kong SJ, Zhang HY, and Chen XM. UGT1A1*6 polymorphisms are correlated with irinotecan-induced neutropenia: A systematic review and meta-analysis. *Cancer Chemother Pharmacol* 80(1);2017:135–49.

Zheng HT, Peng ZH, Li S, and He L. Loss of heterozygosity analyzed by single nucleotide polymorphism array in cancer. *World J Gastroenterol* 11(43);2005:6740–4.

Zhu L, and Fang J. The structure and clinical roles of MicroRNA in colorectal cancer. *Gastroenterol Res Pract* 2016;2016:1360348.

Zlobec I, Bihl MP, Schwarb H, Terracciano L, and Lugli A. Clinicopathological and protein characterization of BRAF- and K-RAS-mutated colorectal cancer and implications for prognosis. *Int J Cancer* 127(2);2009:367–80.

Zlobec I, Kovac M, Erzberger P et al. Combined analysis of specific KRAS mutation, BRAF and microsatellite instability identifies prognostic subgroups of sporadic and hereditary colorectal cancer. *Int J Cancer* 127;2010:2569–75.

Zumwalt TJ, Wodarz D, Komarova NL et al. Aspirin induced chemoprevention and response kinetics are enhanced by PIK3CA mutations in colorectal cancer cells. *Cancer Prev Res (Phila)* 10(3);2017:208–18.

Zurek M, Altschmied J, Kohlgruber S, Ale-Agha N, and Haendeler J. Role of telomerase in the cardiovascular system. *Genes (Basel)* 7(6);2016:29.

6

Precision medicine in prostate cancer

NIGEL P. MURRAY

INTRODUCTION

With the changing demographics of the world population and increasing life expectancy, prostate cancer has become the most common nonskin cancer in developed countries. In the United Kingdom, it is the most common cancer in men, representing 24% of all new cancer cases. For example, 37,051 cases were registered in 2008 with a lifetime risk of 1 in 9 (Office for National Statistics, Cancer Statistics Registrations, UK, 2008) and this figure increased to 40,331 in 2015 (Office for National Statistics, Cancer Statistics Registrations, UK, 2017). In the United States an estimated 218,890 men were newly diagnosed with prostate cancer in 2007 with a lifetime risk of 1 in 6 (National Cancer Institute Surveillance Epidemiology and End Results Program, Cancer Stat Facts: Cancer of Prostate, 2006), however in 2016 the number of new cases registered had decreased to 180,890 (National Cancer Institute Surveillance Epidemiology and End Results Program, Cancer Stat Facts: Cancer of Prostate, 2016). This decrease may be the result of the United States Preventive Services Task Force (USPSTF) recommendations of 2012, whereby

it concluded that the evidence was insufficient to assess the balance of benefits and harms of prostate cancer screening using the serum prostate-specific antigen (PSA) in men younger than 75 years, and recommended against screening in men older than 75 years (Lin et al., 2011).

Worldwide, an estimated 899,000 men were diagnosed with prostate cancer in 2008, and more than two-thirds are diagnosed in developed countries (Ferlay et al., 2008). The highest rates are in Australia/New Zealand, Western and Northern Europe, and North America, largely because the prostate specific antigen (PSA) testing and subsequent biopsy has become widespread in these regions (Table 6.1).

As can be seen in Table 6.1, the number of cases of prostate cancer reported to the World Health Organization (WHO) has increased in all world regions.

Within the United States there are significant differences between racial groups, with the incidence of prostate cancer being 50% higher in African Americans than for white Americans, while rates for Asian Americans are 40% lower than for white Americans. The 2001–2005 age standardized

Table 6.1 Number of cases and deaths reported to the WHO by region for 2008 and 2012

Estimated number	Cases 2008	Deaths 2008	Cases 2012	Deaths 2012
World	899,000	258,000	1,095,000	307,000
More developed regions	644,000	136,000	742,000	142,000
Less developed regions	255,000	121,000	353,000	165,000
WHO Africa region (AFRO)	34,000	24,000	52,000	37,000
WHO Americas region (PAHO)	334,000	76,000	413,000	85,000
WHO East Mediterranean region (EMRO)	12,000	9,000	19,000	12,000
WHO Europe region (EURO)	379,000	94,000	420,000	101,000
WHO Southeast Asia region (SEARO)	28,000	19,000	39,000	25,000
WHO Western Pacific region (WPRO)	109,000	33,000	153,000	46,000
IARC membership (22 countries)	611,000	128,000	791,000	157,000
United States of America	186,000	28,000	233,000	30,000
China	33,000	14,000	47,000	23,000
India	14,000	10,000	19,000	12,000
European Union	323,000	71,000	345,000	72,000

incidences were 249/100,000, 157/100,000, and 94/100,000 for African Americans, white Americans, and Asian Americans, respectively (Ries, 2005).

The risk of prostate cancer rises steeply with age, with the highest rates occurring in the 75–79 year old age group. In the United Kingdom, the incidence is 155/100,000 men aged 55–59 years, 510/100,000 for the group 65–69 years, and 751/100,000 by 75–79 years (Office for National Statistics, Cancer Statistics registrations, UK, 2010). In the United States the median age at diagnosis for prostate cancer is 67 years, with 0.6% of cases between 35 and 44 years, 9.1% between 45 and 54 years, 30.7% between 55 and 64 years, 35.3% between 65 and 74 years, 19.9% between 75 and 84 years, and 4.4% for 85-plus years of age (National Cancer Institute Surveillance Epidemiology and End Results Program, Cancer Stat Facts: Cancer of Prostate, 2006) (Figure 6.1).

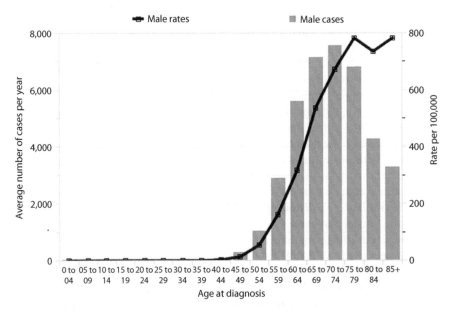

Figure 6.1 Age at diagnosis. (From Office for National Statistics, Cancer Statistics Registrations: Registrations of cancer diagnosed in 2010, England. Series MB1 London; National Statistics.)

However, studies published using postmortem data have shown that approximately half of all men in their 50s have histological evidence of prostate cancer, which in men over 80 rises to 80% (Sakr et al., 1996; Burford et al., 2008), but only 1 in 26 (3.8%) of these men will ultimately die from their cancer. In other words, more men will die with their prostate cancer than from it; this is important when considering population screening of asymptomatic men (Frankel et al., 2003; Selley et al., 1997).

THE PROSTATE-SPECIFIC ANTIGEN (PSA) CONTROVERSY FOR PROSTATE CANCER SCREENING

PSA is a glycoprotein produced almost exclusively by the epithelial of the prostate gland. Serum levels may be elevated due to increased PSA production or architectural distortions in the prostate gland that allow greater PSA access to the circulation. Only approximately 30% of men with a serum PSA >4.0 ng/mL, a standard cutoff point to determine the need for a prostate biopsy, have cancer confirmed on biopsy. Many false positives are attributed to benign prostatic hyperplasia or to subclinical prostatic inflammation (Nadler et al., 1995). It has been published that the cutoff value of 4.0 ng/mL has a sensitivity of 46% with respect to the identification of cases of prostate cancer than would occur over a 10-year period. However, the specificity of 91% in this population with a mean age of 63 years fell to 54% in men over 80 years as a result of the increase of benign prostatic hyperplasia (Gann et al., 1995). Moreover, two studies have shown that biopsies taken from men with serum PSA values of 2.5–4.0 ng/mL detected cancer in 12%–23% of cases (Babaian et al., 2001; Catalona et al., 1997). Thus there is no cutoff point for PSA that determines if there is or is not prostate cancer.

To confound matters, the benefits of screening are controversial. A screening test should ideally detect clinically significant cancer, which if not treated would increase mortality and/or morbidity, and not detect those cancers that clinically would not cause harm to the patient. Two large-scale studies designed to address this question from the United States and Europe reported conflicting results. The United States study reported by Andriole et al. (2009), compared a screening group with annual serum and digital rectal examination (DRE) and a group of usual care that sometimes included screening. In this control group, at the sixth year 52% of patients underwent screening tests. They concluded that there was a 22% increase in the rate of prostate cancer diagnosis in the screening group as compared to the control group; however there was no reduction in prostate cancer mortality during the first seven years of the trial. In contrast the European study showed a reduction of 20% in the rate of death from prostate cancer among men between the ages of 55 and 69 years at study entry (Schroder et al., 2009).

Serum total PSA is prostate specific; however, it is increased in benign diseases, such as hyperplasia and prostatitis (Bozeman et al., 2002; Punglia et al., 2006). As such, 10% to 20% of men aged between 50 and 70 years will have a raised PSA. However, only 25% of those with a serum total PSA of 4–10 ng/mL will be found to have a biopsy positive for cancer (Smith et al., 1997). Furthermore the Prostate Cancer Prevention Trial (Thompson et al., 2003) reported that 39.2% of men with a PSA of 2.1–3.0 ng/mL; 27.7% of men with a PSA of 1.1–2.0 ng/mL; and 1.3% of men with a PSA of <1.0 ng/mL had an end-of-trial prostate biopsy with foci of adenocarcinoma. In other words, 38% of men with prostate cancer have a PSA <4.0 ng/mL and 70% of men with a PSA >4.0 ng/mL do not have cancer.

In addition, the frequency of men having an elevated PSA and benign biopsy is country dependent (Belbase et al., 2013) and may be significantly different between rural and metropolitan populations in the same country (Lalitha et al., 2012).

Not all prostate cancers need treatment. It has been estimated that of screen-detected prostate cancers, 23%–42% are overtreated (Draisma et al., 2009). Active surveillance (AS) is a recognized initial treatment option for men with early stage low-grade prostate cancer. The option to delay or avoid definitive therapy avoids or minimizes patient morbidity without compromising long-term outcomes in appropriately selected patients (Dall'Era et al., 2008; Warlick et al., 2006). However, active surveillance requires repeat biopsies, often yearly, with the patient thus assuming the risks of biopsy. Men with clinically insignificant prostate cancers that were never destined to have symptoms or to affect their life expectancy may not benefit from knowing that they have the "disease." The detection of clinically insignificant prostate cancer could be considered as an adverse effect of the prostate biopsy. As such, there is considerable anxiety and distress found in men undergoing active surveillance (van den Bergh et al., 2009).

Thus for every 100 men with an elevated PSA between 4–10 ng/mL, only approximately 14 will have a clinically significant prostate cancer detected, or 86 men underwent a biopsy with the associated risks for a benign disease. A prostate biopsy is not without risks; infection and hemorrhage being the main potentially serious side effects, with a 30-day complication rate of 3.7%, especially in older patients (Anastasiadis et al., 2014). Thus avoiding unnecessary biopsies is a worthwhile aim if the number of clinically significant cancers detected is not prejudiced.

Therefore, any new screening test developed to detect prostate cancer has to improve on the standard serum PSA test; in other words detect clinically significant prostate cancer and not indolent cancer.

GENETIC SCREENING AND PERSONALIZED MEDICINE

That there is a genetic component to prostate cancer risk has been shown by a family history of prostate cancer. Evidence for this genetic link was reported in the 1960s (Woolf, 1960). Later, two large meta-analyses of studies published between 1966 and 2002 (Goh et al., 2012; Johns and Houlson, 2003) concluded that a positive family history was significantly associated with an increased risk of developing prostate cancer. The estimated relative risk varied between studies, being between 1.93 and 2.50. This risk is higher in men with a positive family history in first-degree relatives than second-degree relatives, and is higher still in those with affected brothers rather than fathers, and in those whose relatives were diagnosed with prostate cancer before the age of 60 years and decreased with age. From monozygotic and dizygotic twin studies (Ahlbom et al., 1997; Page et al., 1997), the concordance rates varied from 20%–27% for monozygotic twins to between 4%–7% for dizygotic twins.

Thus men with a positive family history of prostate cancer have a 1.5- to 2.5-fold increased risk of developing prostate cancer (Johns and Houlson, 2003); however, less than 10% of men with prostate cancer have a family history (Goh et al., 2012).

RARE HIGH PENETRANCE PROSTATE CANCER SUSCEPTIBILITY GENES

To identify chromosomal regions that may contain major prostate cancer susceptibility genes, linkage studies are used in families with multiple affected members. In most inherited cancer syndromes, cases tend to be diagnosed significantly earlier than sporadic cases. In prostate cancer the results from segregation analysis suggests that hereditary cancers are diagnosed at an earlier age, some 6–8 years younger than in sporadic cases (Carter et al., 1992). Linkage studies are hindered by several related problems, the first being a high lifetime incidence of prostate cancer (approximately 15%) and second an estimated 90% of cases are sporadic in nature. This problem is especially relevant with the advent of serum PSA screening in which the incidence rates have increased more than threefold as a result of increased detection.

Polymorphic markers (microsatellite repeats or single nucleotide polymorphisms [SNPs]) across the genome are genotyped in family members, and their cosegregation (linkage) with the presence of prostate cancer is tested using a logarithm of odds score (LOD). A score of over 3 is considered as evidence that the linkage is significant. With the chromosome region identified, sequencing methods are used to pinpoint the susceptibility gene.

In the early years of this century, three genes were identified: Hereditary Prostate Cancer 1 (HCP1) on chromosome 1q24 was reported from families in the United States and Sweden (Carter et al., 1992). The RNASEL gene in this region was identified in 2002 as a susceptibility gene using a combination of fine mapping and direct sequencing (Carpten et al., 2002). RNASEL is a constitutively expressed latent endonuclease, of which two mutations—Met1Ile and Glu265X—have been reported to segregate with prostate cancer (Carpten et al., 2002). The Glu265X mutation has been reported to be associated with prostate cancer in both familial and sporadic forms of prostate cancer (Rokman et al., 2002). One of these variants, the Arg462Gln, has been found in up to 13% of prostate cancer cases in a Finnish family–based case-control study but no association in patients without a family history (Casey et al., 2002). This variant has shown a strong association with disease severity; however, this effect is modified by family history and racial group. In European Americans without a family history, the allele is associated with early stage low-grade disease, differing from those patients with a family history where the allele was associated with aggressive disease (Rennert et al., 2005). Inversely Wang et al. (2002) found an inverse relationship with familial prostate cancer, associated with less aggressive cancer in younger patients and no association with sporadic cancer.

In 2001, using positional cloning and mutation screening of the chromosome 17p ELAC2 was identified in Utah families with hereditary prostate cancer (Tavtigian et al., 2001). A possible role in the control of the cell cycle has been hypothesized as it encodes for a 3' processing endoribonuclease, which interacts with gamma-tubulin, a component of the mitotic cycle (Korver et al., 2003; Takaku et al., 2003). However, mutations in the gene are rare, but sequence analysis has identified two missense changes—Ser217-Leu and Ala541Thr—that have been reported to be associated with prostate cancer (Rebbeck et al., 2000; Tavtigian et al., 2001).

More recently however, there was no reported association between these two variants and prostate cancer (Rennert et al., 2005). Furthermore three meta-analyses showed conflicting results. Severi et al. (2003) found no association with prostate cancer risk in a study of 1,557 patients and data obtained from seven previous studies. Using new data and data from seven previously published reports, Meitz et al. (2002) showed only a moderate risk with an odds ratio of 1.27. In Canadian patients the number of men homozygotes for Ser217-Leu did not differ between cases and controls (8.6% versus 8.5%), although heterozygotic expression was found in 61.8% of cancer cases versus 50.3% of controls and was more common in men with prostatic intraepithelial neoplasia (PIN), 42.3% versus 26.7% (Alder et al., 2003). In Ala541Thr heterozygotes there appeared to be an increase in late onset prostate cancer and PIN (Alder et al., 2003). In a larger study of U.S. males, the Ser217-Leu mutation was found in 32% of cases and 29% of controls, whereas the Ala541Thr variant was found in 4% of both cases and controls (Stanford et al., 2003). They reported that the heterozygote state was associated with less aggressive localized cancer, whereas the homozygote state was associated with Gleason scores ≥7.

A third gene was identified in the region p22–23 of chromosome 8, MSR1 (Xu et al., 2012), and is a member of a family of membrane receptors called scavenger receptors. MSR1 can bind to molecules ranging from bacteria to modified lipoproteins (Platt and Gordon, 2001) and is reported to be commonly deleted in prostate cancer (Lieberfarb et al., 2003). In families with hereditary prostate cancer six missense and one nonsense mutation have been described, and the prevalence of MSR1 mutations is higher in European and African-American members than in case controls (Xu et al., 2002a, 2002b). The most common mutations found in these men were the Arg293X and Ser41Tyr variants (Xu et al., 2002a). The IVS7delTTA homozygous genotype is significantly associated with low-grade sporadic-type prostate cancer, which appears at a later age in European racial groups. This differs from the Arg293X variant, which is associated with high-grade sporadic disease in men younger than 60 years. Rennert et al. (2005) found no significant association of MSR1 variants with familial or sporadic prostate cancer in European or African-American men. MSR1 mutations were found in 4.4% of white men with nonhereditary prostate cancer as opposed to 0.8% of unaffected men. In Afro-American men these frequencies were 12.5% and 1.8%, respectively (Xu et al., 2002a, 2002b).

While the evidence for these three genes is conflicting, the more recent discovery of the prostate cancer susceptibility gene HOXB13 seems to be more clear-cut. More than 200 genes from the chromosomal region 17q21–22 were sequenced from families with hereditary prostate cancer. Probands from four families were found to have a rare but recurrent mutation (G84E) in HOXB13, a homeobox transcription factor important in prostate development (Ewing et al., 2012). The mutation was found in all men with prostate cancer within these four families. In population studies, the heterozygous carrier state in sporadic prostate cancer is increased by a factor of 20. The G84E mutation in a prescreened white Canadian population was more frequent in men with a prostate biopsy positive for cancer, 0.7% versus those with a biopsy negative for cancer 0.1% (Akbari et al., 2012). This signifies that the mutation cosegregates with prostate cancer in hereditary prostate cancer families and is associated with prostate cancer risk in unrelated cases and controls. In the Reduction by

Dutasteride of Prostate Cancer Events (REDUCE) Trial, all 3508 men had an initial negative prostate biopsy and were biopsied after 2 and 4 years of treatment or placebo. The G84E mutation was only detected in Caucasians, with the highest frequency in Northern Europe (1.06%), followed by Western Europe (0.60%) and North America (0.31%). No mutation carrier was observed in Southern or Eastern Europe, Latin America, Australia, and South Africa, which highlights the importance of differences in the genetic load of differing populations. In Caucasians the detected mutation frequency was 0.99% and 0.24% in men with a biopsy positive and negative for cancer, respectively. In those men with a biopsy positive for cancer, the detection frequency was higher in those with a family history of prostate cancer, 4.31% versus 0.34% in those without a family history.

After 4 years of follow-up the prostate cancer detection rate was 53.8% in the 13 men heterozygotic for G84E and 22% among the 3186 noncarriers for a relative risk of 2.45 (Chen et al., 2013). A Swedish-based study of 4693 controls and 5003 prostate cancer cases reported that the G84E mutation was present in 1.3% of the population controls and strongly associated with prostate cancer risk with an odds ratio of 3.4. The strongest associated was reported for young onset prostate cancer (odds ratio 8.6) and hereditary prostate cancer (odds ratio 6.6). As in the REDUCE study, haplotype analysis supported the idea that the G84E mutation is a founder mutation. Carriers for the mutation had an estimated cumulative risk of 33% of developing prostate cancer up to the age of 80 years as compared to 12% in noncarriers (Karlsson et al., 2014). However, although carriers of this mutation were younger at the time of diagnosis and more likely to have a family history of prostate cancer, there was no association with the Gleason score in the surgical piece or pathological stage of the cancer (Beebe-Dimmer et al., 2014). In Danish men undergoing radical prostatectomy, 2.51% were positive for G84E while in a healthy control population the mutation frequency was 0.49%. Differing from the study of Beebe-Dimmer et al. (2014), the authors found that carriers were more likely to have a higher serum PSA level at diagnosis, mean PSA of 19.9 ng/mL versus 13.6 ng/mL. In this context it must be mentioned that the study was conducted between 1997 and 2011. In men with prostate cancer detected in the PSA screening era, the trigger for prostate biopsy is a total serum PSA of 4–10 ng/mL. This Danish study reported an association with higher-grade prostate cancer defined as a Gleason score of $\geq$7 (83.3% versus 60.9%) and positive surgical margins 56.0% versus 28.5%. Risk allele carriers were also more likely to have aggressive disease, defined as a preoperative total PSA $\geq$20 ng/mL, a Gleason score of $\geq$7, and/or the presence of regional or distant disease. However, there was no significant association with biochemical failure after primary treatment (Storebjerg et al., 2016).

COMMON AND LOW-PENETRANCE RISK-ASSOCIATED SINGLE NUCLEOTIDE POLYMORPHISMS

Genetic association studies are designed to identify common and low-penetrance genes in the general population, comparing the allele/genotype frequencies of markers, typically single nucleotide polymorphisms (SNPs), between healthy controls and cancer patients. With the development of high throughput and low-cost genotyping arrays it has become feasible to systematically screen hundreds of SNPs in the genome for their association with disease risk without the need to limit the search to specific genes or chromosomal regions. The Genome-Wide Association Study (GWAS) (http://www.genome.gov/gwastudies/) has identified thousands of SNPs in independent population groups, which are consistently associated with the risk for many complex diseases including prostate cancer. Many of these SNPs are not located in apparent candidate genes or pathways and others are not within genes (Hindorff et al., 2009). Since 2007 when the first two GWASs of prostate cancer were reported, more than 50 prostate cancer risk SNPs have been consistently associated with prostate cancer risk in Caucasians, African Americans, Japanese, and Chinese men from the GWAS and fine mapping of implicated regions. Table 6.2 shows the SNPs associated with prostate cancer risk.

From these studies several important observations can be noted, such as, most of these markers can be consistently reproduced in independent study populations, and few of the SNPs were found in well-known candidate genes and pathways and many are in intergenic regions. Whereas some SNPs indicate an increased risk for prostate cancer in multiple races, some are race specific. The majority

Table 6.2 SNPs associated with prostate cancer risk

SNP	Cytoband	Location (build36)	Risk allele	OR (95%CI)	Population	Reference	# Cases/controls
rs10187424	2p11	85.647.808	A	1.19(1.09–1.32)	Caucasian	Kote-Jarai et al. (2011)	37,250/36,359
rs721048	2p15	62.985.235	A	1.15(1.10–1.21)	Caucasian	Gudmundsson et al. (2008)	10,054/28,879
rs1465618	2p21	43.407.453	A	1.27(1.14–1.43)	Caucasian	Eeles et al. (2009)	21,733/20,655
rs13385191	2p24	20.751.746	G	1.15(1.10–1.21)	Japanese	Takata et al. (2010)	4,584/8,801
rs12621278	2q31	173.019.799	A	1.35(1.10–1.64)	Caucasian	Eeles et al. (2009)	21,733/20,655
rs2292884	2q37.3	238.107.965	G	1.14(1.09–1.19)	Caucasian	Schumacher et al. (2011)	10,140/11,190
rs2055109	3p11.2	87.550.022	C	1.20(1.13–1.29)	Japanese	Akamatsu et al. (2012a,b)	7,141/11,804
rs2660753	3p12	87.193.364	T	1.18(1.06–1.31)	Caucasian	Eeles et al. (2008)	5,122/5,260
rs10934853	3q21	129.521.063	A	1.12(1.08–1.16)	Caucasian	Gudmundsson et al. (2009)	13,774/47,614
rs6763931	3q23	142.585.523	T	1.18(1.07–1.29)	Caucasian	Kote-Jarai et al. (2011)	37,250/36,359
rs10936632	3q26	171.612.796	A	1.14(1.02–1.25)	Caucasian	Kote-Jarai et al. (2011)	37,250/36,359
rs17021918	4q22	95.781.900	C	1.19(1.08–1.32)	Caucasian	Eeles et al. (2009)	21,733/20,655
rs7679673	4q24	106.280.983	C	1.19(1.14–1.37)	Caucasian	Eeles et al. (2009)	21,733/20,655
rs2121875	5p12	44.401.302	G	1.09(0.99–1.21)	Caucasian	Kote-Jarai et al. (2011)	37,250/36,359
rs12653946	5p15	1.948.829	T	1.26(1.20–1.33)	Japanese	Takata et al. (2010)	4,584/8,801
rs130067	6p21	31.226.490	G	1.20(1.07–1.34)	Caucasian	Kote-Jarai et al. (2011)	37,250/36,359
rs1983891	6p21	41.644.405	T	1.15(1.09–1.21)	Japanese	Takata et al. (2010)	4,584/8,801
rs339331	6q22	117.316.745	T	1.22(1.15–1.28)	Japanese	Takata et al. (2010)	4,584/8,801
rs9364554	6q25	160.753.654	T	1.17(1.08–1.26)	Caucasian	Eeles et al. (2008)	5,122/5,260
rs10486567	7p15	27.943.088	G	1.19(0.95–1.49)	Caucasian	Thomas et al. (2008)	5,113/5,121
rs6465657	7q21	97.654.263	C	1.12(1.05–1.20)	Caucasian	Eeles et al. (2008)	5,122/5,260
rs2928679	8p21	23.494.920	A	1.53(1.27–1.83)	Caucasian	Eeles et al. (2009)	21,733/20,655
rs1512268	8p21	23.582.408	A	1.23(1.12–1.35)	Caucasian	Eeles et al. (2009)	21,733/20,655
rs1447295	8q24 (Region 1)	128.554.220	A	2.23(1.58–3.14)	Caucasian	Yeager et al. (2007;	4,296/4,299
rs16901979	8q24 (Region 2)	128.194.098	A	1.79(1.53–2.11)	Caucasian	Gudmundsson et a. (2007)	2,663/5,509
rs6983267	8q24 (Region 3)	128.482.487	G	1.58(1.40–1.78)	Caucasian	Yeager et al. (2007)	4,296/4,299
rs16902094	8q24 (Region 4)	128.389.528	G	1.21(1.15–1.26)	Caucasian	Gudmundsson et a . (2009)	13,774/47,614

(Continued)

Table 6.2 (Continued) SNPs associated with prostate cancer risk

SNP	Cytoband	Location (build36)	Risk allele	OR (95%CI)	Population	Reference	# Cases/controls
rs620861	8q24 (Region 4)	128.404.855	C	1.28(1.16–1.43)	Caucasian	Al Olama et al. (2013)	5,504/5,834
rs10086908	8q24 (Region 5)	128.081.119	T	1.25(1.12–1.37)	Caucasian	Al Olama et al. (2009)	5,504/5,834
rs817826	9q31.2	109.196.121	C	1.41(1.29–1.54)	Chinese	Xu et al. (2012)	4,484/8,934
rs1571801	9q33	123.467.194	T	1.36(1.13–1.63)	Caucasian	Duggan et al. (2007)	1,235/1,599
rs10993994	10q11	51.219.502	T	1.57(1.36–1.81)	Caucasian	Thomas et al. (2008)	5,113/5,121
rs2252004	10q26	122.834.699	G	1.16(1.10–1.22)	Japanese	Akamatsu et al. (2012a,b)	7,141/11,804
rs4962416	10q26	126.686.862	C	1.46(1.22–1.76)	Caucasian	Thomas et al. (2008)	5,113/5,121
rs7127900	11p15	2.190.150	A	1.28(1.14–1.44)	Caucasian	Eeles et al. (2009)	21,733/20,655
rs1938781	11q12	58.671.686	C	1.16(1.11–1.21)	Japanese	Akamatsu et al. (2012a,b)	7,141/11,804
rs12418451	11q13	68.691.995	A	1.36(1.18–1.56)	Caucasian	Zheng et al. (2008)	7,012/4,775
rs10896449	11q13	68.751.243	G	1.41(1.22–1.61)	Caucasian	Thomas et al. (2008)	5,113/5,121
rs10875943	12q13	47.962.277	C	1.18(1.06–1.31)	Caucasian	Kote-Jarai et al. (2011)	37,250/36,359
rs902774	12q13	51.560.171	A	1.17(1.11–1.24)	Caucasian	Schumacher et al. (2011)	10,140/11,190
rs9600079	13q22	72.626.140	T	1.18(1.12–1.24)	Japanese	Takata et al. (2010)	4,584/8,801
rs11649743	17q12	33.149.092	G	1.50(1.26–1.77)	Caucasian	Sun et al. (2009)	9,626/7,337
rs4430796	17q12	33.172.153	A	1.22(1.15–1.30)	Caucasian	Gudmundsson et al. (2007)	3,490/14,345
rs7210100	17q21.32	44.791.748	A	1.51(1.35–1.69)	African American	Haiman et al. (2011)	5,262/6554
rs1859962	17q24	66.620.348	G	1.20(1.14–1.27)	Caucasian	Gudmundsson et al. (2007)	3,490/14,345
rs8102476	19q13	43.427.453	C	1.12(1.08–1.15)	Caucasian	Gudmundsson et al. (2009)	13,774/47,614
rs887391	19q13	46.677.464	T	1.15(1.09–1.21)	Caucasian	Hsu et al. (2009)	9,516/7,252
rs2735839	19q13	56.056.435	G	1.20(1.10–1.33)	Caucasian	Eeles et al. (2008)	5,122/5,260
rs103294	19q13.4	59.489.660	C	1.28(1.21–1.36)	Chinese	Xu et al. (2012)	4,484/8,934
rs9623117	22q13	38.782.065	C	1.18(1.11–1.26)	Caucasian	Sun et al. (2009)	9,626/7,337
rs5759167	22q13	41.830.156	G	1.14(1.04–1.25)	Caucasian	Eeles et al. (2009)	21,733/20,655
rs5945619	Xp11	51.258.412	C	1.19(1.07–1.31)	Caucasian	Eeles et al. (2008), Gudmundsson et al. (2008)	5,122/5,260 (10,054/28,879)
rs5919432	Xq12	66.938.275	A	1.09(1.00–1.20)	Caucasian	Kote-Jarai et al. (2011)	37,250/36,359

of these SNPs are present in the general population with a frequency of 5% or higher, conferring a modest risk increase, commonly a relative risk of 1.1–1.2. However, when considered together they confer a stronger genetic risk for the development of prostate cancer. With regard to ethnic populations and potential race-specific differences, Han et al. (2015) reported that the majority of GWAS identified loci harbor risk alleles that are common and shared across populations. To support this concept, the predictive value of 105 known prostate cancer risk SNPs were used to predict prostate cancer. The genetic risk score (GRS) performed better in the Caucasian and Latin groups than in African-Americans and those from East Asia (Hoffmann et al., 2015). It has been further suggested that race-/ethnic-specific GRS should be used (Na et al., 2013).

Caution is further warranted. Klein et al. (2012), in a study of 891 cases and 2521 controls, were unable to replicate many SNPs and found a significant association with prostate cancer at only 14 of 37 SNPs, with a poor predictive value for the detection of prostate cancer. They also commented that only 25%–60% of reported SNPs were associated with prostate cancer risk and some associations were weak, especially with regard to aggressive disease (Klein et al., 2012).

CLINICAL APPLICATION OF SNPs IN PROSTATE CANCER

Current screening methods, digital rectal examination, total PSA, free percent PSA, PSA density, Prostate Health Index, and differing nomograms have been used to identify men with possible prostate cancer. These tests are used to select men to undergo a prostate biopsy, which is the gold standard for the diagnosis of prostate cancer. As previously mentioned, only about one third of men who undergo a prostate cancer biopsy will have cancer diagnosed and not all will need treatment.

In this context the use of precision medicine has to be applied, in which the additive effect of numerous genes along with environmental determinants create a normal distribution of disease risk in the general population. The risk alleles identified by GWAS studies individually confer a modest increase in the risk of prostate cancer. In men undergoing prostate biopsy for an elevated total PSA, defined as >4.0 ng/mL, or an abnormal digital rectal examination, the presence of prostate

cancer was predicted by the number of genetic variants (Kashyap et al., 2014). Of the 4548 men who underwent biopsy, 1834 (40.3%) had prostate cancer detected. In men with no variant present, the detection rate was 29% increasing to 63% in men with at least seven risk alleles. To assess genetic risk, a population standardized GRS has been developed. This score assesses prostate cancer risk using both the odds ratio and allele frequency of each SNP to assign a risk score, where 1.0 represents the average population risk, <1.0 a lower risk, and >1.0 a higher risk. Studies have classified patients into three risk groups based on the GRS: <0.5 as low risk, 0.5–1.5 as intermediate risk, and >1.5 as high risk. The first prostate cancer risk–associated SNPs identified were associated with a relatively high ability to predict the risk of prostate cancer, with estimated odds ratios ranging from 1.20 to 1.79. With increasing information, the predictive effect of newly determined SNPs is much lower when compared with the original discoveries, with odds ratios in the order of 1.06–1.15. Ren et al. (2013) described a plateau effect, and that incorporating new SNPs had little additional predictive value. However, comparing SNPs used in 2007 with those used in 2013, Krier et al. (2016) reported a 50% change in classification of patients. As such there is no consensus at present of how many SNPs should be incorporated into the GRS.

To date, a family history has been used as a risk factor for prostate cancer detection. However, the family history may change with time and is strongly dependent on other factors such as family size, age of relatives, health care access within the family, and the level of family communication. This is seen in that family history is a relatively poor indicator of prostate cancer risk as measured by area under the receiver operating curve (AUC), reported as being as low as 0.52 (little more than tossing a coin) (Liss et al., 2015). For the GRS to be clinically useful it must provide additional information that cannot be assessed using currently available methods, must be applicable to all men, and outperform current methods of risk assessment.

In the Prostate Cancer Prevention Trial both family history and GRS were used to identify men at higher risk for prostate cancer and the results of the prostate biopsy. Of the participating men, 17% had a positive family history, of whom 29% had prostate cancer diagnosed at biopsy. This was significantly higher than in men without a family history, in whom 23.4% had a cancer detected. When a positive family and/or a GRS of >1.4 were used as selection criteria, 36% of men were classified as high risk and the detection of prostate cancer was reported to be 30.98%. The authors also reported that when the GRS plus family history were used nearly 50% of cases with prostate cancer would be detected, including 45% of high-grade cancers (Chen 2016). In the REDUCE study (Kader et al., 2012), following an initial negative biopsy (which is not a general screening population) the AUC increased from 0.52 (family history alone) to 0.59 (GRS). Although statistically significant, both AUCs are rated as poor in terms of predictive value. A model combining clinical parameters had an AUC of 0.64, which increased to 0.67 when combined with the GRS, both prediction values being considered as acceptable. In the REDUCE trial the authors concluded that among men classified as low or high risk by clinical parameters, the GRS had limited clinical impact in men with a previous negative for cancer prostate biopsy. In those with an intermediate risk, defined as 23% for any prostate cancer and 6% for high-risk prostate cancer, the addition of the GRS had the following affect: In men with a low GRS the detection rate of prostate cancer was 14% with 4% being high-grade cancers. Those men with a high GRS had a detection rate of 35% for any prostate cancer and 14% had a high-grade cancer.

In terms of predicting prostate cancer, the SNPs have a poor predictive value on their own with AUC of between 0.57 and 0.64 (Aly et al., 2011; Hsu et al., 2010; Klein et al., 2012). Nam et al. (2009) and Johansson et al. (2013) found only small increases in the AUC when incorporating SNPs in a general screening population. Using the GRS the AUC was 0.64, whereas for total PSA and percent free PSA the AUC was 0.86. Combining the GRS with serum PSA produced an AUC difference that was statistically significant; the combined AUC was only 0.87 and thus not clinically significant. More recently Gilbert et al. (2015) reported that incorporating known genetic variants did not improve the accuracy of PSA testing to predict prostate cancer at biopsy or to identify patients with high-risk prostate cancer.

ENVIRONMENTAL EFFECTS ON SNPs

In prostate cancer, there is little evidence to link environmental factors on genetic variants and prostate cancer risk. However, it has been shown that some pesticides modify the association between genetic variants on chromosome 8q24 and prostate cancer risk (Koutros et al., 2010). Obesity has been linked to prostate cancer, both for increased risk of developing prostate cancer and a worse prognosis. Two recent studies combined body mass index with the GRS in Chinese patients. Zhang et al. (2015) reported that the predictive value of body mass index was strongly modified by the GRS. In those patients defined as high GRS (>1.4), the body mass index was an independent predictor of prostate cancer, whereas in those with a low GRS (<0.5) the body mass index had no predictive value. Liu et al. (2016) reported that in men with a body mass index of >28 kg/m^2 the presence of the alleles rs6983561 CC and rs16901966 GG increased the odds ratio to 7.66 and 5.33, respectively. In smokers, the presence of the alleles rs7679673 CC + CA and rs12653946 TT increased the odds ratio to 2.77 and 3.11, respectively, whereas in men who consumed alcohol the presence of the allele rs7679673 increased the odds ratio to 4.37. However, apart from a healthy diet and avoiding being overweight, not smoking, limiting alcohol consumption, and the general advice of healthy living, there is no evidence at present of the direct effect on prostate cancer risk.

CONCLUSIONS

The idea that genetic mapping could improve the identification of men destined to develop prostate cancer would appear to be the hallmark of precision medicine. However, in a cancer where the familial or genetic component is of the order of 10%, it would not be surprising that as a population screening test it may not be ideal. To date, the use of GRS based on SNP variants has had limited success on an individual basis. The genetic profile cannot be used to recommend prophylactic treatment nor a prostate biopsy. In the absence of lifestyle interventions or chemoprevention for prostate cancer, the aim of population health care is early detection and adequate treatment. Genetic profiling to date has not found its place in this aspect of cancer detection.

In the clinical situation, an ideal biomarker would identify those men who would need a prostate biopsy to confirm a clinically significant prostate cancer that would need treatment and thus decrease the mortality or morbidity of the disease. It would have to be cost effective and in the real world not require expensive setup costs in terms of equipment and highly trained staff. Thus, it could be implemented in a general district hospital of any country.

For low-frequency, high-penetrance genes, in terms of population screening the cost-benefit ratio does not warrant general testing. However, in high-risk families, defining the genetic marker may be beneficial to identify individual family members who may benefit from PSA testing.

With common low-penetrance SNPs, the fundamental question is how to use the information, and to differentiate between statistical significance and clinical significance. SNP testing does not determine if a patient has prostate cancer, nor does it signify the need for a prostate biopsy. The GRS is not intended to be used as a diagnostic marker for discriminating cancer versus noncancer patients. The GRS should be used as a measurement of genetic susceptibility, similar to the family history. The family history does not require a laboratory test, but neither does it provide a direct measurement of the patient's inherited risk. A genetic score of 1.0 indicates an average risk of developing prostate cancer and is a continuous variable. In the U.S. population, 8% have a twofold higher risk and 2% a threefold higher risk of developing prostate cancer based on SNP analysis.

However, the GRS could be used to stratify men into low-, intermediate-, and high-risk groups. This would then determine at what age PSA screening would be more appropriate. Men with a lower GRS risk group could be screened less frequently or screening started at an older age.

A concern over SNP profiling is that against environmental or lifestyle risk factors it may have little influence on the development of a cancer, or if the modification of environmental or lifestyle factors can change the risk in developing a cancer. The presence of a genetic variant does not signify it will be expressed, nor is there evidence on how other modifiable risk factors may influence its expression.

Furthermore, in identifying patients at risk of aggressive prostate cancer, the use of SNP profiling has not been shown to improve predictive values.

In terms of practical prostate cancer screening, the use of GRS may be statistically significant, but in the day-to-day clinical management of patients it has not proved its potential.

ACKNOWLEDGMENTS

I would like to thank Mrs. Ana Maria Palazuelos for her help in the writing of this manuscript.

CONFLICTS OF INTEREST

There were no conflicts of interest.

REFERENCES

Ahlbom A, Lichtenstein P, Malmstrom H et al., Cancer in twins: Genetic and non-genetic familial risk factors. *J Natl Cancer Inst* 89;1997:287–93.

Akamatsu S, Takata R, Haiman CA et al., Common variants at 11q12, 10q26 and 3p11.2 are associated with prostate cancer susceptibility in Japanese. *Nat Genet* 44;2012a:426–9.

Akamatsu S, Takahashi A, Takata R et al., Reproducibility, performance and clinical utility of a genetic risk prediction model for prostate cancer in Japanese. *PLOS ONE* 2012b;e46454.

Akbari MR, Trachtenberg J, Lee J et al., Association between germline HOXB13 G84E mutation and risk of prostate cancer. *J Natl Cancer Inst* 104;2012:1260–2.

Alder D, Kanji N, Trpkov K et al., HPC2/ELAC2 gene variants associated with incident prostate cancer. *J Hum Genet* 48;2003:634–8.

Al Olama A, Kote-Jarai Z, Scummacher FR et al., A meta-analysis of GWAS to identify prostate cancer susceptibility loci associated with aggressive and non-aggressive disease. *Hum Mol Genet* 22;2013:408–15.

Aly M, Wiklund F, Xu J et al., Polygenic risk score improves prostate cancer risk prediction: Results from the Stockholm-1 cohort study. *Eur Urol* 60;2011:21–8.

Anastasiadis E, van der Meulin J, Emberton M. Hospital admissions after TRUS biopsy of the prostate in men diagnosed with prostate cancer: A database analysis in England. *Int J Urol* 22;2014:181–6.

Andriole GL, Grubb RL, Buys SS et al., Mortality results from a randomized prostate cancer screening trial. *N Eng J Med* 360;2009:1310–9.

Babaian RJ, Johnston DA, Naccarato W et al., The incidence of prostate cancer in a screening population with a serum PSA between 2.5 and 4.0 ng/ml. *J Urol* 165;2001:757–60.

Beebe-Dimmer JL, Isaacs WB, Zuhlke KA et al., Prevalence of the HOXB13 G84E prostate cancer risk allele in men treated with radical prostatectomy. *BJU Int* 113;2014:830–5.

Belbase NP, Agrawal CS, Pokharel PK et al., Prostate cancer screening in a healthy population cohort in Eastern Nepal: An explanatory trial study. *Asian Pac J Cancer Prev* 14;2013:2835–8.

Bozeman CB, Carver BS, Eastham JA, Venable DD. Treatment of chronic prostatitis lowers serum PSA. *J Urol* 167;2002:1723–6.

Burford DC, Kirby M, Austoker J. *Prostate Cancer Risk Management Programme information for Primary Care; PSA testing for asymptomatic men.* 2008, NHS Cancer Screening Programmes: Sheffield.

Carpten J, Nupponen N, Issacs S et al., Germline mutations in the ribonuclease L gene in families showing linkage with HPC-1. *Nat Genet* 30;2002:181–4.

Carter BS, Beaty TS, Steinberg GD et al., Mendelian inheritance of familial prostate cancer. *Proc Natl Acad Sci USA* 89;1992:3367–71.

Casey G, Neville PJ, Plummer SJ et al., Arg462Gln variant is implicated in up to 13% of prostate cancer cases. *Nat Genet* 32;2002:581–3.

Catalona WJ, Smith DS, Ornstein DK. Prostate cancer detection in men with serum PSA concentrations of 2.6–4.0 ng/ml and benign prostate examination: Enhancement with free PSA measurements. *JAMA* 277;1997:1452–5.

Chen H, Liu X, Brendier CB et al., Adding genetic risk score to family history identifies twice as many high risk men for prostate cancer. Results from the Prostate Cancer Prevention Trial. *Prostate* 76;2016:1120–9.

Chen Z, Greenwood C, Isaacs WB et al., The G84E mutation of HOXB13 is associated with increased risk for prostate cancer: Results from the REDUCE trial. *Carcinogenesis* 34;2013:1260–4.

Dall'Era MA, Cooperberg MR, Chan JM et al., 2008. Active surveillance for early stage prostate cancer: Review of the current literature. *Cancer* 112;2008:1650–9.

Draisma G, Etzioni R, Tsodikov A et al., Lead time and overdiagnosis in PSA screening: Importance of methods and context. *J Natl Cancer Inst* 101;2009:374–88.

Duggan D, Zheng SL, Knowlton M et al., Two GWAS of aggressive prostate cancer implicate putative prostate suppressor gene DAB2IP. *J Natl Cancer Inst* 99;2007:1836–44.

Eeles RA, Kote-Jarai Z, Al Olama AA et al., Identification of seven new prostate cancer susceptibility loci through a GWAS. *Nat Genet* 41;2009:1116–21.

Eeles RA, Kote-Jarai Z, Giles GG et al., Multiple newly identified loci associated with prostate cancer susceptibility. *Nat Genet* 40;2008:316–21.

Ewing CM, Ray AM, Lange EM et al., Germline mutations in HOXB13 and prostate cancer risk. *N Eng J Med* 366;2012:141–9.

Ferlay J, Shin HR, Bray F et al., GLOBOCAN 2008 v1.2. Cancer incidence and mortality worldwide. Lyon, France: International Agency for Research on Cancer, 2010. http://globocan.iarc.fr.

Frankel S, Smith GD, Donovan J, Neal D et al., Screening for prostate cancer. *Lancet* 361(9363);2003:1122–8.

Gann PH, Hennekens CH, Stampfer MJ. A prospective evaluation of plasma PSA for detection of prostate cancer. *JAMA* 273;1995:289–94.

Gilbert R, Martin RM, Evans DM et al., Incorporating known genetic variants does not improve the accuracy of PSA testing to identify high risk prostate cancer on biopsy. *PLOS ONE* 2015; doi: 10.1371/journal.pone.0136735.

Goh CI, Schumacher FR, Easton D et al., Genetic variants associated with predisposition to prostate cancer and potential clinical implications. *J Intern Med* 271;2012:353–65.

Gudmundsson J, Sulem P, Gudbjartsson DF et al., GWA and replication studies identify four variants associated with prostate cancer susceptibility. *Nat Genet* 41;2009:1122–6.

Gudmundsson J, Sulem P, Manolescu A et al., GWAS identifies a second prostate cancer susceptibility variant at 8q24. *Nat Genet* 39;2007:631–7.

Gudmundsson J, Sulem P, Rafnar T et al., Common sequence variants on 2p15 and Xp11.22 confer susceptibility to prostate cancer. *Nat Genet* 40;2008:281–3.

Haiman CA, Chen GK, Blot WJ et al., GWAS of prostate cancer in men of African ancestry identifies a susceptibility locus a 17q21. *Nat Genet* 43;2011:570–3.

Han Y, Signorello LB, Strom SS et al., Generalizability of established prostate cancer risk variants in men of African ancestry. *Int J Cancer* 136;2015:1210–7.

Hindorff LA, Sethupathy P, Junkins HA et al., Potential etiologic and functional implications of genome wide association loci for human diseases and traits. *Proc Natl Acad Sci USA* 106;2009:9362–7.

Hoffmann TJ, Van Den Eeden SK, Sakoda LC et al., A large multiethnic genome wide association study of prostate cancer identifies novel risk variants and substantial ethnic differences. *Cancer Discov* 5;2015:878–891.

Hsu FC, Sun J, Wiklund F et al., A novel prostate cancer susceptibility locus at 19q13. *Cancer Res* 69;2009:2720–3.

Hsu FC, Sun J, Zhu Y et al., Comparison of two methods for estimating absolute risk of prostate cancer based on SNPs and family history. *Cancer Epidemiol Biomarkers Prev* 19;2010:1083–8.

Johansson M, Holmstrom B, Hinchcliffe SR et al., Combining 33 genetic variants with prostate specific antigen for prediction of prostate cancer: Longitudinal study. *Int J Cancer* 130;2012:129–37.

Johns LE, Houlston RS. A systemic review and meta-analysis of familial prostate cancer risk. *BJU Int* 91;2003:789–94.

Kader AK, Sun J, Reck BH et al., Potential impact of adding genetic markers to clinical parameters in predicting prostate biopsy outcomes in men following an initial negative biopsy: Findings from the REDUCE Trial. *Eur Urol* 62;2012:953–61.

Karlsson R, Aly M, Clements M et al., A population based assessment of germline HOXB13 G84E mutation and prostate cancer risk. *Eur Urol* 65;2014:169–76.

Kashyap A, Kluzniak W, Wokolorczyk D et al., The presence of prostate cancer at biopsy is predicted by a number of genetic variants. *Int J Cancer* 134;2014:1139–46.

Klein RJ, Hallden C, Gupta A et al., Evaluation of multiple risk-associated SNPSs versus PSA at baseline to predict prostate cáncer in unscreened men. *Eur Urol* 61;2012: 471–7.

Korver W, Guevara C, Chen Y et al., The product of the candidate prostate cancer susceptibility gene ELAC2 interacts with the gamma-tubulin complex. *Int J Cancer* 104;2003:283–8.

Kote-Jarai Z, Olama AA, Giles GG et al., Seven prostate cancer susceptibility loci identified by a multistage GWAS. *Nat Genet* 43;2011:785–91.

Koutros S, Beane-Freeman LE, Berndt SL et al., Pesticide use modifies the association between genetic variants on chromosome 8q24 and prostate cancer. *Cancer Res* 70;2010:9224–33.

Krier J, Barfield R, Green RC et al., Reclassification of genetic based risk predictions as GWAS data accumulate. *Genome Med* 8;2016:20.

Lalitha K, Suman G, Pruthvish S, Mathew A, Murthy NS. Estimation of time trends of incidence of prostate cancer: An Indian scenario. *Asian Pac J Cancer Prev* 13;2012:6245–50.

Lieberfarb ME, Lin M, Lechpammer M et al., Genome wide loss of heterozygosity analysis from laser capture microdissected prostate cancer using single nucleotide polymorphic allele (SNP) arrays and a novel bioinformatics platform of dChipSNP. *Cancer* 63;2003:4781–5.

Lin K, Croswell JM, Koenig H et al., PSA based screening for prostate cancer. An evidence update for the US Preventive Services Task Force Rockville (MD): Agency for Healthcare Research and Quality (US) 2011 Oct. Report N: 12-05160-EF-1.

Liss MA, Xu J, Chen H et al., Prostate genetic score (PGS-33) is independently associated with risk of prostate cancer in the PCLO Trial. *Prostate* 75;2015:1322–8.

Liu M, Shi X, Yang F et al., The cumulative effect of gene-gene and gene-environment interactions on the risk of prostate cancer in Chinese men. *Int J Environ Res Public Health* 13;2016:162.

Meitz JC, Edwards SM, Easton DF et al., HPC2/ELAC2 polymorphisms and prostate cancer risk: Analysis by age of onset of disease. *Br J Cancer* 87;2002:905–8.

Nadler RB, Humphrey PA, Smith DS et al., Effect of inflammation and benign prostatic hyperplasia on elevated serum PSA levels. *J Urol* 154;1995:407–13.

Nam RK, Zhang WW, Trachtenberg J et al., Utility of incorporating genetic variants for early detection of prostate cancer. *Clin Cancer Res* 15;2009:1787–93.

Na R, Liu F, Zhang P et al., Evaluation of reported prostate cancer risk associated SNPs from GWAS of various racial populations in Chinese men. *Prostate* 73;2013:1623–35.

National Cancer Institute Surveillance Epidemiology and End Results Program, Cancer Stat Facts: Cancer of Prostate, 2006. www.seer.cancer.gov/statfacts/html/prost.html, Accessed 2017.

National Cancer Institute Surveillance Epidemiology and End Results Program, Cancer Stat Facts: Cancer of Prostate, 2016. http://seer.cancer.gov, Accessed 2017.

Office for National Statistics, Cancer Statistics Registrations: Registrations of cancer diagnosed in 2008, England. Series MB1 No.39 2010 London; National Statistics.

Office for National Statistics, Cancer Statistics Registrations: Registrations of cancer diagnosed in 2010, England. Series MB1 London; National Statistics.

Office for National Statistics, Cancer Statistics Registrations: Registrations of cancer diagnosed in 2015, England. Series MB1 London; National Statistics, 2017.

Page WF, Braun MM, Partin AW et al., Heredity and prostate cancer: A study of World War II veteran twins. *Prostate* 33;1997: 240–250.

Platt N, Gordon S. Is the class A macrophage scavenger receptor (SR-A) multifunctional? *J Clin Invest* 108;2001:649–54.

Punglia RS, D'Amico AV, Catalona WJ et al., Impact of age, benign prostatic hyperplasia and cancer on PSA level. *Cancer* 106;2006:1507–13.

Rebbeck TR, Walker AH, Zeigler-Johnson C et al., Association of the HPC2/ELAC2 genotypes and prostate cancer. *Am J Hum Genet* 67;2000:1014–9.

Ren S, Xu J, Zhou T et al., Plateau effect of prostate cancer risk associated SNPs in discriminating prostate biopsy outcomes. *Prostate* 73;2013:1824–35.

Rennert H, Zeigler-Johnson CM, Addya K et al., Association of the susceptibility alleles in ELAC2/HPC2, RNASEL/HPC1 and MSR1 with prostate cancer severity in European American and African American men. *Cancer Epidemiol Biomarkers Prev* 14;2005:949–57.

Ries LAG. *SEER Cancer Statistics Review 1975–2002*. 2005. NCI, Bethesda, MD.

Rokman A, Ikonen T, Seppala EH et al., Germline alterations of the RNASEL gene, a candidate HPC1 gene at 1q25 in patients and families with prostate cancer. *Am J Hum Genet* 70;2002:1299–304.

Sakr WA, Grigon DJ, Hass GP et al., Age and racial distribution of prostatic intraepithelial neoplasia. *Eur Urol* 30(2);1996:138–44.

Schroder FH, Hugosson J, Roobol MJ et al., Screening and prostate cancer mortality in a randomized European study. *N Eng J Med* 360;2009:1320–8.

Schumacher FR, Berndt SI, Siddiq A et al., Genome-wide association study identifies new prostate cancer susceptibility loci. *Hum Mol Genet* 20;2011:3867–75.

Selley, S, Donovan J, Faulkner A et al., Diagnosis, management and screening of early localized prostate cancer. *Health Technol Assess* 1(2);1997.

Severi G, Giles GG, Southey MC et al., ELAC2/HPC2 polymorphisms, PSA levels and prostate cancer. *J Natl Cancer Inst* 95;2003:818–24.

Smith DS, Humphrey PA, Catalona WJ. The early detection of prostate carcinoma with PSA. *Cancer* 80;1997:1852–6.

Stanford JL, Sabacan LP, Noonan EA et al., Association of HPC2/ELAC2 polymorphisms with risk of prostate cancer in a population based study. *Cancer Epidemiol Biomarkers Prev* 12;2003:876–81.

Storebjerg TM, Hoyer S, Kirkegaard P et al., Prevalence of the HOXB13 G84E mutation in Danish men undergoing radical prostatectomy and its correlations with prostate cancer risk and aggressiveness. *BJU Int* 118;2016:646–53.

Sun J, Zheng SL, Wiklund F et al., Sequence variants at 22q13 are associated with prostate cancer risk. *Cancer Res* 69;2009:10–15.

Takaku H, Minagawa A, Taskagi M et al., A candidate prostate cancer susceptibility gene encodes tRNA 3' processing endoribonuclease. *Nucleic Acids Res* 31;2003:2272–8.

Takata R, Akamatsu S, Kubo M et al., GWAS identifies five new susceptibility loci for prostate cancer in the Japanese population. *Nat Genet* 42;2010:751–4.

Tavtigian SV, Simard J, Teng DH et al., A candidate prostate cancer susceptibility gene at chromosome 17p. *Nat Genet* 27;2001:321–5.

Thomas G, Jacobs KB, Yeager M et al., Multiple loci identified in a GWAS of prostate cancer. *Nat Genet* 40;2008:310–5.

Thompson IM, Goodman PJ, Tangen CM et al., The influence of finisteride on the development of prostate cancer. *N Eng J Med* 349;2003:215–24.

van den Bergh RC, Essink-Bot M, Roobol MJ et al., Anxiety and distress during active surveillance for early prostate cancer. *Cancer* 115;2009:3868–78.

Wang L, McDonnell SK, Elkins DA et al., Analysis of the RNASEL gene in familial and sporadic prostate cancer. *Am J Hum Genet* 71;2002:116–23.

Warlick C, Trock BJ, Landis P et al., Delayed versus immediate surgical intervention and prostate cancer outcome. *J Natl Cancer Inst* 98;2006:355–7.

Woolf CM. An investigation of the familial aspects of carcinoma of the prostate. *Cancer* 13;1960:739–44.

Xu J, Mo Z, Ye D et al., GWAS in Chinese men identifies two new prostate cancer risk loci at 9q31.2 and 19q13.4. *Nat Genet* 44;2012:1231–5.

Xu J, Zheng SL, Komiya A et al., Common sequence variants of the macrophage scavenger receptor 1 gene are associated with prostate cancer. *Am J Hum Genet* 72;2002a:208–12.

Xu J, Zheng SL, Komiya A et al., Germline mutations and sequence variants of the macrophage scavenger receptor 1 gene are associated with prostate cancer risk. *Nat Genet* 32;2002b:321–5.

Yeager M, Orr N, Hayes RB et al., GWAS of prostate cancer identifies a second risk locus at 8q24. *Nat Genet* 39;2007:645–9.

Zhang GM, Zhu Y, Chen HT et al., Association between the body mass index and prostate cancer at biopsy is modified by genetic risk: A cross sectional analysis in China. *Medicine (Baltimore)* 94;2015:e1603.

Zheng SL, Sun J, Wiklund F et al., Cumulative association of five genetic variants with prostate cancer. *N Eng J Med* 58;2008:910–9.

7

Breast cancer epigenetic targets for precision medicine

RAMONA G. DUMITRESCU

INTRODUCTION

Breast cancer is the most common cancer in women and the second leading cause of cancer death among women in the United States. Due to the improved understanding of molecular defects involved in the initiation and progression of different subtypes of breast cancer (Polyak, 2011), targeted therapeutics have been developed, leading to great advancements for individualized breast cancer patients' treatment.

Among these defects, epigenetic changes involved in altering gene expression play an essential role in cancer development and progression. Recent studies described the existence of breast cancer–specific DNA methylation profiles (Cancer Genome Atlas Network, 2012) that can help in identifying breast cancer specific pathways, which could be modified by targeted interventions, leading to reduced breast cancer incidence. These methylation signatures can also help in refining breast cancer screening to achieve early detection and can help in clinical care by determining personalized cancer treatments and predicting disease-free and overall survival (Terry et al., 2016).

Thus, these epigenetic biomarkers could play an essential role for primary, secondary, and tertiary prevention.

Unfortunately, acquired or intrinsic resistance to targeted therapy has been reported in breast cancer treatment. It has been observed that targeted therapy resistance in breast cancer can be mediated by epigenetic changes (Liu et al., 2016; Walsh et al., 2016). Epigenetic modifications could be important targets for both the prevention and treatment of primary tumors as well as resistant tumors due to the reversible nature of the epigenetic changes. Thus, the study of cancer epigenetic changes could lead to expanding the range of therapeutic options in precision medicine.

In this chapter, several epigenetic targets for breast cancer prevention and treatment with potential important implications for the disease prognosis and survival will be discussed.

Epigenetic targets for breast cancer prevention

Dysregulation of the epigenome in cancer is well recognized to play a crucial role in development of breast cancer (Baylin and Jones, 2011) but it is not well known how early these epigenetic changes take place. Although most studies examined breast tumors to look for these epigenetic events, the early changes in the epigenome are very difficult to be examined and detected. One model used to study the epigenetic remodeling during breast cells' malignant transformation is represented by the human mammary epithelial cells (HMECs) (Hinshelwood and Clark, 2008; Locke and Clark, 2012). These cells derive from breast tissue surgically removed during the reduction mammoplasty procedure in healthy women.

Another related model is represented by the variant HMECs (vHMECs), which are cells undergoing early carcinogenic transformation by going through selection or overcoming senescence barriers. These cells are believed to be a model of basal-like breast carcinogenesis (Hinshelwood et al., 2008; Locke and Clark, 2012). The changes observed in these cells include the promoter hypermethylation of *p16INK4a* tumor suppressor gene, silencing of the transforming growth factor beta gene (*TGFB*), and increased expression of the chromatin methyltransferase EZH2 to name few early events (Hinshelwood et al., 2007). Thus, the HMECs could be a good model to look at for specific changes from normal cells to premalignant cells.

Looking at the gene expression between HMECs and vHMECs, it has been observed that genes differentially expressed in vHMECs have critical roles in cellular growth and proliferation, cellular movement, cell cycle and DNA replication, recombination and DNA repair, and the most significant pathways identified, including the oncogenes *MYC* and *EZH2* (Locke et al., 2015). The same analysis looked at the activity of transcriptions factors and found that NF-kB and E2F as well as stem cell/cancer-associated transcription factor NANOG and ligand-activated transcription factor AHR signaling pathways were activated. The transcription factors that were found by this analysis to be inhibited were p53 and RB (Locke et al., 2015).

Furthermore, differentially methylated regions (DMRs) were identified in early passages of vHMECs. The coordinated DNA hypermethylation of target loci were regulated by key transcriptional factors like p53, AHR, and E2F family members. This together with long-range epigenetic dysregulation is believed to represent the earliest stage of breast malignancy. This finding supports the hypothesis that breast cancer initiates when there is epigenetic disruption of the transcription factor binding that could lead to dysregulation of numerous networks involved in cancer development (Locke et al., 2015).

In the last few years, it has been reported that blood-based DNA methylation markers could be used to evaluate breast cancer risk (Severi et al., 2014; Tang et al., 2016).

It has been shown that a new epigenetics-based system was able to differentiate healthy volunteers from breast cancer patients, with high accuracy (Uehiro et al., 2016). This system is based on 12 novel epigenetic markers that were identified after a methylation array analysis was conducted. The authors suggested that early breast cancer detection based on this system is similar to the mammography screening detection (Uehiro et al., 2016).

Similar results were shown by a different group, when a six-gene methylation panel had a high sensitivity and specificity in breast cancer diagnosis when compared with healthy and benign disease controls. This further suggests that the detection of aberrant methylation of several genes in serum DNA could be a potential diagnosis tool for breast cancer (Shan et al., 2016).

Recent studies linked the "epigenetic clock," or the DNA methylation-based markers of ageing, to the risk of cancer development. In order to identify possible epigenetics biomarkers for breast cancer susceptibility and risk stratification, a new study conducted a DNA methylome analysis in the European Prospective Investigation into Cancer and Nutrition (EPIC) cohort using the Illumina HumanMethylation 450K BeadChip arrays. It was found that higher CpG islands' DNA methylation is significantly associated with postmenopausal breast cancer susceptibility, suggesting that accelerated epigenetic ageing increases risk of breast cancer development (Ambatipudi et al., 2017).

Furthermore, when the advanced DNA methylation (DNAm) age in breast tissue was compared with the DNAm age of peripheral blood tissue from the same individuals, it was observed that DNAm age was highly correlated with chronological age in both breast tissues and peripheral blood. The study showed that breast tissue has a higher epigenetic age relative to peripheral blood, but the difference between the DNAm age in breast and blood is decreasing with advancing age (Sehl et al., 2017).

In addition, several studies showed that another DNA methylation marker, the global hypomethylation is associated with increased breast cancer risk (Choi et al., 2009; Tang et al., 2016).

When the epigenome-wide methylation analysis was conducted in three independent prospective nested case-control studies in relationship to breast cancer risk, it was observed that epigenome-wide hypomethylation is associated with breast cancer risk (van Veldhoven et al., 2015). In addition, decreased average methylation levels were detected in blood samples, years before breast cancer diagnosis. This finding suggests that this genome-wide epigenetic change could be used as a clinical biomarker, with predictive value for breast cancer risk (van Veldhoven et al., 2015).

Epigenetic targets for breast cancer treatments

Personalized medicine in breast cancer treatment would involve tailored drug usage based on the genetic defects identified in the primary tumor (Cho et al., 2012). Some genetic defects are predictive of the response to specific drugs, as in the case of *BRCA1* and *BRCA2* mutations in relation to PARP inhibitors and *PIK3CA* mutations in relation to PIK3 inhibitors. However, in many sporadic breast cancers, several genes are inactivated through DNA methylation changes, and these changes play a role in the tumors' response to breast cancer therapy. Several of these methylation changes with important roles in the personalized breast cancer treatment will be discussed.

THE ROLE OF BRCA1 METHYLATION IN PREDICTING THE TREATMENT RESPONSE

Breast cancers develop in about 5% of *BRCA1* and *BRCA2* mutation carriers, however, somatic mutations of *BRCA1* rarely occur in sporadic breast cancer. Lower than normal rates of BRCA1 expression or loss of BRCA1 expression through epigenetic inactivation was observed in sporadic breast tumors. Thus, epigenetic silencing of *BRCA1* is believed to be an important factor that contributes to breast cancer development. Considering that *BRCA1* methylation is present in about 10%–15% of breast cancer patients (Birgisdottir et al., 2006), it would be extremely valuable to take into account the epigenetic changes, in addition to genetic mutations, so that more women other than mutation carries could benefit from targeted interventions.

Anticancer drugs leading to the formation of double-stranded DNA breaks (DSBs) have been experienced in breast cancer, because this therapy is predicted to be effective in killing BRCA1- or BRCA2-defective cells. For example, PARP inhibitors have been reported to be effective in targeting cancer cells exhibiting errors in the DNA repair mechanisms (Bryant et al., 2005).

The determination of methylation status of *BRCA1* and other genes of the BRCA/homologous recombination (HR) pathway could be used as a predictor of the response of breast tumors to PARP inhibitors and cisplatin therapy (Stefansson and Esteller, 2013; Cai et al., 2014).

Therefore, it was suggested that in triple-negative breast cancers (TNBCs) patients with acquired or inherited defects in the *BRCA1* gene, the use of platinum-based drugs, or PARP inhibitors, for the treatment of these aggressive tumors could have a significant impact on the disease outcome (Stefansson and Esteller, 2013).

BRCA1 loss associated with impaired DNA repair has been observed in triple-negative breast cancers. In fact, when DNA methylation was analyzed in these tumors, *BRCA1* promoter methylation was found to be associated with reduced BRCA1 expression, aggressive phenotype, and shorter overall survival. Thus, *BRCA1* promoter methylation is an important mechanism that leads to functional loss of BRCA1 (Yamashita et al., 2015) and could be a promising biomarker for the prognosis of triple-negative breast cancers (Zhu et al., 2015).

When *BRCA 1* methylation and *TP53* mutations were analyzed in triple-negative breast cancer patients without pathological complete response to taxane-based neoadjuvant chemotherapy, it was observed that *BRCA1* methylation was associated with a significant decrease in disease-free survival, providing important information for breast cancer treatment and prognosis (Foedermayr et al., 2014).

THE ROLE OF ESTROGEN RECEPTOR (ER) METHYLATION IN BREAST CANCER TREATMENT

The expression of estrogen receptor (ER) determines if a breast cancer patient receives endocrine therapy but does not assure a therapeutic response.

More specifically, the loss of ERα expression leads to lack of response to antihormone treatment. In a significant fraction of ER+/PR– and ER–/PR– breast tumors, the loss of expression is a result of epigenetic changes within the *ERα* gene promoter, such as DNA methylation and histone deacetylation (Stearns et al., 2007).

Patients expressing ER and PR proteins respond to hormonal treatment. Therefore, it was hypothesized that reversing the *ER* promoter methylation may help tumors become responsive to tamoxifen therapy and this was amply described in a review by Stearns and his collaborators (Stearns et al., 2007). When ER-negative breast cancer cell lines were treated with HDAC inhibitors and/or demethylating agents like trichostatin A (TSA) and 5-aza-2-deoxycytidine, the reexpression of a functional ER was observed. Similar findings were reported in ER-negative breast cancer cell lines after the administration of an HDAC inhibitor, scriptaid. Also, scriptaid growth inhibition was observed together with increased acetylation of histone tails. Similar to trichostatin A combination treatment, the scriptaid and 5-aza-2-deoxycytidine were more effective in reexpressing ER than either agent alone (Stearns et al., 2007). These findings showed that the administration of HDAC inhibitors can reactivate the ER in ER– breast cancer cell lines and that restores the sensitivity to tamoxifen and then led to cell growth inhibition. Therefore, targeting to reactivate genes that are silenced by promoter hypermethylation in breast cancer with demethylating agents could be an important therapeutic approach.

Furthermore, when a new study looked at the DNA methylation role in the breast cancers response to endocrine therapy, it was found that DNA hypermethylation of the estrogen-responsive enhancers leads to reduced ER binding and reduced expression of genes regulated by the ER (Stone et al., 2015). The study also showed that ER enhancers' hypomethylation plays a critical role in the transition from normal cells to ER-positive cancerous cells that respond to endocrine therapy. These findings highlight the important role of estrogen-regulated enhancers' DNA methylation for the endocrine sensitivity in breast cancer and also for the potential prediction of ER-positive status (Stone et al., 2015).

Recently, a new role for the ERα-associated gene expression regulation was proposed. It has been reported that ERα can silence genes via DNA methylation. More specifically, ERα may direct DNA methylation–associated silencing of a sub-population of basal markers, cancer stem cells, epithelial–mesenchymal transition, and inflammatory and tumor suppressor genes with a potential role in enforcing a phenotype with luminal characteristics (Ariazi et al., 2017). These findings show how important the ER-associated methylation is for the breast cancer phenotype and treatment.

SEVERAL OTHER EPIGENETIC CHANGES AND EPIGENETIC MODULATION IN BREAST CANCER TREATMENT

As mentioned before, promoter methylation of several other key genes has been shown to be critical in breast cancer initiation and progression (Baylin and Jones, 2011).

Recently, Mathe and his group looked if the DNA methylation contributes to the altered gene expression in TNBCs as well as in disease progression from primary breast tumor to lymph node metastasis (Mathe et al., 2016). The authors reported that this whole genome DNA methylation analysis in a TNBC cohort showed an altered methylation of 18 genes associated with lymph node metastasis that contributes to the dysregulation of gene expression changes and is associated with overall survival (Mathe et al., 2016).

In the ER+ breast cancers, the epigenetic-associated overexpression of histone H2B variants was shown to play an important role in the endocrine response of these tumors to treatment (Nayak et al., 2015).

In addition, in the ER+ breast cancers, the *RUNX3* hypermethylation was frequently observed and a higher RUNX3 mRNA expression was associated with better relapse-free survival (Lu et al., 2017). Thus, *RUNX3* methylation could be a therapeutic target for the development of personalized therapy.

Some HDAC inhibitors and DNMT inhibitors have been investigated in preclinical models in breast cancer, and it was shown that drugs that modulate epigenetic changes may be used as single agent or in combination with standard treatments. Some of these preclinical studies have shown that HDAC inhibitors' effect may not be complete or can be reversible if the treatment is stopped. Therefore, the combination therapy of HDAC inhibitors and tamoxifen or retinoic acid could be a better approach. Another combination that was suggested was HDAC inhibitors and DNMT inhibitors or drugs like trastuzumab. Others suggested HDAC inhibitors and chemotherapeutic agents, such as docetaxel and gemcitabine, so that growth inhibition and apoptosis are achieved (Stearns et al., 2007).

For the human epidermal growth factor (HER) 2-positive breast cancers, the standard therapy is targeting the HER2. This therapeutic approach was associated with disease-free survival if a pathological complete response to the therapy is achieved. A new study conducted a genome-wide DNA methylation analysis and identified eight genomic regions differentially methylated in patients with pathological complete response. Among those regions, *HSD17B4* encoding type 4 17β-hydroxysteroid dehydrogenase was most significantly differentially methylated (Fujii et al., 2017). The authors concluded that the pathological complete response of HER2-positive breast cancers to trastuzumab and chemotherapy can be predicted by *HSD17B4* methylation, and this can have important treatment implications for this group of breast cancer patients (Fujii et al., 2017).

Another tumor suppressor gene frequently hypermethylated in breast cancer is *RARβ2* and the silencing of this gene is reinforced by the histone deacetylation at the promoter region. Thus, similarly to *ER* methylation, the reactivation of *RARβ2* could restore RARβ effects in breast cancer. In fact, the treatment with 5-aza-2deoxycytidine and TSA has been reported to restore RARβ2 activity, suggesting that these drugs with effect on chromatin remodeling can reactivate the *RARβ2* gene and overcome the retinoic acid resistance in breast cancer (Stearns et al., 2007).

In a recent study, it was found that the combination of entinostat, ATRA, and doxorubicin (EAD) restored the epigenetically silenced RAR-β expression and resulted in significant tumor regression. The findings in this study suggest that Entinostat mediates Doxorubicin's action on cytotoxicity and differentiation driven by retinoids in order to achieve tumor regression in TNBC (Merino et al., 2016).

Furthermore, several inhibitors of histone deacetylases were considered of a great therapeutic value (Stearns et al., 2007), especially for TNBC patients. A recent study examining the effects of suberoyl anilide hydroxamic acid (SAHA) on TNBCs has revealed that SAHA has great potential to be used as an anticancer agent in the treatment of these type of tumors (Wu et al., 2016).

Moreover, it has previously been observed that triple-negative breast cancer cell lines show heterogeneous responses to the PARP and HDAC inhibitors. When the coadministration of olaparib and SAHA were studied to examine their effect on the growth of TNBC, it was found that cells that expressed functional phosphatase and tensin homolog (PTEN) would favorably respond to the combination therapy (Min et al., 2015).

Recently, carnitine palmitoyl transferase-1A (CPT1A) was identified as a new tumor specific target in human breast cancer (Pucci et al., 2016). It has been suggested before that CPT1A variant 2 product is involved in the epigenetic regulation of cancer survival, cell death escaping, and metastasis pathways. Indeed, the inactivation of CPT1A variant 2 by using small interfering RNAs (siRNAs) led to apoptosis in several breast cancer cell lines. *CPT1A* silencing was associated with reduction of HDAC activity and histone hyperacetylation, leading to upregulated transcription of proapoptotic genes (*BAD, CASP9, COL18A1*) and downmodulation of invasion and metastasis genes (*TIMP-1, PDGF-A, SERPINB2*). Thus, CPT1A has been proposed as a new tumor-specific target, more effective than the well-known aforementioned HDAC inhibitors (Pucci et al., 2016).

Aberrant DNA hypermethylation has been shown to play a critical role in the regulation of renewal and maintenance of cancer stem cells (CSCs), which are targets for cancer initiation by chemical and environmental factors. The administration of a DNA hypermethylation inhibitor like decitabine (DAC) can have an important effect on the chemotherapeutic response and the acquired drug resistance of the CSCs (Li et al., 2015). Based on this hypothesis, *in vitro* studies were conducted to look at the effect of the treatment with nanoparticles loaded with low-dose DAC (NPDAC) combined with nanoparticles loaded with doxorubicin (NPDOX) on the proportion of CSCs with high aldehyde dehydrogenase activity (ALDHhi). It has been observed that this therapeutic approach achieved the highest proportion of apoptotic tumor cells and the lowest proportion of ALDHhi CSCs. This finding suggests that the combined treatment of NPDAC and NPDOX has a critical role in the inhibition of breast cancer growth (Li et al., 2015).

Unfortunately, many cancer cells acquire multidrug resistance (MDR) that was associated with the altered expression of glucosylceramide synthase

(GCS). Liu and his group analyzed the association of methylation at the *GCS* gene promoter with its expression and MDR in invasive ductal breast cancer and found that the *GCS* promoter methylation changes correlate with multidrug resistance in breast cancer (Liu et al., 2016).

In conclusion, the epigenetic changes described play a crucial role in breast cancer initiation, progression, and response to targeted therapies. These epigenetic biomarkers, once validated in clinical studies, can provide essential information for breast cancer risk assessment for predicting the type of breast cancer and the most effective treatment for that tumor type. This is moving the field of precision oncology forward, with a great impact in breast cancer patients' lives.

REFERENCES

Ambatipudi S, Horvath S, Perrier F et al. DNA methylome analysis identifies accelerated epigenetic ageing associated with postmenopausal breast cancer susceptibility. *Eur J Cancer* 75;2017:299–307.

Ariazi EA, Taylor JC, Black MA, Nicolas E, Slifker MJ, Azzam DJ, and Boyd J. A new role for ERα: Silencing via DNA methylation of basal, stem cell, and EMT genes. *Mol Cancer Res* 15(2);2017:152–64.

Baylin SB, and Jones PA. A decade of exploring the cancer epigenome—biological and translational implications. *Nat Rev Cancer* 11(10);2011:726–34.

Birgisdottir V, Stefansson OA, Bodvarsdottir SK, Hilmarsdottir H, Jonasson JG, and Eyfjord JE. Epigenetic silencing and deletion of the BRCA1 gene in sporadic breast cancer. *Breast Cancer Res* 8(4);2006:R38.

Bryant HE, Schultz N, Thomas HD, Parker KM, Flower D, Lopez E, Kyle S, Meuth M, Curtin NJ, and Helleday T. Specific killing of BRCA2-deficient tumours with inhibitors of poly(ADP-ribose) polymerase. *Nature* 434(7035);2005:913–7.

Cai F, Ge I, Wang M, Biskup E, Lin X, and Zhong X. Pyrosequencing analysis of BRCA1 methylation level in breast cancer cells. *Tumour Biol* 35(4);2014:3839–44.

Cancer Genome Atlas Network. Comprehensive molecular portraits of human breast tumours. *Nature* 490(7418);2012:61–70.

Cho SH, Jeon J, and Kim SI. Personalized medicine in breast cancer: A systematic review. *J Breast Cancer* 15(3);2012:265–72.

Choi JY, James SR, Link PA, McCann SE, Hong CC, Davis W, Nesline MK, Ambrosone CB, and Karpf AR. Association between global DNA hypomethylation in leukocytes and risk of breast cancer. *Carcinogenesis* 30(11);2009:1889–97.

Foedermayr M, Sebesta M, Rudas M et al. BRCA-1 methylation and TP53 mutation in triple-negative breast cancer patients without pathological complete response to taxane-based neoadjuvant chemotherapy. *Cancer Chemother Pharmacol* 73(4);2014:771–8.

Fujii S, Yamashita S, Yamaguchi T, Takahashi M, Hozumi Y, Ushijima T, and Mukai H. Pathological complete response of HER2-positive breast cancer to trastuzumab and chemotherapy can be predicted by HSD17B4 methylation. *Oncotarget* 8(12);2017:19039–48.

Hinshelwood RA, and Clark SJ. Breast cancer epigenetics: Normal human mammary epithelial cells as a model system. *J Mol Med (Berl)* 86(12);2008:1315–28.

Hinshelwood RA, Huschtscha LI, Melki J et al. Concordant epigenetic silencing of transforming growth factor-beta signaling pathway genes occurs early in breast carcinogenesis. *Cancer Res* 67(24);2007:11517–27.

Li SY, Sun R, Wang HX, Shen S, Liu Y, Du XJ, Zhu YH, and Jun W. Combination therapy with epigenetic-targeted and chemotherapeutic drugs delivered by nanoparticles to enhance the chemotherapy response and overcome resistance by breast cancer stem cells. *J Control Release* 205;2015:7–14.

Liu J, Zhang X, Liu A et al. Altered methylation of glucosylceramide synthase promoter regulates its expression and associates with acquired multidrug resistance in invasive ductal breast cancer. *Oncotarget* 7(24);2016:36755–66.

Locke WJ, and Clark SJ. Epigenome remodeling in breast cancer: Insights from an early *in vitro* model of carcinogenesis. *Breast Cancer Res* 14(6);2012:215.

Locke WJ, Zotenko E, Stirzaker C, Robinson MD, Hinshelwood RA, Stone A, Reddel RR, Huschtscha LI, and Clark SJ. Coordinated epigenetic remodelling of transcriptional networks occurs during early breast carcinogenesis. *Clin Epigenetics* 7(1);2015:52.

Lu DG, Ma YM, Zhu AJ, and Han YW. An early biomarker and potential therapeutic target of RUNX 3 hypermethylation in breast cancer, a system review and meta-analysis. *Oncotarget* 8(13);2017:22166–74.

Mathe A, Wong-Brown M, Locke WJ, Stirzaker C, Braye SG, Forbes JF, Clark SJ, Avery-Kiejda KA, and Scott RJ. DNA methylation profile of triple negative breast cancer-specific genes comparing lymph node positive patients to lymph node negative patients. *Sci Rep* 6;2016:33435.

Merino VF, Nguyen N, Jin K et al. Combined treatment with epigenetic, differentiating, and chemotherapeutic agents cooperatively targets tumor-initiating cells in triple negative breast cancer. *Cancer Res* 76(7);2016:2013–24.

Min A, Im S, Kim DK et al. Histone deacetylase inhibitor, suberoylanilide hydroxamic acid (SAHA), enhances anti-tumor effects of the poly (ADP-ribose) polymerase (PARP) inhibitor olaparib in triple-negative breast cancer cells. *Breast Cancer Res* 17;2015:33.

Nayak SR, Harrington E, Boone D et al. A role for histone H2B variants in endocrine-resistant breast cancer. *Horm Cancer* 6(5–6);2015:214–24.

Polyak K. Heterogeneity in breast cancer. *J Clin Invest* 121(10);2011:3786–8.

Pucci S, Zonetti MJ, Fisco T, Polidoro C, Bocchinfuso G, Palleschi A, Novelli G, Spagnoli LG, and Mazzarelli P. Carnitine palmitoyl transferase-1A (CPT1A): A new tumor specific target in human breast cancer. *Oncotarget* 7(15);2016:19982–96.

Sehl ME, Henry JE, Storniolo AM, Ganz PA, and Horvath S. DNA methylation age is elevated in breast tissue of healthy women. *Breast Cancer Res Treat* 2017 Mar 31 (Epub ahead of print).

Severi G, Southey MC, English DR, Jung CH, Lonie A, McLean C, Tsimiklis H, Hopper JL, Giles GG, and Baglietto L. Epigenome-wide methylation in DNA from peripheral blood as a marker of risk for breast cancer. *Breast Cancer Res Treat* 148(3);2014:665–73.

Shan M, Yin H, Li J et al. Detection of aberrant methylation of a six-gene panel in serum DNA for diagnosis of breast cancer. *Oncotarget* 7(14);2016:18485–94.

Stearns V, Zhou Q, and Davidson NE. Epigenetic regulation as a new target for breast cancer therapy. *Cancer Invest* 25(8);2007:659–65.

Stefansson OA, and Esteller M. Epigenetic modifications in breast cancer and their role in personalized medicine. *Am J Pathol* 183(4);2013:1052–63.

Stone A, Zotenko E, Locke WJ et al. DNA methylation of oestrogen-regulated enhancers defines endocrine sensitivity in breast cancer. *Nat Commun* 6;2015:7758.

Tang Q, Cheng J, Cao X, Surowy H, and Burwinkel B. Blood-based DNA methylation as biomarker for breast cancer: A systematic review. *Clin Epigenetics* 8;2016:115.

Terry MB, McDonald JA, Wu HC, Eng S, and Santella RM. Epigenetic biomarkers of breast cancer risk: Across the breast cancer prevention continuum. *Adv Exp Med Biol* 882;2016:33–68.

Uehiro N, Sato F, Pu F, Tanaka S, Kawashima M, Kawaguchi K, Sugimoto M, Saji S, and Toi M. Circulating cell-free DNA-based epigenetic assay can detect early breast cancer. *Breast Cancer Res* 18(1);2016:129.

van Veldhoven K, Polidoro S, Baglietto L et al. Epigenome-wide association study reveals decreased average methylation levels years before breast cancer diagnosis. *Clin Epigenetics* 7;2015:67.

Walsh L, Gallagher WM, O'Connor DP, and Ní Chonghaile T. Diagnostic and therapeutic implications of histone epigenetic modulators in breast cancer. *Expert Rev Mol Diagn* 16(5);2016:541–51.

Wu S, Luo Z, Yu PJ, Xie H, and He YW. Suberoylanilide hydroxamic acid (SAHA) promotes the epithelial mesenchymal transition of triple negative breast cancer cells via HDAC8/FOXA1 signals. *Biol Chem* 397(1);2016:75–83.

Yamashita N, Tokunaga E, Kitao H et al. Epigenetic inactivation of BRCA1 through promoter hypermethylation and its clinical importance in triple-negative breast cancer. *Clin Breast Cancer* 15(6);2015:498–504.

Zhu X, Shan L, Wang F et al. Hypermethylation of BRCA1 gene: Implication for prognostic biomarker and therapeutic target in sporadic primary triple-negative breast cancer. *Breast Cancer Res Treat* 150(3);2015:479–86.

8

Precision medicine in ovarian carcinoma

SHAILENDRA DWIVEDI, PURVI PUROHIT, RADHIEKA MISRA, JEEWAN
RAM VISHNOI, APUL GOEL, PUNEET PAREEK, SANJAY KHATTRI, PRAVEEN
SHARMA, SANJEEV MISRA, AND KAMLESH KUMAR PANT

INTRODUCTION

Advancement in preventing, screening, and therapy of ovarian carcinoma has been affected by the fact that ovarian cancer is neither a frequent nor a rare disease. The risk of developing this disease is 1 in 70 and the occurrence is 1 in 2500 for postmenopausal women, patients mostly diagnosed after crossing 50 years of age. It is fifth most common reason of cancer-linked death in women. The probable yearly incidence of this cancer globally is over 200,000 individuals, with about 125,000 deaths (Sankaranarayanan and Ferlay, 2006).

Relentless progress in the knowledge of the natural history of the disease and thorough preliminary staging, along with surgical and chemotherapeutic interventions, has enriched the short-term course of ovarian carcinoma. Yet, despite such developments, most patients revert after primary treatment and succumb to disease advancement. Surgical management of advanced-stage ovarian cancer has long been a central view of genuine management of the patients suffering from this disease, although complete elimination is not possible. Though the order of chemotherapy and surgical intercession is disputed, there is the comprehensive attitude that integration of the two modalities signifies the best primary strategy for women with metastatic disease.

Precision medicine focuses on the designing of the medical treatment to the individual characteristics of each patient. Precision medicine depends on the broad exploration of individual molecular database of genomics, epigenetics, proteomics, and metabolomics, as well as in silico tactics to acquire a thorough knowledge of the association between the control of gene(s) (functional protein) and disease status. This advancement relies on modern cutting-edge technologies of real-time PCR, microarray, and next-generation sequencing (NGS) having capabilities to explore the molecular characteristics in short time.

As we are aware that any drug has its beneficial effects to some individuals while others fail to show its response, now it is established that this fact depends on molecular characteristics of individuals. Each individual or group having an exclusive phenotype (like disease) is mainly due to manifestation of its molecular characteristics, that is, variations (mutations) in gene make up or expression profile, so for targeting any disease, our first goal is to identify such changes and then we can observe the suitability of any drug. Thus we will be able to sub-group them into beneficial or null or negative response groups. The control of gene expression can also take place at the phases of translation and posttranslation modification. Recent cutting-edge in genomics, transcriptome profiling, and epigenetic fingerprinting have been useful to cancer research and can probably be incorporated into a cancer systems biology approach for cultivating cancer medicine.

ADVANCEMENT OF CUTTING EDGE TECHNOLOGY FOR ACCURATE DIAGNOSIS

The arena of oncology is relentlessly expanding mainly due to the utilization of enormous funds and improvements in the basic sciences. Modern molecular diagnostic techniques and tools of genomics, transcriptomics and proteomics have been widely exploited in diagnosis of various cancers. Now automated and more advanced technology like PCR, real-time PCR, etc. have improved the amplification and detection of nucleic acid sequences.

These automated and robotic platforms have delivered precision of not only in assay efficiency but also in quality control of the tests and have added calibration of traditional biomarkers. It is documented since Gregor Mendel that factors are responsible for expression of characters and after the launching of human genome, the picture of genotype to phenotype connection seems more bright and dynamic. The purposes of genes and proteins have been reflected by postgenomics advanced technologies such as expression profiling by DNA microarray, proteomics, and genetic variance analysis, coupled with bioinformatics. Now, genetics has become the dynamic force in medical research and is now ready for amalgamation into medical practice.

Genomics

Genomics is an interdisciplinary arena of science concentrating on genomes. A genome is actually the study of a whole set of DNA within a single cell of an organism, and as such genomics is a division of molecular biology related with the structure, function, evolution, and mapping of genomes. The genetics of humans fundamentally explains their individual characteristics. Delicate variances found in the genetic makeup of a population generate the diverse characters perceived in community.

Further, differences in a genetic frame among a population result into different gene isoforms that may consequently alter gene function producing phenotypic variations in the form of characteristics either protective or more susceptible for any disease. These are normally denoted as mutated genes. Similarly like other several diseases and cancer in ovarian cancer, mutated genes inherited from a parent can make a person more risk or susceptible of developing the disease in their lifespan. Though, inherited mutations only interpret for a small fraction of (20%–25%) all diagnosed cases in ovarian cancer.

According to current predictions of 2017 by The American Cancer Society, about 22,440 women will receive a new diagnosis of ovarian cancer and about 14,080 women will die from ovarian cancer. Thus, it ranks fifth in cancer deaths among women, responsible for more deaths than any other cancer of the female reproductive system. So sporadic, scattered unpredictable gene alterations or deregulated gene expression would be the main sources for maximum ovarian cancer patients.

GENOME AND GENETIC CHARACTERIZATION TOOLS AND TECHNIQUES

In 2003 with the publication of human genome draft, a new revolutionary phase of molecular biotechnology comes into existence, but still the value of conventional tools and techniques is not outdated. Validated mutations still screened by RFLP, probe based method or SYBR green-based approaches by utilizing real time PCR, or by micro-array. As discussed earlier, cancer has a tremendously multifaceted etiology and in the course of the development of the disease several number of mutations can occur. Thus, genomic sequencing has a great value but due to high cost still few laboratories process the sample by real-time polymerase chain reaction (PCR), microarray or by sanger sequencing. Genomic sequencing has made it possible to screen any mutation or variation if it has any preexistence that would make it more susceptible to develop cancer in their lifetime. For instance, mutations found in the BRCA1 and BRCA2 genes upsurge a woman's susceptibility for developing ovarian cancer in her lifetime to approximately 40% and 18%, respectively. Further, these high throughput sequencing platforms have the capacity to screen novel sporadic mutations in a very short time and with more precision.

Genotyping or mutation analysis

Restriction fragment length polymorphism (RFLP) is a technique able to detect variation in homologous DNA sequences that can be screened by the presence of fragments of variable lengths after digestion of the DNA by using precise restriction endonucleases. These fragments of varied length could be separated on the basis of molecular size in agarose gel electrophoresis.

Further, these fragments could be utilized in characterizing unique blotting patterns after hybridization with a complementary probe labeled with some fluorescent dye. RFLP probes are commonly utilized in genome sequence mapping and in variation analysis (genotyping, hereditary disease diagnostics, etc.). These techniques can be generally categorized into three subgroups:

1. RFLP, enzymatic methods—RFLP analysis was traditionally the first technique broadly utilized in investigating the variations in restriction enzyme sites, leading to the gain or loss of restriction events. Later, several enzymatic methods for mutation screening have been developed. These procedures employ the activity of resolvase enzymes T4 endonuclease VII and, in recent times, T7 endonuclease I to digest heteroduplex DNA made by annealing wild-type and mutant DNA. Digested fragments specify the presence and the position of any mutations. One more enzymatic method for mutation exploration is the oligonucleotide ligation assay. In this method, two oligonucleotides are hybridized to complementary DNA fragments at locations of probable mutations. The oligonucleotide primers are customized such that the 3'-end of the one primer is instantly contiguous to the 5'-end of the other primer. Thus, if the first primer matches entirely with the target DNA, then the primers can be ligated by DNA ligase. On the other hand, if a disparity happens at the 3'-end of the first primer, then no ligation products will be found.

2. Electrophoretic-based techniques—This class is well known by several diverse methods planned for detection of known or unknown mutations, centered on the different electrophoretic mobility of the mutant alleles, under denaturing or nondenaturing conditions. Single-strand conformation polymorphism (SSCP) and heteroduplex analysis (HDA) were among the main approaches considered to detect molecular defects in genomic loci. In arrangement with capillary electrophoresis, SSCP and HDA investigation now offer an outstanding, modest, and fast mutation-finding platform with low processing costs and, most interestingly, the capability of easily being automated, thus providing high-throughput analysis of a patient's DNA. Likewise, denaturing and temperature gradient gel electrophoresis (DGGE and TGGE, respectively) can be utilized equally well for mutation screening. In this method, electrophoretic mobility differences between a wild type and mutant allele can be visualized in a gradient of denaturing agents, such as urea and formamide, or of increasing temperature. Finally, a progressively utilized mutation detection technique is the two-dimensional gene scanning, based on two-dimensional electrophoretic separation of amplified DNA fragments, as per their size and

base pair sequence. The latter involves DGGE, following the size separation step.

3. Solid phase-based techniques/hybridization or blotting techniques—These established methods involve the foundation for most of the contemporary mutation screening technologies, because they do not require any extra effort as of being fully automated and therefore are extremely praised for high throughput mutation detection or screening. In the 1970s there was advancement in nucleic acid hybridization techniques, which is centered on the pairing of two complementary nucleotide strands. This pairing is primarily due to envelopment of hydrogen, thus duplex or hybrid results. The hybrids may be consequent of DNA-DNA, RNA-RNA, or DNA-RNA, thus the single-stranded molecule may be DNA or RNA in which one nucleic acid strand (the probe) originates from an organism of recognized identity and the other strand (the target) originates from an unidentified organism to be detected or identified.

A quick, precise, and appropriate method for the discovery of known mutations is reverse dot-blot, originally established by Saiki et al. (1989), and executed for the screening of b-thalassemia mutations. The crux of this technique is the application of oligonucleotides, bound to a membrane, as hybridization targets for amplified DNA. The extra benefit of this technique is that one membrane strip can be applied to screen many various known mutations in a single individual (a one strip-one patient type of assay), the potential of automation, and the ease of analysis of the results, using a classical avidin-biotin system. Though this technique cannot be utilized for the uncovering of unknown mutations, constant expansion has given rise to allele-specific hybridization of amplified DNA (PCR-ASO) on filters and newly extended on DNA oligonucleotide microarrays for high throughput mutation analysis (Gemignani et al., 2002; Chan et al., 2004).

Real-time polymerase chain reaction (PCR): Semiquantitative approach

Higuchi first demonstrated methodological improvement in the form of "real-time PCR" discovered approximately two decades earlier, which is modest, quantitative evaluation for any amplifiable DNA sequence. This technique uses fluorescent labeled probes to screen, approve, and quantify the PCR yields as they are being produced in real time. In real-time PCR, which is poised on three novel characteristics, as temperature cycling happens substantially faster than in standard PCR assays, hybridization of specific DNA probes take place continuously throughout the amplification reaction and a fluorescent dye is attached to the probe and fluoresces only when hybridization occurs. No post-PCR handling of amplified products is needed. The generation of amplified products is perceived automatically by continuous monitoring of fluorescence. In recent years, several commercial automated real-time PCR systems have been available (Light Cycler & TaqMan). In these systems, such as the Light Cycler™ and the Smart Cycler®, the real-time fluorescence monitoring is accomplished by using fluorescent dyes such as SYBR-Green I, which binds nonspecifically to double-stranded DNA produced during the PCR amplification. Others, such as the TaqMan, use florescent probes that bind exactly to amplification target sequences.

Microarrays

A microarray is an assembly of enriched characteristics of a microscopic system, in which normally DNA is hybridized with target molecules for quantitative (gene expression) or qualitative (diagnostic) test of enormous quantities of genes concurrently or to genotype multiple loci of a genome. Each DNA spot contains approximately picomoles (10^{-12} moles) of a specific DNA sequence, known as probes (or reporters). Broadly due to improvements in fabrication, robotics, and bioinformatics, microarray technology has improved in terms of efficiency, resolution power, robustness, sensitivity, and specificity. These enhancements have permitted the transition of microarrays from strictly research bench site to bed site in clinical diagnostic applications. Microarrays can be distinguished on the basis of features such as the nature of the probe, the solid-surface support applied, and the specific method utilized for probe addressing and/or target detection. An effective hybridization episode between the labeled target and the immobilized probe will consequently lead to an increase of fluorescence intensity over a background level, which can be studied by using a fluorescent scanner (Miller, 2009). The entire strength of the signal, from a spot (feature), can be determined by the amount of target sample binding to the probes that exist on that spot. Microarrays use relative quantization in which the intensity of characteristics is compared to the intensity of the same characteristics under a different condition, and the identity of the feature is known by its position. In situ synthesized arrays are exceedingly high-density microarrays that use oligonucleotide probes, of which Gene Chips (Affymetrix, Santa Clara, California) are the most commonly known. In addition to the printed oligonucleotide arrays discussed earlier, the oligonucleotide probes are blended directly on the surface of the microarray, which is typically a 1.2 cm^2 quartz wafer. As in situ synthesized probes are usually short (20–25 bp), multiple probes per target are involved to improve sensitivity, specificity, and statistical accuracy. As with in situ hybridized microarrays, Bead Arrays (Illumina, San Diego, California) offer a spotted substrate for the high-density detection of target nucleic acids. Though, instead of glass slides or silicon wafers as direct substrates, Bead Arrays rely on 3 μm silica beads that arbitrarily self-assemble onto one of two available substrates: the Sentrix Array Matrix (SAM) or the Sentrix Bead Chip (Oliphant et al., 2002; Fan et al., 2006). Nothing like the other array, the special characteristics of Bead Arrays depend on passive transport for the hybridization of nucleic acids. One additional type of array, electronic microarrays, exploits active hybridization via electric fields to control nucleic acid transport. Microelectronic cartridges (NanoChip 400; Nanogen, San Diego, California) modify complementary metal-oxide semiconductor technology for the electronic addressing of nucleic acids.

Traditional methods and next-generation sequencing

Two dissimilar approaches for sequencing DNA were established in 1977: the chain termination method and the chemical degradation method. In 1976–1977, A. Maxam and W. Gilbert developed a DNA sequencing technique based on chemical alteration of DNA and successive cleavage at specific bases. Maxam–Gilbert sequencing swiftly became more recognized, as purified DNA could be utilized directly, though the initial Sanger technique required that every read initiated be cloned for making of single-stranded DNA. Though, with the enhancement of the chain-termination method, Maxam–Gilbert sequencing has dropped out of

favor due to its practical complexity barring its utilization in standard molecular biology kits, wide use of harmful chemicals, and complications with scale-up. Each of four reactions (G, A + G, C, and C + T) is separated into different bands. Thus a series of labeled fragments are produced, from the radiolabeled end to the first cut site in each molecule. The fragments in the four reactions are electrophoresed side by side in denaturing acrylamide gels for size separation. To see the fragments, the gel is visualized onto x-ray film for autoradiography, yielding a series of dark bands each corresponding to a radio labeled DNA fragment, from which the sequence may be inferred, also sometimes known as the "chemical sequencing" method.

Another method, known as the Sanger sequencing method, is centered on the principle that single-stranded DNA molecules that vary in length by just a single nucleotide can be separated from one another by using polyacrylamide gel electrophoresis. The stable laser beam excites the fluorescently labeled DNA bands and the light radiated is noticed by sensitive photodetectors. Automated DNA sequencing is freely automated by a variation of Sanger's sequencing method in which dideoxynucleosides castoff for each reaction is labeled with a differently colored fluorescent tag. Several commercial and noncommercial software packages can trim low-quality DNA traces automatically. DNA sequencing is still painstakingly the golden standard and the final experimental procedure for mutation detection. However, the costs for the initial investment and the difficulties for standardization and interpretation of ambiguous results have restricted its use to basic research laboratories. Because the chain-terminator tactics or Sanger approach is more efficient and this procedure involves fewer toxic chemicals and lower amounts of radioactivity than the method of Maxam and Gilbert, it swiftly became the method of choice.

The declaration of the first draft sequence of the human genome in February 2001 (International Human Genome Sequencing Consortium, 2001; Venter et al., 2001) and then with the genomic sequence of other organisms, molecular biology has moved into a new era with exclusive opportunities and challenges. Technology has improved rapidly in the last two decades and new mutation-detection techniques have been claimed, whereas old methodologies have advanced to fit the increasing need for automated and high-throughput screening.

The chromatographic screening of polymorphic changes of disease-causing mutations by utilizing denaturing high-performance liquid chromatography (DHPLC) is one of the novel technologies that occurred. DHPLC discloses the existence of a genetic variation by the differential retention of homo- and heteroduplex DNA on reversed-phase chromatography under partial denaturation. Single-base substitutions, deletions, and insertions can be identified effectively by ultraviolet (UV) or fluorescence monitoring within 2 to 3 minutes in unpurified PCR products as large as 1.5 kilobases. These characteristics, together with its low cost, make DHPLC one of the most potent techniques for mutational analysis.

HIGH-THROUGHPUT SEQUENCING TECHNOLOGIES

Sanger sequencing and added sequence analysis methods have augmented sequencing outputs by orders of magnitude and driven down per-base sequencing cost considerably (Church, 2006; Roukos, 2010). It is now normally estimated that NGS will permit the in-depth characterization of the cancer cell genome and further improvement in the fields of molecular pathology and personalized medicine for patients with cancer.

Illumina's Genome Analyzer

The massively parallel signature sequencing advanced by Lynx Therapeutics was the second or next-generation approach to DNA sequencing. The elementary Lynx Therapeutics platform was a microsphere (bead)-based system that discovers nucleotides in groups of four via an adapter ligation and adapter decoding strategy using reversible dye terminators (Mardis, 2008). Lynx Therapeutics (Hayward, California) merged with Solexa, which was later acquired by Illumina. This short-read sequencing technique is today incorporated into a fluidic flow cell design (HiSeq and Genome Analyzer systems, Illumina, San Diego, California) with eight individual lanes. The flow cell surface is established with capture oligonucleotide anchors, which hybridize the properly modified DNA segments of a sequencing library generated from a genomic DNA sample.

By a process called bridge amplification, engaged DNA templates are amplified in the flow cell by bending over and hybridizing to a contiguous anchor oligonucleotide primer (Mardis, 2008).

Genuine sequencing is accomplished by hybridizing a primer complementary to the adapter sequence, then cyclically adding DNA polymerase and a mixture of four differently colored fluorescent reversible dye terminators to the captured DNA in the flow cell.

By using this technique, nonaltered DNA fragments and unincorporated nucleotides are washed away, while apprehended DNA fragments are extended one nucleotide at a time. After each nucleotide-coupling cycle takes place, the flow cell is scanned, and digital images are developed to record the locations of fluorescently labeled nucleotide amalgamations. Next to imaging, the fluorescent dye and the terminal 3′ blocker are chemically detached from the DNA before the next nucleotide coupling cycle.

The Illumina method is the most extensively used NGS platform, but it is restricted by a relatively low multiplexing capability (Erlich et al., 2008). The Illumina system has been applied in programs for gene innovation, whole exome analysis, and SNP detection by resequencing (Margulies et al., 2005).

Roche second-generation sequencing

First presented in 2005, the 454 Genome Sequencer FLX Titanium System (Roche, Branford, Connecticut) NGS platform regulated by highly parallel PCR reactions happening in minute emulsions consists of a primer-coated bead with a single captured DNA template covered with the DNA polymerase, oligonucleotide ligation after PCR amplification in an emulsion primers, and nucleoside triphosphates (NTPs) vital for PCR in an oil droplet. PCR amplification results in each bead becoming covered with a single DNA amplicon. The emulsions are cracked, and the DNA-coated beads are loaded onto an array of picoliter wells for the sequencing reaction (Tawfik and Griffiths, 1998).

Pyrosequencing is accomplished over the picoliter well array and the nucleotide additions are visualized and located by a fiberoptic-coupled imaging camera. The system provides longer read lengths than other NGS technologies, a strength of this system (Margulies et al., 2005).

Helicos HeliScope: Single-molecule sequencing

The HeliScope platform includes fragmenting the sample DNA and performing polyadenylation at the 3' ends of the fragments. Denatured polyadenylated strands are occupied by hybridization to poly (dT) oligonucleotides immobilized on a flow-cell surface. This technique was the first process to effectively perform single-molecule sequencing (Braslavsky et al., 2003). The flow cell is cyclically swamped up with the fluorescently labeled deoxynucleoside triphosphates (dNTPs) in the existence of DNA polymerase, which incorporates nucleotides from the oligo-dT primer. The flow cell is imaged at each cycle using a CCD camera, allowing the documentation of the location of each nucleotide incorporation event. As in other systems, the fluorescent label is cut and washed away before each succeeding sequencing cycle. (Braslavsky et al., 2003; Margulies et al., 2005). The HeliScope system is precise enough to provide the most nonbiased DNA sequence, which is its power, although relative to competing NGS platforms, it has relatively high NTP incorporation error rates (Margulies et al., 2005).

SOLiD sequencing: Sequencing-by-ligation approach

SOLiD sequencing (Supported Oligonucleotide Ligation and Detection) is based on DNA ligase–mediated oligonucleotide ligation after PCR amplification in an emulsion format. The primers in SOLiD NGS are progressively offset to allow the adapter bases to be sequenced when utilized in conjunction with the color-space coding for defining the template sequence by deconvolution. Fluorescent signals are taken by CCD camera imaging before enzymatic cleavage of the ligated probes and, after washing, repeating the sequencing process. The SOLiD tactic has been used in applications similar to the Illumina NGS, including whole genome sequencing, whole exome capture, and sequencing and SNP finding. Strengths of the SOLiD approach include reduction in sequencing error rates relative to the Illumina NGS by using two-base encoding. A drawback of the SOLiD system has been its relatively long run times and complex analysis requirements (Margulies et al., 2005).

Ion Torrent sequencing: Ion semiconductor sequencing

This platform was picked up by Life Technologies, which suggests that this postlight sequencing technology has the utmost significant benefit of being the first platform to eliminate the cost and effort

associated with the four-color optical detection, currently used in all other NGS platforms. The Ion Torrent method relies on standard DNA polymerase sequencing with unchanged dNTPs but uses semiconductor-based screening of hydrogen ions liberated during every cycle of DNA polymerization. Every nucleotide incorporated into the budding complementary DNA strand causes the release of a hydrogen ion that is detected by a hypersensitive ion sensor. The initial Ion Torrent system has relatively low parallelism, so it has a tendency to be concentrated on short sequence determination of mutation hot spots throughout the genome.

All NGS platforms have high entry costs, but all also have the probability to dramatically reduce the cost of comprehensive genomic profiling of cancer cells in the forthcoming years. Some techniques suggest speed, such as the 454 Pyro sequencing and Ion Torrent platforms, but, compared with the Illumina and SOLiD platforms, may not be as well suited for clinical somatic tumor DNA sequencing due to their relatively restricted capacities for supporting highly parallel, deep sequencing.

Somatic and germ-line mutation/ variance studied in ovarian carcinoma

Liede et al. (1998) acknowledged in 1 out of 8 of the ovarian cancer cases, the 185delAG mutation in the BRCA1 gene (113705.0003) segregated with the cancer. Liede et al., established that site-specific ovarian cancer families perhaps signify a variant of the breast-ovarian cancer syndrome and have characteristic mutation in either BRCA1 or BRCA2.

Stratton et al. (1999) showed a population-based study to conclude the involvement of germ-line mutations, they reported 2 out of 101 women with invasive ovarian cancer had germ-line mutations in the MLH1 gene (120436), and no germ-line mutations were recognized in any of the other genes explored, including BRCA1, the ovarian cancer cluster region (nucleotides 3139-7069) of BRCA2, and MSH2. This study concluded that germ-line mutations in BRCA1, BRCA2, MSH2, and MLH1 add to only a lesser of cases of early-onset epithelial ovarian cancer (Stratton et al., 1999).

Cesari et al. (2003) recognized the whole PARK2 gene (602544) within a Loss of Heterozygosity (LOH) region on chromosome 6q25-q27. LOH investigation of 40 malignant breast and ovarian tumors showed a shared minimal region of loss, including the markers D6S305 (50%) and D6S1599 (32%), both of which are located within the PARK2 gene. Further, they found two somatic truncating deletions in the PARK2 gene (see, e.g., 602544.0016) in 3 of 20 ovarian cancers. This report proved that PARK2 may work as a tumor suppressor gene (Cesari et al., 2003).

Sellar et al. (2003) demonstrated that D11S4085 on 11q25 is positioned in the second intron of the OPCML gene (600632). OPCML was commonly somatically disabled in epithelial ovarian cancer tissue by allele loss and by CpG island methylation. OPCML behaves like a tumor suppressor gene as proved both in vitro and in vivo. A somatic missense mutation from an individual with epithelial ovarian cancer represents perfect proof of loss of function (600632.0001) (Sellar et al., 2003).

The Cancer Genome Atlas Project has completed whole exome sequencing on ovarian cancer (Cancer Genome Atlas Research Network 2011). Screening DNA from 316 high-grade serous ovarian cancer patients and compared with normal for 19,356 somatic mutations (about 61 per tumor) were interpreted. High-grade serous ovarian cancer was recognized by TP53 mutations in almost all tissues (96%). BRCA1 and BRCA2 were mutated in 22% of tumors, as the study considered a mixed type of germ-line and somatic mutations. Additional significant mutated genes, including NF1, RB1, FAT3, CSMD3, GABRA6, and CDK12, occurred in 2%–6% of cases. Mutational analysis also presented that mutations in BRAF, PIK3CA, KRAS, and NRAS may be central forces in high-grade serous carcinoma. For instance, clear cell types have few TP53 mutations but have recurrent ARID1A and PIK3CA mutations. Although CTNNB1, ARID1A, and PIK3CA mutations were commonly seen in endometrioid ovarian cancer histology, KRAS mutations were prevalent in mucinous types (Wiegand et al., 2010).

Bell et al. (2011) completed a study based on 316 HGS-OvCa samples and compared with normal samples for each individual. The study was based on captured 180,000 exons from 18,500 genes totaling 33 megabases of nonredundant sequence. Enormously parallel sequencing by using the Illumina GAIIx platform or ABI SOLiD 3 platform (80 sample pairs) obtained 14 gigabases per sample bases in total. TP53 was mutated in 303 of 316 samples. BRCA1 and BRCA2 had germ-line mutations in 9% and 8% of cases, respectively, and showed somatic mutations in a further 3% of cases. Further, this group also demonstrated the presence of other mutated genes: RB1, NF1, FAT3, CSMD3, GABRA6, and CDK12.

Recently a study reported 11,479 somatic mutations in the 142 fresh TCGA cases. These mutations were manually reviewed, resulting in a total of 27,280 mutations in 429 cases. TP53, NF1, RB1, CDK12(CRKRS), and BRCA14, as well as the novel SMG, KRAS. BRCA2 and RB1CC1 were reported significantly associated. This group also identified 4 NRAS mutations; 3 NF2 mutations; and 3, 8, and 10 mutations in the identified tumor suppressor genes: ATR, ATM, and APC, respectively. Somatic truncation mutations were also detected in histone modifier genes including ARID1A, ARID1B, ARID2, SETD2, SETD4, SETD6, JARID1C, MLL, MLL2, and MLL3 as well as the DNA excision repair gene ERCC6 (Kanchi et al., 2014). Tables 8.1 and 8.2 have shown the various mutations as characterized by various researchers.

Table 8.1 Mutations predicted and verified by various studies

Gene	Number of mutations proposed	Number of mutations verified	Reference
TP53	302	294	Bell et al. (2011), Nick et al. (2015)
BRACA1	11	10	Bell et al. (2011), Couch et al. (2013)
BRACA2	10	10	Bell et al. (2011), Kanchi et al. (2014)
CSMD3	19	19	Bell et al. (2011), Kanchi et al. (2014)
NF1	13	13	Bell et al. (2011), Sangha et al. (2008)
CDK12	9	9	Bell et al. (2011), Dong, Lu, and Lu (2016)
FAT3	19	18	Bell et al. (2011), Kanchi et al. (2014)
GABRA6	6	6	Bell et al. (2011), Kanchi et al. (2014)
RB1	6	6	Bell et al. (2011), Kanchi et al. (2014)

Table 8.2 Prevalence of somatic mutations in epithelial ovarian cancer

Gene mutation	Epithelial ovarian cancer overall	High grade serous (type II)	Low grade serous (type I)	Clear cell (type I)	Endometrioid (type I)	Mucinous (type I)
BRAF	11% (Kurman and Shih, 2011)	<1% (Bell et al., 2011)	24%–33% (Nakayama et al., 2006; Singer et al., 2003)	1% (Kuo et al., 2009; Zannoni et al., 2016)	24% (Singer et al., 2003)	50%–75% (Gemignani et al., 2003)
KRAS	11% Kurman and Shih, (2011)	<1% (Bell et al., 2011)	33% (Nakayama et al., 2006; Singer et al., 2003)	<1%–7% (Kuo et al., 2009; Singer et al., 2003)	<1% (Singer et al., 2003)	50%–75% (Gemignani et al., 2003)
PIK3CA	6.7% (Levine et al., 2005; Campbell et al., 2005)	<1% (Bell et al., 2011)	5% (Nakayama et al., 2006; Wu et al., 2013)	20%–33% (Kuo et al., 2009)	20%–33% (Cho and Shih, 2009; Samuels and Waldman, 2010)	Rare
PTEN	20% (Kurman and Shih, 2011)	<1% mutation (Bell et al., 2011)	20% (Landen et al., 2008)	<1%–5% (Kuo et al., 2009; Willner et al., 2007)	20%–31% (Kurman and Shih, 2011; Willner et al., 2007; McConechy et al., 2014)	Rare (Obata et al., 1998)

Transcriptomics

For proper functioning of our biological system, genes expression is needed in precise quality and quantity so that smooth processing of all activities could be maintained. Modification or change may happen in the controlling region of the particular gene so the product of that gene may be abnormal in quantity, thus deregulation of governing function may occur. Commonly this abrupt regulation may manifest in two ways—nonsense-mediated decay (NMD) or ubiquitin-mediated decay—and may cause carcinogenesis. A few studies suggested that this upregulation of gene expression may advance the stage of the cancer (Dwivedi, Goel, Khattri, et al., 2015; Dwivedi, Goel, Mandhani, et al., 2015; Dwivedi, Singh, et al., 2015)

Exploring a large sample size of cancer specimen can establish a gene expression database that can be used not only for molecular subtyping but also for screening the advancement of the disease, which ultimately could be very fruitful in management of these patients (Dwivedi et al., 2013). Currently, one tactic broadly utilized to assess the gene expression in terms of copy number, that is, a semiquantitative approach by the real-time PCR. In this process, first, isolation of total RNA is required from any sample like blood or tissue and then converted into cDNA by reverse transcriptase. Thus the converted cDNA is quantified by using a fluorescently labeled probe or the SYBR green-based method.

This method permits researchers to evaluate expression of genes related to pathways or pathogenesis of patients, and further help to compare the data with healthy individuals in a very short time. For example, one study looked at gene expression by quantifying the mRNAs of the genes that have a role in the developmental and hormonal regulation of the transmembrane TJ protein, occludin (OCLN), and the cytoplasmic TJ proteins, TJ protein 1 (TJP1) and cingulin (CGN) in bovine granulosa cells (GC) and theca cells (TC) and found altered gene expression of these genes during ovarian cancer (Zhang et al., 2017).

An additional standard technique for transcriptome profiling is microarray assays. Microarray exploration means extracting the total RNA from samples and transforming it into cDNA as real-time PCR. The microarrays are dependent on in situ hybridization of complementary nucleotide strands unlike gene polymerization in real-time PCR. DNA spots are fabricated on the microarray surface and every spot includes a custom deliberated DNA sequence that works as a probe for specific gene detection. The sample has fluorescently labeled cDNA and when this is combined with their corresponding spots on the microarray, a fluorescent signal is radiated based on the quantity of cDNA bound to the probe DNA. Microarray assays can be accomplished in a swift and economical manner that proves it a potent tool for medical transcriptome profiling of patient tumor biopsies and can also detect patient samples for distinctive subtype-specific gene subtypes that can assist in forecast therapy response, tumor progression, and patient prognosis.

Recently, a study by Carrarelli et al. (2017) on myostatin expression compared endometrium in benign (endometriosis, polyps) and malignant (endometrial adenocarcinoma) patients and with healthy women during the menstrual cycle. As myostatin is a growth factor member of the transforming growth factor β superfamily, which is recognized to play main roles in cell proliferation and differentiation. The current data demonstrated for the first time the expression of myostatin in healthy endometrium and aggravated copy number expression in endometriosis and endometrial cancer, proposing myostatin involvement in human endometrial physiology and related pathologies.

Similarly, Wei et al. (2017) reported high expression of ITGA6 (Integrin subunit alpha 6) in 287 ovarian cancer patients of the TCGA cohort, which was significantly associated with poorer progression-free survival. This study provided the basis of drug resistance in ovarian cancer, and integrins could be a probable biomarker for prognosis of ovarian cancer. Table 8.3 shows the recent studies that proved significant gene expression in ovarian carcinoma.

Epigenetic (DNA methylation analysis)

As mentioned earlier, mutations, copy number, and gene expression play a role in development and progression of ovarian carcinoma, but these factors alone are not responsible for carcinogenesis mechanism. So, it is now believed that epigenetic modifications have a role in carcinogenesis. Epigenetics can be defined as the potent permanent and inheritable alteration in gene expression that does not affect DNA sequence but alter the morphology of the gene or chromosome. Epigenetic modifications to the genome generally take place during normal cell cycle regulation. More notably, a study has revealed these modifications tend to occur more commonly than mutations. Epigenetic modifications among cancer cells effect abnormal gene expression via three process: (1) DNA methylation, (2) histone modifications, and (3) noncoding microRNAs. These modifications have shown a significant association with initiation and progression of ovarian cancers.

Two major types of epigenetic controls generally exist: the first ensues at the gene level and a second that arises at the chromosomal level. The first is recognized as genomic methylation in which "methyl" groups is added to definite locus of genes, which can either have active or inactive gene expression. DNA methylation happens among cytosine residues in cytosine–guanine (CpG) dinucleotides, which are frequently dispersed in the CpG-rich regions mentioned as "CpG islands." This type of methylation is accomplished by DNA methyltransferases (DNMTs), which are a family of enzymes that assist to transfer methyl groups onto DNA. In humans, DNMTs are classified into two groups: DNMT1 and DNMT3. Alterations in DNA methylation regulating gene expression are very common and appear in both normal and cancerous cells. As per assumptions, about 80% of CpG dinucleotides in the human genome get methylated during the lifespan (Nguyen et al., 2014).

EPIGENETIC ALTERATIONS IN CANCER

DNA hypermethylation has a role in gene silencing, whereas DNA hypomethylation is known for alteration in gene expression. Both have been reported in malignancy and cancer cells. Commonly, hypermethylated CpG regions within the DNA cause silencing of tumor suppressor genes, demolishing the cell's capability to repair DNA damage, control cell growth, and inhibit proliferation. Whereas, DNA hypomethylation participates in oncogenesis when formerly silenced oncogenes become transcriptionally activated. Furthermore, DNA hypomethylation has a role in triggering

Table 8.3 Recent study on exploration of gene expression in ovarian carcinoma

Gene	Expression	Diagnostic/prognostic/ therapeutics	Types of study	Reference
E74-like factor 5 (ELF5)	Decreased	Therapeutics (Gene Therapy)	Human sample study	Yan et al. (2017)
Receptor for advanced glycation end products (RAGE)	Increased	Diagnostic	Human sample study	Rahimi et al. (2017)
RAP80 (HR-pathway-related gene)	Decreased	Poor prognosis	Human sample study	Romeo et al. (2017)
Musashi-2 (MSI2)	Increased	Therapeutics (paclitaxel resistance), poor prognosis	Human sample study	Lee et al. (2016)
DNA-PKcs, Akt3, and p53	Increased	Poor prognosis	Human sample study	Shin et al. (2016)
Aberrant ALK (anaplastic lymphoma kinase)	Present in ovarian serous carcinoma	Prognosis	Human sample study	Tang et al. (2016)
YAP (autophagy related)	Present	Therapeutics (cisplatin-resistant; protective)	C13 K cell line	L. Xiao et al. (2016)
Urothelial carcinoma associated 1 (UCA1)	Increased	Poor prognosis and therapeutics	Human sample study	Zhang et al. (2016)
Cyclin Y (CCNY)	Increased	Therapeutics	Cell line	H. Liu et al. (2016)
Spalt-like transcription factor 4 (SALL4)	Increased	Poor prognosis	Cell line	Yang, Xie, and Ding (2016)
Carnitine palmitoyltransferase 1A (CPT1)	Increased	Poor prognosis, therapeutics	Cell line	Shao et al. (2016)

latent transposons and thus chromosomal instability take place in specific pericentromeric satellite regions (Sapiezynski et al., 2016).

EPIGENETIC ROLE IN OVARIAN CANCER

Epigenetic modifications have shown their potent promise to be utilizing as biomarkers not for early diagnosis but also to screen the advancement of the disease and prognosis. Epigenetic databases derived from the patient samples along with mutations and gene expression could be used in better management of the cancer. Table 8.4 summarizes the recent studies on epigenetic changes associated with ovarian cancer.

Micro-RNA expression in ovarian cancer

The exclusive biomolecule micro-RNA (miRNA) was revealed by Victor Ambros et al., in 2011

while working on lin-14 gene, which regulates the development of nematodes. Mature miRNA is a class of endogenous, noncoding, single-stranded small RNA, which is composed of about 20–22 nucleotides (Lehrbach and Miska, 2008). miRNA is involved in several physiological processes and is expressed aberrantly in many pathological conditions. These aberrant expressions of miRNA are meticulously linked to the manifestation, development, progression, diagnosis, and prognosis of the human disease. The role for miRNAs in cancer was initially reported from the laboratory of Carlo Croce. A bicistronic miRNA cluster comprising *miR-15a* and *miR-16* at chromosome 13q14 was detected to be mutated, deleted, or have reduced expression in chronic lymphocytic leukemia. Afterward, germ-line mutations in *miR-15a/-16* were screened and it was proved that both of these target antiapoptotic *BCL-2* mRNA (Calin et al., 2006).

Table 8.4 Recent study on exploration of epigenetic changes in ovarian carcinoma

Gene	Methylation/ acetylation	Diagnostic/prognostic/ therapeutics	Reference
GATA	Hypermethylation	Better prognosis	Bubancova et al. (2017)
miR-128-2	Methylation	Progression of the cancer	Jang et al. (2017)
RGS2 (regulator of G-protein signaling 2)	Methylated	Therapeutics (chemoresistance)	Cacan (2017)
4-1BB ligand (4-1BBL/ CD157) and OX-40 ligand (OX-40L/CD252)	Histone deacetylation and DNA methylation	Therapeutics (chemoresistance)	Cacan (2017)
Solute carrier family 6, member 12 (SLC6A12)	Methylation	Poor prognosis and advancement of the cancer	Sung et al. (2017)
HNF1B	Methylation	Poor prognosis and advancement of the cancer	Ross-Adams et al. (2016)
TBX15	Promoter methylation	Therapeutics	Gozzi et al. (2016)
P16INK4a	Methylation	Risk, diagnosis	Xiao et al. (2016)
DAPK1 and SOX1	Promoter methylation	Risk, diagnosis	Kaur et al. (2016)
Cadherin 13 (CDH13), Dickkopf WNT signaling pathway inhibitor 3 (DKK3) and Forkhead box L2 (FOXL2)	Promoter hypermethylation	Diagnostic and advancement of the disease	Y. Xu et al. (2016)

The alterations in miRNA expression can ensue at both the DNA and RNA level. Modifications in Dicer1/Ago2 expression effects global changes in miRNA expression, though change of specific miRNAs in cancer is also very frequent. This may happen via one of many mechanisms, like by germline mutation, deletion, or promoter methylation (Calin et al., 2005). In recent years, several novel micro-RNA, as shown in Table 8.5, have been deciphered and added into database, suggesting their immense potential in ovarian cancer diagnosis, prognosis, and therapeutics.

Proteomics

The proteome is defined as the array of proteins expressed in a particular cell at a fixed set of conditions. In the human proteome more than a million proteins are produced. Initially, proteomic equipment has progressed from gel-based techniques (one- and two-dimensional SDS PAGE) to now gel-free techniques of mass spectrometry.

Reversed-phase protein array (RPPA) is a high-throughput antibody-based method that offers higher sensitivity, quantification, and multiplexing abilities than traditional immunoassay. The Cancer Genome Atlas (TCGA) utilized the RPPA practice in many tumor types with numerous proteins screened by RPPA method, but there were restrictions due to unavailability of antibodies to detect isoform or phosphorylated antibodies with various distinct functions. So, mass spectrometry (MS) is now emerging as a technology of preference for screening various proteins. Currently, electrospray ionization-MS and matrix-assisted laser desorption/ionization (MALDI)-MS are the main systems utilized in exploring protein profiling, screening of posttranslational modifications, and meticulous quantification. Nowadays, mass spectrometry–based proteomics created many milestones in terms of sensitivity, robustness, and consistency.

Additionally, the advancement of quantitative tactics has opened new vistas for exploring

Table 8.5 Recent study on exploration of miRNA expression and their role in ovarian carcinoma

miRNA	Increased/ decreased expression	Involved mechanisms/ pathway	Diagnostic/ prognostic/ therapeutics	Reference
miR-125b	Repression	Cell migration by TGF-β	Therapeutic target	Yang et al. (2017)
let-7a and miR-30c	Deregulation	Increased expression of High-mobility group AT-hook 2 protein (HMGA2) i	Therapeutic target	Agostini et al. (2017)
miR-130b-3p	Increased	Cytidine monophosphate kinase inhibition by the TGF-β signaling pathway	Therapeutic target	Zhou et al. (2017)
miR-27b	Increased	Suppressed ovarian cancer cell migration and invasion by binding with VE-cadherin	Therapeutic target	Liu et al. (2017)
miR-196b	Increased	Invasion activities in recurrent epithelial ovarian carcinoma by regulating the HOXA9 gene	Prognosis/ therapeutic target	Chong et al. (2017)
miR-429	Increased	Suppress zinc finger E-box binding homeobox 1 (ZEB1) and increased cisplatin sensitivity	Prognosis/ therapeutic target	Zou et al. (2017)
miR-490-3p	Increased	Increased cisplatin (CDDP) sensitivity by downregulating ABCC2 expression	Prognosis/ therapeutic target	J. Tian et al. (2017)
let-7e	Decreased	Activation of BRCA1 and Rad51 expression and enhancement of DSB repair, which in turn results in cisplatin resistance	Therapeutic target	Xiao et al. (2017)
miR-221	Increased	Promotes cell proliferation by targeting the apoptotic protease activating factor-1	Diagnostic/ prognostic/ therapeutics	Li et al. (2017)
miR-28-5p	Increased	Progression of ovarian cancer cell cycle, proliferation, migration and invasion, inhibited apoptosis, and induced the process of EMT through inhibition of N4BP1	Diagnostic/ prognostic/ therapeutics	J. Xu et al. (2017)

the protein differential expression, and posttranscriptional and posttranslational modifications in diverse conditions in an endeavor to comprehend the functional significances of modified gene expression. Quantitative proteomics has perceived major revolutions in precision and relative quantification methods, including spectral counting, stable isotope labeling by amino acid in cell culture, isotope-coded affinity tags, and isobaric tags for relative and absolute quantification (iTRAQ). Proteomic methods are now exclusively applied in several areas of ovarian cancer research not only in deciphering mechanism and characterization of the biomarkers (diagnostics and prognostics) but also in searching biomolecules involved in resistance of the therapy (Sapiezynski et al., 2016).

The ovarian cancer patients do not exhibit any specific symptoms during disease initiation and so when they are diagnosed the disease has progressed into advanced stage. Currently for investigation in the early stage, a biomarker with higher sensitivity and specificity is warranted. Any standard

Table 8.6 Current progress in proteomics-based biomarker

Subtype	Markers	Method	Reference
Serous	Wilm's tumor 1 (WT-1)	MALDI-TOF	Zhu et al. (2006)
	Ras-related protein (Rab-3D)	MALDI-TOF	Zhu et al. (2006)
	Mesothelin	LC-MS/MS	Tian et al. (2011)
	ICAM3, CTAG2, p53, STYXL1, PVR, POMC, NUDT11, TRIM39, UHMK1, KSR1, and NXF3	Protein microarrays	Katchman et al. (2017)
Endometroid	Estrogen receptor α (ERα)	RPPA	Sereni et al. (2015)
Clear cell	Annexin-A4 (ANXA4)	2-DE, MALDI-TOF	Toyama et al. (2012), Zhu et al. (2006)
	Napsin A	Immuno-histochemistry and tissue microarrays	Alshenawy and Radi (2017), Skirnisdottir et al. (2013)
	Phosphoserine aminotransferase (PSAT1)	2-DE	Toyama et al. (2012)
Mucinous	Serpin B5 (SPB5)	2-DE	Toyama et al. (2012)
	FOXA1	Immunohistochemistry	Karpathiou et al. (2017)
	REG4	Immunohistochemistry tissue microarrays	Lehtinen et al. (2016)
	CEA5	LC-MS/MS	Tian et al. (2011)
	CEA6	LC-MS/MS	Tian et al. (2011)

biomarker can be any DNA, RNA, or protein that may show its presence in body fluids like blood, urine, and saliva. Several biomarkers have been characterized for a few types of cancer, but ovarian cancer fails to show robustness that correlate with ovarian tumor formation and progression.

One such well-known biomarker is the CA-125 glycoprotein. Several researchers have claimed that several patients with ovarian cancer showed increased levels of CA-125 (Sapiezynski et al., 2016). In fact, various results showed that approximately 78% (70%–90%) of patients with ovarian cancer have raised levels of this glycoprotein. Thus due to its higher specificity in comparison to other known biomarkers, this biomarker is recommended initially to presymptomatic women, though the strength of CA-125 as a biomarker for screening ovarian cancer is disputed. There are several other conditions where the levels of CA-125 shoots up like in all inflammatory diseases, cirrhosis, and liver diseases; diabetes mellitus; and in cancers of endometrial, fallopian, lung cancers. False-positive outcomes weaken CA-125 to be used as a standard biomarker. Therefore, there is need of some other potent biomarker that could differentiate ovarian cancer from normal conditions

with a higher specificity and sensitivity (Elzek and Rodland, 2015). Table 8.6 summarizes the current progress in proteomics-based biomarkers that showed some potency to be a better biomarker.

CURRENT SCENARIO IN OVARIAN CANCER THERAPY

Ovarian cancer remains one of the most lethal gynecological cancers with an estimated 225,000 women being diagnosed with it every year (Ferlay et al., 2015). The survival rate in ovarian cancer, however, is 40% over a period of five years. This is because most of the patients present at a later stage of the disease. The current treatment modality for epithelial ovarian cancer is cytoreductive/interval debulking surgery (IDS) followed by platinum-based chemotherapy (Banerjee and Kaye, 2013).

Cytoreductive surgery and interval debulking surgery (IDS)

Cytoreductive surgery/tumor debulking surgery is done with the idea that removal of as much cancer as possible shall be useful for the patient, even when complete resection is not possible. The primary debulking surgery can be challenging in patients of advanced (stage III and IV) ovarian cancer. It is often a high-risk preposition, with extensive bowel resection and major blood loss and even risk of morbidity as reported in 2015 Cochrane reviews comparing conventional, primary surgery with secondary IDS in ovarian cancer patients.

IDS is a secondary surgery that is performed after two to four cycles of neoadjuvant chemotherapy (NAC) or induction chemotherapy to remove the bulk of the tumor, and followed by adjuvant chemotherapy of the same type. Since most patients of ovarian cancer present in advance stages, IDS is the surgery of choice. The term neoadjuvant chemotherapy (NAC) describes the administration of chemotherapy when primary debulking surgery is not feasible, and only a biopsy is done for histologic diagnosis. If there is some tumor response after few cycles of chemotherapy, then the secondary surgery can be done before proceeding with further chemotherapy cycles (Tangjitgamol et al., 2010).

Chemotherapy in ovarian cancer

Ovarian cancer although not chemocurable is chemoresponsive. The 1960s and '70s saw the use of single alkylating agents with little success. Gradually there was the development of combination chemotherapy, where initially a combination of cisplatin and cyclophosphamide was used, and currently platinum compounds with paclitaxel are the treatment of choice. Platinum compounds are the most active cytotoxic agents currently used for the treatment of ovarian cancer. According to the American Cancer Society (ACS), the standard approach is the combination of a platinum compound, such as cisplatin or carboplatin, and a taxane, such as paclitaxel (Taxol®) or docetaxel (Taxotere®). For IV chemotherapy, carboplatin is the choice of drug over cisplatin due to its fewer side effects and being as effective. Several other drugs can produce regression of epithelial ovarian cancers, including pegylated liposomal doxorubicin (PLD), gemcitabine, and topotecanin combined with paclitaxel and/or carboplatin with two or three drugs in combination (Bast, 2011).

Another important arm of chemotherapy in ovarian cancer is hyperthermic intraperitoneal chemotherapy (HIPEC). Since epithelial ovarian cancer has the tendency to disseminate into the peritoneal cavity and remain confined to the peritoneum and intra-abdominal viscera, HIPEC can be useful in these patients. Improved long-term results can be achieved in highly selected patients using cytoreductive surgery (CRS), in combination with intraoperative hyperthermic intraperitoneal chemotherapy (HIPEC) (Martín-Cameán et al., 2016). Currently intraperitoneal cisplatin is validated in phase III trials but with increased nonhematologic toxicity; intraperitoneal carboplatin therapy phase III trial is underway (Teo, 2014).

However, the advances in chemotherapy are marred with a significant risk of recurrence and resistance to therapy making ovarian cancer difficult to cure. Hence, there is an urgent need to develop smarter treatment options.

Precision medicine in ovarian cancer

To improve outcomes of ovarian cancer, translational research must be prioritized and accelerated. For this it is important to identify multiple molecular abnormalities in human ovarian cancer that can be used for earlier diagnosis or used as therapeutic targets. Besides, the need of the hour is the enhancement of predictive models and biomarkers, developing therapeutic agents, and assays targeting molecular abnormalities. Targeted/personalized therapy is the latest approach for treatment of ovarian cancer. It utilizes the molecular profile of a patient's cancer for designing an efficacious treatment plan. A variety of targeted therapeutic approaches have been utilized for the management of ovarian cancer. The significant ones are discussed in the subsequent sections.

ANTIANGIOGENIC THERAPY

Angiogenesis is crucial for tumor development and progression. Clinical outcomes in ovarian cancer are worsened with increased angiogenesis. Angiogenesis inhibition has been the choicest therapeutic target for tumor control since the 1970s. Vascular endothelial growth factor (VEGF) is the major component involved in angiogenesis (Sapiezynski et al., 2016). VEGF overexpression leads to increased angiogenesis, and its association with ovarian cancer development and peritoneal dissemination is well established.

The VEGF/VEGF receptor (VEGFR) signaling pathway is the most promising angiogenic target.

There are two major strategies for inhibition of angiogenesis in ovarian cancer tumor treatment.

Inhibition of vascular endothelial growth factor (VEGF) ligand with antibodies or soluble receptors

Inhibition of VEGF receptor with tyrosine kinase inhibitors

Elevated expression of the VEGF ligands and receptors promotes malignant progression and correlates with poor prognosis in EOC. There are many drugs directed toward the VEGF/VEGFR signaling pathway like bevacizumab, trebananib, pazopanib, nintedanib, cediranib, and aflibercept.

Inhibition of VEGF ligand with antibodies or soluble receptors

Bevacizumab

Bevacizumab was developed by Roche and is a recombinant humanized monoclonal antibody capable of binding to all VEGF isoforms, thus inhibiting the binding of VEGF to its receptor and the downstream signaling (Conteduca et al., 2014). It is one of the best antiangiogenic agents since its action leads to the reduction in vascularization and normalization of the tumor microenviornment without any cytotoxicity. Bevacizumab has been used in combination with carboplatin and paclitaxel (Conteduca et al., 2014).

There have been two major clinical trials on bevacizumab, namely, international collaboration on ovarian neoplasms 7 (ICON-7) and gynecologic oncology group study 218 (GOG-218). Both these trials are in phase III. The ICON-7 trial enrolled 1528 women randomly to receive carboplatin and paclitaxel every 3 weeks for 6 cycles, or the same regimen plus bevacizumab every 3 weeks during chemotherapy, followed by 12 cycles or until unacceptable toxicity or disease progression. The objectives were to measure progression-free survival (PFS), overall survival (OS), response to treatment, toxicity, and quality of life. Final analysis with a 49-month median follow-up, showed an increase in PFS in the high-risk-of-progression group, with an increase of 5.5 months (16.0 vs. 10.5 months; HR = 0.73 (95% CI = 0.61–0.88); $p = 0.001$) compared to placebo group without bevacizumab. Further there was an increase of 9.4 months in OS in the high-risk-of-progression

group (39.7 vs. 30.3 months; HR = 0.78 (95% CI = 0.63–0.97); $p = 0.03$).

The ICON-7 trial however also reported certain side effects that need to be considered before making a decision on a combination of bevacizumab with cytotoxic drugs. The treatment with bevacizumab was associated with an increase in grade 1 mucocutaneous bleeding, grade 2 or higher acute hypertension (18% vs. 2%), grade 3 or higher thromboembolic events (7% vs. 3%), and gastrointestinal perforation (10 cases vs. 3 cases). The questionnaire-based quality of life scores showed that continuation of treatment with bevacizumab led to a small but clinically significant decline in quality of life compared to standard chemotherapy (Bermejo et al., 2016).

The second trial was the GOG-218 trial which randomly recruited 1873 women patients at stage III or IV of ovarian cancer. The study participants had undergone cytoreduction surgery and were divided into three groups according to their drug combination. All patients received carboplatin AUC 6 and paclitaxel 175 mg/m^2 every 3 weeks for 6 cycles and the study treatment. The three study groups then received drugs as follows: Group I received placebo every 3 weeks from cycle 2 to cycle 22; group II received bevacizumab every 21 days from cycle 2 to cycle 6 followed by placebo from cycle 7 to cycle 22; and group III received bevacizumab at the same dose from cycle 2 to cycle 22. The objectives were the same as ICON-7 trial.

The results showed the arm with bevacizumab (bevacizumab throughout) compared with the standard chemotherapy arm (placebo arm) showed a statistically significant increase in PFS. However, in the bevacizumab initiation group there was no increase in PFS. OS was similar in the three groups: 39.3, 38.7, and 39.7 months, respectively, with no statistically significant differences (Bermejo et al., 2016). Compared to the ICON-7 trial, the GOG-218 trial had just one significant side effect, that is, grade 2 hypertension in bevacizumab as compared to the placebo group.

Avastin use was studied in the platinum-resistant epithelial ovarian cancer study (AURELIA). It is a phase III trial where the combination therapy of paclitaxel, topotecan, or liposomal doxorubicin is being used with bevacizumab. The addition of bevacizumab showed an improved median PFS as well as OS (16.6 months)

(Sapiezynski et al., 2016). Bevacizumab is yet to be approved by the U.S. Food and Drug Administration (FDA) as the first line of treatment in ovarian cancer, although it is approved as the first line treatment in combination therapy.

VEGF signaling involves the tyrosine kinases and inhibition of these tyrosine kinases have proven to be another promising area of ovarian cancer treatment. One such inhibitor is pazopanib, an orally administered multikinase inhibitor of VEGFR-1/-2/-3 and of platelet-derived growth factor receptor (PDGFR)-α/-β and of c-Kit (Sapiezynski et al., 2016).

Pazopanib has been studied in combination therapy in phase I/II trials to evaluate the safety and efficacy of paclitaxel (175 mg/m^2) plus carboplatin (AUC 5 [group A] or AUC 6 [group B]) once every 3 weeks for up to 6 cycles, with either 800 or 400 mg per day of pazopanib. However, drug limiting toxicities were encountered in 4 out of 6 patients receiving pazopanib in combination therapy at low dose (400 mg) and higher dose (800 mg), which included gastrointestinal perforations and myelotoxicities (Friedlander et al., 2010).

Further pazopanib was also studied as monotherapy in patients with recurrent epithelial ovarian/fallopian tube/primary peritoneal cancer. The study was a nonrandomized, multicentric phase II trial (VEG104450; NCT00281632). The primary objective of the study was to study the response rate (determined by normalization of CA-125 levels) and secondary objectives were overall response and PFS. These patients were treated with pazopanib (800 mg OD) until progression or toxicity. At the end of 113 days the overall response rate (as per CA 125) was seen in 18% patients (Friedlander et al., 2010). The overall response rate in monotherapy paved the way for maintenance therapy of epithelial ovarian, fallopian tube, or primary peritoneal cancer, stages II to IV with no progression of disease postsurgery. This increased PFS compared with the placebo (17.9 vs. 12.3 months, HR = 0.77; 95% CI = 0.64–0.91; p = 0.0021). Grades 3 and 4 adverse events were observed, including hypertension (30.8%), neutropenia (9.9%), transaminase elevation (9.4%), diarrhea (8.2%), fatigue (2.7%), thrombocytopenia (2.5%), and palmoplantar erythrodysesthesia (1.9%) in the pazopanib arm. Although the maintenance therapy improved median PFS, OS has not yet been shown. Thus the use of pazopanib is not currently recommended in ovarian cancer management.

Some other significant antiangiogenic therapeutics include trebananib (a peptibody that inhibits angiopoietin 1 and 2), nintedanib (inhibits VEGFR-1, VEGFR-2, and VEGFR-3; PDGFR α and β; and FGFR-1, FGFR-2, and FGFR-3), cediranib (inhibits (VEGFR 1, VEGFR2, and VEGFR3), and c-Kit and aflibercept/VEGF trap (binds to and neutralizes all forms of VEGF-A and VEGF-B and inhibits placental growth factor [PGF] activation) (Friedlander et al., 2010).

Invasive ovarian cancer tumors cover a spectrum from low-grade to high-grade malignancy, with differences in their morphologic, histologic, and clinical features. Further genomic analysis has revealed that low-grade tumors and low-malignancy potential (LMP) tumors are a separate class of tumors as compared to high-grade malignancy (HGM) tumors due to a difference in their mutated genes (high *TP53*, its downstream effector *CDKN1A*, activators of p53, such as *PPM1A*, and decreased levels of inhibitors of p53, *UBE2D1* and *ADNP* in LMP tumors versus high expression of *PDCD4*, *E2F3*, *MCM4*, *CDC20*, and *PCNA* in HGM tumors (Bast and Mills 2010).

Based on the gene expression profile, two types of ovarian cancer can be defined (Cerrato et al., 2016):

Type I—Low-grade and borderline serous cancers, endometrioid, mucinous and clear-cell tumors with frequent mutations in PTEN, PI3 K catalytic subunit-α (PIK3CA), KRAS, BRAF, and b-catenin (CTNNB1) genes

Type II—High-grade serous carcinomas, mixed malignant mesodermal tumors, carcinosarcomas, and undifferentiated cancers with high genomic instability and up to 80% of patients' TP53 is mutated and is characteristic of BRCA1 and BRCA2 mutations (fallopian tubes and the peritoneum cancer)

Unfortunately, these differences in the tumor heterogeneity do not make any change in the treatment regimen of the veritable types of tumors. Thus currently, with advances in the knowledge on the molecular basis of tumor genesis, new drugs are under trial targeting various mutated genes. Some of the significant ones are discussed in the text to follow.

DRUGS TARGETING MUTATIONS IN OVARIAN CANCER PRECISION MEDICINE

Poly(ADP ribose) polymerase (PARP) inhibitors

Poly(ADP-ribose) polymerase (PARP) is a crucial enzyme involved in the base excision repair pathway of DNA repair mechanism. The *PARP* family has 17 structurally similar proteins. Cancer research has been mainly focusing on the *PARP1*. *PARP1*s like SSBs detect DNA strand interruptions and promote the NAD$^+$-dependent synthesis of poly(ADP-ribose) (PAR). Poly(ADP-ribosylation) of histones cause their release from DNA and allow access of chromatin more repair components (Cerrato et al., 2016). *PARP* inhibition causes single-stranded breaks in the DNA that eventually collapse and form double-stranded breaks during replication. Double-strand break repairs are carried out by homologous recombination mediated by *BRCA* or by error susceptible nonhomologous end joining, causing genetic instability and strand breakage. PARP inhibition is based upon this very theory of causing genetic instability and shall be very effective in ovarian cancer patients with *BRCA* mutations. This is because in individuals with *BRCA* deficiencies, double-stranded breaks shall be repaired by nonhomologous end joining, which usually results in cell death. Some of the significant PARP inhibitors utilized with positive results and involved in further trials are summarized in Table 8.7. The results so far received in different clinical trials suggest them to be potentially useful in treating patients with *BRCA1* and *BRCA2* mutations.

MDM–TP53 inhibitor: Nutlin-3 alpha

The *TP53* tumor suppressor gene is the most frequently altered gene in human cancers including most high- and low-grade ovarian cancers and MDM2 is the main negative regulator of p53, regulating p53 through ubiquitin-dependent degradation (Zanjirband et al., 2016). Another important strategy to treat low-grade ovarian cancer could be to exploit the persistence of wild-type TP53 in most low-grade cancers. Meijer et al. (2013) reported that in wild-type p53-expressing ovarian, colon, and lung cancer cell lines and in an ex vivo model of human ovarian cancer, Nutlin-3 enhanced p53, p21, MDM2, and DR5 surface expression. They further concluded that Nutlin-3

is a potent enhancer of D269H/E195R-induced apoptosis in wild-type p53-expressing cancer cells. These results were replicated by Crane et al. (2015) who further showed it in 15 ovarian cancer cell lines of varying histologic subtypes and demonstrated apoptosis in sensitive cell lines treated with Nutlin-3a.

Recently Zanjirband et al. (2016) analyzed combination therapy of Nutlin-3 with cisplatin. They demonstrated in ovarian cancer cell lines that Nutlin-3 or RG7388 effect in combination with cisplatin was additive to or synergistic in a p53-dependent manner, resulting in increased p53 activation, cell cycle arrest, and apoptosis, associated with increased p21WAF1 protein and/or caspase-3/7 activity compared to cisplatin alone.

Folate and folate receptor antagonists

Folate is an essential component required by rapidly dividing cells since it is important for DNA synthesis and helps promote cell division. Targeting folate and folate receptors can prove to be potentially useful in tumor growth regression. The folate dependent drugs can be categorized as

- Folate antagonists
- Folate receptor inhibitors
- Folate-conjugated therapeutic agents

Folate antagonists

Many folate antagonists have been studied to date. The most prominent is aminopterin and its successor molecule methotrexate. Both of these competitively inhibit the enzyme folate reductase. Besides folate reductase, another enzymatic target is thymidylate synthase, which is regulated by suicidal inhibition of fluorouracil-5, but has had limited success partly due to toxicity and patient tolerance issues. Recently drugs like pemetrexed and raltitrexed inhibiting thymidylate synthase have been used in ovarian cancer combination therapy. Pemetrexed in combination with bevacizumab showed median patient PFS improved to 7.9 months and median OS increased to 25.7 months (Sapiezynski et al., 2016).

Folate receptor inhibitors

The reports of overexpression of folate receptors (FR-α) in epithelial ovarian cancer (80% high) and its correlation with staging of ovarian cancer,

Table 8.7 PARPP inhibitors and their trials

PARP inhibitor	Target	Therapy type	Trial completed	Results	Ongoing trials
Olaparib (Sapiezynski et al., 2016)	BRCA-deficient ovarian tumors	Monotherapy and combination	Phase II	Patients with BRCA mutation had longer PFS	Phase III trials SOLO
Rucaparib (Sapiezynski et al., 2016)	BRCA-mutated cells, cells with low-level expression of HRR-associated genes, platinum-sensitive, relapsed high-grade epithelial ovarian cancer, fallopian tube cancer, peritoneal cancer	Monotherapy	Ongoing	Results awaited in 2017	Phase II trials ARIEL2 and 3
Veliparib (Sapiezynski et al., 2016)	Patients with mutation in the BRCA 1 and/or 2 gene	Monotherapy and combination	Phase II trial	Effective as monotherapy and has acceptable side effects	Phase II trials ongoing for efficacy
Niraparib (Sehouli et al., 2016)	BRCA mutation carriers and patients with sporadic cancer	Monotherapy	Phase I trial	Phase I trial showed 300 mg as maximum tolerable limit	Phase II and III trials (NOVA) for efficacy and tolerability in monotherapy for platinum-sensitive recurrent ovarian carcinoma
Talazoparib (Sehouli et al., 2016)	1. In vitro studies HRR-deficient cells (BRCA1, BRCA2, PTEN gene defects) 2. BRCA-associated ovarian and peritoneal carcinoma	Monotherapy	Phase I	Tumor response (RECIST and/or CA-125) was shown at BMN 673 doses of 100–1100 μg daily in 11 out of 17 patients	Phase II trials for efficacy in ovarian cancer ongoing

makes FR-α as potential drug targets. Clinical trials have been conducted using farletuzumab as

- Combination with cisplatin and taxanes (phase II)
- Combination with carboplatin and taxane in platinum-sensitive ovarian cancer (phase III)

The results of a phase II trial showed that the combinational therapy improved the overall tumor response rate and farletuzumab was well tolerated by patients. However, phase III trial results failed to reach the primary PFS endpoint and raised concerns about its clinical efficacy in patients with recurrent ovarian cancer (Sapiezynski et al., 2016).

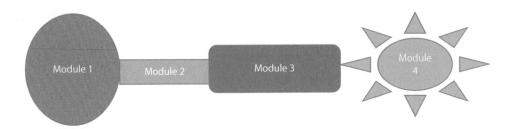

Figure 8.1 Structure of a typical folate-conjugated drug.

Folate-conjugated therapeutic agents

A typical folate drug conjugate contains four modules (Figure 8.1), namely,

Module 1—Pteroic acid

Module 2—A glutamic acid residue juxtaposed to module 1

Module 3—Has cleavable bonds

Module 4—Has drug moiety

Vintafolide, a folate molecule conjugated with desacetylvinlastine hydrazine, has entered phase II and phase III trials. Vintafolide can deliver chemotherapy to FR expressing cells, as it is a conjugate of folate and the chemotherapeutic agent desacetylvinblastine monohydrazide. It has been studied in combination therapy of patients who have undergone less than three chemotherapeutic cycles in the phase II PRECEDENT study. It is given in combination with PEGylated liposomal doxorubicin (PLD) for recurrent platinum-resistant ovarian tumors. The combination therapy was clearly effective with median PFS 5 months versus 2.7 months in monotherapy (Sapiezynski et al., 2016).

Human epidermal growth factor receptor (HER) antagonists

Most of the high-grade ovarian cancer tumors show defects with the signaling pathways, 67% showed RB1 pathway defects, 45% had defects in PI3K and/or Ras signaling, and 22% had defects with notch signaling (Bast and Mills 2010). Similarly *ERBB2* is a proto-oncogene coding for human epidermal growth factor receptor (*HER*). It is associated with high expression in advanced epithelial ovarian cancer in some rare cases. Currently there are two HER antagonists being used in cancer therapeutics: trastuzumab (Herceptin®) and pertuzumab humanized monoclonal antibodies. Pertuzumab prevents HER2 dimerization and inhibits multiple HER-mediated pathways. It has been tolerated well in two platinum-resistant ovarian cancer cohorts (61 and 62 patients) in a phase II trial. It was well tolerated with a RR of 4.3% in heavily pretreated OC patients (Gordon et al., 2006). A phase III trial is under way in combination with gemcitabine, paclitaxel, or gemcitabine, and the other arm a placebo in combination with gemcitabine, paclitaxel, or gemcitabine (Sapiezynski et al., 2016).

Estrogen receptor antagonists

Estrogen receptors are a potential therapeutic target in ovarian cancer owing to their expression in almost 60% of these patients. Besides, hormone replacement therapy has lesser side effects compared to the cytotoxic therapy, making it an attractive treatment modality to explore. Both estrogen receptor α (ERα) and ERβ get expressed in ovarian cancer cells. In clinical trials with tamoxifen, the overall response was only 13% but with fulvestrant it was 38%. Tamoxifen, however, was used in a limited fashion, that is, it was given only with chemotherapy for only 36 weeks (shorter duration than that recommended for breast cancer use in this setting) and most patients had residual disease. Further phase III trial with tamoxifen was also conducted versus thalidomide. The thalidomide group had a worse median and overall survival with a higher risk of death (HR = 1.76, 95% CI = 1.16–2.68) and had higher toxicity. But the study did not have a control arm (Simpkins et al., 2013). Recently Chan et al. (2014) reported that targeting ER subtypes may enhance the response to hormonal treatment in women with ovarian cancer. They used highly modulated 1, 3-bis (4-hydroxyphenyl)-4-methyl-5-[4-(2 -piperidinylethoxy)phenol]-1H-pyrazole dihydrochloride (MPP) (ERα antagonist) or 2,3-bis(4-hydroxy-phenyl)-propionitrile (DPN) (ERβ agonist) significantly suppressed cell growth in ovarian cancer cell lines SKOV3 and OV2008.

Table 8.8 Aromatase inhibitors and their clinical trials

Aromatase inhibitor	Target	Therapy type	Trial completed	Results
Anastrozole (Sapiezynski et al., 2016; del Carmen et al., 2003)	Patients with ovarian, peritoneal, or fallopian tube carcinoma	Monotherapy	Phase II	Well-tolerated oral agent but with minimal tumoricidal activity in women with recurrent/persistent Müllerian cancer
Letrozole (Sapiezynski et al., 2016; Smyth et al., 2007)	Previously treated ER-positive ovarian cancer patients with a rising CA125	Monotherapy	Phase II, Phase III trial in CA breast not in OC	26% patients had a PFS of >6 months, and 5% patients had a PFS of ≥2 years; minimal toxicity from letrozole in this patient group
Exemestane (Sapiezynski et al., 2016)	Refractory ovarian cancer patients with prior chemotherapy of platinum and taxane	Monotherapy	Phase II	Stable disease 36% and well tolerated with low toxicity

To date there has been low success with ER antagonists in ovarian cancer, however, it still remains a lucrative target of research.

Aromatase inhibitors

The enzyme aromatase is an estrogen synthase and can be a potential target in ER-positive tumors. Some of the significant aromatase inhibitors used in ovarian cancer are summarized in Table 8.8. Aromatase inhibitors remain an active area of research in ovarian cancer, owing to fewer large prospective studies. Although the results of the existing clinical trials are not clinically as much valuable, the use of these drugs in combination therapy shall be an interesting preposition to watch out for.

FUTURE PROSPECTS

The future of precision medicine in ovarian cancer treatment looks exciting. With the advances in the field of genomics and integration of basic science and clinical databases, novel therapeutics can be developed targeted to specific oncoproteins responsible for multidrug resistance, tumor progression and antiapoptotic cellular defense in ovarian cancer cells, or getting overexpressed in tumor cells compared to surrounding tissue. One such upcoming field is the nanotechnology-based delivery of therapeutics. One such method has

been discussed by (Sapiezynski et al., 2016). They used nanotechnology-based targeted delivery systems (Figure 8.2). This is true personalized treatment, where tumor proteins of a particular patient are identified and targeted using nanotechnology-based targeted delivery systems. The systems contain only one protein inhibitor (siRNA) to suppress one targeted protein or an anticancer drug. These nanotechnology-based targeted delivery systems can be used in any combination with each other.

The authors report encouraging in vitro and in vivo results at the preliminary stage, since their data showed that such a personalized treatment approach is much more effective when compared with a standard treatment protocol. In the mice model they selected five genes that were overexpressing (*BCL2*, *MDR1*, *CD44*, *MMP9*, *PGR*) and caused metastases. The nanotechnology-based targeted delivery systems were designed accordingly. The systems (siRNAs) were delivered along with paclitaxel using a dendrimer-based delivery system. The combination therapy reduced the expression of the target genes. This is because of significantly enhanced cell death induction, imposed tumor shrinkage, and preventing the development of intraperitoneal metastases.

Another nanotechnology-based treatment regimen for ovarian cancer has been designed in mice models for delivery of folate conjugates. It

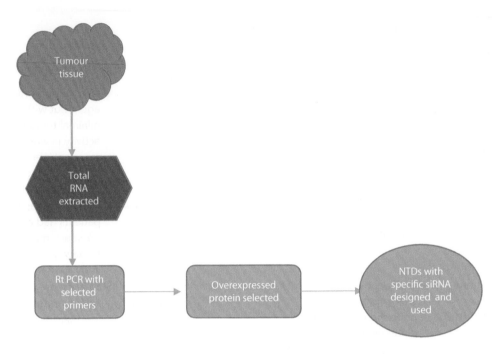

Figure 8.2 Protocol for personalized treatment using nanotechnology-based targeted delivery systems.

has been reported that nanoceria (NCe), nanoparticles of cerium oxide when conjugated to folic acid (NCe-FA), increased the cellular NCe internalization and inhibited cell proliferation. A combination therapy of NCe-FA and cisplatinum lowered the tumor burden significantly, and NCe-FA significantly reduced proliferation and angiogenesis in the xenograft mouse model (Hijaz et al., 2016).

In times to come, the exciting field of precision medicine shall become a useful arena not just for research but the advances shall translate into clinical practice benefitting hundreds of ovarian cancer patients with early diagnosis and effective treatment.

REFERENCES

Agostini A, Brunetti M, Davidson B et al. Genomic imbalances are involved in miR-30c and let-7a deregulation in ovarian tumors: Implications for HMGA2 expression. *Oncotarget* 8(13);2017:21554–60.

Alshenawy HA, and Radi DA. Napsin-A, a possible diagnostic marker for differentiating clear cell ovarian carcinoma from other high-grade ovarian carcinomas. *Appl Immunohistochem Mol Morphol* 2017 March. doi:10.1097/PAI.0000000000000510.

Ambros V. MicroRNAs and developmental timing. *Curr Opin Genet Dev* 21;2011:511–7.

Banerjee S, and Kaye SB. New strategies in the treatment of ovarian cancer: Current clinical perspectives and future potential. *Clin Cancer Res* 19(5);2013:961–68.

Bast RC. Molecular approaches to personalizing management of ovarian cancer. *Ann Oncol* 22(Suppl 8);2011:viii5–viii15.

Bast RC, and Mills GB. Personalizing therapy for ovarian cancer: BRCAness and beyond. *J Clin Oncol* 28(22);2010:3545–48.

Bell D, Berchuck A, Birrer M et al. Integrated genomic analyses of ovarian carcinoma. *Nature* 474(7353);2011:609–15.

Bermejo MA, Alonso L, Rey Iglesias ME, Pérez López A, Montes F, and García Mata J. Antiangiogenic therapy in epithelial ovarian cancer. In: *Gynecologic Cancers: Basic Sciences, Clinical and Therapeutic Perspectives*. Farghaly SA (Ed). InTech. 2016. doi:10.5772/60822.

Braslavsky I, Hebert B, Kartalov E, and Quake SR. Sequence information can be obtained from single DNA molecules. *Proc Natl Acad Sci USA* 100;2003:3960–4.

Bubancova I, Kovarikova H, Laco J, Ruszova E, Dvorak O, Palicka V, and Chmelarova M.

Next-generation sequencing approach in methylation analysis of HNF1B and GATA4 Genes: Searching for biomarkers in ovarian cancer. *Int J Mol Sci* 18(2);2017:pii.

Cacan E. Epigenetic regulation of RGS2 (Regulator of G-protein signaling 2) in chemo-resistant ovarian cancer cells. *J Chemother* 29(3);2017:173–8.

Calin GA, Ferracin M, Cimmino A et al. A microRNA signature associated with prognosis and progression in chronic lymphocytic leukemia. *N Engl J Med* 353(17);2005:1793–801.

Calin GA, Garzon R, Cimmino A, Fabbri M, and Croce CM. MicroRNAs and leukemias: How strong is the connection? *Leuk Res* 30(6);2006:653–55.

Campbell IG, Russell SE, and Phillips WA. PIK3CA mutations in ovarian cancer. *Clin Cancer Res* 11(19 Pt 1);2005:7042; author reply 7042–3.

Cancer Genome Atlas Research Network. Integrated genomic analyses of ovarian carcinoma. *Nature* 474(7353);2011:609–15.

Carrarelli P, Funghi L, Ciarmela P, Centini G, Reis FM, Cruz CD, Mattei A, Vannuccini S, and Petraglia F. Deep infiltrating endometriosis and endometrial adenocarcinoma express high levels of myostatin and its receptors messenger RNAs. *Reprod Sci* 24(12);2017:1577–82.

Cerrato A, Morra F, and Celetti A. Use of poly ADP-ribose polymerase [PARP] inhibitors in cancer cells bearing DDR defects: The rationale for their inclusion in the clinic. *J Exp Clin Cancer Res* 35(1);2016:179.

Cesari R, Martin ES, Calin GA et al. Parkin, a gene implicated in autosomal recessive juvenile parkinsonism, is a candidate tumor suppressor gene on chromosome 6q25-q27. *Proc Natl Acad Sci USA* 100(10);2003:5956–61.

Chan K, Wong MS, Chan TK, and Chan V. A thalassaemia array for Southeast Asia. *Br J Haematol* 124(2);2004:232–9.

Chan KK-L, Leung TH-Y, Chan DW, Wei N, Lau GT-Y, Liu SS, Siu MK-Y, and Ngan HY-S. Targeting estrogen receptor subtypes (ERα and ERβ) with selective ER modulators in ovarian cancer. *J Endocrinol* 221(2);2014:325–36.

Cho KR, and Shih IM. Ovarian cancer. *Annu Rev Pathol* 4;2009:287–313.

Chong GO, Jeon H-S, Han HS, Son JW, Lee YH, Hong DG, Park HJ, Lee YS, and Cho YL. Overexpression of microRNA-196b accelerates invasiveness of cancer cells in recurrent epithelial ovarian cancer through regulation of homeobox A9. *Cancer Genomics Proteomics* 14(2);2017:137–41.

Church GM. The personal genome project. *Mol Syst Biol* 2005; 1:2005.0030.

Conteduca V, Kopf B, Burgio SL, Bianchi E, Amadori D, and De Giorgi U. The emerging role of anti-angiogenic therapy in ovarian cancer (review). *Int J Oncol* 44(5);2014:1417–24.

Couch FJ, Wang X, McGuffog L et al. Genome-wide association study in BRCA1 mutation carriers identifies novel loci associated with breast and ovarian cancer risk. *PLoS Genet* 9(3);2013:e1003212.

Crane EK, Kwan S-Y, Izaguirre DI, Tsang YTM, Mullany LK, Zu Z, Richards JS, Gershenson DM, and Wong K-K. Nutlin-3a: A potential therapeutic opportunity for TP53 wild-type ovarian carcinomas. *PLOS ONE* 10(8);2015:e0135101.

del Carmen MG, Fuller AF, Matulonis U, Horick NK, Goodman A, Duska LR., Penson R, Campos S, Roche M, and Seiden MV. Phase II trial of anastrozole in women with asymptomatic müllerian cancer. *Gynecol Oncol* 91(3);2003:596–602.

Dong A, Lu Y, and Lu B. Genomic/epigenomic alterations in ovarian carcinoma: Translational insight into clinical practice. *J Cancer* 7(11);2016:1441–51.

Dwivedi S, Goel A, Khattri S, Mandhani A, Sharma P, Misra S, and Pant KK. Genetic variability at promoters of IL-18 (pro-) and IL-10 (anti-) inflammatory gene affects susceptibility and their circulating serum levels: An explorative study of prostate cancer patients in North Indian populations. *Cytokine* 74(1);2015:117–22.

Dwivedi S, Goel A, Mandhani A, Khattri S, Sharma P, Misra S, and Pant KK. Functional genetic variability at promoters of pro-(IL-18) and anti-(IL-10) inflammatory affects their mRNA expression and survival in prostate carcinoma patients: Five year follow-up study. *Prostate* 75(15);2015:1737–46.

Dwivedi S, Shukla KK, Gupta G, and Sharma P. Non-invasive biomarker in prostate carcinoma: A novel approach. *Indian J Clin Biochem* 28(2);2013:107–9.

Dwivedi S, Singh S, Goel A, Khattri S, Mandhani A, Sharma P, Misra S, and Pant KK. Pro-(IL-18) and anti-(IL-10) inflammatory promoter genetic variants (intrinsic factors) with tobacco exposure (extrinsic factors) may influence susceptibility and severity of prostate carcinoma: A prospective study. *Asian Pac J Cancer Prev* 16(8);2015:3173–81.

Elzek MA, and Rodland KD. Proteomics of ovarian cancer: Functional insights and clinical applications. *Cancer Metastasis Rev* 34(1);2015:83–96.

Erlich H, Valdes AM, Noble J et al. HLA DR-DQ haplotypes and genotypes and type 1 diabetes risk: Analysis of the type 1 diabetes genetics consortium families. *Diabetes* 57(4);2008:1084–92. doi: 10.2337/db07-1331.

Fan J-B, Gunderson KL, Bibikova M et al. Illumina universal bead arrays. *Methods Enzymol* 410;2006:57–73.

Ferlay J, Soerjomataram I, Dikshit R, Eser S, Mathers C, Rebelo M, Parkin DM, Forman D, and Bray F. Cancer incidence and mortality worldwide: Sources, methods and major patterns in GLOBOCAN 2012. *Int J Cancer* 136(5);2015:E359–86.

Friedlander M, Hancock KC, Rischin D, Messing MJ, Stringer CA, Matthys GM, Ma B, Hodge JP, and Lager JJ. A phase II, open-label study evaluating pazopanib in patients with recurrent ovarian cancer. *Gynecol Oncol* 119(1);2010:32–7.

Gemignani F, Perra C, Landi S, Canzian F, Kurg A, Tonisson N, Galanello R, Cao A, Metspalu A, and Romeo G. Reliable detection of beta-thalassemia and G6PD mutations by a DNA microarray. *Clin. Chem.* 48;2002:2051–4.

Gordon MS, Matei D, Aghajanian C et al. Clinical activity of pertuzumab (rhuMAb 2C4), a HER dimerization inhibitor, in advanced ovarian cancer: Potential predictive relationship with tumor HER2 activation status. *J Clin Oncol* 24(26);2006:4324–32.

Gozzi G, Chelbi ST, Manni P, Alberti L, Fonda S, Saponaro S, Fabbiani L, Rivasi F, Benhattar J, and Losi L. Promoter methylation and downregulated expression of the TBX15 gene in ovarian carcinoma. *Oncol Lett* 12(4);2016:2811–19.

Hijaz M, Das S, Mert I et al. Folic acid tagged nanoceria as a novel therapeutic agent in ovarian cancer. *BMC Cancer* 16(March);2016:220.

International Human Genome Sequencing Consortium. Initial sequencing and analysis of the human genome. *Nature* 409;2001:860–921.

Jang K, Kim M, Gilbert CA, Simpkins F, Ince TA, and Slingerland JM. VEGFA activates an epigenetic pathway upregulating ovarian cancer-initiating cells. *EMBO Mol Med* 9(3);2017:304–18.

Kanchi KL, Johnson KJ, Lu C et al. Integrated analysis of germline and somatic variants in ovarian cancer. *Nat Commun* 5;2014:3156.

Karpathiou G, Venet M, Mobarki M, Forest F, Chauleur C, and Peoc'h M. FOXA1 is expressed in ovarian mucinous neoplasms. *Pathol* 49(3);2017:271–76.

Katchman BA, Chowell D, Wallstrom G, Vitonis AF, LaBaer J, Cramer DW, and Anderson KS. Autoantibody biomarkers for the detection of serous ovarian cancer. *Gynecol Oncol* 146(1);2017:129–36.

Kaur M, Singh A, Singh K, Gupta S, and Sachan M. Development of a multiplex MethyLight assay for the detection of DAPK1 and SOX1 methylation in epithelial ovarian cancer in a North Indian population. *Genes Genetic Syst* 91(3);2016:175–81.

Kuo K-T, Mao T-L, Jones S et al. Frequent activating mutations of PIK3CA in ovarian clear cell carcinoma. *Am J Pathol* 174(5);2009:1597–601.

Kurman RJ, and Shih IM. Molecular pathogenesis and extraovarian origin of epithelial ovarian cancer—Shifting the paradigm. *Hum Pathol* 42(7);2011:918–31.

Landen CN, Birrer MJ, and Sood AK. Early events in the pathogenesis of epithelial ovarian cancer. *J Clin Oncol* 26(6);2008:995–1005.

Lee J, An S, Choi YM et al. Musashi-2 is a novel regulator of paclitaxel sensitivity in ovarian cancer cells. *Int J Oncol* 49(5);2016:1945–52.

Lehrbach NJ, and Miska EA. Functional genomic, computational and proteomic analysis of C. Elegans microRNAs. *Brief Funct Genomics Proteomic* 7(3);2008:228–35.

Lehtinen L, Vesterkvist P, Roering P et al. REG4 is highly expressed in mucinous ovarian cancer: A potential novel serum biomarker. *PLOS ONE* 11(3);2016:e0151590.

Levine DA, Bogomolniy F, Yee CJ, Lash A, Barakat RR, Borgen PI, and Boyd J. Frequent mutation of the PIK3CA gene in ovarian and breast cancers. *Clin Cancer Res* 11(8);2005:2875–8.

Li J, Li Q, Huang H, Li Y, Li L, Hou W, and You Z. Overexpression of miRNA-221 promotes cell proliferation by targeting the apoptotic protease activating factor-1 and indicates a poor prognosis in ovarian cancer. *Int J Oncol* 2017 March. doi:10.3892/ijo.2017.3898.

Liede A, Tonin PN, Sun CC, Serruya C, Daly MB, Narod SA, and Foulkes WD. Is hereditary site-specific ovarian cancer a distinct genetic condition? *Am J Med Genet* 75(1);1998:55–8.

Liu H, Shi H, Fan Q, and Sun X. Cyclin Y regulates the proliferation, migration, and invasion of ovarian cancer cells via Wnt signaling pathway. *Tumour Biol* 37(8);2016:10161–75.

Liu W, Lv C, Zhang B, Zhou Q, and Cao Z. MicroRNA-27b functions as a new inhibitor of ovarian cancer-mediated vasculogenic mimicry through suppression of VE-cadherin expression. *RNA* 23(7);2017:1019–27.

Mardis ER. Next-generation DNA sequencing methods. *Annu Rev Genomics Hum Genet.* 9;2008:387–402. doi: 10.1146/annurev.genom.9.081307.164359.

Margulies M, Egholm M, Altman WE et al. Genome sequencing in microfabricated high-density picolitre reactors. *Nature.* 437(7057);2005:376–80.

Martín-Cameán M, Delgado-Sánchez E, Piñera A, Diestro MD, De Santiago J, and Zapardiel I. The role of surgery in advanced epithelial ovarian cancer. *Ecancermedicalscience* 10;2016:666.

McConechy MK, Ding J, Senz J et al. Ovarian and endometrial endometrioid carcinomas have distinct CTNNB1 and PTEN mutation profiles. *Mod Pathol* 27(1);2014:128–34.

Meijer A, Kruyt FAE, van der Zee AGJ, Hollema H, Le P, ten Hoor KA, Groothuis GMM, Quax WJ, de Vries EGE, and de Jong S. Nutlin-3 preferentially sensitises wild-type p53-expressing cancer cells to DR5-selective TRAIL over rhTRAIL. *Br J Cancer* 109(10);2013:2685–95.

Miller MB, and Tang YW. Basic concepts of microarrays and potential applications in clinical microbiology. *Clinical Microbiology Reviews* 22(4);2009:611–33.

Nakayama K, Nakayama N, Kurman RJ, Cope L, Pohl G, and Samuels Y. Sequence mutations and amplification of PIK3CA and AKT2 genes in purified ovarian serous neoplasms. *Cancer Biol Ther* 5(7);2006:779–85.

Nguyen HT, Tian G, and Murph MM. Molecular epigenetics in the management of ovarian cancer: Are we investigating a rational clinical promise? *Front Oncol* 4;2014:71.

Nick AM, Coleman RL, Ramirez PT, and Sood AK. A framework for a personalized surgical approach to ovarian cancer. *Nat Rev Clin Oncol* 12(4);2015:239–45.

Obata K, Morland SJ, Watson RH, Hitchcock A, Chenevix-Trench G, Thomas EJ, and Campbell IG. Frequent PTEN/MMAC mutations in endometrioid but not serous or mucinous epithelial ovarian tumors. *Cancer Res* 58(10);1998:2095–7.

Oliphant A, Barker DL, Stuelpnagel JR, and Chee MS. BeadArray technology: Enabling an accurate, cost-effective approach to high-throughput genotyping. *Biotechniques* 32;2002:S56.

Rahimi F, Karimi J, Goodarzi MT, Saidijam M, Khodadadi I, Razavi ANE, and Nankali M. Overexpression of receptor for advanced glycation end products (RAGE) in ovarian cancer. *Cancer Biomark* 18(1);2017:61–8.

Romeo M, Karachaliou N, Chaid I, Queralt C, De Aguirre I, Del Carmen Gómez M, Sanchez-Ronco M, Radua J, Ramírez JL, and Rosell R. Low RAP80 mRNA expression correlates with shorter survival in sporadic high-grade serous ovarian carcinoma. *Int J Biol Markers* 32(1);2017:e90–95.

Ross-Adams H, Ball S, Lawrenson K, Halim S, Russell R, and Wells C. HNF1B variants associate with promoter methylation and regulate gene networks activated in prostate and ovarian cancer. *Oncotarget* 7(46);2016:74734–46.

Roukos DH. Personalized cancer diagnostics and therapeutics. *Expert Rev Mol Diagn.* 9(3);2009:227–9.

Saiki RK, Walsh PS, Levenson CH, and Erlich HA. Genetic analysis of amplified DNA with immobilized sequence-specific oligonucleotide probes. *Proc Natl Acad Sci USA* 86(16);1989:6230–4.

Samuels Y, and Waldman T. Oncogenic mutations of PIK3CA in human cancers. *Curr Top Microbiol Immunol* 347;2010:21–41.

Sangha N, Wu R, Kuick R et al. Neurofibromin 1 (NF1) defects are common in human ovarian serous carcinomas and co-occur with TP53 mutations. *Neoplasia (New York, N.Y.)* 10(12);2008:1362–72, following 1372.

Sankaranarayanan R, and Ferlay J. Worldwide burden of gynaecological cancer: The size of the problem. *Best Pract Res Clin Obstet Gynaecol* 20(2);2006:207–25.

Sapiezynski J, Taratula O, Rodriguez-Rodriguez L, and Minko T. Precision targeted therapy of ovarian cancer. *J Control Release* 243(December);2016:250–68.

Sehouli J, Braicu EI, and Chekerov R. PARP inhibitors for recurrent ovarian carcinoma: Current treatment options and future perspectives. *Geburtshilfe Und Frauenheilkunde* 76(2);2016:164–69.

Sellar GC, Watt KP, Rabiasz GJ et al. OPCML at 11q25 is epigenetically inactivated and has tumor-suppressor function in epithelial ovarian cancer. *Nat Genet* 34(3);2003:337–43.

Sereni MI, Baldelli E, Gambara G et al. Functional characterization of epithelial ovarian cancer histotypes by drug target based protein signaling activation mapping: Implications for personalized cancer therapy. *Proteomics* 15(2–3);2015:365–73.

Shao H, Mohamed EM, Xu GG, Waters M, Jing K, and Ma Y. Carnitine palmitoyltransferase 1A functions to repress FoxO transcription factors to allow cell cycle progression in ovarian cancer. *Oncotarget* 7(4);2016:3832–46.

Shin K, Kim KH, Yoon MS, Suh DS, Lee JY, Kim A, and Eo W. Expression of interactive genes associated with apoptosis and their prognostic value for ovarian serous adenocarcinoma. *Adv Clin Exp Med* 25(3);2016:513–21.

Simpkins F, Garcia-Soto A, and Slingerland J. New insights on the role of hormonal therapy in ovarian cancer. *Steroids* 78(6);2013:530–37.

Singer G, Oldt R, Cohen Y, Wang BG, Sidransky D, Kurman RJ, and Shih I-M. Mutations in BRAF and KRAS characterize the development of low-grade ovarian serous carcinoma. *J Natl Cancer Inst* 95(6);2003:484–86.

Skirnisdottir I, Bjersand K, Akerud H, and Seidal T. Napsin A as a marker of clear cell ovarian carcinoma. *BMC Cancer* 13(November);2013:524.

Smyth JF, Gourley C, Walker G et al. Antiestrogen therapy is active in selected ovarian cancer cases: The use of letrozole in estrogen receptor-positive patients. *Clin Cancer Res* 13(12);2007:3617–22.

Stratton JF, Thompson D, Bobrow L et al. The genetic epidemiology of early-onset epithelial ovarian cancer: A population-based study. *Am J Hum Genet* 65(6);1999:1725–32.

Sung HY, Yang SD, Park AK, Ju W, and Ahn JH. Aberrant hypomethylation of solute carrier family 6 member 12 promoter induces metastasis of ovarian cancer. *Yonsei Med J* 58(1);2017:27–34.

Tang S, Yang F, Du X, Lu Y, Zhang L, and Zhou X. Aberrant expression of anaplastic lymphoma kinase in ovarian carcinoma independent of gene rearrangement. *Int J Gynecol Pathol* 35(4);2016:337–47.

Tangjitgamol S, Manusirivithaya S, Srijaipracharoen S et al. Endometrial cancer in Thai women: Clinicopathological presentation and survival. *Asian Pac J Cancer Prev* 11;2010:1267–72.

Tawfik DS, Griffiths AD. Man-made cell-like compartments for molecular evolution. *Nat Biotechnol* 16;1998:652–6.

Teo MCC. Update on the management and the role of intraperitoneal chemotherapy for ovarian cancer. *Curr Opin Obstet Gynecol* 26(1);2014:3–8.

Tian J, Xu Y-Y, Li L, and Hao Q. MiR-490-3p sensitizes ovarian cancer cells to cisplatin by directly targeting ABCC2. *Am J Transl Res* 9(3);2017:1127–38.

Tian Y, Yao Z, Roden RBS, and Zhang H. Identification of glycoproteins associated with different histological subtypes of ovarian tumors using quantitative glycoproteomics. *Proteomics* 11(24);2011:4677–87.

Toyama A, Suzuki A, Shimada T et al. Proteomic characterization of ovarian cancers identifying annexin-A4, phosphoserine aminotransferase, cellular retinoic acid-binding protein 2, and serpin B5 as histology-specific biomarkers. *Cancer Sci* 103(4);2012:747–55.

Venter JC, Adams MD, Myers EW et al. The sequence of the human genome. *Science* 291;2001:1304–51.

Wei L, Yin F, Zhang W, and Li L. STROBE-compliant integrin through focal adhesion involve in cancer stem cell and multidrug resistance of ovarian cancer. *Medicine* 96(12);2017:e6345.

Wiegand KC, Shah SP, Al-Agha OM et al. ARID1A mutations in endometriosis-associated ovarian carcinomas. *N Engl J Med* 363;2010:1532–43.

Willner J, Wurz K, Allison KH, Galic V, Garcia RL, Goff BA, and Swisher EM. 2007. Alternate molecular genetic pathways in ovarian carcinomas of common histological types. *Hum Pathol* 38(4);607–13.

Wu R, Baker SJ, Hu TC, Norman KM, Fearon ER, and Cho KR. Type I to type II ovarian carcinoma progression: Mutant Trp53 or Pik3ca confers a more aggressive tumor phenotype in a mouse model of ovarian cancer. *Am J Pathol* 182(4);2013:1391–99.

Xiao L, Shi X-Y, Zhang Y, Zhu Y, Zhu L, Tian W, Zhu B-K, and Wei Z-L. YAP induces cisplatin resistance through activation of autophagy in human ovarian carcinoma cells. *OncoTargets and Therapy* 9;2016:1105–14.

Xiao M, Cai J, Cai L, Jia J, Xie L, Zhu Y, Huang B, Jin D, and Wang Z. Let-7e sensitizes epithelial ovarian cancer to cisplatin through repressing DNA double strand break repair. *J Ovarian Res* 10(1);2017:24.

Xu J, Jiang N, Shi H, Zhao S, Yao S, and Shen H. miR-28-5p promotes the development and progression of ovarian cancer through inhibition of N4BP1. *Int J Oncol* 2017 March. doi:10.3892/ijo.2017.3915.

Xu Y, Li X, Wang H, Xie P, Yan X, Bai Y, and Zhang T. Hypermethylation of CDH13, DKK3 and FOXL2 promoters and the expression of EZH2 in ovary granulosa cell tumors. *Mol Med Rep* 14(3);2016:2739–45.

Yan H, Qiu L, Xie X, Yang H, Liu Y, Lin X, and Huang H. ELF5 in epithelial ovarian carcinoma tissues and biological behavior in ovarian carcinoma cells. *Oncol Rep* 37(3);2017:1412–8. doi: 10.3892/or.2017.5418.

Yang L, Zhang X, Ma Y, Zhao X, Li B, and Wang H. Ascites promotes cell migration through the repression of miR-125b in ovarian cancer. *Oncotarget* 8(31);2017:51008–15.

Yang M, Xie X, and Ding Y. SALL4 is a marker of poor prognosis in serous ovarian carcinoma promoting invasion and metastasis. *Oncol Rep* 35(3);2016:1796–806.

Zanjirband M, Edmondson RJ, and Lunec J. Pre-clinical efficacy and synergistic potential of the MDM2-p53 antagonists, nutlin-3 and RG7388, as single agents and in combined treatment with cisplatin in ovarian cancer. *Oncotarget* 7(26);2016:40115–34.

Zannoni GF, Improta G, Pettinato A, Brunelli C, Troncone G, Scambia G, and Fraggetta F. Molecular status of PI3KCA, KRAS and BRAF in ovarian clear cell carcinoma: An analysis of 63 patients. *J Clin Pathol* 69(12);2016:1088–92.

Zhang L, Cao X, Zhang L, Zhang X, Sheng H, and Tao K. UCA1 overexpression predicts clinical outcome of patients with ovarian cancer receiving adjuvant chemotherapy. *Cancer Chemother Pharmacol* 77(3);2016:629–34.

Zhang T, Xu J, Deng S et al. Core signaling pathways in ovarian cancer stem cell revealed by integrative analysis of multi-marker genomics data. *PLoS ONE* 13(5);2018:e0196351. https://doi.org/10.1371/journal.pone.0196351

Zhou D, Zhang L, Sun W, Guan W, Lin Q, Ren W, Zhang J, and Xu G. Cytidine monophosphate kinase is inhibited by the TGF-β signalling pathway through the upregulation of miR-130b-3p in human epithelial ovarian cancer. *Cell Signal* 35;2017:197–207.

Zhu Y, Wu R, Sangha N, Yoo C, Cho KR, Shedden KA, Katabuchi H, and Lubman DM. Classifications of ovarian cancer tissues by proteomic patterns. *Proteomics* 6(21);2006:5846–56.

Zou J, Liu L, Wang Q, Yin F, Yang Z, Zhang W, and Li L. Downregulation of miR-429 contributes to the development of drug resistance in epithelial ovarian cancer by targeting ZEB1. *Am J Transl Res* 9(3);2017:1357–68.

9

Precision medicine in myelodysplastic syndromes

OTA FUCHS

INTRODUCTION

Myelodysplastic syndromes (MDS) are a family of clonal hematopoietic stem cell malignancies characterized by ineffective hematopoiesis, peripheral cytopenias, frequent karyotypic abnormalities, and risk of transformation to acute myeloid leukemia (AML) (Nimer, 2008; Tefferi and Vardiman, 2009; Sekeres, 2010; Adés et al., 2014; Bejar and Steensma, 2014; Pellagatti and Boultwood, 2015; Kennedy and Ebert, 2017; Shastri et al., 2017). The incidence of MDS is approximately 3–5 cases per 100,000 population per year with 30 cases per 100,000 population per year in patients after the age of 70 (Phekoo et al., 2006; Ma et al., 2007; Germing et al., 2008; Rollison et al., 2008; Cogle et al., 2011; Neukirchen et al., 2011; Rodger and Morison, 2012; Avgerinou et al., 2013; Otrock et al., 2015; Lubeck et al., 2016). Thus, MDS are mainly a disease of the elderly with a sharp increase in incidence in the age decade above 70 years. Approximately 10,000 to 15,000 new cases are diagnosed annually in the United States of America. The overall incidence of MDS is higher in males than in females (1.5–2:1).

MDS is rare in children, where it is more frequently associated with monosomy of chromosome 7. Adolescents with monosomy 7 frequently have (72%) GATA2 mutations and the high risk for progression to advanced disease (Wlodarski et al., 2016). Therefore, timely hematopoietic stem cell transplantation must be considered in these cases. MDS in children and young and middle aged adults is frequently associated with underlying genetic predisposition syndromes (bone marrow failure such as Fanconi anemia and dyskeratosis congenita) and nonsyndromic familial MDS/AML predisposition caused by mutations in *GATA2, RUNX1, CEBPA, TERT/TERC* (telomerase subunit genes), *ANKRD26* (ankyrin repeat domain 26), *ETV6* (gene for ETS family transcriptional repressor, variant 6), and *SRP72* (signal recognition particle 72) genes (Owen et al., 2008; Godley, 2014; West et al., 2014; Babushok and Bessler, 2015; Churpek et al., 2015; Babushok et al., 2016; Bannon and DiNardo, 2016; Koeffler and Leong, 2017).

The majority of MDS cases occur *de novo* (80%–90%), whereas 10%–20% of MDS cases are secondary, therapy-related MDS. The etiology of *de novo* MDS is unclear. Cumulative exposure to environmental agents (benzene, ionizing radiation, tobacco smoke pesticides, insecticides, and possibly hair dyes or extremely low frequency magnetic fields) and chronic inflammation, genetic differences in leukemogen susceptibility and metabolism, and hematopoietic stem cell genomic senescence may contribute to disease pathogenesis in *de novo* cases (Bowen, 2013). Therapy-related MDS arise as a result of chemotherapy, radiation therapy, or a combination of these therapies. Therapy-related MDS have worse outcomes than *de novo* MDS (Zeidan et al., 2017). Therapy-related MDS have more complex karyotypes and mutations in p53. The revised version of the World Health Organization (WHO) classification combines therapy-related MDS (t-MDS) and therapy-related AML (t-AML) in the one entity of therapy-related myeloid neoplasms (t-MNs) (Heuser, 2016; Ganser and Heuser, 2017). Treatment options include best supportive care, azacitidine, or intensive chemotherapy/allogeneic transplant according to genetic risk profile of patients with t-MNs (Ganser and Heuser, 2017). Median overall survival (OS) was 14 months and 5-year OS was 13.8%. Improved outcomes were seen with allogeneic transplantation. Clonal hemopoiesis of indeterminate potential (CHIP) is an age-associated genetic event linked to increased risk of primary hematological malignancies. Patients with cancer who have CHIP are at increased risk of developing therapy-related myeloid neoplasms (Gillis et al., 2017; Takahashi et al., 2017).

METHODS FOR THE DIAGNOSIS AND TREATMENT OF MYELODYSPLASTIC SYNDROMES (MDS)

Cytogenetic analysis

The most important diagnostic and prognostic method in patients with MDS is conventional karyotyping (Haase et al., 2007; Haase, 2008; Tiu et al., 2011; Giagounidis and Haase, 2013; Bacher et al.,

2015). Conventional karyotyping is performed on metaphase cells (a stage in cell division when chromosomes are condensed) with Giemsa stain resulting in a bending pattern of light and dark stripes, known as G-banding. The patterns are specific, allowing to identify each chromosome. Clonal chromosome abnormalities can be found in roughly 50% of *de novo* MDS patients and in up to 90% of MDS patients with therapy-related or secondary disease. The remaining MDS patients with normal karyotype can have molecular abnormalities (Bejar et al., 2011; Haferlach, 2012; Itzykson et al., 2013; Nybakken and Bagg, 2014; Bejar, 2015; Kennedy and Ebert, 2017).

The most common MDS abnormality is the interstitial deletion of chromosome 5/5q– or del(5q)/, del(7q)/monosomy 7 and trisomy 8 (Table 9.1), but a number of other chromosome abnormalities occur in MDS (Table 9.1). Karyotyping is critical, not only for MDS patients but also for patients with unexplained cytopenias in the absence of morphologically identifiable dysplasia in the differential diagnosis. Trisomy 8, del(20q), and loss of chromosome Y without other alterations are not presumptive evidence of MDS, because these cytogenetic abnormalities also occur in other myeloid neoplasms. The presence of three or more cytogenetic abnormalities are associated with poor prognosis (Table 9.2).

In comparison to chromosome banding analysis, fluorescence *in situ* hybridization (FISH) has some advantages. FISH can be performed on nondividing cells but is restricted according the respective FISH probe used in analysis. Therefore, FISH is not suitable for an initial analysis where we need comprehensive view on the karyotype. On the other hand, FISH is valuable for the verification of a suspected anomaly like del(5q).

MDS is a dynamic disease with a high probability for a karyotype evolution. Thus, the karyotype should be monitored repeatedly during the course of the disease, mainly from peripheral blood (immunomagnetically enriched circulating CD34+ cells) due to ethical reasons.

Importance of the morphological evaluation of bone marrow slides and peripheral blood films for the diagnosis of MDS

The morphological evaluation of peripheral blood films and bone marrow slides remains the essential

Table 9.1 Frequency of chromosome abnormalities in MDS

| Abnormality | Total (% of all MDS cases) | % of the individual abnormality | | |
		Isolated	With one additional abnormality	As part of complex abnormalities
5q−	15.1	47.0	17.0	36.0
−7/7q−	11.1	37.5	13.5	49.0
+8	8.4	46.8	21.4	31.8
−18/18q−	3.8	3.8	2.6	93.6
20q−	3.6	48.6	13.5	37.8
−5	3.3	1.4	5.8	92.8
−Y	2.8	70.7	8.6	20.7
+21	2.2	11.1	40.0	48.9
−17/17p−	2.0	2.4	2.4	95.2
inv/t(3q)	2.0	39.0	19.5	41.5
−13/13q−	1.9	12.5	15.0	72.5
+1/+1q	1.8	8.1	16.2	75.7
−21	1.6	9.1	12.1	78.8
+11	1.4	21.4	14.3	64.3
−12	1.3	0	7.7	92.3
12p−	1.2	28.0	24.0	48.0
t(5q)	1.2	25.0	12.5	62.5
11q−	1.1	34.8	17.4	47.8
9q−	1.1	34.8	13.0	52.2
t(7q)	1.1	27.3	27.3	45.0
−20	1.1	0	0	100.0

Source: Adapted from Haase D, *Ann Hematol* 87;2008:515–26.
Note: Total study cohort consisted of 2072 MDS patients.

Table 9.2 Cytogenetic abnormalities in *de novo* MDS with prognostic implications

| Prognostic subgroup | Cytogenetic abnormality | Frequency (%) | Median survival (years) | Progression to AML (years) | Hazard ratio | |
					OS	AML
Very good	−Y, del(11q)	4	5.4	NR	0.7	0.4
Good	Normal, del(5q), del(12q), del(20q); double abnormality including del(5q)	72	4.8	9.4	1.0	1.0
Intermediate	del(7q), +8, +19, i(17q); any other single or double independent clones	13	2.7	2.5	1.5	1.8
Poor	−7, inv(3), t(3q) or del(3q); double abnormality including −7 or del(7q); complex (3 abnormalities)	4	1.5	1.7	2.3	2.3
Very poor	Complex (>3 abnormalities)	7	0.7	0.7	3.8	3.6

Source: Adapted from Nybakken GE, and Bagg A. *J Mol Diagn* 16;2014:145–58.
Note: NR, not reached; OS, overall survival.

method of MDS diagnosis, and it needs a considerable amount of expertise and training. In addition, this method is cheap, rapid, and universally available. A number of cytological abnormalities (see Table 9.3) need to be considered.

Histopathology

Bone marrow trephine biopsies, examined by experienced hematopathologists are important mainly in cases of hypoplastic MDS or fibrotic MDS, where cytomorphology is of limited value for the diagnosis and prognosis of MDS. Hematopathologists use immunohistochemistry for CD34+ cells and in del(5q) MDS as well as for *TP53* overexpression caused by *TP53* mutation.

Classification of MDS patients

The initial French-British-American (FAB) classification system of MDS was published in 1982 (Bennett et al., 1982) and was later refined to the International Prognostic Scoring System (IPSS) (Greenberg et al., 1997) and to the WHO prognostic scoring system (WPSS) (Malcovati et al., 2005, 2007; Vardiman et al., 2009; Senent et al., 2013; Table 9.4). The new revised IPSS (IPSS-R)

integrated marrow cytogenetic subset, marrow blast percentage, and depth of cytopenias (hemoglobin, platelet and absolute neutrophil count) and was published in 2012 (Greenberg et al., 2012). Validation of WPSS for MDS and comparison with IPSS-R has been recently described (Della Porta et al., 2015). Two other prognostic systems for MDS subgroups (MD Anderson lower risk MDS prognostic scoring system; chronic myelomonocytic leukemia (CMML) prognostic scoring system) exist (Jonas and Greenberg, 2015; Zcidan et al., 2015). The 2016 revision to the WHO classification of myeloid neoplasms (Arber et al., 2016; Lichtman, 2016; Table 9.5) introduced refinements in morphological interpretation and cytopenia assessment and addressed the influence of recent developments in the molecular pathogenesis of MDS.

Mandatory for the diagnosis of MDS is the presence of dysplasia in >10% of cells within one or more cell lineages or presence of >15% ring sideroblasts or presence of MDS-associated cytogenetic abnormalities. The WHO proposal has raised some concern regarding minimal diagnostic criteria particularly in patients with normal karyotype without robust morphological markers of dysplasia (such as sideroblasts or excess of blasts).

Table 9.3 Cytological abnormalities in MDS

Peripheral blood

Granulopoiesis: pseudo-Pelger cells, abnormal chromatin clumping, hypo-/degranulation, an increase in the number of immature leukocytes (left shift)

Platelets: giant platelets, anisometry of platelets

Red cells: anisocytosis, poikilocytosis, dimorphic erythrocytes, polychromasia, hypochromasia, megalocytes (large nonnucleated red blood cells), basophilic stippling, presence of nucleated erythroid precursors, tear drop cells, ovalocytes (red blood cells with pale centers), fragmentocytes (fragmented or split red cells)

Bone marrow

Cellularity of the marrow

Erythropoiesis: megaloblastoid changes, multinuclearity, nuclear budding, nuclear bridges, atypical mitoses, sideroblastosis, ring sideroblasts, periodic acid-Schiff (PAS) positive red cell precursors

Megakaryocytes: micromegakaryocytes, mononuclear megakaryocytes, dumbbell-shaped nuclei, hypersegmentation, multinuclearity with multiple isolated nuclei

Granulopoiesis: left shift, increased medullary blast count, Auer rods or Auer bodies, hypo-/degranulation, pseudo-Pelger cells, nuclear anomalies (e.g., hypersegmentation, abnormal chromatin clumping), deficiency of myeloperoxidase, increase and morphological abnormality of monocytes

Source: Adapted from Giagounidis A, and Haase D. *Best Pract Res Clin Hematol* 26;2013:337–53.

Table 9.4 World Health Organization MDS classification and criteria, 2008

Classification	Blood findings	Bone marrow findings
Refractory cytopenia with unilineage dysplasia (RCUD)	Unicytopenia or bicytopenia No or rare blasts (<1%)	Unilineage dysplasia: >10% of cells in one myeloid lineage
Refractory anemia (RA), refractory neutropenia (RN), refractory thrombocytopenia (RT)	Anemia Neutropenia Thrombocytopenia	<5% blasts; <15% of erythroid precursors are ring sideroblasts
Refractory anemia with ring sideroblasts (RARS)	Anemia No blasts	>15% of erythroid precursors are ring sideroblasts Erythroid dysplasia only; <5% blasts
Refractory cytopenia with multilineage dysplasia (RCMD)	Cytopenia(s) No or rare blasts (<1%) No Auer rods <1 × 10⁹/L monocytes	Dysplasia in >10% of the cells in two or more myeloid lineages <5% blasts in marrow No Auer rods ±15% ring sideroblasts
Refractory anemia with excess blasts-1 (RAEB-1)	Cytopenia(s) <5% blasts No Auer rods <1 × 10⁹/L monocytes	Unilineage or multilineage dysplasia; 5%–9% blasts No Auer rods
Refractory anemia with excess blasts-2 (RAEB-2)	Cytopenia(s) 5%–19% blasts Auer rods± <1 × 10⁹/L monocytes	Unilineage or multilineage dysplasia; 10%–19% blasts Auer rods±
Myelodysplastic syndrome–unclassified (MDS-U)	Cytopenias <1% blasts	Unequivocal dysplasia in <10% of cells in one or more myeloid cell lines when accompanied by a cytogenetic abnormality considered as presumptive evidence for a diagnosis of MDS <5% blasts
MDS associated with isolated del(5q)	Anemia Usually normal or increased platelet count No or rare blasts (<1%)	Normal to increased megakaryocytes with hypolobated nuclei <5% blasts Isolated del(5q) cytogenetic abnormality No Auer rods

Source: Adapted from Nybakken GE, and Bagg A. *J Mol Diagn* 16;2014:145–58.

Flow cytometry (FCM) immunophenotyping has been proposed as a tool to improve the evaluation of marrow dysplasia. Immunophenotyping is an accurate method for quantitative and qualitative evaluation of hematopoietic cells. MDS have been found to have abnormal expression of several cellular antigens. Although no single immunophenotypic parameter has been proven to be diagnostic of MDS, combinations of such parameters into scoring systems have been shown to discriminate MDS from other cytopenias with high sensitivity and acceptable specificity. When morphology and cytogenetics are indeterminate, an abnormal phenotype determined by FCM can help to establish a definitive diagnosis of MDS (Cremers et al., 2016; Della Porta and Picone, 2017).

Table 9.5 World Health Organization MDS classification and criteria, 2016 revision

Classification	Blood findings	Bone marrow findings
MDS with single lineage dysplasia (MDS-SLD)	Peripheral blood (PB) blasts <1%	Bone marrow (BM) blasts <5%
MDS with multilineage dysplasia (MDS-MLD)	PB blasts <1%	BM blasts <5%
MDS with ring sideroblasts and with single lineage dysplasia (MDS-RS-SLD)	PB blasts <1%	BM blasts <5%
MDS with ring sideroblasts and with multilineage dysplasia (MDS-RS-LD)	PB blasts <1%	BM blasts <5%
MDS with isolated del(5q)	PB blasts <1%	BM blasts <5%
MDS with excess blasts (MDS-EB-1)	PB blasts 2%–4%	BM blasts 5%–9%
MDS with excess blasts (MDS-EB-2)	PB blasts 5%–19%	BM blasts 10%–19%
MDS, unclassifiable (MDS-U) with 1% blood blasts	PB blasts = 1%	BM blasts <5%
MDS, unclassifiable (MDS-U) with single lineage dysplasia and pancytopenia	PB blasts <1%	BM blasts <5%
MDS, unclassifiable (MDS-U) based on defining cytogenetic abnormality	PB blasts <1%	BM blasts <5%
Refractory cytopenia of childhood	PB blasts <2%	BM blasts <5%

Source: Adapted from Arber DA et al., *Blood* 127;2016:2391–405.

Note: Cytopenia is defined as hemoglobin <10 g/dL, platelet count <100 × 10^9/L, and absolute neutrophil count <1.8 × 10^9/L. Rarely, MDS may be present with mild anemia or thrombocytopenia above these levels. PB monocytes must be <1 × 10^9/L. Isolated del(5q) is defined as del(5q) alone or del(5q) with 1 additional abnormality except −7 or del(7q). Cases with >15% ring sideroblasts by definition have significant erythroid dysplasia and are classified as MDS-RS-LD.

Somatic mutations of individual genes as clinically relevant biomarkers in MDS

Great developments in molecular technologies (high-throughput next-generation sequencing) helped to elucidate differential roles of mutations in MDS. It is estimated that one or more mutated genes can be identified in over 90% of MDS cases (Bejar et al., 2011; Haferlach, 2012; Itzykson et al., 2013; Papaemmanuil et al., 2013; Haferlach et al., 2014; Nybakken and Bagg, 2014; Bejar, 2015; Kennedy and Ebert, 2017). The terms "driver" and "passenger" mutation have been used. A driver mutation is causally implicated in oncogenesis. It has conferred growth advantage on the cancer cell and has been positively selected in the microenvironment of the tissue in which the cancer arises. A driver mutation need not be required for maintenance of the final cancer (although it often is), but it must have been selected at some point along the lineage of cancer development. Driver mutations have equivalent prognostic significance, whether clonal or subclonal, and leukemia-free survival deteriorated steadily as numbers of driver mutations increased. Thus, analysis of oncogenic mutations in a large cohort of patients illustrates the interconnection between the cancer genome and disease biology.

A passenger mutation has not been selected, has not conferred clonal growth advantage, and has therefore not contributed to cancer development. Passenger mutations are found within cancer genomes because somatic mutations without functional consequences often occur during cell division. Thus, a cell that acquires a driver mutation will already have biologically inert somatic mutations within its genome. These will be carried along in the clonal expansion that follows and therefore will be present in all cells of the final cancer.

Frequency and impact of chosen genetic mutations identified in MDS are shown in Table 9.6. The most frequently mutated MDS genes are *SF3B1*, *TET2*, *ASXL1*, *SRSF2*, *DNMT3A*, *RUNX1*, *U2AF1*, *ZRSR2*, *STAG2*, *TP53*, *NRAS*, and *EZH2*. No single gene is mutated in the majority of MDS cases and

Table 9.6 Frequency and impact of chosen genetic mutations identified in MDS

Gene mutation	Prognostic impact	Estimated frequency in MDS	Additional findings
SF3B1	Favorable	20%–25%	Highly associated with ring sideroblasts
TP53	Poor	8% (found in 20% of 5q– MDS)	Decreased response to lenalidomide in 5q– MDS, poor outcome post alloSCT, associated with complex karyotype, higher BM blasts, and thrombocytopenia
EZH2	Poor	5%	Identifies lower risk MDS with aggressive course
ETV6	Poor	2%–3%	
RUNX1	Poor	9%–10%	
ASXL1	Poor	15% (found in 40% of CMML)	
NRAS	Poor	4%–8%	
DNMT3A	Poor	10%–15%	Poor outcome post alloSCT
SRSF2	Poor	12%	Occurrence with TET2 mutation is associated with monocytosis
PTPN11	Poor	1%	
U2AF1	Poor	7%–10%	
TET2	Unclear	20%–30%	Improved response to azacitidine, poor outcome post alloSCT, occurrence with SRSF2 mutation is associated with monocytosis

Source: Adapted from Lee EJ et al., Blood Rev 30;2016:1–10.

most genes are mutated in fewer than 5% of cases. Nearly two-thirds of MDS patients carry a splicing factor mutation and more than half have a mutation in an epigenetic regulator. Several additional pathways are frequently affected including hematopoietic transcription factors, tyrosine kinase signaling, cohesin genes, and DNA damage response and repair enzymes.

The clinical impact of selected genetic lesions has been studied but only in a limited number of MDS patients in serially collected samples (Walter et al., 2013). To better understand MDS progression in terms of gene mutations, a large number of serially collected samples is necessary. Clonal architecture is derived from the allelic burden of driver mutations. During MDS progression, the number of mutations, their diversity, and clone sizes increase (Makishima et al., 2017). Makishima et al. (2017) compared high-risk and low-risk MDS samples. Mutations in 10 genes were enriched in high-risk MDS, including GATA2, NRAS, KRAS, IDH2, TP53, RUNX1, STAG2, ASXL1, ZRSR2, and TET2. MDS patients with these mutations might be simply associated with secondary AML (sAML) evolution from MDS. SF3B1 mutations

were strongly enriched in low-risk MDS, compared to high-risk MDS. MDS patients with SF3B1 mutations are less likely to progress to sAML. SF3B1 mutations may define a distinct subset of MDS with ring sideroblasts (Malcovati et al., 2015; Malcovati and Cazzola, 2016).

Mutations in other genes (e.g., DNMT3A and U2AF1) are not enriched in specific MDS subtypes. These mutations represent founder or ancestral mutations that initiate the early stage of MDS, rather than secondary mutations involved in MDS progression. Mutations of TP53, EZH2, RUNX1, ETV6, and ASXL1 are associated with greater risk than predicted by the IPSS and the IPSS-R (Bejar et al., 2011). Further studies found additional mutated genes (NRAS, CBL, PRPF8, PTPN11, and NF1) with adverse MDS prognosis independent of the IPSS-R (Papaemmanuil et al., 2013; Haferlach et al., 2014). One or more of these adverse mutations can be found in over one-third of MDS patients. Therefore, we routinely underestimate disease risk using conventional analysis (Bejar, 2017). Complex karyotype MDS patients have about 50% TP53 mutations and have the worst overall outcomes, even after treatment.

Table 9.7 Clinical criteria for ICUS, CHIP, and MDS

	Nonclonal ICUS		CHIP	Lower Risk MDS	Higher Risk MDS
Clonality	−	+	+	+	+
Dysplasia	−	−	−	+	+
Cytopenia	+	−	+ (CCUS)	+	+
BM blast %	<5%	<5%	<5%	<5%	<19%
Overall risk	Very low	Very low	Low (?)	Low	High

Source: Adapted from Steensma DP et al. *Blood* 126;2015:9–16.
Abbreviation: ICUS, idiopathic cytopenia of undetermined significance; CHIP, clonal hematopoiesis of indeterminate potential; MDS, myelodysplastic syndrome; CCUS, clonal cytopenia of undetermined significance.

MDS patients with complex karyotype who do not carry *TP53* mutations have better outcomes approaching patients without complex karyotype (Bejar and Haferlach, 2014). Furthermore, *TP53* mutations are present in approximately 20% of MDS patients with isolated del(5q) and disturb treatment with immunomodulatory or cereblon-binding drug lenalidomide and have worse survival (Kulasekararaj et al., 2013; Saft et al., 2014; Zhang et al., 2017). The DNA hypomethylating agents (HMAs) azacitidine and decitabine (Takahashi et al., 2016) or allogeneic hematopoietic stem cell transplantation (Bejar et al., 2014; Brierley and Steensma, 2016; de Witte et al., 2017) do not abrogate the negative prognostic effect of *TP53* mutations. The overall survival of MDS patients was significantly lower in the presence of *TP53* mutations, especially when these mutations are combined with poor cytogenetics. Next-generation sequencing is a valuable technology, but may not be economically feasible for routine use. Alternatively, immunohistochemistry is fast, reproducible, and cost effective for routine laboratory use. The relationship between p53 expression and *TP53* mutation is known and the immunohistochemical pattern of p53 can be used as a measure of *TP53* mutation burden and adverse clinical outcome in MDS and sAML (Saft et al., 2014; McGraw et al., 2016).

Several studies have demonstrated that somatic mutations in the blood of unselected individuals in the premalignant state with a frequency that increases with age (Steensma et al., 2015; Jan et al., 2017; Makishima et al., 2017). These individuals have CHIP and the majority of these persons with CHIP will never develop a hematologic disorder. Patients who do not meet the diagnostic criteria for MDS have idiopathic cytopenias of undetermined significance (ICUS). ICUS is a heterogeneous group that includes a clonal subset. Clinical criteria for ICUS, CHIP, and MDS are shown in Table 9.7.

Therapies for MDS

Guidelines for the diagnosis and management of adult myelodysplastic syndromes were published (Killick et al., 2014; Tefferi and Garcia-Manero, 2015; Greenberg et al., 2017). Erythropoietic stimulating agents (ESAs) remain the first-line treatment in most cases of non-del(5q) low-risk MDS, where anemia is the major clinical problem (Santini, 2016; Almeida et al., 2017). Lenalidomide is used in low-risk MDS with del(5q) (List et al., 2014; Komrokji and List, 2016). Lenalidomide was also used in lower-risk patients with MDS without 5q deletion after failure of ESAs (Raza et al., 2008; Sibon et al., 2012; Santini et al., 2016). Clinical and genetic markers with potential for predicting responsiveness to lenalidomide are shown in Table 9.8 (Stahl and Zeidan, 2017b).

Best supportive care (BSC) was considered the primary standard treatment for high-risk MDS, except for patients younger than 65 years of age with a compatible donor for allogeneic HSCT following myeloablative or nonmyeloablative conditioning regimens. Recently, however azacitidine and decitabine as hypomethylating agents have become the standard approach for older high-risk MDS patients who are not amenable to allogeneic HSCT and have largely replaced chemotherapy. Azacitidine provides a survival advantage in high-risk MDS patients, but decitabine does not for a variety of reasons. Guadecitibine (SGI-110) is a second-generation HMA and a dinucleotide of decitabine and deoxyguanosine (Nebbioso et al.,

Table 9.8 Clinical and genetic markers with potential to predict responsiveness to lenalidomide

Marker	Prediction of response
Clinical markers (Negoro et al., 2016; Stahl and Zeidan, 2017b)	
Sex	Negative association with male sex
Platelet and neutrophil count	Negative associations with thrombocytopenia and neutropenia
IPSS-R	Negative association with high/very high risk
Erythropoietic gene expression signature (Ebert et al., 2008a)	A total of 47 genes involved in erythroid differentiation are more highly expressed in nonresponders to lenalidomide In MDS without del(5q); patients who respond to lenalidomide have evidence of defective erythroid differentiation, but it failed to predict responders in the larger MDS-005 trial
Mutations (Negoro et al., 2016; Stahl and Zeidan, 2017b)	
U2AF1 mutations	Negatively associated in del(5q) and non-del(5q) cohort
TP53 mutations	Negatively associated in del(5q) cohort
DEAD box RNA helicase mutations	Positively associated in non-del(5q) cohort
High level of full length cereblon mRNA (Jonasova et al., 2015)	Positively associated in del(5q) cohort with response to lenalidomide
Polymorphism A/G located at −29 nucleotides upstream of the transcription start site of the CRBN gene located on chromosome 3 at nucleotide 3179748; NC_000003.12 (Sardnal et al., 2013)	A higher distribution of G alleles is a biomarker of response to lenalidomide in non-del(5q) cohort
Low expression of NPM1 before treatment with lenalidomide (Chesnais et al., 2014)	Negatively associated with response to lenalidomide in non-del(5q) cohort
Immune environment (Epling-Burnette et al., 2013)	Higher percentage of CD28 negative T cells negatively associated with response to lenalidomide in non-del(5q) in non-del(5q) MDS
CD28 negative T cells	

Source: Adapted from Stahl M, Zeidan AM, Cancer 2017b, doi: 10.1002/cncr.30585.

2015; Navada and Silverman, 2016). *TET2*-mutated MDS have a higher probability of responding to these hypomethylating agents (Itzykson et al., 2011). *U2AF1* mutation was significantly associated with nonresponse to azacitidine, which was consistent in multivariate analysis. *DNMT1, DNMT3A, RAS*, and *TP53* were independent predicting factors of shorter overall survival (Jung et al., 2016). Combinations of azacitidine and lenalidomide (DiNardo et al., 2015) or azacitidine and histone deacetylase inhibitors vorinostat, entinostat, or pracinostat (dual targeting of aberrant epigenetic processes) are also in clinical trials (Ornstein et al., 2015; Garcia-Manero et al., 2017; Stahl and Zeidan, 2017a). However, there is a problem due to higher toxicity in some cases. These agents have improved both survival and quality of life, but results overall remain poor (Zeidan et al., 2016; Bhatt and Blum, 2017). Patients in the highest risk groups had a median OS from diagnosis of 11–16 months and should be considered for experimental approaches. Combination of HMA with Bcl-2 (B-cell lymphoma-2 inhibitors, e.g., venetoclax/ABT-199, GDC-0199/ or ABT-737) (Bogenberger et al., 2014;

Jilg et al., 2016), IDH2 inhibitor (AG-221) (Dang et al., 2017; Medeiros et al., 2017), CD33 antibody-drug conjugates (Daver et al., 2016), Fc-engineered CD33 antibody BI 836858 (Eksioglu et al., 2017), CD16xCD33 bispecific killer cell engager (BiKE) (Gleason et al., 2014), and hedgehog inhibitors like sonidegib (erismodegib, LDE255) and glasdegib (PF-04449913) (Medler et al., 2015; Zou et al., 2015; Hanna and Shevde, 2016) are in ongoing trials.

It will be a better way to develop new therapies for MDS to move promising new agents into an earlier treatment in the course of disease, instead of conducting randomized trials in end-stage patients (Blum, 2016; Zahr et al., 2016). Allogeneic transplantation (alloHSCT) remains the only curative therapy for MDS patients (de Witte et al., 2017). Recent findings of a phase III randomized trial comparing reduced-intensity conditioning (RIC) regimens with less toxicity and high-intensity preparative regimens (myeloablastive conditioning [MAC]) have shown that overall survival was higher with MAC, but this was not statistically significant. RIC resulted in lower treatment-related toxicity but higher relapse rates compared with MAC (Scott et al., 2017). MDS patients older than 65 years are today more frequently transplanted and this treatment is covered as participation in research. All fit patients without comorbidities should be considered for HSCT when the disease-related factors allow the recommendation. Somatic mutations in *ASXL1*, *RUNX1*, *RAS*, *JAK2*, and *TP53* were associated with unfavorable outcomes and shorter survival after allogeneic HSCT (Della Porta et al., 2016; Lindsley et al., 2017).

New knowledge about molecular and cellular mechanisms of MDS as basis for new therapy

Advances in supportive care are connected with advances in MDS molecular biology and pathogenesis research (Gill et al., 2016). A new target has emerged in treating MDS-related anemia. Defects in erythroid differentiation lead to increased production of erythropoietin without effective hemoglobin synthesis. Growth differentiation factor GDF11, an important erythropoiesis regulator, accumulates and can be inhibited by transforming growth factor β (TGF-β) superfamily ligand trap strategies (Dussiot et al., 2014; Suragani et al., 2014). Luspatercept is an activin receptor antagonist that

functions as a ligand trap for GDF11 and other TGF-β family ligands to suppress Smad2/3 activation and to increase hemoglobin synthesis (Mies et al., 2016; Almeida et al., 2017).

Thrombocytopenia is commonly seen in MDS patients, and bleeding complications are a major cause of morbidity and mortality. Thrombocytopenia is an independent factor for decreased survival and has been incorporated in newer prognostic scoring systems. The mechanisms of thrombocytopenia are multifactorial and involve a differentiation block of megakaryocytic progenitor cells, leading to dysplastic, hypolobated, and microscopic appearing megakaryocytes or increased apoptosis of megakaryocytes and their precursors. Dysregulated thrombopoietin (TPO) signaling and increased platelet destruction through immune or nonimmune mechanisms are frequently observed in MDS. The clinical management of patients with low platelet counts remains challenging and approved chemotherapeutic agents, such as lenalidomide and azacytidine, can also lead to a transient worsening of thrombocytopenia. Platelet transfusion was the only supportive treatment option for clinically significant thrombocytopenia. The TPO receptor agonists romiplostim and eltrombopag have shown clinical activity in clinical trials in MDS (Li et al., 2016; Oliva et al., 2017).

Rigosertib (ON01910.Na) is an inhibitor of the phosphoinositide 3-kinase and polo-like kinase pathways (a multikinase inhibitor). Rigosertib has demonstrated therapeutic activity for patients with high-risk MDS in clinical trials. Rigosertib-induced apoptosis blocked the cell cycle at the G2/M phase and subsequently inhibited the proliferation of CD34+ cells from MDS, while it minimally affected the normal CD34+ cells. Rigosertib acted also via the activation of the p53 signaling pathway. Bioinformatics analysis based on gene expression profile and flow cytometry analysis revealed the abnormal activation of the Akt-PI3K, Jak-STAT, and Wnt pathways in high-grade MDS, while the p38 MAPK, SAPK/JNK, and p53 pathways were abnormally activated in low-grade MDS. Rigosertib could markedly inhibit the activation of the Akt-PI3K and Wnt pathways, whereas it activated the SAPK/JNK and p53 pathways in high-grade MDS (Chung, 2016; Garcia-Manero et al., 2016).

In recent years, much research has been aimed at the pathogenetic role of haploinsufficient genes located on the deleted region of the long arm of chromosome 5 (Pellagatti and Boultwood, 2015). Ebert et al. (2008b) and Dutt et al. (2011) described impaired ribosome biosynthesis due to *RPS14* (ribosomal protein 14 of the small ribosome subunit) gene haploinsufficiency leads to the E3 ubiquitin ligase HDM2 (human homologue to mouse double minute 2, major negative regulator of p53) inactivation by free ribosomal proteins, particularly RPL11. HDM2 degradation drives p53-mediated apoptosis of erythroid cells carrying the del(5q) aberration. This p53-mediated apoptosis of erythroid cells is a key effector of hypoplastic anemia in MDS patients with del(5q) (Dutt et al., 2011). *RPS14* haploinsufficiency causes a block in erythroid differentiation mediated by calprotectin (the heterodimeric S100 calcium-binding proteins S100A8 and S100A9) (Schneider et al., 2016). Diminutive somatic deletions in the 5q region were also found in some of these patients. These deletions were not identified by fluorescence *in situ* hybridization or cytogenetic testing but by single nucleotide polymorphism array genotyping (Vlachos et al., 2013). Therefore, some patients originally considered as MDS patients without del(5q) can have a phenotype of atypical 5q– syndrome and can be sensitive to lenalidomide therapy. Low RPS14 expression in 50%–70% MDS patients without del(5q) confers a higher apoptosis rate of nucleated erythrocytes and predicts prolonged survival (Czibere et al., 2009; Wu et al., 2013).

What is the mechanism of lenalidomide in del(5q)MDS based on what has been achieved and elucidated to date? Lenalidomide stabilizes E3 ubiquitin ligase HDM2, thereby accelerating p53 degradation (Wei et al., 2009, 2013). Lenalidomide inhibits phosphatases PP2a and Cdc25c (coregulators of cell cycle in which genes are very commonly deleted in del(5q) MDS) with consequent G2 arrest of del(5q) MDS progenitors and their apoptosis. PP2a and Cdc25c inhibition by lenalidomide suppress HDM2 autoubiquitination and subsequent degradation. Thus, lenalidomide has been shown to not only reverse apoptosis within the erythroid compartment, but also directly induce apoptosis of the myeloid clone in del(5q) MDS (Gandhi et al., 2006; Matsuoka et al., 2010). Lenalidomide upregulates expression of other two haploinsufficient genes located on chromosome 5q, genes for microRNAs (miR-145 and miR-146a) (Venner et al., 2013). These miRs are involved in

the Toll-like receptor pathway, IL-6 induction, and regulation of megakaryopoiesis (Starczynowski et al., 2010). Ito et al. (2010) discovered that thalidomide (founding member of IMiDs) binds CRBN in the terminal C-region (parts of exons 10 and 11 of the *CRBN* gene code in this IMiD binding region). Several researchers confirmed CRBN as the target of lenalidomide in multiple myeloma (MM), lymphoma, chronic lympocytic leukemia, and del(5q) MDS (Ito and Handa, 2016). CRBN is an ubiquitously expressed 51kDa protein with a putative role in cerebral development, especially memory and learning (Chang and Stewart, 2011; Kim et al., 2016).

We (Jonasova et al., 2015) have found that del(5q) MDS patients (the so-called 5q minus syndrome) have higher levels of full-length CRBN mRNA than other patients with lower risk MDS, linking higher levels of a known lenalidomide target cereblon (CRBN) and an MDS subgroup known to be especially sensitive to lenalidomide.

CRBN is a member and substrate receptor of the cullin 4 RING E3 ubiquitin ligase complex (CRL4). CRBN recruits substrate proteins to the CRL4 complex for ubiquitination and the subsequent degradation in proteasomes. IMiDs binds to CRBN in CRL4 complex and block normal endogenous substrates (CRBN and the homeobox transcription factor MEIS2 in MM) from binding to CRL4 for ubiquitination and degradation (Fischer et al., 2014). After IMID binding to CRBN, CRL4 complex is recruiting transcription factors Ikaros (IKZF1) and Aiolos (IKZF3) for ubiquitination and degradation in MM cells (Krönke et al., 2014). Degradation of these transcription factors explains lenalidomide's growth inhibition of MM cells and increased interleukin-2 (IL-2) release from T cells. However, it is unlikely that degradation of IKZF1 and IKZF3 accounts for lenalidomide's activity in MDS with del(5q). Krönke et al. (2015) identified a novel target casein kinase1A1 (CSNK1A1) by quantitative proteomics in the myeloid cell line KG-1. CSNK1A1 is encoded in the del(5q) commonly deleted region and the gene is haploinsufficient. Lenalidomide treatment leads to increased ubiquitination of the remaining CSNK1A1 and decreased protein abundance. CSNK1A1 negatively regulates β-catenin, which drives stem cell self-renewal, and *CSNK1A1* haploinsufficiency causes the initial clonal expansion in patients with the del(5q) MDS and contributes to the pathogenesis of del(5q) MDS. The further inhibition of CSNK1A1 in del(5q) MDS is associated with del(5q) failure and p53 activation. The inhibition of CSNK1A1 reduced RPS6 phosphorylation, induced p53 expression and growth inhibition, and triggered the myeloid differentiation program. TP53-null leukemia did not respond to CSNK1A1 inhibition, strongly supporting the importance of the p53 expression for the yield of CSNK1A1 inhibition. *CSNK1A1* mutations have been recently found in 5%–18% of MDS patients with del(5q) (Schneider et al., 2014). These mutations are associated similarly to the effect of *TP53* mutations with rise to a poor prognosis in del(5q) MDS (Smith et al., 2015). Other studies did not find impact of *CSNK1A1* mutations on lenalidomide treatment in del(5q) MDS (Heuser et al., 2015; Negoro et al., 2016).

While CSNK1A1 is the CRL4^{CRBN} target in del(5q) MDS, CRL4^{CRBN} targets in lower risk non-del(5q) remain to be determined. The mechanism of action of lenalidomide is still unclear in non-del(5q) MDS cells. Recent evidence shows that lenalidomide directly improves erythropoietin receptor (EPOR) signaling by EPOR upregulation mediated by a posttranscriptional mechanism (Basiorka et al., 2016a). Lenalidomide stabilizes the EPOR protein by inhibition of the E3 ubiquitin ligase RNF41 (ring finger protein 41, also known as neuregulin receptor degradation protein-1 [Nrdp1] and fetal liver ring finger [FLRF]) (Basiorka et al., 2016a) and induces lipid raft assembly to enhance EPOR signaling in MDS erythroid progenitors (McGraw et al., 2012, 2014).

After failure of ESAs, lenalidomide yields red blood cell transfusion independence in 20%–30% of lower-risk non-del(5q) MDS. Indeed, several observations suggest an additive effect of ESA and lenalidomide in this situation (Komrokji et al., 2012; Toma et al., 2016). Basiorka et al. reported activation of the NLRP3 inflammasome in MDS (Basiorka et al., 2016b; Sallman et al., 2016). NRLP3 drives clonal expansion and pyroptotic cell death. Independent of genotype, MDS hematopoietic stem and progenitor cells (HSPCs) overexpress inflammasome proteins. Activated NLRP3 complexes direct then activation of caspase-1, generation of interleukin-1β (IL-1β) and IL-18, and pyroptotic cell death. Mechanistically, pyroptosis is triggered by the alarmin S100A9 that is found in excess in MDS HSPCs and bone marrow

plasma. Further, like somatic gene mutations, S100A9-induced signaling activates NADPH oxidase (NOX) and increasing levels of reactive oxygen species (ROS). ROS initiates cation influx, cell swelling, and β-catenin activation. Knockdown of NLRP3 or caspase-1, neutralization of S100A9, and pharmacologic inhibition of NLRP3 or NOX suppress pyroptosis, ROS generation, and nuclear β-catenin in MDSs and are sufficient to restore effective hematopoiesis. Thus, alarmins and founder gene mutations in MDSs cause a common redox-sensitive inflammasome circuit. They are new candidates for therapeutic intervention.

The effects of lenalidomide in non-del(5q) are thought to be secondary to modulation of the immune system. Hyperactivated T cells inhibit hematopoiesis. Immunosuppressive therapies with antithymocyte globulin alone and in combination with prednisone or cyclosporine show response rates between 25% and 40% (Haider et al., 2016; Stahl and Zeidan, 2017b). Several trials with various combinations of lenalidomide are shown in Table 9.9.

THE CONCEPT OF PERSONALIZED OR PRECISION MEDICINE

The concept of personalized or precision medicine was introduced in the State of the Union speech given by U.S. President Barack Obama in 2015. He announced the precision medicine initiatives that would enable development of individualized patient care in order to improve patient outcomes. Initial attempts to analyze the human genome started with cytogenetic analyses and chromosome banding. Chromosomal analysis can only visualize the human genome at 0.5%–1% resolution. To further improve the detection of genomic abnormalities, chip and array techniques were developed. Single nucleotide polymorphisms, frequently called SNPs (pronounced "snips"), are the most common type of genetic variation among people. Each SNP represents a difference in a single DNA building block, called a nucleotide. For example, a SNP may replace the nucleotide cytosine (C) with the nucleotide thymine (T) in a certain stretch of DNA. SNPs occur normally throughout a person's DNA. They occur once in every 300 nucleotides on average, which means there are roughly 10 million SNPs in the human genome. Most commonly, these variations are found in the DNA between genes.

Table 9.9 List of clinical trials investigating combinations of lenalidomide

Combination partner	Mechanism of action	Clinical trial number	Phase	MDS population
Cyclosporine-A	Immune-suppressive drug	NCT00840827	Phase 2	Low, intermediate-1; non-del(5q)
Prednisone	Immune-suppressive drug	NCT01133275	Phase 2	Low, intermediate-1; non-del(5q)
Romiplostim	TPO agonist	NCT00418665	Phase 2	All MDS
Eltrombopag	TPO agonist	NCT02928419 NCT01772420	Phase 2	Low, intermediate-1; with and without del(5q)
Ezatiostat	Small molecule glutathione analog inhibitor of glutathione S-transferase leads to activation of Jun kinase	NCT01062152	Phase 1–2	Low, intermediate-1; non-del(5q)
Clofarabine	Chemotherapy	NCT01629082	Phase 1–2	Higher-risk MDS
Melphalan	Chemotherapy	NCT00744536	Phase 2	Higher-risk MDS
Bortezomib	Proteasome inhibitor	NCT02312102 NCT00580242	Phase 1	Higher-risk MDS
Lintuzumab	Anti-CD33 monoclonal antibody	NCT00502112	Phase 1	All MDS

Source: Adapted from Stahl M, Zeidan AM, Cancer 2017b, doi: 10.1002/cncr.30585.

They can act as biological markers, helping scientists locate genes that are associated with disease. When SNPs occur within a gene or in a regulatory region near a gene, they may play a more direct role in disease by affecting the gene's function. Most SNPs have no effect on health or development. Some of these genetic differences, however, have proven to be very important in the study of human health. Researchers have found SNPs that may help predict an individual's response to certain drugs, susceptibility to environmental factors such as toxins, and risk of developing particular diseases. SNPs can also be used to track the inheritance of disease genes within families.

The first DNA sequencing effort was developed in the 1970s when fragments of DNA were cloned and amplified by polymerase chain reaction (PCR). This method enabled reading of sequences about 1000 bp in length. New genomic technologies and mainly advances in bioinformatics helped to develop next-generation sequencing (NGS), whole genome sequencing, whole exome sequencing, and targeted deep sequencing of specific mutations (Platzbecker and Fenaux, 2015; Nazha and Sekeres, 2017; Sheikine et al., 2017). Since the completion of the Human Genome Project, a 2.91 billion base pair (bp) consensus sequence of the euchromatic portion of the human genome was generated by the whole genome shotgun sequencing method and was published in 2001 (Venter et al., 2001). The genomic landscape of MDS was analyzed by these methods (Bejar et al., 2011; Haferlach, 2012; Itzykson et al., 2013; Papaemmanuil et al., 2013; Haferlach et al., 2014; Nybakken and Bagg, 2014; Bejar, 2015; Kennedy and Ebert, 2017). The impact of genomic data on diagnosis, prognosis, and therapy of individual MDS patients was described in this chapter.

FUTURE OPPORTUNITIES

Genomic and clinical data can be used to further understand the biology and pathophysiology of MDS, a group of heterogeneous and complex hematologic disorders primarily found within the older population. A risk-adapted treatment strategy is necessary in this disease with a highly variable clinical course. However, this personalized approach, including performance status, EPO level, ferritin level, prognostic scoring systems (good risk or poor risk), cytogenetics (especially presence of del(5q) for targeted treatment with lenalidomide), and mutations, is not supported by prospective randomized trials. Although limitations exist in the application of precision medicine in MDS, mutation profiling analyses will further improve the classification of MDS and discover new therapeutic biomarkers. Recently, great progress has been done in the study of bone marrow (BM) microenvironment in MDS (Bulycheva et al., 2015; Cogle et al., 2015; Rankin et al., 2015; Balderman et al., 2016; Glenthøj et al., 2016; Pleyer et al., 2016; Korn and Méndez-Ferrer, 2017). This development will extend our current limited therapeutic portfolio by detection of new therapeutic targets for targeted therapies. These therapies will offer in the near future truly personalized approaches (Platzbecker and Fenaux, 2015). Accumulating evidence suggests that the altered BM microenvironment in general, and in particular the components of the stem cell niche, including mesenchymal stem and progenitor cells (MSPCs) and their progeny, play a pivotal role in the evolution and propagation of MDS. MSPCs have potent immunomodulatory capacities and communicate with diverse immune cells, but also interact with various other cellular components of the microenvironment as well as with normal and leukemic stem and progenitor cells. Moreover, compared to normal MSPCs, MSPCs in MDS and AML often exhibit altered gene expression profiles, an aberrant phenotype, and abnormal functional properties (Table 9.10). These alterations contribute to the "reprogramming" of the stem cell niche into a disease-permissive microenvironment where an altered immune system, abnormal stem cell niche interactions, and an impaired growth control lead to disease progression. Common signaling pathways, such as PI3K/AKT and WNT/β-catenin, regulate multiple MSPC properties, including proliferation, differentiation, and cell-cell interaction. Impairment of these signaling pathways was found in stromal cells of BM microenvironment in MDS. This finding highlights a potential new strategy for treating some myeloid disorders. Inhibition of the Wnt/β-catenin pathway by pyrvinium in the bone marrow niche prevented the development of MDS in the Apcdel/+ MDS mouse model (Falconi et al., 2016). Pyrvinium, an anthelmintic (against infections with parasitic worms) drug approved by the U.S. Food and Drug Administration (FDA), has

Table 9.10 Frequent alterations of mesenchymal stem and progenitor cells (MSPCs) in MDS

Alterations
Cytogenetic aberrations in MDS-MSPCs (chromosomes 1 and 7 are more frequently involved)
Lower expression of *Dicer1, DROSHA, AURKA, AURKB* genes in MDS-MSPCs
Altered immunophenotype in MDS-MSPCs; decreased CD44 and CD49e, CD90, CD104, and CD105 expression, increased CXCL12 expression
Impaired proliferation and differentiation capacity of MDS-MSPCs
Impaired cytokine production, including IL-32 by MDS-MSPCs
Dysregulation of Wnt signaling pathway in MDS-MSPCs
Impaired hematopoietic stem and progenitor cells (HSPC) support by MDS-MSPCs

Source: Adapted from Bulycheva E et al., *Leukemia* 29;2015:259–68.

been identified as a potent inhibitor of the Wnt/β-catenin pathway. It stimulates casein kinase 1α (CK1α), which is part of the β-catenin destruction complex where it phosphorylates β-catenin at serine 45, priming the subsequent phosphorylation of β-catenin by glycogen synthase kinase 3β (GSK3β). These phosphorylations mark β-catenin for proteasomal degradation. This is one of the central regulating events controlling the Wnt signaling pathway. Furthermore, CK1 has been shown to additionally regulate Wnt signaling through phosphorylation of lymphocyte enhancer factor-1 (LEF-1) and β-catenin leading to disruption of the LEF-1/β-catenin transcription complex.

ACKNOWLEDGMENTS

This work was supported by a research project for conceptual development of research organization (00023736; Institute of Hematology and Blood Transfusion, Prague) from the Ministry of Health of the Czech Republic.

CONFLICT OF INTEREST

There are no commercial or other associations in connection with this review article.

REFERENCES

Adés L, Itzkson R, and Fenaux P. Myelodysplastic syndromes. *Lancet* 383;2014:2239–52.

Almeida A, Fenaux P, List AF, Raza A, Platzbecker U, and Santini V. Recent advances in the treatment of lower-risk non-del(5q) myelodysplastic syndromes (MDS). *Leuk Res* 52;2017:50–7.

Arber DA, Orazi A, Hasserjian R, Thiele J, Borowitz MJ, Le Beau MM, Bromfield CD, Cazzola M, and Vardiman JW. The 2016 revision to the World Health Organization classification of myeloid neoplasms and acute leukemia. *Blood* 127;2016:2391–405.

Avgerinou C, Alamanos Y, Zikos P et al. The incidence of myelodysplastic syndrome in Western Greece is increasing. *Ann Hematol* 92;2013:877–87.

Babushok DV, and Bessler M. Genetic predisposition syndromes: When should they be considered in the work-up of MDS? *Best Pract Res Clin Hematol* 28;2015:55–68.

Babushok DV, Bessler M, and Olson TS. Genetic predisposition to myelodysplastic syndrome and acute myeloid leukemia in children and young adults. *Leuk Lymphoma* 57;2016:520–36.

Bacher U, Schanz J, Braulke F, and Haase D. Rare cytogenetic abnormalities in myelodysplastic syndromes. *Mediterr J Hematol Infect Dis* 7;2015:e2015034.

Balderman SE, Li AJ, Hoffman CM et al. Targeting of the bone marrow misroenvi-ronment improves outcome in a murine model of myelodysplastic syndrome. *Blood* 127;2016:616–25.

Bannon SA, and DiNardo CD. Hereditary predisposition to myelodysplastic syndrome. *Int J Mol Sci* 17;2016:838.

Basiorka AA, McGraw KL, De Ceuninck L et al. Lenalidomide stabilizes the erythropoietin receptor by inhibiting the E3 ubiquitin ligase RNF41. *Cancer Res* 76;2016a:3531–40.

Basiorka AA, McGraw KL, Eksioglu EA et al. The NLRP3 inflammasome functions as a driver of the myelodysplastic syndrome phenotype. *Blood* 128;2016b:2960–75.

Bejar R. Myelodysplastic syndromes diagnosis: What is the role of molecular testing? *Curr Hematol Malig Rep* 10;2015:282–91.

Bejar R. Implications of molecular genetic diversity in myelodysplastic syndromes. *Curr Opin Hematol* 24;2017:73–8.

Bejar R, and Haferlach T. TP53 mutation status divides MDS patients with complex karyotypes into distinct prognostic risk groups: Analysis of combined datasets from the International Working Group for MDS-Molecular Prognosis Committee. *Blood* 124;2014:532a.

Bejar R, and Steensma DP. Recent developments in myelodysplastic syndromes. *Blood* 124;2014:2793–803.

Bejar R, Stevenson K, Abdel-Wahab O et al. Clinical effect of point mutations in myelodysplastic syndromes. *N Engl J Med* 364;2011:2496–506.

Bejar R, Stevenson K, Caughey B et al. Somatic mutations predict poor outcome in patients with myelodysplastic syndrome after hematopoietic stem-cell transplantation. *J Clin Oncol* 32;2014:2691–8.

Bennett JM, Catovsky D, Daniel MT, Flandrin G, Galton DA, Gralnick HR, and Sultan C. Proposals for the classification of the myelodysplastic syndromes. *Br J Haematol* 51;1982:189–99.

Bhatt G, and Blum W. Making the most of hypomethylating agents in myelodysplastic syndrome. *Curr Opin Hematol* 24;2017:79–88.

Blum W. Why the stagnation in effective therapy for MDS? *Best Pract Res Clin Hematol* 29;2016:334–40.

Bogenberger JM, Kornblau SM, Pierceall WE et al. BCL-2 family proteins as 5-azacytidine sensitizing targets and determinants of response in myeloid malignancies. *Leukemia* 28;2014:1657–65.

Bowen DT. Occupational and environmental etiology of MDS. *Best Pract Res Clin Hematol* 26;2013:319–26.

Brierley CK, and Steensma DP. Allogeneic stem cell transplantation in myelodysplastic syndromes: Does pretransplant clonal burden matter? *Curr Opin Hematol* 23;2016:167–74.

Bulycheva E, Rauner M, Medyouf H, Theurl I, Bornhäuser M, Hofbauer LC, and Platzbecker U. Myelodysplasia is in the niche: Novel concepts and emerging therapies. *Leukemia* 29;2015:259–68.

Chang XB, and Stewart AK. What is the functional role of the thalidomide binding protein cereblon? *Int J Biochem Mol Biol* 2;2011:287–94.

Chesnais V, Renneville A, Toma A et al. NPM1 expression level and a CRBN polymorphism are able to predict the rate of response to lenalidomide in non del(5q) lower risk MDS patients resistant to erythropoiesis-stimulating agents: The GFM experience. *Blood* 124;2014:533.

Chung SS. Novel therapeutic strategies in MDS: Do molecular genetics help. *Curr Opin Hematol* 23;2016:79–87.

Churpek JE, Pyrtel K, Kanchi KL et al. Genomic analysis of germ line and somatic variants in familial myelodysplasia/acute myeloid leukemia. *Blood* 126;2015:2484–90.

Cogle CR, Craig BM, Rollison DE, and List AF. Incidence of myelodysplastic syndromes using a novel claims-based algorithm: High number of uncaptured cases by cancer registries. *Blood* 117;2011:7121–5.

Cogle CR, Saki N, Khodadi E, Li J, Shahjahani M, and Azizidoost S. Bone marrow niche in the myelodysplastic syndromes. *Leuk Res* 39;2015:1020–7.

Cremers EMP, Westers TM, Alhan C, Cali C, Wondergem MJ, Poddighe PJ, Ossenkoppele GJ, and van de Loosdrecht AA. Multiparameter flow cytometry is instrumental to distinguish myelodysplastic syndromes from non-neoplastic cytopenias. *Eur J Cancer* 54;2016:49–56.

Czibere A, Bruns I, Junge B, Singh R, Kobbe G, Haas R, and Germing U. Low RPS14 expression is common in myelodysplastic syndromes without 5q- aberration and defines a subgroup of patients with prolonged survival. *Haematologica* 94;2009:1453–5.

Dang L, Yen K, and Attar EC. IDH mutations in cancer and progress toward development of targeted therapeutics. *Ann Oncol* 27;2017:599–608.

Daver N, Kantarjian H, Ravandi F et al. A phase II study of decitabine and gemtuzumab ozogamicin in newly diagnosed and relapsed acute myeloid leukemia and high-risk myelodysplastic syndrome. *Leukemia* 30;2016:268–73.

Della Porta MG, Galli A, Bacigalupo A et al. Clinical effects of driver somatic mutations on the outcomes of patients with myelodysplastic syndromes treated with allogeneic hematopoietic stem-cell transplantation. *J Clin Oncol* 34;2016:3627–37.

Della Porta MG, and Picone C. Diagnostic utility of flow cytometry in myelodysplastic syndromes. *Mediterr J Hematol Infect Dis* 9;2017:e2017017.

Della Porta MG, Tuechler H, Malcovati L et al. Valadation of WHO classification-based Prognostic Scoring System (WPSS) for myelodysplastic syndromes and comparison with the revised International Prognostic Scoring System (IPSS-R). A study of the International Working Group for Prognosis in Myelodysplasia (IWG-PM). *Leukemia* 29;2015:1502–13.

de Witte T, Bowen D, Robin M et al. Allogeneic hematopoietic stem cell transplantation for MDS and CMML: Recommendations from an international expert panel. *Blood* 129;2017:1753–62.

DiNardo CD, Daver N, Jabbour E et al. 2015. Sequential azacitidine and lenalidomide in patients with high-risk myelodysplastic syndromes and acute myeloid leukaemia: A single-arm, phase 1/2 study. *Lancet Haematol* 2: e12–e20.

Dussiot M, Maciel TT, Fricot A et al. An activin receptor IIA ligand trap corrects ineffective erythropoiesis in beta-thalassemia. *Nat Med* 20;2014:398–407.

Dutt S, Narla A, Lin K et al. Haploinsufficiency for ribosomal protein genes causes selective activation of p53 in human erythroid progenitor cells. *Blood* 117;2011:2567–76.

Ebert BL, Galili N, Tamayo P et al. An erythroid differentiation signature predicts response to lenalidomide in myelodysplastic syndrome. *PLOS Med* 5;2008a:e35.

Ebert BL, Pretz J, Bosco J et al. Identification of RPS14 as a 5q- syndrome gene by RNA interference screen. *Nature* 451;2008b:335–9.

Eksioglu EA, Chen X, Heider K-H et al. *Leukemia* 2017, doi: 10.1038/leu.2017.21.

Epling-Burnette PK, Han Y, Mailloux AW et al. Novel predictor of lenalidomide response in non-del(5q) MDS reveals linkage to molecular mechanism: First characterization of T-cell function in cereblon homozygous deficient mice. *Blood* 122;2013:748.

Falconi G, Fabiani E, Fianchi L et al. Impairment of PI3K/AKT and WNT/β-catenin pathways in bone marrow mesenchymal stem cells isolated from patients with myelodysplastic syndromes. *Exp Hematol* 44;2016:75–83.

Fischer ES, Böhm K, Lydeard JR et al. Structure of the DDB1-CRBN E3 ubiquitin ligase in complex with thalidomide. *Nature* 512;2014:49–53.

Gandhi AK, Kang J, Naziruddin S, Parton A, Schafer PH, and Stirling D. Lenalidomide inhibits proliferation of Namalwa CSN.70 cells and interferes with Gab1 phosphorylation and adaptor protein complex assembly. *Leuk Res* 30;2006:849–58.

Ganser A, and Heuser M. Therapy-related myeloid neoplasms. *Curr Opin Hematol* 24;2017:152–8.

Garcia-Manero G, Fenaux P, Al-Kali A et al. Rigosertib versus best supportive care for patients with high-risk myelodysplastic syndromes after failure of hypomethylating drugs (ONTIME): A randomised, controlled, phase 3 trial. *Lancet Oncol* 17;2016:496–508.

Garcia-Manero G, Montalban-Bravo G, Berdeja JG et al. Phase 2, randomized, double-blind study of pracinostat in combination with azacitidine in patients with untreated, higher-risk myelodysplastic syndromes. *Cancer* 123;2017:994–1002.

Germing U, Aul C, Niemeyer CM, Haas R, and Bennett JM. Epidemiology, classification and prognosis of adults and children with myelodysplastic syndromes. *Annu Hematol* 87;2008:691–9.

Giagounidis A, and Haase D. Morphology, cytogenetics and classification of MDS. *Best Pract Res Clin Hematol* 26;2013:337–53.

Gill H, Leung AYH, and Kwong Y-L. Molecular and cellular mechanisms of myelodysplastic syndrome: Implications on targeted therapy. *Int J Mol Sci* 17;2016:440.

Gillis NK, Ball M, Zhang Q et al. Clonal haemopoiesis and therapy-related myeloid malignancies in elderly patients: A proof-of-concept, case-control study. *Lancet Oncol* 18;2017:112–21.

Gleason MK, Ross JA, Warlick ED et al. CD16xCD33 bispecific killer cell engager (BiKE) activates NK cells against primary MDS and MDSC CD33+ targets. *Blood* 123;2014:3016–26.

Glenthøj A, Ørskov AD, Hansen JW, Hadrup SR, O'Connel C, and Grønbaek K. 2016. Immune mechanisms in myelodysplastic syndrome. *Int J Mol Sci* 17: 944.

Godley LA. Inherited predisposition to acute myeloid leukemia. *Semin Hematol* 51;2014:306–21.

Greenberg P, Cox C, LeBeau MM et al. International scoring system for evaluating prognosis in myelodysplastic syndromes. *Blood* 89;1997:2079–88.

Greenberg PL, Stone RM, Al-Kali A et al. Myelodysplastic syndromes, Version 2. 2017, NCCN Clinical Practice Guidelines in Oncology. *J Natl Compr Canc Netw* 15;2017:60–87.

Greenberg PL, Tuechler H, Schanz J et al. Revised international prognostic scoring system for myelodysplastic syndromes. *Blood* 120;2012:2454–65.

Haase D. Cytogenetic features in myelodysplastic syndromes. *Ann Hematol* 87;2008:515–26.

Haase D, Germing U, Schanz J et al. New insights into the prognostic impact of the karyotype in MDS and correlation with subtypes: Evidence from a core dataset of 2124 patients. *Blood* 110;2007:4385–95.

Haferlach T. Molecular genetics in myelodysplastic syndromes. *Leuk Res* 36;2012:1459–62.

Haferlach T, Nagata Y, Grossmann V et al. Landscape of genetic lesions in 944 patients with myelodysplastic syndromes. *Leukemia* 28;2014:241–7.

Haider M, Al Ali N, Padron E, Epling-Burnette P, Lancet J, List A, and Komrokji R. 2016. Immunosuppressive therapy: Exploring an underutilized treatment option for myelodysplastic syndrome. *Clin Lymphoma Myeloma Leuk* 16(Suppl): S44–8.

Hanna A, and Shevde LA. Hedgehog signaling: Modulation of cancer properties and tumor microenvironment. *Mol Cancer* 15;2016:24.

Heuser M. Therapy-related myeloid neoplasms: Does knowing the origin help to guide treatment? *Hematology Am Soc Hematol Educ Program* 2016(1);2016:24–32.

Heuser M, Meggendorfer M, Cruz MM et al. Frequency and prognostic impact of casein kinase 1A1 mutations in MDS patients with deletion of chromosome 5q. *Leukemia* 29;2015:1942–5.

Ito T, Ando H, Suzuki T, Ogura T, Hotta K, Imamura Y, Yamaguchi Y, and Handa H. Identification of a primary target of thalidomide teratogenicity. *Science* 327;2010:1345–50.

Ito T, and Handa H. Cereblon and its downstream substrates as molecular targets of immunomodulatory drugs. *Int J Hematol* 104;2016:293–9.

Itzykson R, Kosmider O, Cluzeau T et al. Impact of TET2 mutations on response rate to azacitidine in myelodysplastic syndromes and low blast count acute myeloid leukemias. *Leukemia* 25;2011:1147–52.

Itzykson R, Kosmider O, and Fenaux MD. Somatic mutations and epigenetic abnormalities in myelodysplastic syndromes. *Best Pract Res Clin Hematol* 26;2013:355–64.

Jan M, Ebert BL, and Jaiswal S. Clonal hematopoiesis. *Semin Hematol* 54;2017:43–50.

Jilg S, Reidel V, Müller-Thomas C et al. Blockade of BCL-2 proteins efficiently induces apoptosis in progenitor cells of high-risk myelodysplastic syndromes patients. *Leukemia* 30;2016:112–23.

Jonas BA, and Greenberg PL. MDS prognostic scoring systems—Past, present, and future. *Best Pract Res Clin Haematol* 28;2015:3–13.

Jonasova A, Bokorova R, Polak J et al. High level of full-length cereblon mRNA in lower risk myelodysplastic syndrome with isolated 5q deletion is implicated in the efficacy of lenalidomide. *Eur J Haematol* 95;2015:27–34.

Jung SH, Kim JY, Yim SH et al. Somatic mutations predict outcomes of hypomethylating therapy in patients with myelodysplastic syndrome. *Oncotarget* 23;2016:55264–75.

Kennedy JA, and Ebert BL. Clinical implications of genetic mutations in myelodysplastic syndrome. *J Clin Oncol* 35;2017:968–74.

Killick SB, Carter C, Culligan D et al. Guidelines for the diagnosis and management of adult myelodysplastic syndromes. *Br J Haematol* 164;2014:503–25.

Kim HK, Ko TH, Nyamaa B et al. Cereblon in health and disease. *Pflugers Arch* 468;2016:1299–309.

Koeffler HP, and Leong G. Preleukemia: One name, many meanings. *Leukemia* 31;2017:534–42.

Komrokji RS, Lancet JE, Swern AS, Chen N, Paleveda J, Lush R, Saba HI, and List AF. Combined treatment with lenalidomide and epoetin alfa in lower-risk patients with myelodysplastic syndrome. *Blood* 120;2012:3419–24.

Komrokji RS, and List AF. Short- and long-term benefits of lenalidomide treatment in patients with lower-risk del(5q) myelodysplastic syndromes. *Annals Oncol* 27;2016:62–8.

Korn C, and Méndez-Ferrer S. Myeloid malignancies and the microenvironment. *Blood* 129;2017:811–22.

Krönke J, Fink EC, Hollenbach PW et al. Lenalidomide induces ubiquitination and degradation of CK1α in del(5q) MDS. *Nature* 523;2015:183–8.

Krönke J, Udeshi ND, Narla A et al. Lenalidomide causes selective degradation of IKZF1 and IKZF3 in multiple myeloma cells. *Science* 343;2014:301–5.

Kulasekararaj AG, Smith AE, Mian SA et al. 2013. TP53 mutations in myelodysplastic syndrome are strongly correlated with aberrations of chromosome 5, and correlate with adverse prognosis. *Br J Haematol* 160: 660–72.

Lee EJ, Podoltsev N, Gore SD, and Zeidan AM. The evolving field of prognostification and risk stratification in MDS: Recent developments and future directions. *Blood Rev* 30;2016:1–10.

Li W, Morrone K, Kambhampati S, Will B, Steidl U, and Verma A. Thrombocytopenia in MDS: Epidemiology, mechanisms, clinical consequences and novel therapeutic strategies. *Leukemia* 30;2016:536–44.

Lichtman MA. The World Health Organization revisits the classification of the myelodysplastic syndromes: Improvement and insufficiencies. *Blood Cells Mol Dis* 60;2016:12–5.

Lindsley RC, Saber W, Mar BG et al. Prognostic mutations in myelodysplastic syndrome after stem-cell transplantation. *N Engl J Med* 376;2017:536–47.

List AF, Bennett JM, Sekeres MA, Skikne B, Fu T, Shammo JM, Nimer SD, Knight RD, and Giagounidis A. Extended survival and reduced risk of AML progression in erythroid-responsive lenalidomide-treated patients with lower-risk del(5q) MDS. *Leukemia* 28;2014:1033–40.

Lubeck DP, Danese M, Jennifer D, Miller K, Richhariya A, and Garfin PM. Systematic literature review of the global incidence and prevalence of myelodysplastic syndrome and acute myeloid leukemia. *Blood* 128;2016:5930.

Ma X, Does M, Raza A, and Mayne ST. Myelodysplastic syndromes: Incidence and survival in the United States. *Cancer* 109;2007:1536–42.

Makishima H, Yoshizato T, Yoshida K et al. Dynamics of clonal evolution in myelodysplastic syndromes. *Nature Genet* 49;2017:204–12.

Malcovati L, and Cazzola M. Recent advances in understanding of myelodysplastic syndromes with ring sideroblasts. *Br J Haematol* 174;2016:847–58.

Malcovati L, Germing U, Kuendgen A et al. Time-dependent prognostic scoring system for predicting survival and leukemic evolution in myelodysplastic syndromes. *J Clin Oncol* 25;2007:3503–510.

Malcovati L, Karimi M, Papaemmanuil E et al. *SF3B1* mutation identifies a distinct subset of myelodysplastic syndrome with ring sideroblasts. *Blood* 126;2015:233–41.

Malcovati L, Porta MG, Pascutto C et al. Prognostic factors and life expectancy in myelodysplastic syndromes classified according to WHO criteria: A basis for clinical decision making. *J Clin Oncol* 23;2005:7594–603.

Matsuoka A, Tochigi A, Kishimoto M et al. Lenalidomide induces cell death in an MDS-derived cell line with deletion of chromosome 5q by inhibition of cytokinesis. *Leukemia* 24;2010:748–55.

McGraw KL, Basiorka AA, Johnson JO et al. Lenalidomide induces lipid raft assembly to enhance erythropoietin receptor signaling in myelodysplastic syndrome progenitors. *PLOS ONE* 9;2014:e114249.

McGraw KL, Fuhler GM, Johnson JO, Clark JA, Caceres GC, Sokol L, and List AF. Erythropoietin receptor signaling is membrane raft dependent. *PLOS ONE* 7;2012:e34477.

McGraw KL, Nguyen J, Komrokji RS et al. Immunohistochemical pattern of p53 is a measure of *TP53* mutation burden and adverse clinical outcome in myelodysplastic syndromes and secondary acute myeloid leukemia. *Haematologica* 101;2016:e320–3.

Medeiros BC, Fathi AT, DiNardo CD, Pollyea DA, Chan SM, and Swords R. Isocitrate dehydrogenase mutations in myeloid malignancies. *Leukemia* 31;2017:272–81.

Medler CJ, Waddell JA, and Solimando DA. Cancer chemotherapy update. *Gefitinib and sonidegib. Hosp Pharm* 50;2015:868–72.

Mies A, Hermine O, and Platzbecker U. Activin receptor II ligand traps and their therapeutic potential in myelodysplastic syndromes with ring sideroblasts. *Curr Hematol Malig Rep* 11;2016:416–24.

Navada SC, and Silverman LR. The safety and efficacy of rigosertib in the treatment of myelodysplastic syndromes. *Expert Rev Anticancer Ther* 16;2016:805–10.

Nazha A, and Sekeres MA. Precision medicine in myelodysplastic syndromes and leukemias: Lessons from sequential mutations. *Annu Rev Med* 68;2017:127–37.

Nebbioso A, Benedetti R, Conte M, Iside C, and Altucci L. Genetic mutations in epigenetic modifiers as therapeutic targets in acute myeloid leukemia. *Expert Opin Ther Targets* 19;2015:1187–202.

Negoro E, Radivoyevitch T, Polprasert C et al. Molecular predictors of response in patients with myeloid neoplasms treated with lenalidomide. *Leukemia* 30;2016:2405–9.

Neukirchen J, Schoonen WM, Strupp C, Gattermann N, Aul C, Haas R, and Germing U. Incidence and prevalence of myelodysplastic syndromes: Data from the Düsseldorf MDS-registry. *Leuk Res* 35;2011:1591–6.

Nimer SD. Myelodysplastic syndromes. *Blood* 111;2008:4841–51.

Nybakken GE, and Bagg A. The genetic basis and expanding role of molecular analysis in the diagnosis, prognosis, and therapeutic design for myelodysplastic syndromes. *J Mol Diagn* 16;2014:145–58.

Oliva EN, Alati C, Santini V et al. Eltrombopag versus placebo for low-risk myelodysplastic syndromes with thrombocytopenia (EQoL-MDS): Phase 1 results of a single-blind, randomised, controlled, phase 2 superiority trial. *Lancet Haematol* 4;2017:e127–36.

Ornstein MC, Mukherjee S, and Sekeres MA. More is better: Combination therapies for myelodysplastic syndromes. *Best Pract Res Clin Hematol* 28;2015:22–31.

Otrock ZK, Chamseddine N, Salem ZM et al. A nationwide non-interventional epidemiological data registry on myelodysplastic syndromes in Lebanon. *Am J Blood Res* 5;2015:86–90.

Owen C, Barnett M, and Fitzgibbon J. Familial myelodysplasia and acute myeloid leukaemia—A review. *Br J Haematol* 140;2008:123–32.

Papaemmanuil E, Gerstung M, Malcovati L et al. Clinical and biological implications of driver mutations in myelodysplastic syndromes. *Blood* 122;2013:3616–27.

Pellagatti A, and Boultwood J. The molecular pathogenesis of the myelodysplastic syndromes. *Eur J Haematol* 95;2015:3–15.

Phekoo KJ, Richards MA, Mϕller H, and Schey SA. The incidence and outcome of myeloid malignancies in 2,112 adult patients in southeast England. *Haematologica* 91;2006:1400–4.

Platzbecker U, and Fenaux P. Personalized medicine in myelodysplastic syndromes: Wishful thinking or already clinical reality? *Haematologica* 100;2015:568–71.

Pleyer L, Valent P, and Greil R. Mesenchymal stem and progenitor cells in normal and dysplastic hematopoiesis-masters of survival and clonality? *Int J Mol Sci* 17;2016:1009.

Rankin EB, Narla A, Park JK, Lin S, and Sakamoto KM. Biology of the bone marrow microenvironment and myelodysplastic syndromes. *Mol Genet Metab* 116;2015:24–8.

Raza A, Reeves JA, Feldman EJ et al. Phase 2 study of lenalidomide in transfusion-dependent, low-risk and intermediate-1 risk myelodysplastic syndromes with karyotypes other than deletion 5q. *Blood* 111;2008:86–93.

Rodger EJ, and Morison IM. Myelodysplastic syndrome in New Zealand and Australia. *Intern Med J* 42;2012:1235–42.

Rollison DE, Howlader N, Smith MT, Strom SS, Merritt WD, Ries LA, Edwards BK, and List AF. Epidemiology of myelodysplastic syndromes and chronic myeloproliferative disorders in the United States, 2001-2004, using data from the NAACCR and SEER programs. *Blood* 112;2008:45–52.

Saft L, Karimi M, Ghaderi M et al. p53 protein expression independently predicts outcome in patients with lower-risk myelodysplastic syndromes with del(5q). *Haematologica* 99;2014:1041–9.

Sallman DA, Cluzeau T, Basiorka AA, and List A. Unraveling the pathogenesis of MDS: The NLRP3 inflammasome and pyroptosis drive the MDS phenotype. *Front Oncol* 6;2016:151.

Santini V. Treatment of low-risk myelodysplastic syndromes. *Hematology Am Soc Hematol Educ Program* 2016;2016:462–9.

Santini V, Almeida A, Giagounidis A et al. Randomized phase III study of lenalidomide versus placebo in RBC transfusion-dependent patients with lower-risk non-del(5q) myelodysplastic syndromes and ineligible for or refractory to erythropoiesis-stimulating agents. *J Clin Oncol* 34;2016:2988–96.

Sardnal V, Rouquette A, Kaltenbach S et al. A G polymorphism in the *CRBN* gene acts as a biomarker of response to lenalidomide in low/int-1 risk MDS without del(5q). *Leukemia* 27;2013:1610–3.

Schneider RK, Ademá V, Heckl D et al. Role of casein kinase 1A1 in the biology and targeted therapy of del(5q) MDS. *Cancer Cell* 26;2014:509–20.

Schneider RK, Schenone M, Ferreira MV et al. Rps14 haploinsufficiency causes a block in erythroid differentiation mediated by S100A8 and S100A9. *Nat Med* 22;2016:288–97.

Scott BL, Pasquini MC, Logan BR et al. Myeloablative versus reduced-intensity hematopoietic cell transplantation for acute myeloid leukemia and myelodysplastic syndromes. *J Clin Oncol* 35;2017:1154–61.

Sekeres MA. Are we nearer to curing patients with MDS? *Best Pract Res Clin Haematol* 23;2010:481–7.

Senent L, Arenillas L, Luño E, Ruiz JC, Sanz G, and Florensa L. Reproducibility of the World Health Organization 2008 criteria for myelodysplastic syndrromes. *Haematologica* 98;2013:568–75.

Shastri A, Will B, Steidl U, and Verma A. Stem and progenitor cell alterations in myelodysplastic syndromes. *Blood* 129;2017:1586–94.

Sheikine Y, Kuo FC, and Lindeman NI. Clinical and technical aspects of genomic diagnostics for precision oncology. *J Clin Oncol* 35;2017:929–33.

Sibon D, Cannas G, Baracco P et al. Lenalidomide in lower-risk myelodysplastic syndromes with karyotypes other than deletion 5q and refractory to erythropoiesis-stimulating agents. *Br J Haematol* 156;2012:619–25.

Smith AE, Kulasekararaj AG, Jiang J et al. CSNK1A1 mutations and isolated del(5q) abnormality in myelodysplastic syndrome: A retrospective mutational analysis. *Lancet Haematol* 2;2015:e212–221.

Stahl M, and Zeidan AM. Hypomethylating agents in combination with histone deacetylase inhibitors in higher risk myelodysplastic syndromes: Is there a light at the end of the tunnel. *Cancer* 123;2017a:911–4.

Stahl M, Zeidan AM. Lenalidomide use in myelodysplastic syndromes: Insights into the biologic mechanisms and clinical applications. *Cancer* 2017b. doi: 10.1002/cncr.30585.

Starczynowski DT, Kuchenbauer F, Argiropoulos B et al. Identification of miR-145 and miR-146a as mediators of the 5q- syndrome phenotype. *Nat Med* 16;2010:49–58.

Steensma DP, Bejar R, Jaiswal S, Lindsley RC, Sekeres MA, Hasserjian RP, and Ebert BL. Clonal hematopoiesis of indeterminate potential and its distinction from myelodysplastic syndromes. *Blood* 126;2015:9–16.

Suragani RN, Cadena SM, Cawley SM et al. Transforming growth factor-β superfamily ligand trap ACE-536 corrects anemia by promoting late-stage erythropoiesis. *Nat Med* 20;2014:408–14.

Takahashi K, Patel K, Bueso-Ramos C et al. Clinical implications of *TP53* mutations in myelodysplastic syndromes treated with hypomethylating agents. *Oncotarget* 7;2016:1472–87.

Takahashi K, Wang F, Kantarjian H, Doss D, Khanna K, Thompson E, and Zhao L. Prelekaemic clonal haemopoiesis and risk of therapy-related neoplasms: A case-control study. *Lancet Oncol* 18;2017:100–11.

Tefferi A, and Garcia-Manero G. CME Information: Annual clinical updates 2015: Myelodysplastic syndromes. *Am J Hematol* 90;2015:831–41.

Tefferi A, and Vardiman JW. Myelodysplastic syndromes. *N Engl J Med* 361;2009:1872–85.

Tiu RV, Visconte V, Traina F, Schwandt A, and Maciejewski JP. Updates in cytogenetics and molecular markers in MDS. *Curr Hematol Malig Rep* 6;2011:126–35.

Toma A, Kosmider O, Chevret S et al. Lenalidomide with or without erythropoietin in transfusion-dependent erythropoiesis-stimulating agent-refractory lower-risk MDS without 5q deletion. *Leukemia* 30;2016:897–905.

Vardiman JW, Thiele J, Arber DA et al. The 2008 revision of the World Health Organization (WHO) classification of myeloid neoplasms and acute myeloid leukemia: Rationale and important changes. *Blood* 114;2009:937–51.

Venner CP, Woltosz JW, Nevill TJ et al. Correlation of clinical response and response duration with miR-145 induction by lenalidomide in CD34+ cells from patients with del(5q) myelodysplastic syndrome. *Haematologica* 98;2013:409–13.

Venter JC, Adams MD, Myers EW et al. The sequence of the human genome. *Science* 291;2001:1304–51.

Vlachos A, Farrar JE, Atsidaftos E et al. Diminutive somatic deletions in the 5q region lead to a phenotype atypical of classical 5q-syndrome. *Blood* 122;2013:2487–90.

Walter MJ, Shen D, Shao J et al. Clonal diversity of recurrently mutated genes in myelodysplastic syndromes. *Leukemia* 27;2013:1275–82.

Wei S, Chen X, McGraw K, Zhang L, Komrokji R, Clark J, and Caceres G. Lenalidomide promotes p53 degradation by inhibiting MDM2 auto-ubiquitination in myelodysplastic syndrome with chromosome 5q deletion. *Oncogene* 32;2013:1110–20.

Wei S, Chen X, Rocha K et al. A critical role for phosphatase haplodeficiency in the selective suppression of deletion 5q MDS by lenalidomide. *Proc Natl Acad Sci USA* 106;2009:12974–9.

West AH, Godley LA, and Churpek JE. Familial myelodysplastic syndrome/acute myeloid leukemia syndromes: A review and utility for translational investigations. *Annals NY Acad Sci* 1310;2014:111–8.

Wlodarski MW, Hirabayashi S, Pastor V et al. Prevalence, clinical characteristics, and prognosis of GATA2-related myelodysplastic syndromes in children and adolescents. *Blood* 127;2016:1387–97.

Wu L, Li X, Xu F, Zhang Z, Chang C, and He Q. Low RPS14 expression in MDS without 5q– aberration confers higher apoptosis rate of nucleated erythrocytes and predicts prolonged survival and possible response to lenalidomide in lower risk non-5q– patients. *Eur J Haematol* 90;2013:486–93.

Zahr AA, Ramirez CB, Wozney J, Prebet T, and Zeidan AM. New insights into the pathogenesis of MDS and the rational therapeutic opportunities. *Expert Rev Hematol* 9;2016:377–88.

Zeidan AM, Al Ali N, Barnard J et al. Comparison of clinical outcomes and prognostic utility of risk stratification tools in patients with therapy-related vs *de novo* myelodysplastic syndromes: Report on behalf of the MDS Clinical Research Consortium. *Leukemia* 2017. doi: 10.1038/leu. 2017.33.

Zeidan AM, Sekeres MA, Garcia-Manero G et al. Comparison of risk stratification tools in predicting outcomes of patients with higher-risk myelodysplastic syndromes treated with azanucleotides. *Leukemia* 30;2016:649–57.

Zeidan AM, Sekeres MA, Wang XF et al. Comparing the prognostic value of risk stratifying models for patients with lower-risk myelodysplastic syndromes: Is one model better? *Am J Hematol* 90;2015:1036–40.

Zhang L, McGraw KL, Sallman DA, and List AF. The role of p53 in myelodysplastic syndromes and acute myeloid leukemia: Molecular aspects and clinical implications. *Leuk Lymphoma* 58;2017:1777–90.

Zou J, Zhou Z, Wan L, Tong Y, Qin Y, Wang C, and Zhou K. Targeting the sonic Hedgehog-Gli1 pathway as a potential new therapeutic strategy for myelodysplastic syndromes. *PLOS One* 10;2015:e0136843.

10

Precision medicine in acute myeloid leukemia

OTA FUCHS

INTRODUCTION

Acute myeloid leukemia (AML) is a genetically heterogeneous, malignant clonal disease and the most common acute leukemia in adults (about 80% of cases in this group) (Yamamoto and Goodman, 2008). GLOBOCAN estimates the worldwide total leukemia incidence of AML for 2012 to be 351,965 with an age-standardized rate (ASR) per 100,000 of 4.7, a 5-year prevalence of 1.5%, and a M (male):F (female) ratio of about 1.4 (GLOBOCAN 2014). The incidence of AML ranges from 3 to 5 cases per 100,000 population in the United States. In 2015, 20,830 new cases of AML were expected to occur in the United States, including 12,370 in males and 8,100 in females, and over 10,000 patients died from this disease (De Kouchkovsky and Abdul-Hay, 2016; Kansal, 2016; Siegel et al., 2015). The incidence of AML increases with age, from about 1.3 per 100,000 population in patients less than 65 years old, to 12.2 cases per 100,000 population in

those over 65 years. Prognosis in the older patients who account for the majority of new cases remains poor (Shah et al., 2013). Even with current treatments, as much as 70% of patients 65 years or older will die of their disease within 1 year of diagnosis (Meyers et al., 2013). AML is characterized by clonal expansion of abnormal hematopoietic progenitor cells and impaired production of normal blood cells (Estey and Dohner, 2006; Jan et al., 2017; Prada-Arismendy et al., 2017). A high relapse rate for patients with AML is still a major barrier to the long-term survival of these patients (Rowe, 2016).

The first reports of an unknown condition with a "milky blood" were from Scottish physician Peter Cullen (1811). French physician Alfred Velpeau defined the leukemia-associated symptoms (1825) followed by French physician Alfred Donné who detected amaturation arrest of the white blood cells (1844). British physician John Hughes Bennett named the disease "leucocythemia," based on the microscopic accumulation of purulent leucocytes

(1845). The same year, a famous German physician Rudolf Virchow defined a reversed white and red blood cell balance. He introduced the disease as "leukemia" in 1847. British physician Henry Fuller performed and described the first microscopic diagnosis of a leukemia patient during life (1846) (Kampen, 2012; Thomas, 2013).

The purpose of this review is to describe some of the recent major advances that occurred due to the impact of genomics, mainly through the new genomic technologies, in the diagnostic classification, risk stratification, and treatment of AML. New genomic technologies and mainly advances in bioinformatics helped to develop next-generation sequencing (NGS), whole genome sequencing, whole exome sequencing, and targeted deep sequencing of specific mutations (Eisfeld et al., 2017; Kohlmann et al., 2014; Lin et al., 2017; Nazha et al., 2016; Wang et al., 2016; White and DiPersio, 2014).

METHODS FOR THE DIAGNOSIS AND TREATMENT OF ACUTE MYELOID LEUKEMIA (AML)

Cytogenetic analysis

Cytogenetic analysis is an important and mandatory component of AML diagnosis. Cytogenetics remains the most important disease-related prognostic factor and powerful independent prognostic indicator in AML (Ferrara et al., 2008; Foran, 2010; Grimwade and Mrózek, 2011; Röllig et al., 2011). Until now, more than 100 balanced chromosomal rearrangements (translocations, insertions, and inversions) have been identified and cloned. These chromosomal rearrangements are critical initiating events in the pathogenesis of AML. Karyotype analysis identifies biologically distinct subsets of AML that differ in their response to therapy and treatment outcome (Mrózek et al., 2004; Smith et al., 2011).

Cytogenetic abnormalities used in the WHO classification (Vardiman et al., 2009) are shown in Table 10.1. Approximately 10% of adult AML and 20% of childhood AML are classified as having core binding factor (CBF) leukemia with balanced chromosomal rearrangements that disrupt gene *RUNX1* (also known as *CBFA2* or *AML1*), which plays a critical role in hematopoiesis and leukemogenesis (Lam and Zhang, 2012; Marcucci, 2006;

Table 10.1 Cytogenetic abnormalities used in the WHO classification (2008) of AML

Cytogenetic abnormalities used to define entities within the WHO category of AML with recurrent genetic abnormalities
t(8;21)(q22;q22); *RUNX1-RUNX1T1*
inv(16)(p13.1q22) or t(16;16)(p13;q22); *CBFB-MYH11*
t(15;17)(q22;q12-21); *PML-RARA*
t(9;11)(p22;q23); *MLLT3-MLL*
t(6;9)(p23;q34); *DEK-NUP214*
inv(3)(q21q26.2) or t(3;3)(q21;q26.2); *RPN1-EVI1*
t(1;22)(p13;q13); *RBM15-MKL1*
Cytogenetic abnormalities sufficient to diagnose the WHO category of AML with myelodysplasia-related changes
Complex karyotype
(defined as 3 or more unrelated abnormalities, none of which can be a translocation or inversion associated with AML with recurrent genetic abnormalities)
Unbalanced abnormalities
  −7 or del(7q)
  −5 or del(5q)
  i(17q), an isochromosome for long arm of chromosome 17 or t(17p)
  −13 or del(13q)
  del(11q)
  del(12p) or t(12p)
  del(9q)
  idic(X)(q13), isodicentric X chromosome
Balanced abnormalities
  t(11;16)(q23;p13.3)
  t(3;21)(q26.2;q22.1)
  t(1;3)(p36.3;q21.1)
  t(2;11)(p21;q23)
  t(5;12)(q33;p12)
  t(5;7)(q33;q11.2)
  t(5;17)(q33;p13)
  t(5;10)(q33;q21)
  t(3;5)(q25;q34)

Metzeler and Bloomfield, 2017; Mrózek et al., 2008; Sangle and Perkins, 2011; Utsun and Marcucci, 2015; Yamagata et al., 2005). Patients with two specific, clonal, recurring cytogenetic abnormalities t(8;21)(q22;q22), inv(16)(p13.1q22) or t(16;16) (p13.1q22) are called CBF AML. Compared to

Table 10.2 Variations in cytogenetic risk in patients with acute myeloid leukemia. (United Kingdom Medical Research Council cytogenetic group assignment 2010)

Risk group assignment	Cytogenetic abnormality	Comments
Favorable	t(15;17)(q22;q12-21)	Irrespective of additional cytogenetic abnormalities
	t(8;21)(q22;q22)	
	inv(16)(p13.1q22) or t(16;16)(p13.1q22)	
Intermediate	Other noncomplex entities, not classified as favorable or adverse	
	Normal karyotype	
Adverse	abn(3q) [excluding t(3;5)(q21–25;q31–35)], inv(3)(q21q26) or t(16;16)(p13;q22), add(5q), del(5q), −5, −7, add(7q), del(7q), t(6;11)(q27;q23), t(10;11)(p11–13;q23), other t(11;q23) [excluding t(9;11)(p21–22;q23) and t(11;19)(q23;p13)], t(9;22)(q34;q11), −17, abn(17p)	
	Complex (>4 unrelated abnormalities)	

Note: Table is based on multivariable analysis conducted in 5876 adults (16–59 years) treated in the United Kingdom Medical Research Council AML10, 12, and 15 trials (Grimwade et al., 2010).

other cytogenetic AML subgroups, CBF AML is considered a more favorable subset of AML (Table 10.2). CBF AML results in the formation of hybrid fusion genes called *RUNX1-RUNX1T1* (also known as *AML1-ETO*) and *CBFB-MYH11*, which can be quantified in patients before, during, and after the therapy, including stem cell transplantation (Yin et al., 2012). Favorable karyotype patients have a good prognosis with complete remission rates exceeding 90%, a five-year survival of at least 65%, and relapse rates too low and salvage rates too high to benefit from routine use of allograft in first complete remission (Smith et al., 2011).

Patients with CBF AML and mutations in the *KIT* gene (exon 17) have a higher relapse risk (Ayatollahi et al., 2017; Paschka et al., 2006; Shimada et al., 2006). The mechanism underlying *KIT* gene mutations that adversely affects the prognosis involves phosphorylation of the KIT receptor after physiologic binding of KIT ligand, which activates downstream pathways supporting cell proliferation and survival (Lennartsson et al., 2005).

About 10%–20% of AML patients have adverse cytogenetics, which includes monosomal

karyotype (monosomies of chromosomes 5 and/or 7(-5/-7) (Brands-Nijenhuis et al., 2016; Kayser et al., 2012; Perrot et al., 2011), abnormalities of 3q/abn(3q) (Hinai and Valk, 2016; Lugthart et al., 2010), or a complex karyotype often associated with TP53 alterations (Rücker et al., 2012). Complex karyotype anomalies (clinically defined as having an adverse outcome) have at least three to five anomalies and do not contain the balanced chromosomal translocations/inversions defining a favorable prognosis, such as t(15;17) or t(8;21) (Blum et al., 2017). The monosomal karyotype entity is defined as two or more autosomal monosomies or combination of one monosomy with structural abnormalities. These patients are usually older and had MDS or were treated by chemotherapy or radiation. They have therapy-related AML (tAML) (Heuser, 2016). Complete remission rates are around 60% and a five-year survival about 10%. Due to this very poor prognosis with current therapies, an allograft in first complete remission or an experimental treatment approach may be justified (Fang et al., 2011; Paun and Lazarus, 2012). However, patients with single monosomy, more than 10% of cells with normal metaphase, the absence of del(17p), and achievement of complex remission after induction therapy were prognostic factors for better overall survival (Jang et al., 2015).

The mixed lineage leukemia gene (*MLL*, also known as *ALL-1* or *HRX*), located on chromosome 11q23, encodes a histone methyltransferase and is frequently rearranged in AML. Recurrent translocations of *MLL* are generally considered to confer a poor prognosis (Meyer et al., 2009; Pigneux et al., 2015). Fusion proteins with more than 60 partners were characterized and these partners also have a significant effect on disease outcome (Coenen et al., 2011). Other entities that have been associated with poor prognosis are t(6;9)(p23;q34), t(9;22)(q34;q11), and 17p deletions (Smith et al., 2011).

AML with isolated trisomies for chromosomes 4, 8, 11, and 21 should, with exception of trisomy for chromosome 13, be classified as intermediate risk (Lazarevic et al., 2017).

Since cytogenetics provides the framework for current risk stratification schemes in AML with a major role in determining whether patients are candidates for allogenic transplant in first remission, it is important to have clearly defined risk groups not only for direction of patient management but also for the comparison of data between different clinical results (Smith et al., 2011).

Morphology

The categories of AML with t(8;21)(q22:q22), inv(16)(p13.1:q22) or t(16;16)(p13.1:q22), and t(15;17)(q22:q12) are considered AML without regard to blast count. Identification of 20% or more blasts in the blood or bone marrow is required in most categories of AML. The morphology of the blasts often correlates with cytogenetic abnormalities. For example, AML with t(8:21) often has large blasts with numerous azurophilic granules, which may occasionally be very large (pseudo-Chediak-Higashi type), and paranuclear clearing or hofs. Long, thin Auer rods are frequently seen and eosinophils may be increased. AML with inv(16) has blasts with myelomonocytic features and increased number of eosinophils at all stages of maturation (Weinberg et al., 2017b).

AML with t(15;17), also known as acute promyelocytic leukemia (APL), is characterized by abnormal promyelocytes with bilobed or indented nuclei and moderate cytoplasm, generally with abundant cytoplasmic granules and Auer rods that may be in bundles ("faggot cells").

It is now necessary to use a combination of morphology with other techniques as immunocytochemistry or flow cytometry to confirm a diagnosis. These latter techniques are essential for the diagnosis of the newer described entities of AML with minimal differentiation (FAB-M0) and acute megakaryoblastic leukemia (FAB-M7). Cytogenetics and fluorescent in situ hybridization techniques are also very important for diagnosis as in FAB-M3 (promyelocytic leukemia) but also for detecting those myeloid leukemias that are associated with a favorable or unfavorable response (Bacher et al., 2005; Krause 2000).

Flow cytometry

Flow cytometric immunophenotyping is an accurate and objective method for quantitative and qualitative evaluation of hematopoietic cells. Compared with AML-not otherwise specified (AML-NOS), granulocytic cells in AML with myelodysplasia-related changes (AML-MRC) had higher CD33 expression but lower CD45, CD11b, and CD15. Monocytes in AML-MRC had lower expression of CD14, CD56, and CD45. Morphologic dysplasia was associated with significantly lower granulocytic forward scatter, side scatter, and CD10 but higher CD33 expression. Our results suggest that the workup of AML cases should include flow cytometric assessment of granulocytes and monocytes. This analysis can aid a morphologic impression of multilineage dysplasia in distinguishing AML-MRC from AML-NOS, especially in cases with limited maturing myeloid cells (Weinberg et al., 2017a).

Classification of AML patients

To assist with patient diagnosis and management, classifications, such as the French-American-British (FAB) (Bennett et al., 1976, 1985) and the World Health Organization (WHO) classification (Arber et al., 2016; Döhner et al., 2010; Vardiman et al., 2009; Table 10.3), were developed to define clinically relevant disease subtypes. FAB classification is still widely used in clinical setting that groups AML into eight subgroups (M0-M7) based on its degree of differentiation and morphology.

Table 10.3 WHO classification (2016) of acute myeloid leukemia and related neoplasms

---

Acute myeloid leukemia with recurrent genetic abnormalities
    AML with t(8;21)(q22;q22); *RUNX1-RUNX1T1*
    AML with inv(16)(p13.1q22) or t(16;16)
      (p13;q22); *CBFB-MYH11*
    APL with t(15;17)(q22;q12); *PML-RARA*
    AML with t(9;11)(p22;q23); *MLLT3-MLL*
    AML with t(6;9)(p23;q34); *DEK-NUP214*
    AML with inv(3)(q21q26.2) or t(3;3)
      (q21;q26.2); *GATA2, MECOM*
    AML (megakaryoblastic) with t(1;22)(p13;q13);
      RBM15-MKL1
    Provisional entity: *AML* with *BCR-ABL1*
    AML with mutated *NPM1*
    AML with biallelic mutations of *CEBPA*
    Provisional entity: AML with mutated *RUNX1*
Acute myeloid leukemia with myelodysplasia-related changes
Therapy-related myeloid neoplasma
Acute myeloid leukemia, not otherwise specified
    AML with minimal differentiation
    AML without maturation
    AML with maturation
    Acute myelomonocytic leukemia
    Acute monoblastic/monocytic leukemia
      Pure erythroid leukemia
      Acute megakaryoblastic leukemia
    Acute basophilic leukemia
    Acute panmyelosis with myelofibrosis
Myeloid sarcoma
Myeloid proliferations related to Down syndrome
    Transient abnormal myelopoiesis (TAM)
    Myeloid leukemia associated with Down syndrome
Blastic plasmacytoid dendritic cell neoplasm
Acute leukemias of ambiguous lineage

Treatment for all subtypes of AML, except the M3 subtype, acute promyelocytic leukemia (APL), involves combination chemotherapy and a possible hematopoietic stem cell transplantation as a part of consolidation therapy (Burnett, 2012; Paun and Lazarus, 2012). In APL, the balanced t(15;17) translocation rearranges the promyelocytic leukemia gene PML on chromosome 15 with retinoic acid receptor α (RARA) located on chromosome 17. APL with fusion gene PML-RARA is treated with molecularly targeted therapy such as the differentiation-inducing agent all-trans retinoic acid (ATRA) and arsenic trioxide (ATO), especially in cases where ATRA plus anthracycline-based chemotherapy cannot be used (Baljevic et al., 2011; Lo-Coco et al., 2008).

A new provisional entity AML with mutated *RUNX1* (excluding cases with myelodysplasia-related changes) was added (Gadzik et al., 2016; Haferlach et al., 2016). Myeloid neoplasms with germ line predisposition is also a new category that was added to the WHO classification (Babushok et al., 2016; Churpek and Godley, 2016; Table 10.4). The molecular basis of AML with inv(3)(q21q26.2) or t(3;3)(q21;q26.2) was revisited showing that repositioning of a *GATA2* enhancer element leads to overexpression of the *MECOM* (*EVI1*) gene and to haploinsufficiency of GATA2 (Gröschel et al., 2014).

Table 10.4 WHO classification of AML with germ-line predisposition and guide for molecular genetic diagnostics

---

Classification
Myeloid neoplasms with germ-line predisposition without a preexisting disorder or organ dysfunction
    AML with germ-line *CEBPA* mutation
    Myeloid neoplasms with germ line *DDX41* mutation
Myeloid neoplasms with germ-line predisposition and preexisting platelet disorders
    Myeloid neoplasms with germ line *RUNX1* mutation
    Myeloid neoplasms with germ line *ANKRD26* mutation
    Myeloid neoplasms with germ line *ETV6* mutation
Myeloid neoplasms with germ-line predisposition and other organ dysfunction
    Myeloid neoplasms with germ line *GATA2* mutation
    Myeloid neoplasms associated with bone marrow failure syndromes

Prognostic stratification significance of gene mutations in intermediate-risk AML

Approximately 45% of adult patients with AML have normal karyotype (cytogenetically normal AML; CN-AML patients) and are usually classified as an intermediate risk group. Cytogenetic analysis is uninformative for these patients because they have vastly different clinical outcomes within this intermediate risk group. Analysis of specific genetic abnormalities, not detectable by cytogenetics, and quantification of expression of specific genes with prognostic significance are valuable tools for more accurate risk stratification and clinical outcome prediction of these CN-AML patients (Al-Issa and Nazha, 2016; Bullinger et al., 2017; Döhner et al., 2017; Papaemmanuil et al., 2016; Sanders and

Valk, 2013). Functional categories of genes commonly affected in AML are presented in Table 10.5. Genetic abnormalities were correlated with clinical characteristics and outcome of AML patients. The European Leukemia Net (ELN) published a recommendation where a three-group classification system (favorable, intermediate, and adverse) was used for ELN genetic risk stratification (Bullinger et al., 2017; Döhner et al., 2017; Table 10.6).

Current therapy of AML and experimental therapies

Current therapy has not changed substantially in recent years. For decades, the standard therapy has been initial treatment with cytarabine and an anthracycline, the so-called $7 + 3$ induction regimen (a seven-day continuous intravenous

Table 10.5 Functional categories of genes commonly affected in AML

Functional category	Selected gene members	Role in leukemogenesis
Signaling genes	Kinases (e.g., *FLT3*, *KIT*), phosphatases (e.g., *PTPN1*), or RAS family members (e.g., *KRAS*, *NRAS*)	Activated signaling confers a proliferative advantage through RAS/RAF, JAK/STAT, and PI3K/AKT signaling pathways
DNA methylation-associated genes	*DNMT3A*, *TET2*, *IDH1*, *IDH2*	Deregulated DNA methylation and transcriptional deregulation of leukemia-relevant genes
Myeloid transcription factor (TF) gene fusion mutations	TF fusions [t(8;21), inv(16)/t(16;16)]	Aberrant TF function results in transcriptional deregulation and impaired hematopoietic differentiation
	TF mutations (*RUNX1*, *CEBPA*)	Aberrant TF function results in transcriptional deregulation and impaired hematopoietic differentiation
Chromatin-modifying genes	Mutations (e.g., *ASXL1*, *EZH2*) or *KMT2A* fusions	Deregulation of chromatin modification (e.g., methylation of histones H3 and H2A) as well as *KMT2A* fusion-driven impairment of methyltransferases lead to transcriptional deregulation
Nucleophosmin (*NPM1*) gene	*NPM1*	Mutations of *NPM1* result in aberrant cytoplasmic localization of NPM1 and NPM1-interacting proteins
Tumor-suppressor genes	*TP53*, *WT1*, *PHF6*	Mutations lead to transcriptional deregulation
Spliceosome-complex genes	*SRSF2*, *SF3B1*, *U2AF1*, *ZRSR2*	Mutations lead to impaired spliceosome function and deregulated RNA processing
Cohesin-complex genes	*STAG2*, *RAD21*	Mutations may lead to impaired chromosome segregation and impact transcriptional regulation

Table 10.6 2017 ELN risk stratification by genetics

Risk category	Genetic lesion
Favorable	t(8,21)(q22;q22.1); *RUNX1-RUNX1T1*
	inv(16)(p13.1q22) or t(16;16) (p13.1;q22); *CBFB-MYH11*
	Mutated *NPM1* without *FLT3*-ITD or with *FLT3*-ITD^{low}
	Biallelic mutated *CEBPA*
Intermediate	Mutated *NPM1* and *FLT3*-ITD^{high}
	Wild type *NPM1* without *FLT3*-ITD or with *FLT3*-ITD^{low} (without adverse-risk gene mutations)
	t(9;11)(p21.3;q23.3); *MLLT–KMT2A*
	Cytogenetic abnormalities not classified as favorable
Adverse	t(6;9)(p23;q34.1); *DEK-NUP214*
	t(v;11q23.3); *KMT2A* rearranged
	t(9;22)(q34.1; q11.2); *BCR-ABL1*
	inv(3)(q21q26.2) or t(3;3) (q21;q26.2); *GATA2, MECOM* (*EVI1*)
	−5 or del(5q); −7; −17/abn(17p)
	Complex karyotype, monosomal karyotype
	Wild-type *NPM1* and *FLT3*-ITD^{high}
	Mutated *RUNX1*
	Mutated *ASXL1*
	Mutated *TP53*

cytarabine infusion and 3 daily doses of daunorubicin), followed by higher doses of cytarabine as consolidation. This approach is curative in about 40% of younger patients (age range between 18 and 60 years) overall, with patients in some of the more favorable subgroups enjoying cure rates of approximately 60%. The outcomes are worst in older AML patients, who represent the majority of individuals with AML and in patients with AML after a prior diagnosis of MDS (secondary AML).

As a combination of cytarabine and daunorubicin represented the long-standing but inadequate standard of care for induction chemotherapy, in vitro studies analyzed the optimal synergistic ratio of both compounds. These studies identified a 5:1 molar ratio of cytarabine:daunorubicin to be ideal (Mayer et al., 2006) and led to the development of a liposomal formulation to these drugs in this ideal molar ratio within 100 nanometer liposomes (Tardi et al., 2009). This product named CPX-351 was well tolerated and had better efficacy and improved response rates compared to standard administration but without an improvement in event-free survival or overall survival. However, complete remission was nearly doubled in patients with secondary AML (Sallman and Lancet, 2017). These encouraging results led to phase 3 study (NCT01696084) in high-risk AML patients (patients with secondary AML, therapy-related AML, but also AML with WHO-defined MDS cytogenetic abnormalities). The median of overall survival was 9.56 months in patients treated with CPX-351 compared to 5.95 months with 7 + 3. These data supported CPX-351 as a new standard therapy for older patients with secondary AML.

A randomized trial found that fludarabine and high dose cytarabine plus granulocyte colony-stimulating factor (G-CSF; FLAG) plus idarubicin not only produced a lower relapse rate than daunorubicin-cytarabine with or without etoposide, but was associated with more deaths in remission resulting in similar overall survival (Burnett et al., 2013).

New therapies are needed that can target and eradicate resistant subclones early in the disease course. An example is the multikinase inhibitor midostaurin, which when added to 7 + 3 induction has been found to benefit in patients with *FLT3*-mutated AML (Gallogly and Lazarus, 2016). Sorafenib (a tyrosine kinase inhibitor), crenolanib (a FLT3 inhibitor with activity against FLT3-D835 TKD mutation), gilteritinib (ASP-2215, a potent inhibitor of both *FLT3*-ITD and *FLT3*-TKD mutations), and a second-generation *FLT3*-ITD inhibitor quizaritinib (formerly known as AC220) were also used in combination with chemotherapy (De Kouchkovsky and Abdul-Hay, 2016; Grunwald and Levis, 2015). Terminal myeloid differentiation in vivo is induced by FLT3 inhibition in FLT3/ITD AML (Sexauer et al., 2012). Unfortunately, owing to the resistance, the responses to FLT3 inhibitors are not durable and last just 3–6 months. For patients harboring *FLT3*-ITD mutations, allogeneic hematopoietic stem cell transplantation (alloHSCT) is preferable, and autologous HSCT is used mainly for patients with favorable or intermediate-risk cytogenetics in first remission (Tamamyan et al., 2017).

In AML, targeted inhibitors of both IDH1 and IDH2 have been characterized in preclinical models (Chaturvedi et al., 2013; Losman et al., 2013; Wang et al., 2013). The conversion of glutamine to the α-ketoglutarate (α-KG) in cells occurs in two steps. Glutamine is first converted to glutamate and ammonia by glutaminase. Glutamate is subsequently converted to α-KG by either glutamate pyruvate transaminase or glutamate oxaloacetate transaminase (Fathi et al., 2015). Glutamine is the main source of α-KG in *IDH*-mutant leukemic blasts. Therefore, the depletion of glutamine or interruption of its metabolism are both selectively harmful for AML cells with mutated *IDH* (Emadi et al., 2014). The promising results of these preclinical studies warranted early-phase clinical trials (Dang et al., 2017; Fathi et al., 2015; Medeiros et al., 2017; Table 10.7).

Both FLT3 and IDH inhibitors can be used as a bridge to transplant. Current AML management relies largely on intensive chemotherapy and allogeneic hematopoietic stem cell transplantation (HSCT), at least in younger patients who can tolerate such intensive treatments. Safer allogeneic HSCT procedures allow a larger proportion of patients to achieve durable remission. Improved identification of patients at relatively low risk of relapse should limit their undue exposure to the risk of HSCT in first remission. The optimal treatment of AML in first complete remission is uncertain (Buccisano and Walter, 2017). Current consensus, based on cytogenetic risk, recommends myeloablative allogeneic

HSCT for poor-risk but not for good-risk AML. Which patients should be offered allogeneic HSCT and whether this should be in first complete remission or after relapse is a major therapeutic dilemma (Gale et al., 2014). The quantification of measurable minimal residual disease (MRD) in first complete remission is very important (Buccisano and Walter, 2017). A recent single-center analysis showed that patients with positive MRD at the time of transplantation have poor outcomes even if myeloablative conditioning is able to reduce residual amounts of leukemia cells (Zhou et al., 2016). Allogeneic HSCT, autologous transplantation, and consolidation chemotherapy are considered of equivalent benefit for intermediate-risk AML.

In patients not eligible for intensive chemotherapy treatment, hypomethylating agents (HMAs) have become a frequent alternative treatment choice. Dysregulated epigenetic mechanisms play an important role in the pathogenesis of AML. The term *epigenetics* refers to changes in gene expression that are not caused by changes in DNA sequence. Cytosine methylation in DNA, modifications of histone proteins, or RNA-associated gene silencing play a role in epigenetic mechanisms (Wouters and Delwel, 2016). Epigenetic dysregulation in AML is widespread and cannot be explained by recurrent somatic mutations alone. Two HMAs currently approved for clinical use in AML are 5-azacytidine (azacitidine) and 5-aza-2′-deoxycytidine

Table 10.7 Clinical trials targeting IDH mutations in AML

Inhibitor	Target	Phase	Identifier	Patient population
AG-120	*IDH1*	1	NCT02074839	>18 years of age with *IDH1* gene-mutated, advanced hematologic malignancy including relapsed and/or primary refractory AML
				>60 years of age with *IDH1* gene-mutated, untreated AML who are not candidates for standard therapy due to age, performance status, and/or adverse risk factors
AG-221	*IDH2*	1	NCT1915498	>18 years of age with *IDH2* gene-mutated, advanced hematologic malignancy
CB-839	Glutaminase	1	NCT02071927	> 18 years of age with relapsed/refractory AML after at least one prior treatment regimen
Erwinaze	Serum glutamine	1	NCT02283190	> 18 years of age with *IDH1* or *IDH2* gene-mutated relapsed/refractory AML to first-line therapy, with or without additional subsequent therapy

(decitabine). Both are pyrimidine analogs that function as inhibitors of DNA methyltransferases (Table 10.8). Both agents reverse aberrant DNA hypermethylation and restore expression of critical tumor-suppressor genes. Decitabine is incorporated into DNA, whereas azacitidine is incorporated into RNA.

Guadecitibine (SGI-110) is a second-generation HMA and a dinucleotide of decitabine and deoxyguanosine (Nebbioso et al., 2015). A number of

Table 10.8 Examples of epigenetic targeted therapy in AML

Class of epigenetic regulator	Target	Compound	Preclinical or clinical use
DNA methyltransferase (DNMT)	DNMTs	Azacitidine	Approved for clinical use
		Decitabine	Approved for clinical use
		Novel inhibitors	Preclinical and clinical use
Regulator of methylation	IDH1, IDH2	Inhibitors of mutant IDH1/2	See Table 10.7
Histone lysine acetyltransferase	CREBBP (CBP) EP300 (p300)	CREBBP inhibitor, EP300 inhibitor	Preclinical use Preclinical use
Histone deacetylase (HDAC)	HDACs	HDAC inhibitors	Several clinical trials ongoing, often in combination with other treatment modalities, e.g., with DNMT inhibitors, Clinical Trials.gov identifiers NCT01617226 and NCT00867672; with conventional chemotherapy NCT01802333 or in conjuction with allogeneic stem cell transplantation NCT01451268
Histone acetyl reader	Bromodomain-containing proteins (BET proteins)	BET inhibitors	Several clinical trials ongoing with compounds, including OTX-015 (NCT01713582), CPI-0610 (NCT02308761), GSK525762 (NCT01943851)
Histone lysine Methyltransferase	EZH2		EZH2 inhibitors Preclinical
	MLL complexes	DOT1L inhibitors	Clinical trial with compound EPZ-5676 (NCT01684150); inhibitors of MLL-menin preclinical interface Inhibitors of MLL-LEDGF preclinical interface
Histone lysine demethylase	LSD1	LSD1 inhibitors	Clinical trial with compound GSK2879552 (NCT02177612) and tranylcypromine in combination with tretinoine (NCT02261779)
	Jumonji family of histone demethylases (KDMs)	Small molecules targeting and selectively inhibiting Jumonji histone demethylase activity	Preclinical
Histone arginine methyltransferase	PRMTs	PRMT inhibitors	Preclinical

DNMT inhibitors have been reported, but most of them are nucleoside analogs that can lead to toxic side effects and lack specificity. By combining docking-based virtual screening with biochemical analyses, a novel compound, DC_05, was identified (Chen et al., 2014). DC_05 is a non-nucleoside DNMT1 inhibitor with low micromolar IC50 values and significant selectivity toward other AdoMet-dependent protein methyltransferases. Through a process of similarity-based analog searching, compounds DC_501 and DC_517 were found to be more potent than DC_05. These three potent compounds significantly inhibited cancer cell proliferation.

Histone methylome (HMT) is another major epigenetic determinant in gene expression and is frequently deregulated in AML, especially in MLL-rearranged leukemia (Tsai and So, 2017). The first HMT inhibitor targeting DOTL1, EPZ4777, and its second-generation derivative, EPZ5676, have been developed and tested for suppressing MLL leukemia (Daigle et al., 2011, 2013). Both compounds showed selective inhibitory effects on H3K79 methylation and cells bearing MLL fusions, leading to the first clinical trial of HMT inhibitors in AML. Protein arginine N-methyltransferase 1 (PRMT1) inhibitor AMI-408 targets H4R3 methyltransferase, leading to repression of MLL fusion targets (Cheung et al., 2016). Enhancer of zeste 2 polycomb repressive complex 2 subunit (EZH2) inhibitors DZNep and UNC1999 target H3K27 methyltransferase, leading to derepression of polycomb targets (Xu et al., 2015; Zhou et al., 2011).

Bromodomain and extraterminal (BET) proteins bind to acetylated lysines in histones and control gene expression. ABBV-075 is a potent and selective BET family bromodomain inhibitor that recently entered Phase I clinical trials. Apoptosis induced by ABBV-075 was mediated in part by modulation of the intrinsic apoptotic pathway, exhibiting synergy with the BCL-2 inhibitor venetoclax in preclinical models of AML (Bui et al., 2017).

Histone deacetylase inhibitors (HDACIs) have been studied in clinical trials (Table 10.8). The response rates for single-agent HDACIs in AML have been relatively low. Therefore, most investigations of HDACIs have involved combination with other agents, particularly the hypomethylating agents decitabine and azacitidine (Bose and Grant, 2015; Shafer and Grant, 2016), proteasome inhibitors, cyclin-dependent kinase inhibitors, tyrosine kinases inhibitors, beta-catenin inhibitor (Fiskus et al., 2015), G/M checkpoint abrogators, and chemotherapy (Bose and Grant, 2015).

Mutations of the spliceosome machinery range from <1% to 90% in secondary AML compared to <1% to 10.5% in de novo AML (Mohamed et al., 2014; Saez et al., 2017; Yoshida et al., 2011). Mutations in splicing factor 3 subunit b1 (SF3b1), U2 small nuclear RNA auxiliary factor 1 (U2AF1), serine arginine-rich splicing factor 2 (SRSF2), and U2 small nuclear RNA auxiliary factor with zinc finger CCCH-type (ZRSR2) occur in AML. A number of natural products derived from distinct species of bacteria have been found to target the SF3B component of the spliceosome and demonstrate potent antitumor activities. One of the first to be identified was FR901464, a fermentation product from *Pseudomonas*, which has been used as a structural model for the synthesis of several stable chemical analogs of similar or greater potency. In particular, meayamycin B, a soluble synthetic derivative, demonstrates important potential for development as a novel therapy for AML treatment (Wojtuszkiewicz et al., 2014). Meayamycin B inhibits the SF3B1 subunit and can shift alternative splicing of *MCL-1* to promote expression of the proapoptotic *MCL-1s* isoform. Although SF3B1 is among the most commonly mutated splice factors both in MDS and AML, these mutations are not located in the putative binding region of spliceosome inhibitors like meayamycin B, suggesting that they could be effective therapeutic options for patients with these mutations (de Necochea-Campion et al., 2016). Further inhibitors (17S-FD-895, pladienolide B, and sudemycins) were also tried (Shirai et al., 2017; Zhou and Chng 2017).

CD33 antibody-drug conjugates such as gemtuzumab ozogamicin and vadastuximab talirine (SGN-CD33A) (Burnett et al., 2012; Daver et al., 2016; Minagawa et al., 2016; Table 10.9); Hedgehog inhibitors like sonidebig (erismodegib, LDE255), glasdegib (PF-04449913), and vismodegib; neddylation and subsequent ubiquitination inhibitors (pevonedistat); and Wnt signaling pathway inhibitors (Fukushima et al., 2016; Hanna and Shevde, 2016; Ma et al., 2015; Medler et al., 2015; Swords et al., 2017) together with standard chemotherapy are in ongoing trials. Further therapeutic approaches including NF-κB signaling pathway inhibition (Bosman et al., 2016; Fuchs 2010; Siveen et al., 2017; Zhou et al., 2015), PI3 K/AKT/mTOR signaling

Table 10.9 Trials for antibody-based targeting AML stem cells

AML stem cell antigen	Description	Antibody	Trials
CD123	IL-3 receptor α chain	SGN-123A, antibody-drug conjugate	NCT02848248
		JNJ-56022473 dual-specific Antibody with CD16	NCT02472145
		SL-401, antibody toxin chimeric	NCT02113982
		Protein using diphtheria toxin	NCT02270463
		XmAb14045, dual-specific Antibody with CD3	NCT02730312
		MGD006, dual specific antibody with CD3	NCT02152956
		KHK2823, IgG 1 antibody	NCT02181699
CLL-1	C-type lectin-like molecule		
CD33	Receptor of myeloid cells, member of the Ig superfamily	SGN-CD33A, antibody-drug conjugate	NCT02326584 NCT02785900 NCT02312037 NCT02614560
		IMGN779, antibody drug conjugate	NCT02674763
		Gemtuzumab ozogamicin, antibody-drug conjugate	NCT01869803
CD44	Receptor for hyaluronic acid		
CD47	Receptor for signal Regulatory protein α (SIRP α)	Hu5F9-G4, anti-CD47 Monoclonal antibody	NCT02678338
		TTI-621, antibody against CD47	NCT02663518
		Binding domain of SIRP α	
		INBRX-103 (CC-90002), Anti-CD47 monoclonal antibody	NCT02367196 NCT02641002
CD96	Tactile, member of Ig superfamily		
CD93	Marker of a nonquiescent AML stem cell population In MLL-rearranged AML		
CD25	IL-2 receptor α chain		NCT02588092

pathway inhibition, (Brenner et al., 2016; Fuchs 2011; Hauge et al., 2016) and targeting the S100A8/S100A9-TLR4-ERK/JNK/p38 pathway (Laouedj et al., 2017; Tamburini, 2017) were studied. Cell division cycle 25 (CDC25) protein phosphatases inhibition had antiproliferative effects on primary human AML cells for a subset of patients identified by gene expression profiling (Brenner et al., 2017).

Immunotherapies, which have shown promise in the treatment of hematologic malignancies, have the potential to target AML through pathways that are distinct and complementary to current approaches (Lichtenegger et al., 2015). A second-generation CD33-specific chimeric antigen receptor capable of redirecting cytolytic effector T cells against leukemic cells was prepared. CD33 is expressed in approximately 90% of AML cases and has demonstrated utility as a target of therapeutic antibodies. Chimeric antigen receptor (CAR)-modified T cells efficiently killed leukemia cell lines and primary tumor cells in vitro. The antileukemia effect was CD33-specific, mediated through T-cell effector functions, and displayed tumor lysis at effector:target ratios as low as 1:20. Furthermore, the CD33-redirected T cells were effective in vivo, preventing the development of leukemia after prophylactic administration and delaying the progression of established disease in mice. These data provide preclinical validation of the effectiveness of a second-generation anti-CD33 chimeric antigen receptor therapy for AML, and support its continued development as a clinical therapeutic (O'Hear et al., 2015; Minagawa et al., 2016). An alternative antibody-based immunotherapeutic strategy is a novel class of bispecific T-cell-engaging antibodies (BiTEs) targeting CD33 antigen on AML cells and the CD3e component of the T-cell receptor complex (AMG 330) (Lichtenegger et al., 2015).

Therapeutic targeting of AML stem cells

The leukemia stem cells (LSCs) are characterized by their unlimited self-renewal, repopulating potential and long residence in a quiescent state of G_0/G_1 phase. LSCs are considered to have a pivotal role in the relapse and refractory of AML. Therefore, new therapeutic strategies to target LSCs with limited toxicity toward the normal hematopoietic population is critical for the ultimate curing of AML. Ongoing research works with natural products like parthenolide (a natural plant extract derived compound) and its derivatives that have the ability to target multiple pathways that regulate the self-renewal, growth, and survival of LSCs point to ways for a possible complete remission in AML (Siveen et al., 2017). AML stem cells are not only in the quiescent phase of the cell cycle (G0) and can be also found in cycling cells. Their frequency increases as a function of chemotherapy and subsequent relapse. The first clearly defined AML stem

cells–specific antigen was interleukin-3 receptor α (CD123) within $CD34^+/CD38^-$ compartment (Jordan et al., 2000). Multiple additional immunophenotypic differences that may distinguish AML stem cells from normal hematopoietic stem cells were subsequently identified and can be used for therapeutic intervention using antibody-based targeting strategies (Table 10.9) (Pollyea and Jordan, 2017).

FAMILIAL PREDISPOSITION TO AML

AML is not only sporadic but also familial MDS/AML predisposition caused by specific mutations has been recognized. The myeloid neoplasms with genetic predisposition represent a new category in the revised 2016 World Health Organization classification. According to the new classification, these disorders are subdivided based on the clinical and genetic features, including myeloid neoplasms with germ-line predisposition alone, or with preexisting platelet disorder, cytopenias, or other organ failures. The predisposing genetic factors include mutations in RUNX1, CEBPA, GATA2, ANKRD26 (ankyrin repeat domain 26), ETV6, (gene for ETS family transcriptional repressor, variant 6), DDX41, TERC, or TERT (telomerase subunit genes), and SRP72 (signal recognition particle 72 gene) (Babushok et al., 2016; Churpek et al., 2015; Godley, 2014; Owen et al., 2008; West et al., 2014). The genes affected in familial AML are important regulators of hemopoiesis and are frequently implicated in leukemogenesis, providing deeper insight into the understanding of normal and malignant hemopoiesis. Despite the growing knowledge of germ-line predisposing events in the background of familial myeloid malignancies, the germ-line genetic component is still unknown in a subset of these pedigrees, suggesting the presence of so-far-unidentified predisposing mutations.

FIRST MODELS FOR PREDICTING OUTCOME FOR INDIVIDUAL AML PATIENTS

Papaemmanuil et al. (2016) studied whole exome sequences from 111 myeloid cancer genes in diagnostic leukemia samples from 1540 patients with AML who were undergoing intensive treatment in three prospective clinical trials of the

German-Austrian AML Study Group. They identified driver point mutations and combined these data with the clinical database to generate a comprehensive knowledge bank. The 231 predictor variables were derived from these data and they include fusion genes, copy number alterations, point mutations, gene–gene interactions, demographic features, clinical risk factors, and treatment. Various validation strategies were used to assess the accuracy of predictions. The model built from these three prospective clinical trials was tested on an independent AML cohort from the United States (TCGA) (Cancer Genome Atlas Research Network, 2013). The model for predicting outcomes developed from these trials is considerable more complex (Gerstung et al., 2017) than the ELN genetic scoring system (Döhner et al., 2010). The individual risk in this AML cohort was continuous, with no cutoff points for stratification. More detailed survival estimates confirmed known ELN risk groups. However, the predicted survival of one-third of patients deviated more than 20% from the ELN stratification. Gerstung et al. (2017) found that clinical and demographic factors (patient age, performance status and blood counts) substantially influenced rates of early death, including death in remission. Genomic features, such as copy number changes, fusion genes, or driver point mutations, most strongly impacted the dynamics of disease remission and relapse. The model for predicting outcome was also used for the estimation for individual AML patients whether to choose HSCT (allograft) or chemotherapy in first remission. This choice is very important because a transplantation-associated mortality rate is 20%–25% and the risk of graft versus host disease is nonnegligible. Gerstung et al., estimated that 12% (124/995) of patients aged 18–60 years would have an improvement in survival of three years after HSCT in first complete remission as compared to standard chemotherapy. Only 28 from these 124 patients were identified as having "adverse risk" by current criteria. Personally tailored management decisions could reduce the number of HSCT in AML patients by 20%–25% while maintaining the same survival rates.

CONCLUSION

Recent advances in AML biology and its genetic landscape due to the continuously increasing application of genomic techniques have led to major advances in our knowledge of the pathogenesis of AML and should finally lead to more subset-specific AML therapies. These therapies will be ideally tailored to each patient with AML (Dombret and Gardin, 2016). Current advances in nonacute promyelocytic leukemia lack innovation. However, modification and intensification of doses and schedules of standard cytotoxic drugs were improved and great progress in HSCT techniques was achieved. Distinct AML subsets were characterized, but therapy remained uniform. New effective agents and their combination and also combined with standard chemotherapy are still under investigation and under clinical trials. Calculations show that databases require information from thousands of patients for accurate decision support. Clinical and genomic knowledge banks and models for predicting outcome of AML patients will be in the future considerably more complex than those currently used in clinical practice.

Routine clinical use of MRD testing requires further refinements and standardization of assays and reported results. MRD results will be used not only as tool for therapy decision making but also as an alternate end point for drug development and regulatory approval (Hourigan et al., 2017). Currently, survival is the end point used as evidence for the clinical benefit of new drugs or new drug combinations for the therapy of AML.

Research in the last few years revealed alterations in the microenvironment of the hematopoietic stem cells (the niche compartment) in AML. Leukemic cells can remodel the niche into a permissive environment favoring the expansion of leukemic stem cells and progression of disease (Korn and Méndez-Ferrer, 2017). The microenvironment also mediates chemotherapy-resistance pathways. Therefore, the understanding of niche-controlled resistance is important for combined treatment targeting not only the mutated cells but also the microenvironment of leukemic cells.

ACKNOWLEDGMENTS

This work was supported by a research project for conceptual development of research organization (00023736; Institute of Hematology and Blood Transfusion, Prague) from the Ministry of Health of the Czech Republic.

CONFLICT OF INTEREST

There are no commercial or other associations in connection with this review.

REFERENCES

Al-Issa K, Nazha A. Molecular landscape in acute myeloid leukemia: where do we stand in 2016. *Cancer Biol Med* 13;2016:474–82.

Arber DA, Orazi A, Hasserjian R et al. The 2016 revision to the World Health Organization classification of myeloid neoplasms and acute leukemia. *Blood* 127;2016:2391–405.

Ayatollahi H, Shajiei A, Sadeghian MH, Sheikhi M, Yazdandoust E, Ghazanfarpour M, Shams SF, Shakeri S. Prognostic importance of C-KIT mutations in core binding factor acute myeloid leukemia: A systematic review. *Hematol Oncol Stem Cell Ther* 10;2017:1–7.

Babushok DV, Bessler M, Olson TS. Genetic predisposition to myelodysplastic syndrome and acute myeloid leukemia in children and young adults. *Leuk Lymphoma* 57;2016:520–36.

Bacher U, Kern W, Schnittger S, Hiddemann W, Schoch C, Haferlach T. Further correlations of morphology according to FAB and WHO classification to cytogenetics in de novo acute myeloid leukemia: A study on 2,235 patients. *Ann Hematol* 84;2005:785–91.

Baljevic M, Park JH, Stein E, Douer D, Altman JK, Tallman MS. Curing all patients with acute promyelocytic leukemia: Are we there yet? *Hematol Oncol Clin North Amer* 25;2011:1215–33.

Bennett JM, Catovsky D, Daniel MT, Flandrin G, Galton DA, Gralnick HR, Sultan C. Proposals for the classification of the acute leukemias. French-American-British (FAB) co-operative group. *Brit J Haematol* 33;1976:451–58.

Bennett JM, Catovsky D, Daniel MT, Flandrin G, Galton DA, Gralnick HR, Sultan C. Proposed revised criteria for the classification of acute myeloid leukemia. A report of the French-American-British cooperative group. *Ann Int Med* 10;1985:620–25.

Blum S, Greve G, Lübert M. Innovative strategies for adverse karyotype acute myeloid leukemia. *Curr Opin Hematol* 24;2017:89–98.

Bose P, Grant S. Rational combinations of targeted agents in AML. *J Clin Med* 4;2015:634–64.

Bosman MC, Schuringa JJ, Vellenga E. Constitutive NF-κ activation in AML: Causes and treatment strategies. *Crit Rev Oncol Hematol* 98;2016:35–44.

Brands-Nijenhuis AV, Labopin M, Schouten HC et al. Monosomal karyotype as an adverse prognostic factor in patients with acute myeloid leukemia treated with allogeneic hematopoietic stem-cell transplantation in first complete remission: A retrospective survey on behalf of the ALWP of the EBMT. *Haematologica* 101;2016:248–55.

Brenner AK, Andersson Tvedt TH, Bruserud Ø. The compexity of targeting PI3K-AKT-mTOR signalling in human acute myeloid leukaemia: The importance of leukemic cell heterogeneity, neighbouring mesenchymal stem cells and immunocompotent cells. *Molecules* 21;2016:E1512.

Brenner AK, Reikvam H, Rye KP, Hagen KM, Lavecchia A, Bruserud Ø. CDC25 inhibition in acute myeloid leukemia: A study of patient heterogeneity and the effects of different inhibitors. *Molecules* 22;2017:E446.

Buccisano F, Walter RB. Should patients with acute myeloid leukemia and measurable residual disease be transplanted in first complete remission? *Curr Opin Hematol* 24;2017:132–8.

Bui MH, Lin X, Albert DH S et al. Preclinical characterization of BET family bromodomain inhibitor ABBV-075 suggests combination therapeutic strategies. *Cancer Res* 2017. doi: 10.1158/0008-5472.CAN-16-1793.

Bullinger L, Döhner K, Döhner H. Genomics of acute myeloid leukemia diagnosis and pathways. *J Clin Oncol* 2017. doi: 10.1200/JCO.2016.71.2208.

Burnett AK. New induction and postinduction strategies in acute myeloid leukemia. *Curr Opin Hematol* 19;2012:76–81.

Burnett AK, Russell NH, Hills RK et al. Optimization of chemotherapy for younger patients with acute myeloid leukemia: Results of the medical research council AML15 trial. *J Clin Oncol* 31;2013:3360–8.

Burnett AK, Russell NH, Hills RK et al. Addition of gemtuzumab ozogamicin to induction chemotherapy improves survival in older patients with acute myeloid leukemia. *J Clin Oncol* 30;2012:3924–31.

Cancer Genome Atlas Research Network. Genomic and epigenomic landscapes of adult *de novo* acute myeloid leukemia. *N Engl J Med* 368;2013:2059–74.

Chaturvedi A, Araujo Cruz MM, Jyotsana N et al. Mutant IDH1 promotes leukemogenesis in vivo and can be specifically targeted in human AML. *Blood* 122;2013:2877–87.

Chen S, Wang Y, Zhou W et al. Identifying novel selective non-nucleoside DNA methyltransferase 1 inhibitors through docking-based virtual screening. *J Med Chem* 57;2014:9028–41.

Cheung N, Fung TK, Zeisig BB et al. Targeting aberrant epigenetic networks mediated by PRMT1 and KDM4C in acute myeloid leukemia. *Cancer Cell* 29;2016:32–48.

Churpek JE, Godley LA. How I diagnose and manage individuals at risk for inherited myeloid malignancies. *Blood* 128;2016:1800–13.

Churpek JE, Pyrtel K, Kanchi KL et al. Genomic analysis of germ line and somatic variants in familial myelodysplasia/acute myeloid leukemia. *Blood* 126;2015:2484–90.

Coenen EA, Zwaan CM, Meyer C, Marschalek R, Pieters R, van der Veken LT, Beverloo HB, van den Heuvel-Eibrink MM. KIAA1524: A novel MLL translocation partner in acute myeloid leukemia. *Leuk Res* 35;2011:133–5.

Daigle SR, Olhava EJ, Therkelsen CA et al. Potent inhibition of DOT1L as treatment of MLL-fusion leukemia. *Blood* 122;2013:1017–25.

Daigle SR, Olhava EJ, Therkelsen CA et al. Selective killing of mixed lineage leukemia cells by a potent small-molecule DOT1L inhibitor. *Cancer Cell* 20;2011:53–65.

Dang L, Yen K, Attar EC. IDH mutations in cancer and progress toward development of targeted therapeutics. *Ann Oncol* 27;2017:599–608.

Daver N, Kantarjian H, Ravandi F et al. A phase II study of decitabine and gemtuzumab ozogamicin in newly diagnosed and relapsed acute myeloid leukemia and high-risk myelodysplastic syndrome. *Leukemia* 30;2016:268–73.

De Kouchkovsky I, Abdul-Hay M. Acute myeloid leukemia: a comprehensive review and 2016 update. *Blood Cancer J* 6;2016:e441.

de Necochea-Campion R, Shouse GP, Zhou Q, Mirshahidi S, Chen CS. Aberrant splicing and drug resistance in AML. *J Hematol Oncol* 9;2016:85.

Döhner H, Estey EH, Amadori S et al. Diagnosis and management of acute myeloid leukemia in adults: recommendations from an international expert panel, on behalf of the European LeukemiaNet. *Blood* 115;2010:453–74.

Döhner H, Estey EH, Grimwade D et al. Diagnosis and management of AML in adults: 2017 ELN recommendations from an international expert panel. *Blood* 129;2017:424–47.

Dombret H, Gardin C. An update of current treatments for adult acute myeloid leukemia. *Blood* 127;2016:53–61.

Eisfeld AK, Mrózek K, Kohlschmidt J et al. The mutational oncoprint of recurrent cytogenetic abnormalities in adult patients with de novo acute myeloid leukemia. *Leukemia* 2017. doi: 10.1038/leu.2017.86.

Emadi A, Jun SA, Tsukamoto T, Fathi AT, Minden MD, Dang CV. Inhibition of glutaminase selectively suppresses the growth of primary acute myeloid leukemia cells with IDH mutation. *Exp Hematol* 42;2014:247–51.

Estey E, Dohner H. Acute myeloid leukemia. *Lancet* 368;2006:1894–907.

Fang M, Storer B, Estey E, Othus M, Zhang L, Sandmaier BM, Appelbaum FR. Outcome of patients with acute myeloid leukemia with monosomal karyotype who undergo haematopoietic cell transplantation. *Blood* 118;2011:1490–4.

Fathi AT, Wander SA, Faramand R, Emadi A. Biochemical, epigenetic, and metabolic approaches to target IDH mutations in acute myeloid leukemia. *Semin Hematol* 52;2015:165–71.

Ferrara F, Palmieri S, Leoni F. Clinically useful prognostic factors in acute myeloid leukemia. *Crit Rev Oncol Hematol* 66;2008:181–93.

Fiskus W, Sharma S, Saha S et al. Pre-clinical efficacy of combined therapy with novel beta-catenin antagonist BC2059 and histone deacetylase inhibitor against AML cells. *Leukemia* 29;2015:1267–78.

Foran JM. New prognostic markers in acute myeloid leukemia: Perspective from the clinic.

Hematology Am Soc Hematol Educ Prog 2010;2010:47–55.

Fuchs O. Transcription factor NF-κB inhibitors as single therapeutic agents or in combination with classical chemotherapeutic agents for the treatment of hematologic malignancies. Curr Mol Pharmacol 3;2010:98–122.

Fuchs O. Promising activity of mammalian target of rapamycin inhibitors in hematologic malignancies therapy. Curr Sign Transd Ther 61;2011:44–54.

Fukushima N, Minami Y, Kakiuchi S, Kuwatsuka Y, Hayakawa F, Jamieson C, Kiyoi H, Naoe T. Small-molecule Hedgehog inhibitor attenuates the leukemia-initiation potential of acute myeloid leukemia cells. Cancer Sci 107;2016:1422–9.

Gadzik VI, Teleanu V, Papaemmanuil E et al. RUNX1 mutations in acute myeloid leukemia are associated with distinct clinicpathologic and genetic features. Leukemia 30;2016:2160–8.

Gale RP, Wiernik PH, Lazarus HM. Should persons with acute myeloid leukemia have a transplant in first remission? Leukemia 28;2014:1949–52.

Gallogly MM, Lazarus HM. Midostaurin: An emerging treatment for acute myeloid leukemia patients. J Blood Med 7;2016:73–83.

Gerstung M, Papaemmanuil E, Martincorena I et al. Precision oncology for acute myeloid leukemia using a knowledge bank approach. Nat Genet 49;2017:332–40.

GLOBOCAN 2014: Estimated Cancer Incidence, Mortality and Prevalence Worldwide, 2012. www.globocan.iarc.fr. Accessed September 21, 2014.

Godley LA. Inherited predisposition to acute myeloid leukemia. Semin Hematol 51;2014:306–21.

Grimwade D, Hills RK, Moorman AV et al. Refinement of cytogenetic classification in acute myeloid leukemia: Determination of prognostic significance of rare recurring chromosomal abnormalities among 5876 younger adult patients treated in the United Kingdom Medical Research Council trials. Blood 116;2010:354–65.

Grimwade D, Mrózek K. Diagnostic and prognostic value of cytogenetics in acute myeloid leukemia. Hematol Oncol Clin North America 25;2011:1135–61.

Gröschel S, Sanders M, Hoogenboezem R et al. A single oncogenic enhancer rearrangement causes concomitant EVI1 and GATA2 deregulation in leukemia. Cell 157;2014:369–81.

Grunwald MR, Levis MJ. FLT3 tyrosine kinase inhibition as a paradigm for targeted drug development in acute myeloid leukemia. Semin Hematol 52;2015:193–9.

Haferlach T, Stengel A, Eckstein C, Perglerová K, Alpermann T, Kern W, Haferlach C, Meggendorfer M. The new provisional WHO entity "RUNX1 mutated AML" shows specific genetics but no prognostic influence of dysplasia. Leukemia 30;2016:2109–12.

Hanna A, Shevde LA. Hedgehog signaling: modulation of cancer properties and tumor microenvironment. Mol Cancer 15;2016:24.

Hauge M, Bruserud Ø, Hatfield KJ. Targeting of cell metabolism in human acute myeloid leukemia-more than targeting of isocitrate dehydrogenase mutations and PI3 K/AKT/mTOR signaling? Eur J Haematol 96;2016:211–21.

Heuser M. Therapy-related myeloid neoplasms: Does knowing the origin help to guide treatment? Hematology Am Soc Hematol Educ Program 2016(1);2016:24–32.

Hinai AA, Valk PJ. Aberrant EVI1 expression in acute mzeloid leukemia. Brit J Haematol 172;2016:870–8.

Hourigan CS, Gale RP, Gormley NJ, Ossenkoppele Gj, Walter RB. Measurable residual disease testing in acute myeloid leukaemia. Leukemia 2017. doi: 10.1038/leu.2017.113.

Jang JE, Min YH, Yoon J et al. Single monosomy as a relatively better survival factor in acute myeloid leukemia patients with monosomal karyotype. Blood Cancer J 5;2015:e358.

Jan M, Ebert BL, Jaiswal S. Clonal hematopoiesis. Semin Hematol 54;2017:43–50.

Jordan CT, Upchurch D, Szilvassy SJ et al. The interleukin-3 receptor alpha chain is a unique marker for human acute myelogenous leukemia stem cells. Leukemia 14;2000:1777–84.

Kampen KR. The discovery and early understanding of leukemia. Leuk Res 36;2012:6–13.

Kansal R. Acute myeloid leukemia in the era of precision medicine: recent advances in diagnostic classification and risk stratification. *Cancer Biol Med* 13;2016:41–54.

Kayser S, Zucknick M, Döhner K et al. Monosomal karyotype in adult acute myeloid leukemia: prognostic impact and outcome after different treatment strategies. *Blood* 118;2012:551–8.

Kohlmann AA, Bacher U, Schnittger S, Haferlach T. Perspective on how to approach molecular diagnostics in acute myeloid leukemia and myelodysplastic syndromes in the era of next generation sequencing. *Leuk Lymphoma* 55;2014:1725–34.

Korn C, Méndez-Ferrer S. Myeloid malignancies and the microenvironment. *Blood* 129;2017:811–22.

Krause JR. Morphology and classification of acute myeloid leukemias. *Clin Lab Med* 20;2000:1–16.

Lam K, Zhang DE. RUNX1 and RUNX1-ETO: Roles in hematopoiesis and leukemogenesis. *Front Biosci* 17;2012:1120–39.

Laouedj M, Tardif MR, Gil L, Raquil MA, Lachhab A, Pelletier M, Tessier PA, Barabé F. S100A9 induces differentiation of acute myeloid leukemia cells through TLR4. *Blood* 129;2017:1980–90.

Lazarevic V, Rosso A, Juliusson G et al. Incidence and prognostic significance of isolated trisomies in adult acute myeloid leukemia: A population-based study from the Swedish AML registry. *Eur J Haematol* 98;2017:493–500.

Lennartsson J, Jelacic T, Linnekin D, Shivakrupa R. Normal and oncogenic forms of the receptor tyrosine kinase kit. *Stem Cells* 23;2005: 16–43.

Lichtenegger FS, Krupka C, Köhnke T, Subklewe M. Immunotherapy for acute myeloid leukemia. *Semin Hematol* 52;2015:207–14.

Lin PH, Li HY, Fan SC et al. A targeted next-generation sequencing in the molecular risk stratification of adult acute myeloid leukemia: Implications for clinical practice. *Cancer Med* 6;2017:349–60.

Lo-Coco F, Ammatuna E, Montesinos P, Sanz MA. Acute promyelocytic leukemia: Recent advances in diagnosis and management. *Semin Oncol* 35;2008:401–9.

Losman JA, Looper RE, Koivunen P et al. (R)-2-hydroxyglutarate is sufficient to promote leukemogenesis and its effect is reversible. *Science* 339;2013:1621–5.

Lugthart S, Gröschel S, Baverloo HB et al. Clinical, molecular, and prognostic significance of WHO type inv(3)(q21q26.2)/t(3;3)(q21;q26.2) and various other 3q abnormalities in acute myeloid leukemia. *J Clin Oncol* 28;2010:3890–98.

Ma S, Yang LL, Niu T et al. SKLB-677, an FLT3 and Wnt/beta-catenin signaling inhibitor, displays potent activity in models of FLT3-driven AML. *Sci Rep* 5;2015:15616.

Marcucci G. Core binding factor acute myeloid leukemia. *Clin Adv Hematol Oncol* 4;2006:339–41.

Mayer LD, Harasym TO, Tardi PG et al. Ratiometric dosing of anticancer drug combinations: Controlling drug ratios after systemic administration regulates therapeutic activity in tumor-bearing mice. *Mol Cancer Ther* 5;2006:1854–63.

Medeiros BC, Fathi AT, DiNardo CD, Pollyea DA, Chan SM, Swords R. Isocitrate dehydrogenase mutations in myeloid malignancies. *Leukemia* 31;2017:272–81.

Medler CJ, Waddell JA, Solimando DA. Cancer chemotherapy update. Gefitinib and sonidegib. *Hosp Pharm* 50;2015:868–72.

Metzeler KH, Bloomfield CD. Clinical relevance of RUNX1 and CBFB alterations in acute myeloid leukemia and other hematological disorders. *Adv Exp Med Biol* 962;2017:175–99.

Meyer C, Kowarz E, Hofmann J et al. New insights to the *MLL* recombinome of acute leukemias. *Leukemia* 23;2009:1490–9.

Meyers J, Yu Y, Kaye JA, Davis KL. Medicare fee-for-service enrollees with primary acute myeloid leukemia: An analysis of treatment patterns, survival, and healthcare resource utilization and costs. *Appl Health Econ Health Policy* 11;2013:275–86.

Minagawa K, Jamil MO, Al-Obaidi M et al. In vitro pre-clinical validation of suicide gene modified anti-CD33 redirected chimeric antigen receptor T-cells for acute myeloid leukemia. *PLOS ONE* 11;2016:e0166891.

Mohamed AM, Thénoz M, Solly F, Balsat M, Mortreux F, Wattel E. How mRNA is mis-spliced in acute myelogenous leukemia (AML)? *Oncotarget* 5;2014:9534–45.

Mrózek K, Heerema N, Bloomfield CD. Cytogenetics in acute leukemia. *Blood Rev* 18;2004:115–36.

Mrózek K, Marcucci G, Paschka P, Bloomfield CD. Advances in molecular genetics and treatment of core-binding factor acute myeloid leukemia. *Currt Opin Oncol* 20;2008:711–8.

Nazha A, Zarzour A, Al-Issa K et al. The complexity of interpreting genomic data in patients with acute myeloid leukemia. *Blood Cancer J* 6;2016:e510.

Nebbioso A, Benedetti R, Conte M, Iside C, Altucci L. Genetic mutations in epigenetic modifiers as therapeutic targets in acute myeloid leukemia. *Expert Opin Ther Targets* 19;2015:1187–1202.

O'Hear C, Heiber JF, Schubert I, Fey G, Geiger TL. Anti-CD33 chimeric antigen receptor targeting of acute myeloid leukemia. *Haematologica* 100;2015:336–44.

Owen C, Barnett M, Fitzgibbon J. Familial myelodysplasia and acute myeloid leukaemia: A review. *Br J Haematol* 140;2008:123–32.

Papaemmanuil E, Gerstung M, Bullinger L et al. Genomic classification and prognosis in acute myeloid leukemia. *N Engl J Med* 374;2016:2209–21.

Paschka P, Marcucci G, Ruppert AS et al. Adverse prognostic significance of KIT mutations in adult acute myeloid leukemia with inv(16) and t(8;21): A Cancer and Leukemia Group B Study. *J Clin Oncol* 24;2006:3904–11.

Perrot A, Luquet I, Pigneux A et al. Dismal prognostic value of monosomal karyotype in elderly patients with acute myeloid leukemia: A GOELAMS study of 186 patients with unfavorable cytogenetic cytogenetic abnormalities. *Blood* 118;2011:679–85.

Paun O, Lazarus HM. Allogeneic hematopoietic cell transplation for acute myeloid leukemia in first complete remission: Have the indications changed? *Curr Opin Hematol* 19;2012: 95–101.

Pigneux A, Labopin M, Maertens J et al. Outcome of allogeneic hematopoietic stem-cell transplantation for adult patients with AML and 11q23/MLL rearrangement (MLL-r AML). *Leukemia* 29;2015:2375–81.

Pollyea DA, Jordan CT. Therapeutic targeting of acute myeloid leukemia stem cells. *Blood* 129;2017:1627–35.

Prada-Arismendy J, Arroyave JC, Röthlisberger S. Molecular biomarkers in acute myeloid leukemia. *Blood Rev* 31;2017:63–76.

Röllig C, Bornhäuser M, Thiede C et al. Long-term prognosis of acute myeloid leukemia according to the new genetic risk classification of the European LeukemiaNet recommendations: Evaluation of the proposed reporting system. *J Clin Oncol* 29;2011:2758–65.

Rowe JM. AML in 2016: Where we are now? *Best Pract Res Clin Hematol* 29;2016:315–9.

Rücker FG, Schlenk RF, Bullinger L et al. TP53 alterations in acute myeloid leukemiawith complex karyotype correlate with specific copy number alterations, monosomal karyotype, and dismal outcome. *Blood* 119;2012:2114–21.

Saez B, Walter MJ, Graubert TA. Splicing factor gene mutations in hematologic malignancies. *Blood* 129;2017:1260–9.

Sallman DA, Lancet JE. What are the most promising new agents in acute myeloid leukemia? *Curr Opin Hematol* 24;2017:99–107.

Sanders MA, Valk PJM. The evolving molecular genetic landscape in acute myeloid leukaemia. *Curr Opin Hematol* 20;2013:79–85.

Sangle NA, Perkins SL. Core-binding factor acute myeloid leukemia. *Arch Pathol Lab Med* 135;2011:1504–09.

Sexauer A, Perl A, Yang X et al. Terminal myeloid differentiation in vivo is induced by FLT3 inhibition in FLT3/ITD AML. *Blood* 120;2012:4205–14.

Shafer D, Grant S. Update on rational targeted therapy in AML. *Blood Rev* 30;2016:273–83.

Shah A, Andersson TM, Rachet B, Bjorkholm M, Lambert PC. Survival and cure of acute myeloid leukaemia in England, 1971–2006: A population-based study. *Br J Haematol* 162;2013:509–516.

Shimada A, Taki T, Tabuchi K et al. KIT mutations, and not FLT3 internal tandem duplication, are strongly associated with a poor prognosis in pediatric acute myeloid leukemia with t(8;21): A study of the Japanese Childhood AML Cooperative Study Group. *Blood* 107;2006:1806–9.

Shirai CL, White BS, Tripathi M et al. Mutant U2AF1-expressing cells are sensitive to pharmacological modulation of the spliceosome. *Nat Commun* 8;2017:14060.

Siegel RL, Miller KD, Jemal A. Cancer statistics, 2015. *CA Cancer J Clin* 65;2015:5–29.

Siveen KS, Uddin S, Mohammad RM. Targeting acute myeloid leukemia stem cell signaling by natural products. *Mol Cancer* 16;2017:13.

Smith ML, Hills RK, Grimwade D. Independent prognostic variables in acute myeloid leukaemia. *Blood Rev* 25;2011:39–51.

Swords RT, Watts J, Erba HP et al. Expanded safety analysis of pevonedistat, a first-in-class NEDD8-activating enzyme inhibitor, in patients with acute nyeloid leukemia and myelodysplastic syndromes. *Blood Cancer J* 7;2017:e520.

Tamamyan G, Kadia T, Ravandi F et al. Frontline treatment of acute myeloid leukemia in adults. *Crit Rev Oncol Hematol* 110;2017:20–34.

Tamburini J. S100 proteins in AML: Differentiation and beyond. *Blood* 129;2017:1893–4.

Tardi P, Johnstone S, Harasym N et al. In vivo maintenance of synergistic cytarabine:daunorubicin ratios greatly enhances therapeutic efficacy. *Leuk Res* 33;2009:129–39.

Thomas X. First contributors in the history of leukemia. *World J Hematol* 2;2013:62–70.

Tsai CT, So CWE. Epigenetic therapies by targeting aberrant histone methzlome in AML: Molecular mechanism, current preclinical and clinical development. *Oncogene* 36;2017:1753–9.

Utsun C, Marcucci G. Emerging diagnostic and therapeutic approaches in core binding factor acute myeloid leukaemia. *Curr Opin Hematol* 22;2015:85–91.

Vardiman JW, Thiele J, Arber DA et al. The 2008 revision of the World Health Organization (WHO) classification of myeloid neoplasms and acute leukemia: Rationale and important changes. *Blood* 114;2009:937–51.

Wang B, Liu Y, Hou G et al. Mutational spectrum and risk stratification of intermediate-risk acute myeloid leukemia patients based on next-generation sequencing. *Oncotarget* 7;2016:32065–78.

Wang F, Travins J, DeLaBarre B et al. Targeted inhibition of mutant IDH2 in leukemia cells induces cellular differentiation. *Science* 2013;340:622–6.

Weinberg OK, Hasserjian RP, Li B, Pozdnyakova O. Assessment of myeloid and monocytic dysplasia by flow cytometry in de novo AML helps define an AML with myelodysplasia-related changes category. *J Clin Pathol* 70;2017a:109–115.

Weinberg OK, Sohani AR, Bhargava P, Nardi V. Diagnostic work-up of acute myeloid leukemia. *Am J Hematol* 92;2017b:317–21.

West AH, Godley LA, Churpek JE. Familial myelodysplastic syndrome/acute myeloid leukemia syndromes: A review and utility for translational investigations. *Annals NY Acad Sci* 1310;2014:111–8.

White RS, DiPersio JF. Genomic tools in acute myeloid leukemia: From the bench to the bedside. *Cancer* 120;2014:1134–44.

Wojtuszkiewicz A, Assaraf YG, Jansen G, Koide K, Bressin RK, Basu U, Sonneveld E. Spliceosome inhibitor meayamycin B as a novel potential chemotherapeutic agent in ALL and AML. *Blood* 124;2014:924.

Wouters BJ, Delwel R. Epigenetics and approaches to targeted epigenetic therapy in acute myeloid leukemia. *Blood* 127;2016:42–52.

Xu B, On DM, Ma A et al. Selective inhibition of EZH2 and EZH1 enzymatic activity by a small molecule suppresses MLL-rearranged leukemia. *Blood* 125;2015:346–57.

Yamagata T, Maki K, Mitani K. RUNX1/AML1 in normal and abnormal hematopoiesis. *Inter J Hematol* 82;2005:1–8.

Yamamoto JF, Goodman MT. Patterns of leukemia incidence in the United States by subtype and demographic characteristics, 1997–2002. *Cancer Causes Control* 19;2008:379–90.

Yin JAL, O'Brien MA, Hills RK, Daly SB, Wheatley K, Burnett AK. Minimal residual disease monitoring by quantitative RT-PCR in core binding factor AML allows risk stratification and predicts relapse: Results of the United Kingdom MRC AML-15 trial. *Blood* 120;2012:2826–35.

Yoshida K, Sanada M, Shiraishi Y et al. Frequent pathway mutations of splicing machinery in myelodysplasia. *Nature* 478;2011:64–9.

Zhou J, Bi C, Cheong LL et al. The histone methyltransferase inhibitor, DZNep, up-regulates TXNIP, increases ROS production, and targets leukemia cells in AML. *Blood* 118;2011:2830–9.

Zhou J, Ching YQ, Chng W-J. Aberrant nuclear factor-kappa B activity in acute myeloid leukemia: From molecular pathogenesis to therapeutic target. *Oncotarget* 6;2015:5490–500.

Zhou J, Chng W-J. Aberrant RNA splicing and mutations in spliceosome complex in acute myeloid leukemia. *Stem Cell Investig* 4;2017:6.

Zhou Y, Othus M, Araki D et al. Pre- and post-transplant quantification of measurable ("minimal") residual disease via multiparameter flow cytometry in adult acute myeloid leukemia. *Leukemia* 30;2016:1456–64.

PART 2

Precision medicine in NCDs

PART 2

Precision medicine in NIDDx

11

Precision medicine in coronary artery disease

MELVIN GEORGE, LUXITAA GOENKA, AND SANDHIYA SELVARAJAN

BACKGROUND

Coronary artery disease (CAD) continues to be a leading cause of mortality and morbidity worldwide, despite the significant improvement and development in its diagnosis and treatment. CAD is a multifactorial disease that can have differing clinical presentations (Figure 11.1). Around 17.5 million people have died due to CAD worldwide, which represents 31% of the overall global deaths. Among these deaths, 7.4 million were due to coronary heart disease (CHD) and 6.7 million were due to stroke (Krishnan, 2012). CAD decreases the quality of life (QOL) among individuals and contributes to several comorbidities. Moreover, treatment of CAD exceeds $100 billion each year (Heidenreich et al., 2011). As the treatment of CAD is becoming expensive each year, it is necessary to place greater emphasis on cost containment and improve outcome measures.

CURRENT DIAGNOSTIC AND THERAPEUTIC APPROACHES FOR CORONARY ARTERY DISEASE (CAD)

Presently there are several diagnostic and therapeutic strategies available for the detection and treatment of CAD. The assessment of CAD

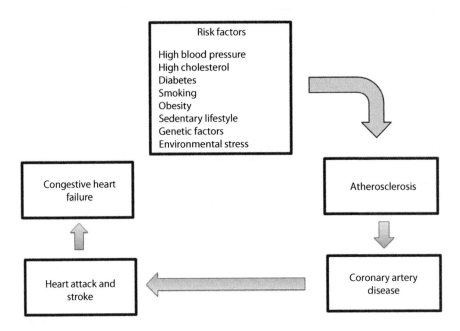

Figure 11.1 Risk factors for coronary artery disease.

includes both invasive and noninvasive coronary assessment. Invasive coronary angiography (CAG) has been the gold standard diagnostic approach for establishing the presence, location, and severity of CAD. Invasive CAG is recommended for patients with high risk of CAD according to patient history, electrocardiogram, and biomarkers of cardiac injury. However, this technique is invasive and expensive and is associated with small but definite risk of harm. The diagnostic approaches utilized for patients with low to intermediate risk of CAD include noninvasive diagnostic approaches, such as exercise tests, radioisotope myocardial perfusion scanning, coronary calcification score, multidetector cardiac computed tomography (MDCT), magnetic resonance (MR) imaging, and stress echocardiography (Kohsaka and Makaryus, 2008). The advances in the current therapeutic strategies for CAD have shown a tremendous decrease in the mortality and morbidity rate among CAD patients. Basic research and clinical trials have evaluated the safety and efficacy of statins in reducing the cholesterol levels among CAD patients, which is one of the major risk factors for the disease (Shepherd et al., 1995; Downs et al., 1998; Emami et al., 2017). Further chronic total occlusions (CTO) are present among some CAD patients; thereby, advances in percutaneous coronary intervention (PCI) strategies have

shown a significant improvement in the treatment of CTO (Loh and Waksman, 2012; Gao and Chen, 2016). Further, the diagnostic approaches are associated with risks of exposure to radiation, contrast-induced anaphylaxis, and acute kidney injury (Einstein et al., 2007, 2010; Rhee et al., 2012). Even though the current diagnostic and therapeutic approaches have been successful in the detection and treatment of CAD, the present approaches have their own pros and cons. The present strategies pose an economic burden on health care systems worldwide. The diagnosis of CAD is highly variable among patients presenting with symptoms of suspected CAD. Only 10% of the chest pain cases evaluated by physicians results in the diagnosis of CAD, despite the augmentation of the expensive diagnostic approaches (Klinkman et al., 1994). Hence, it has become necessary to improvise and develop precise newer diagnostic and therapeutic strategies for the diagnosis, treatment, and management of CAD.

ROLE OF GENETIC, ENVIRONMENTAL, AND LIFESTYLE RISK FACTORS IN PRECISION MEDICINE

The three primary components of precision medicine are genetic information, lifestyle, and

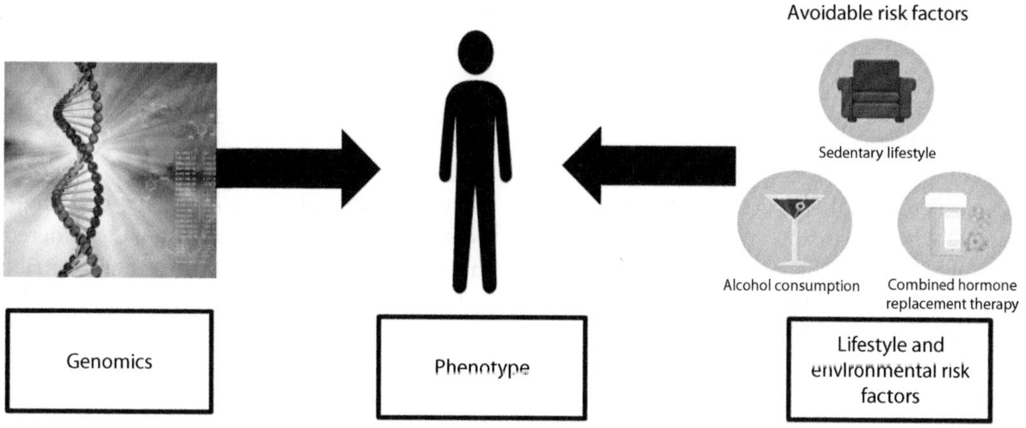

Figure 11.2 Association between genomics, lifestyle, and environmental risk factors with precision medicine.

environmental factors (Figure 11.2). Further, CAD is a known complex disease that is influenced by inherited genetic risk, and environmental and lifestyle factors, such as diets, saturated fats, smoking, and sedentary lifestyle (Tenconi et al., 1992; Alizadeh et al., 2017). These factors play a key role in the pathogenesis and treatment of disease. Hence, identifying the link between these factors and the disease will guide physicians in understanding the disease and develop precise diagnostic, therapeutic, and preventive approaches.

Genetic risk factors

Genetic factors contribute to the risk of CAD and several developments have occurred in the past decade. The evolution of genome-wide association studies (GWAS) over the last 10 years has played a key role in investigating the genetic architecture of complex human disease. Evidence shows that the GWAS represent a powerful approach in the identification of genes involved in complex human diseases such as CAD, cancer, and others (Sayols-Baixeras et al., 2014). GWAS have so far been able to identify approximately 50 genetic variants associated with CAD. Although the application of this strategy in clinical practice is challenging, genetic variants can be used to determine the new therapeutic targets and personal genetic information in order to improve the prognosis and treatment of disease. The development in this field so far has been able to provide insight into the pathophysiology of CAD, which is the beginning of the

individualized treatment strategies (Kessler et al., 2016). Several GWAS have been conducted to identify the single nucleotide polymorphisms (SNPs) on chromosomes that have an association with the susceptibility to CAD. The discovery of pro-protein convertase subtilisin/kexin type 9 (PCSK9) inhibition, which is involved in LDL-C receptor recycling, has led to the discovery and development of alirocumab and evolocumab. These molecules are currently being studied for the treatment of hypercholesterolemia along with statin therapy, which is one of the major risk factors for CAD (Postmus et al., 2013).

The Coriell Personalized Medicine Collaborative (CPMC) research study was done to investigate the potential contributions of common genetic factors to the screening, management, and prevention of complex diseases including CAD. The impact of personalized risk reports including genetic and nongenetic risk factors for CAD on heart health behaviors was assessed. A total of 683 CPMC participants who received personalized CAD risk reports, including genetic risk, family history risk, and self-reported nongenetic risks based on smoking and diabetes status, were evaluated in the study. It was observed that subjects with increased genetic risk of CAD were significantly more likely to report increases in heart health behaviors after viewing their personalized risk report. Thus, the study indicated that subjects who were aware of their genetic risk were highly motivated to increase their heart health behaviors (Scheinfeldt et al., 2016).

Thus, the genetic data of patients can provide a good platform for the diagnosis, prognosis, and treatment of CAD.

Lifestyle risk factors

In the recent decade, lifestyle risk factors have been associated with increasing risk of CAD and have increasingly become the center point of research interest worldwide (Brotman et al., 2007). In previous studies, it had been demonstrated that healthy lifestyle habits have reduced the disease and mortality rates, and the sociodemographic parameters such as gender, age, marital status, economic level, and paid employment correlates with healthy lifestyle (Kromhout et al., 2002).

A meta-analysis was conducted to examine the combined association of job strain and lifestyle with the risk of CAD. The population attributable risk was calculated for three exposures: job strain, unhealthy lifestyle, and their combination. The results of the meta-analysis showed that the risk of CAD was high among participants who had an unhealthy lifestyle and who reported job strain. However, the risk of CAD was half among participants with job strain and healthy lifestyle. Further, the 10-year incidence rate of CAD subjects with job strain and a healthy lifestyle was 53% lower than the incidence among subjects with job strain and an unhealthy lifestyle (Kivimäki et al., 2013). A retrospective study was conducted to evaluate the impact of various daily lifestyle indicators, including the dinner satiety rate, tobacco use, heavy alcohol use, sleep pattern, anxiety, and exercise, among CHD patients who had undergone PCI. The multivariate logistic analysis showed that dinner satiety, tobacco use, heavy alcohol use, and exercise significantly impacted CHD ($p < 0.05$). The clinical composite end point of target lesion revascularization, defined as repeated PCI and coronary artery bypass grafting (CABG), was observed in 55% of the cases between 3 and 5 years of the follow-up period. Hence, it was concluded that poor lifestyle may increase the in-stent restenosis rate, increase the progress of original lesion, and enhance the development of new lesions among patients with CHD (Wan et al., 2015).

In conclusion, lifestyle factors can play a vital role in the prognosis and diagnosis of CAD. Precision medicine has the ability to integrate patient's lifestyle to diagnose and classify the disease process, in order to provide custom-tailored therapeutic solutions.

Environmental and societal influences

Environmental risk factors and socioeconomic status have an influence on CAD. The environmental risk factor is vital but is considered as an underestimated risk factor that contributes to the development and severity of CAD (Shah et al., 2016). It is known that the heart and vascular system are highly susceptible to a variety of environmental agents, such as ambient air pollution and the metals arsenic, cadmium, and lead. The exposure to these environmental risk factors contributes to advancement in diseases and mortality through initiation of the pathophysiological processes associated with CAD inclusive of blood-pressure control, carbohydrate and lipid metabolism, vascular function, and atherogenesis (Chow et al., 2009). A study was conducted to analyze the interaction of a genetic variant of the CDKN2A/B-rs10811661 gene locus with cardiovascular risk factors and environmental exposures like (e.g., diet and physical activity) among 1165 individuals. Genotyping was carried out using the TaqMan real-time PCR-based method. There was a significant association between CDKN2A-rs10811661 polymorphism with cardiovascular risk factors and dyslipidemia in a nondiabetic population. Thus, low energy diet and good physical activity may relieve the unfavorable effects of T allele of CDKN2A/B locus (Mehramiz et al., 2016). Hence, the environmental risk factors contribute to the development of CAD among the population. The analysis of the environmental factors will enable physicians to identify the cause of CAD and provide better therapeutics for CAD patients.

PERSONALIZED MEDICINE IN CAD

The 21st century is poised to develop more targeted and personalized medicines for disease detection and treatment. Available therapies may be effective for some individuals but be associated with treatment failures or toxicities in others. "PM (precision medicine) is defined as a novel approach that can be characterized as molecular, immunologic, and functional endotyping of the disease, leading to personalized care, with the patient's engagement in decision making as to the treatment process

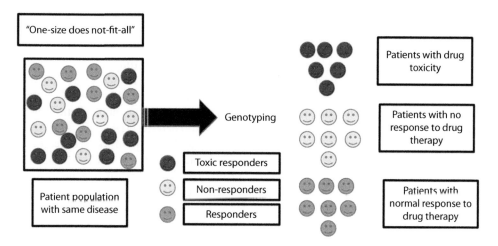

Figure 11.3 Differential response to drug therapy.

and consideration of predictive and preventive aspects of treatment" (Śliwczynski and Orlewska, 2016). Precision medicine is a newly emerging approach, which will evaluate patient's variability in their environmental, lifestyle, and genetic risk factors individually for the prevention and treatment of CAD. CAD is a heritable complex disease caused due to the manifold interactions of different lifestyle, genetic, and environmental risk factors, leading to the change in the molecular outlook of vascular and metabolic tissues to accelerate atherosclerosis (McGrath and Ghersi, 2016). Advancements in technologies, such as genome sequencing, big data analysis, and electronic health records, have enabled the health care to focus on the individual patient approach than the one-size-fits-all approach (Figure 11.3). Hence, the goal of

precision medicine is to integrate the genetic, lifestyle, and environmental information in order to classify the populations into subgroups based on their susceptibility and response to the disease (Duarte and Spencer, 2016). This will enable physicians to optimize the therapy for each individual based on their respective diagnostic testing. The difference between traditional therapy and precision medicine is illustrated in Figure 11.4.

Genomic-based gene expression score (GES)

The utility of genomic-based, personalized medicine may assist physicians' medical decision making, including the need for further cardiac testing. The genomic-based gene expression score (GES)

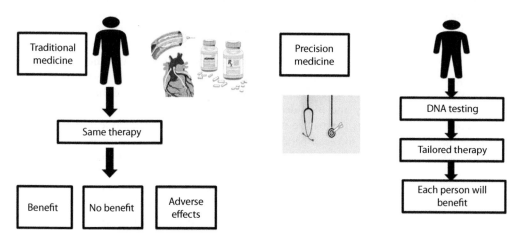

Figure 11.4 Traditional versus precision medicine.

will provide an assessment of the present state of obstructive CAD by quantifying the gene expression changes associated with arthrosclerosis. The GES is calculated by the commercially available, Medicare-covered, validated, quantitative diagnostic test, which measures the expression levels of 23 genes from the peripheral blood to determine the likelihood of obstructive CAD, as diagnosed by CAG (at least one vessel with $\geq$50% angiographic coronary artery stenosis). The raw algorithm scores are calculated from the median expression values for the 23 algorithm genes, grouped in the six terms, four sex-independent and two sex-specific. The scores are then converted to a 0–40 scale and patients with lower GES are at a lower risk of underlying obstructive coronary disease (Elashoff et al., 2011).

Studies evaluating the gene expression score

A prospective trial (Investigation of a Molecular Personalized Coronary Gene Expression Test on Primary Care Practice Pattern, clinical trial) (PREDICT) was conducted to see if the GES results would reduce the diagnostic uncertainty among patients presenting with symptoms suggestive of obstructive CAD and would lead to a change in making medical decisions, including the need for further cardiac testing. The primary outcome measure was the change in the diagnostic plan, which was classified into the following categories (in hierarchical order).

1. No further cardiac testing or treatment
2. Change in lifestyle or medical therapy
3. Stress testing (with or without imaging) or computed tomography/coronary angiography
4. Invasive CAG

The investigators observed a change in diagnostic plan following GES testing in 58% (n = 145) of patients, between the preliminary and final decision. The secondary outcome was to analyze the change in the diagnostic pattern for each patient independently. Out of the 145 patients, on whom the diagnostic plan was altered, it was observed that most patients demonstrated a decrease (n = 93) than an increase (n = 52) in the intensity of testing. In order to observe if the change in the diagnostic pattern had an effect on the patient outcomes, major adverse cardiac events (MACE)

were recorded during the 30-day follow-up. During the follow-up, only 0.4% of MACE were detected (Herman et al., 2014).

A multicenter study was conducted to validate, the diagnostic accuracy of a 23-gene expression-based classifier for the assessment of obstructive CAD in nondiabetic patients. The Diamond–Forrester (DF) risk score comprising age, sex, and chest pain was selected to assess the added value of GES to other clinical factors. The primary outcome measure was the receiver-operating characteristic (ROC) curve area for algorithm score prediction of disease status. The ROC curves were estimated for the gene expression algorithm score, the DF risk score, a combined model of algorithm score and DF risk score, myocardial perfusion imaging, and a combined model of algorithm score and imaging. The primary outcome measure AUC was 0.70 ± 0.02, (P < 0.001), being independently significant in both male (P < 0.001) and female (P < 0.001) subsets. The ROC also showed higher AUC for the algorithm score and DF risk score combination when compared to the DF risk score alone (AUC 0.72 versus 0.66, P = 0.003). Thus, the noninvasive whole blood test, based on gene expression and demographics, might be useful for the assessment of obstructive CAD in nondiabetic patients without known CAD (Rosenberg et al., 2014).

However, one of the limitations of the PREDICT study was that the study population was angiographically referred and the accuracy of GES among patients with lower prevalence of CAD population was unknown. The Coronary Obstruction Detection by Molecular Personalized Gene Expression (COMPASS) study was designed in such a way, where symptomatic nondiabetic patients referred for myocardial perfusion imaging (MPI) were included. Thus, the COMPASS study assessed the GES and MPI performance among a low-risk population by minimizing the selection bias. The prespecified primary end point was the GES ROC characteristics to segregate $\geq$50% stenosis, in order to identify patients with obstructive CAD. It was observed that the area under the ROC curve for GES was 0.79 (95% confidence interval, 0.73–0.84; P < 0.001). The sensitivity and specificity of the GES were 89% and 52%, respectively; the negative and positive predictive values were 96% and 24%, respectively. Thus, GES was found to be a highly significant predictor of obstructive CAD by ROC analysis. Moreover, during the 6-month follow-up it was observed that the majority of the cardiovascular events and revascularizations (27 of 28%, 96%) occurred among patients with GES > 15 (Thomas et al., 2013).

The two multicentric trials, the PREDICT and COMPASS, validated that GES could be used to discriminate patients with and without CAD. However, it is important to look at the reproducibility of the technology, dependence of test performance on ethnicity, and the schedule of testing. Hence, a study was conducted where the testing of more than 1500 patients from the PREDICT study was done, inclusive of their ethnicities and the variation observed in the GES serial testing for one year among 192 patients from the COMPASS study. From the PREDICT study, subjects who completed the discovery and development phases were included in the present study. To determine the sample and process stability, GES was measured for the validation set (n = 648) from the PREDCIT study, whose samples were stored at −20°C for approximately 5 years. There was no significant change in the test performance measured by ROC analysis, between the PREDICT and the present study (area under the curve [AUC] = 0.70 for both, $N = 501$). The analysis of the 138 non-Caucasian and 1364 Caucasian patients also showed a similar performance (AUCs = 0.72 vs. 0.70). Thus, the results demonstrated the sample and GES analytical stability, but the extent of biological variation over time on a per patient basis was not analyzed. Hence, subjects who either had invasive angiograms or research CT-angiograms to determine the coronary anatomy were included from the COMPASS study. The top four sites in the COMPASS study was approached and blood was drawn from these subjects approximately after one year of the study entry. A slight increase in the mean score from 15.9 to 17.3, corresponding to a 2.5% increase in obstructive CAD likelihood, was observed. In conclusion, the changes in the cardiovascular medications did not show significant change in GES score (Daniels et al., 2014).

The CAD test has been extensively evaluated in the two multicentric trials. It is known that CAD has a destructive effect on the QOL and the cost for diagnosing CAD is expensive. Thus, a simple GES blood test might help and can be a valuable tool to discriminate patients with obstructive CAD. Hence, this test could be implemented in daily

clinical practice if it continues to show good results among larger and more diverse cohorts (Vargas et al., 2013).

PHARMACOGENOMICS AND PHARMACOGENETICS

There is an extensive variability in the response to drugs between the patients in terms of both benefits and adverse effects leading to treatment failure or adverse reactions in most of the cases. The massive development in the field of pharmacogenetics and pharmacogenomics has led to a greater understanding of genetics for the prognosis and treatment of CAD, but the implementation of this in daily clinical practice is limited. The genetic mutations that influence the drug metabolisms can be explored in a targeted single-gene manner (pharmacogenetics) or in a whole-genome manner (pharmacogenomics). Pharmacogenomics aims to develop a rationale for the optimization of drug therapy, with respect to patients' genotype, so as to ensure the maximum efficacy with minimal side effects (Figure 11.5). Pharmacogenetics is defined as the science of identifying the genetic differences on the metabolic pathways that can affect individual's response to drugs both in terms of both therapy and adverse effects. This field combines pharmacology and genomics, so as to develop medications that are designed to react with the patient's genetic makeup. The factors that influence the response of patients to each drug include drug dose, adherence to a dosing regimen, drug–drug interactions, genetic variation in genes encoding for drug metabolizing enzymes, and transporter proteins. The genetic variation will have an impact on the pharmacokinetics of the drug through the regulation of drug absorption, distribution, metabolism, or elimination. Since the drugs will have standard doses, individuals with this type of genetic variation would experience adverse drug reactions due to the concentration of drug that will either be toxic or nonefficacious (Chambliss and Chan, 2016).

Antiplatelet pharmacogenomics

The adjuvant therapy of aspirin and clopidogrel has substantially reduced the risk of myocardial infarction, stent thrombosis, and death among patients with acute coronary syndromes and those receiving

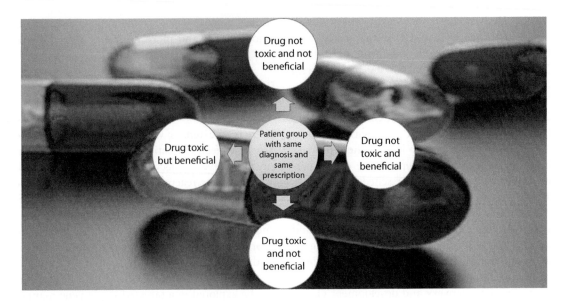

Figure 11.5 Pharmacogenomics in precision medicine.

coronary stents (Yusuf et al., 2001; Steinhubl et al., 2002). However, in several studies it was observed that thousands of patients inclusive of acute coronary syndrome patients experienced genetic resistance to clopidogrel. The presence of the genetic mutation resulted in reduced clopidogrel active metabolite formation, attenuated antiplatelet effect, and there was a three-fold increase in the risk of stent thrombosis, myocardial infarction, and death (Mehta et al., 2001; Collet et al., 2009).

A GWAS was done to confirm the gene variant *CYP2C19*2* genotype to be associated with diminished platelet response to clopidogrel treatment and poorer cardiovascular outcomes. The platelet aggregation was measured at baseline and within 1 h after the last dose of clopidogrel on day 7 and about 400,000 SNPs was simultaneously assessed using GWAS. The *CYP2C19* variant demonstrated an increased risk (345%) for stent thrombosis, along with risk of myocardial infarction and death (Shuldiner et al., 2009). Clopidogrel is one of the highly prescribed drugs to prevent stent thrombosis, myocardial infarction, and death among acute coronary syndrome patients; a number of alternative therapies are available for individuals who are resistant to the drug. The alternative therapies include P2Y12 receptor blockade, prasugrel and cilostazol. Studies have been conducted to evaluate the pharmacodynamic response of switching acute coronary syndrome patients with high posttreatment platelet reactivity (HPPR) from clopidogrel to prasugrel

and cilostazol therapy. The studies demonstrated that alternative therapies reduced the rate of HPPR among these patients (Jeong et al., 2009; Angiolillo et al., 2010). A systematic review and meta-analysis was conducted to investigate the role of *CYP2C19*2* polymorphism in relation to adverse clinical outcome in CAD patients who were undergoing clopidogrel treatment. The outcome measure MACE was defined as any cardiovascular event (fatal and nonfatal myocardial infarction, stroke, unstable angina), death from cardiovascular causes, ischemic recurrences, stent thrombosis, and death from any causes occurred during the follow-up. The results of the meta-analysis demonstrated that the *CYP2C19*2* polymorphism was associated with increased risk of MACE and stent thrombosis (Sofi et al., 2011).

Statin pharmacogenetics

The adherence of statin is often limited due to certain side effects. The *SLCO1B1*5* variant is a major risk factor for statin-specific side effects. However, the effect of *SLCO1B1*5* genotype-guided statin therapy (GGST) is unknown. A pilot study was done to see if *SLCO1B1*5* therapy may improve perceptions and adherence to statins. The primary outcome was the change in the Beliefs about Medications Questionnaire (BMQ) that measured patients' perceptions about needs for statins and concerns about adverse effects. The GGST patients demonstrated a higher "necessity" (pre: 15.1 ± 3.8

versus post: 15.6 ± 4.0, p = 0.24) and lower "concern" (pre: 15.2 ± 4.2 versus post: 14.7 ± 4.3, p = 0.24) at four months compared to baseline. On examining the individual components of the necessity and concerns domains, the greatest change in BMQ was observed in the "need for statin to prevent sickness" (pre: 2.9 ± 0.9 versus post: 3.3 ± 0.9, p < 0.001) and the "concern for statin to disrupt life" (pre: 3.2 ± 1.4 versus post: 2.8 ± 1.2, p = 0.006). The secondary outcome measure was to see if the GGST was associated with improved statin adherence, statin prescribing behavior, and LDL-c, compared to concurrent control subjects receiving standard of care. It was observed that the GGST patients had more statin prescriptions (p < 0.001), higher statin use (p < 0.001), and greater decrease in LDL-c (p = 0.059) during follow-up (Li et al., 2014).

Warfarin pharmacogenetics

Warfarin still remains to be the mainstay for oral anticoagulation despite the recent approval of newer agents. However, warfarin also remains to be one of the challenging medications to be managed. Presently, warfarin ranks as one of the leading drug-related causes of serious adverse events leading to hospitalizations. The challenges associated with the usage of warfarin includes its narrow therapeutic index and wide interpatient variability in its dosage in order to achieve the required optimal anticoagulation. The optimal dosage to achieve therapeutic anticoagulation varies among patients from 0.5 to 11 mg/day or higher. The inappropriate dosage of warfarin leads to increased risk of bleeding and thromboembolic complications (Johnson and Cavallari, 2013). Several GWAS and candidate gene data has shown that genotype contributes to the interpatient variability in warfarin dose requirements (Aithal et al., 1999; Parra et al., 2015). CYP2C9 and VKORC1, which encode for vitamin K epoxide reductase, are major genes that influence the warfarin pharmacokinetics and pharmacodynamics, respectively (Johnson et al., 2011; Giri et al., 2014). Clinical trials were conducted to observe the effect of genotype-guided warfarin dosing among patients, but the trial results were inconsistent (Anderson et al., 2007; Caraco et al., 2008; Kimmel et al., 2013). Two large multicentric comparative studies were conducted to support the effectiveness of genotype-guided warfarin therapy.

A national, prospective, comparative effectiveness study was done to compare the 6-month incidence of hospitalization, including those due to bleeding or thromboembolism among patients receiving warfarin genotyping versus a matched historical control group. During the 6-month follow-up, the patients who received the warfarin genotyping had 31% fewer hospitalizations overall and 28% fewer hospitalizations for bleeding or thromboembolism compared to the control group (Epstein et al., 2010). Another randomized and clinical effectiveness trial was conducted to compare the two pharmacogenetic algorithms and standard care for individualizing warfarin dosing (CoumaGen-II). The primary outcome measure was percentage of out-of-range international normalized ratios (INR) at 1 and 3 months and percentage of time in therapeutic range. The study results demonstrated that the genotype-guided therapy, regardless of the algorithm used, was superior to the standard dosing. This was in terms of reduction in the percent of out-of-range INRs and the percent of INRs ≥4 or ≤1.5 at 3 months (Anderson et al., 2012). A recent meta-analysis was conducted to assess the effect of pharmacogenetics-based warfarin dosing in patients initiating warfarin therapy. The primary outcome measure was the time within the therapeutic range (TTR) and the secondary outcome measures were the time to maintenance dose and time to first INR, an INR greater than 4, adverse events, major bleeding, thromboembolism, and death from any cause. The results of the study showed that the pharmacogenetics-based dosing (PD) did not improve the TTR compared to conventional dosing (CD). However, PD significantly shortened the time to maintenance dose and the time to first therapeutic INR. Further, the PD significantly reduced the risk of adverse events and major bleeding, but there was no change in the percentage of INR greater than 4, the risk of thromboembolic events and death from any cause (Shi et al., 2015).

Presently the website WarfarinDosing.org is available to help clinicians to begin warfarin therapy by evaluating the required therapeutic dose among patients who are new to warfarin therapy. The evaluation is based on the patient's clinical and demographic factors. The genotypes of two genes—cytochrome P450 2C9 (CYP2C9) and vitamin K epoxide reductase (VKORC1)—are also evaluated.

OTHER EXAMPLES

Currently, angiotensin-converting enzyme (ACE) inhibitors are recommended in clinical guidelines for the secondary prevention among patients with stable CAD. However, the CAD population is comprised of a heterogeneous group and the absolute risk of cardiovascular complications varies among the individuals. A study was done to integrate clinical and pharmacogenetic determinants in an ultimate combined risk prediction model. The clinical, genetic, and outcome data were used from the 8726 stable CAD patients participating in the EUROPA/PERGENE trial of perindopril versus placebo. Three SNPs (rs275651 and rs5182 in the angiotensin II type I receptor gene and rs12050217 in the bradykinin type I receptor gene) were utilized to construct the pharmacogenetic risk score (ranging between 0 and 6 points). The primary end point was cardiovascular mortality, nonfatal myocardial infarction, or resuscitated cardiac arrest. There was a reduction in the absolute risk from 1.2% to 7.5% in the 73.5% of patients with PGXscore of 0 to 2. Thereby, the estimated annual numbers to be treated ranged from as low as 29 (clinical risk score ≥10 and PGXscore of 0) to 521 (clinical risk score ≤6 and PGXscore of 2). Further, the data recommended that long-term perindopril prescription in patients with a PGX score of 0 to 2 is cost-effective. Hence, the combination of phenotypical and genetic information demonstrated a very wide range of gradients of absolute treatment benefit (Oemrawsingh et al., 2016). Studies were also done to determine whether ACE insertion/deletion (I/D) polymorphism was associated with the severity CAD and its progression/regression in response to fluvastatin therapy among the Lipoprotein and Coronary Atherosclerosis Study (LCAS) population. The study results demonstrated that the ACE (I/D) polymorphism is associated with the response of plasma lipids and coronary atherosclerosis to treatment with fluvastatin. The subjects with (D/D) mutation experienced a greater reduction in LDL-C, a higher rate of regression and a lower rate of progression of CAD (Marian et al., 2000).

Development of advanced medications

In the future, the pharmaceutical companies can design drugs based on proteins, enzymes, and RNA molecules, which are associated with genes and different diseases. Thus, drugs that can target the more specific disease can be manufactured. This will not only maximize the effectiveness of the drug but can also minimize the damage caused to the nearby cells.

Development of improved and safer drugs

More efficient and safer drugs can be prescribed for the patients. The physicians would be able to analyze the patient's genetic profile and prescribe the appropriate drugs from the onset itself. Thus, unnecessary therapy and adverse effects can be avoided for the patient.

Usage of appropriate doses of drug

The appurtenant dosage of drug can be prescribed to patients based on their individual genetic makeup rather than the traditional method of treatment. Thus, this approach would maximize the therapy value and decrease the possibility of drug overdose.

Precocious screening for disease

Having a prior knowledge about each person's genetic code will enable the subjects to make the required changes in their lifestyle and environment. This would avoid or cut short the severity of genetic disease. Moreover, having an advanced knowledge about the disease susceptibility will provide careful monitoring and the treatment could be provided at the appropriate stage to obtain the maximum benefits of the therapy.

Reduction in the overall cost of health care

The overall decrease in the following domains leads to an overall decrease in the cost of health care.

1. Decrease in the number of adverse drug reactions.
2. Decrease in the number of failed drug trials.
3. Decrease in the time taken for the approval of the drug.
4. Decrease in the length of time the patients are on medications.

5. Decrease in the number of medications the patient has to take to provide improved therapy.

Disadvantages

LIMITED ALTERNATIVE THERAPY

The few therapies that are approved can only be utilized for the treatment of a particular condition. If the identified genetic mutations will prevent the usage of drug, the patient might be left with no other substitute treatment.

COST CONCERNS

The discovery and development of pharmacogenomic drugs will be expensive and thus only a small proportion of population might get the benefit.

EDUCATION OF HEALTH CARE PROVIDERS

Health care providers must have a better understanding regarding genomics. Hence, all the health care providers must be properly trained and educated.

In conclusion, the field of pharmacogenomics is still scanty and its use is currently limited. However, investigations are underway for new approaches. Advancement in the field of genomic science has assisted clinicians in prescribing medications to patients based on their individual genetic data. Thus, the overall screening, therapeutic, and management cost for patients can be reduced (Patil, 2015).

PROTEOMICS

Proteomics is defined as a new biological approach to study the large-scale expression, function, and interaction of the complement of proteins on a large scale as a result of biological processes and perturbations. In the recent years, novel biomarkers have been evaluated for several cardiovascular diseases such as CAD and heart failure. These biomarkers have shown promising results for the prediction of cardiovascular outcomes. For these biomarkers to become useful and be available for personalized medicine, the following criteria need to be fulfilled. The biomarkers must be easy to measure, must provide new information, and must help clinicians for the management of patients (Stratz et al., 2012).

The advancements in the high throughput mass spectrometry for analyzing the clinical samples have lead to the discovery of an outline of human proteome. The collection of this data forms the base for the comparative analysis of clinical proteomics. The comparison of the patient sample with the existing database allows scientists to associate the changes in proteome with particular disease states. Hence, the translation of the aforementioned findings in the clinic and hospital settings will allow mass spectral analysis of patient proteomes. This personalized proteomic analysis has the potential to stimulate the diagnostic accuracy by reducing the cost and time. Thus, this proteomic screening will allow early detection of the presence and severity of disease, which allows for a quick response from medical personnel and improved patient outcome (Duarte and Spencer, 2016). Many studies have been conducted to evaluate the levels of proteins of both normal and disease state for a successful diagnosis, prognosis, and therapy (Sabatine et al., 2005; Parguiña et al., 2010).

Proteomic studies

A study was conducted to evaluate the modifications in the plasma protein map during unstable angina (UA) and AMI using proteomics. The protein plasma levels were quantified among patients with AMI (n = 11) and UA (n = 9). The control group was comprised of age-matched volunteers. The proteins that were analyzed included alpha-1-antitrypsin (AAT), apolipoprotein A-I, fibrinogen gamma chain, immunoglobulin gamma heavy chain, and albumin. The main finding of this study was that different AAT isoforms change in plasma during an acute coronary syndrome (ACS). Seven different AAT isoforms were seen in the plasma of control samples. However, the AAT isoform 1 was absent in the ACS samples. In the study, there was also a significant reduction of isoforms 5, 6, and 7 in the AMI samples when compared with UA samples. The fibrinogen gamma chain 1 and 2 were increased in AMI patients compared to UA patients. Five apolipoprotein A-I isoforms were also identified, but these were reduced in plasma from AMI patients with respect to UA patients. The γ-immunoglobulin heavy chains were identified and were found to be increased in the plasma among ACS patients. Thus, the proteomic analysis will help in the mapping of the protein isoforms that are expressed in plasma during an ACS (Mateos-Cáceres et al., 2004).

Another study was done that combined pharmacogenetics and iTRAQ-coupled LC-MS/MS pharmacoproteomics to analyze the plasma protein profiles in 53 patients. This approach identified a significantly higher level of transthyretin (TTR) precursor among patients receiving low dose of warfarin but not in those on a high dose of warfarin. For the validation of the results, real-time RT-PCR, western blotting, and human IL-6 ELISA assay were done. Thus, the study demonstrated that variations in TTR levels were associated with specific dosing regimens of warfarin. Moreover, in addition to pharmacogenomic biomarkers presently emerging as an aid in the dosing of warfarin, the pharmacoproteomics biomarkers such as TTR may also be investigated for the dosing regimens of warfarin. Hence, the combined approach of pharmacogenomics and pharmacoproteomics can be applied for other target-based therapies in matching a particular marker in a subgroup of patients (Saminathan et al., 2010).

Limitations of proteomics

Even though the field of proteomics is important in cardiovascular medicine, there are several challenges with respect to this field. The first being the advantages and disadvantages of the different technological platforms being utilized for the evaluation of cardiovascular proteome. The second challenge is to identify whether a combination or a one-platform approach is better. The third is with respect to the protocol, that is, whether a common protocol should be for blood sample handling and processing. The fourth challenge is the influence of standard cardiovascular drugs on the blood proteome. The fifth challenge is the method used to integrate current protein data with biomarker data from different investigators (Arab et al., 2006).

CHALLENGES IN THE IMPLEMENTATION OF PRECISION MEDICINE

The main concept of precision medicine is to emphasize the scientific and technological innovations so as to enable the physicians to alter or modify the prognosis of disease, its diagnosis, and

the treatment of each individual patient based on their individual data. Even though studies have demonstrated that the application of personalized medicine has provided positive results, its usage in daily clinical practice comes with many challenges. Being a multidisciplinary approach to science, it requires compliance with many components. The components are advancements in technology, modifications in the health care infrastructure and medical practice, improvements in the efficiency and quality of health care, and ethical and legal issues regarding the storage of genomic information in medical records for a long time (Lee et al., 2012).

Another limitation of personalized medicine is the translation of molecular and genomic advances into clinical available means. The Human Genome Strategy Group (HGSG) report and the whole genome sequencing by the Foundation for Genomics and Population Health (PHG Foundation) outlined many of the challenges involved in the implementation of genomics to test the health care system as a whole (Burton et al., 2012). Although the research has been successful in the identification of the genetics that has the ability to predict the susceptibility of the disease like the interaction of risk factors, the method utilized is not simple and small. It is necessary to develop basic scientific evidence that supports clinical interpretation of the genome based data. This in turn requires the standardization of the databases of the normal and pathogenic variations linked to analytical tools (Lee et al., 2012).

An appropriate development of laboratory setup is vital (i.e., biobanking, coordinated efforts, operational and informatics support, standards, genomic technologies, core laboratories, economies of scale) and the bioinformatical infrastructures is also necessary (Lee et al., 2012). Good clinical and practice guidelines are a must, as there must be a proper understanding and transparency between the physicians and patients. The information regarding what and what will not be analyzed and why testing has to be done should be communicated to the patients. The confidentiality of the patient data has to be protected during the usage of genetics to predict the risks for the patients and relatives. The clinicians should also be aware of the fact that although some of the patients might be interested to know about their genetic inheritance, others may not be willing to know (Burton et al., 2012).

One of the other challenges of the genomic approach is the readiness and willingness of physicians to the application of genomic technologies for maximum effect so as to incorporate genomic medicine into mainstream clinical practice. The clinicians who are working in the field of clinical research should be aware of the advancements in the field of genomic medicine. Some of the clinicians do not consider or recognize the field of genetics in clinical practice, as they perceive that genetic conditions are rare and untreatable. Hence, education and training for all physicians are important. The collaboration between physicians and researchers will help in the understanding of inherited disorders within various areas of clinical medicine (Burton et al., 2012).

FUTURE CONSIDERATIONS

In conclusion, the large-scale genomic association studies have the ability to identify the genes and genetic regions that can screen the risk factors for disease. However, further research and efforts have to be taken to avoid the ascertainment bias among the clinical research participants with wider inclusion criteria and follow-up to find the functional variants in order to eliminate the need for tag SNPs (Scheinfeldt et al., 2016). Although, there is incredible developments in the field of genomic medicine, efforts are still required in the field of precision medicine. Some of the technical challenges are data collection, processing, storing, and sharing. Some are related to the broad scientific, clinical, social, and ethical ramifications (Chambliss and Chan, 2016).

Although, data collection is taking place across the hospitals and communities in traditional ways; the data collection is also being done via social media analytics and wearable devices. The procedure for data collection has improved through the adoption of systematic formats for electronic health records. However, improvements need to be made in data processing and integration. The storage and access to high volume data is difficult, and efforts have to be taken to enable the physicians, policy makers, and academics to access the databases. During the development of the precision approach, efforts have to be taken to accommodate the ethical and transparent use of data (Chambliss and Chan, 2016; Laper et al., 2016).

CONCLUSION

Precision medicine appears to have tremendous potential in the field of cardiovascular medicine. The emergence and advances in precision medicine will make it simpler for clinicians to tailor the drug treatment in accordance with the patient's genetic, environmental, and lifestyle factors. The approaches that are currently being studied in precision medicine include pharmacogenomics, pharmacogenetics, proteomics, and GES. The studies have shown inconsistent results leading to the requirement of more multicentric and well-powered studies. Thus, precision medicine is an emerging field and further research and efforts are required for it to be translated into daily clinical practice.

REFERENCES

Aithal GP, Day CP, Kesteven PJ, and Daly AK. Association of polymorphisms in the cytochrome P450 CYP2C9 with warfarin dose requirement and risk of bleeding complications. *Lancet (London, England)* 353(9154);1999:717–19.

Alizadeh S, Djafarian K, Alizadeh H, Mohseni R, and Shab-Bidar S. Common variants of vitamin D receptor gene polymorphisms and susceptibility to coronary artery disease: A systematic review and meta-analysis. *J Nutrigenet Nutrigenomics* 10(1–2);9–2017:18.

Anderson JL, Benjamin DH, Stevens SM et al. Randomized trial of genotype-guided versus standard warfarin dosing in patients initiating oral anticoagulation. *Circulation* 116(22);2007:2563–70.

Anderson JL, Horne BD, Stevens SM et al. A randomized and clinical effectiveness trial comparing two pharmacogenetic algorithms and standard care for individualizing warfarin dosing(CoumaGen-II). *Circulation* 125(16);2012:1997–2005.

Angiolillo DJ, Saucedo JF, Deraad R, Frelinger AL, Gurbel PA, Costigan TM, Jakubowski JA, Ojeh CK, Effron MB, and SWAP Investigators. Increased platelet inhibition after switching from maintenance clopidogrel to prasugrel in patients with acute coronary syndromes: Results of the SWAP (SWitching Anti Platelet) study. *J Am Coll Cardiol* 56(13);2010:1017–23.

Arab S, Gramolini AO, Ping P, Kislinger T, Stanley B, van Eyk J, Ouzounian M, MacLennan DH, Emili A, and Liu PP. Cardiovascular proteomics: Tools to develop novel biomarkers and potential applications. *J Am Coll Cardiol* 48(9);2006:1733–41.

Brotman DJ, Golden SH, and Wittstein IS. The cardiovascular toll of stress. *Lancet (London, England)* 370(9592);2007:1089–1100.

Burton H, Cole T, and Lucassen AM. Genomic medicine: Challenges and opportunities for physicians. *Clin Med (Lond, England)* 12(5);2012:416–19.

Caraco Y, Blotnick S, and Muszkat M. CYP2C9 Genotype-guided warfarin prescribing enhances the efficacy and safety of anticoagulation: A prospective randomized controlled study. *Clin Pharm Ther* 83(3);2008:460–70.

Chambliss, AB, and Chan DW. Precision medicine: From pharmacogenomics to pharmacoproteomics. *Clin Proteomics* 13;2016:25.

Chow CK, Lock K, Teo K, Subramanian SV, McKee M, and Yusuf S. Environmental and societal influences acting on cardiovascular risk factors and disease at a population level: A review. *Int J Epidemiol* 38(6);2009:1580–94.

Collet J-P, Hulot J-S, Pena A et al. Cytochrome P450 2C19 Polymorphism in young patients treated with clopidogrel after myocardial infarction: A cohort study. *Lancet (London, England)* 373(9660);2009:309–17.

Daniels SE, Beineke P, Rhees B, McPherson JA, Kraus WE, Thomas GS, and Rosenberg S. Biological and analytical stability of a peripheral blood gene expression score for obstructive coronary artery disease in the PREDICT and COMPASS studies. *J Cardiovasc Transl Res* 7(7);2014:615–22.

Downs JR, Clearfield M, Weis S, Whitney E, Shapiro DR, Beere PA, Langendorfer A, Stein EA, Kruyer W, and Gotto AM. Primary prevention of acute coronary events with lovastatin in men and women with average cholesterol levels: Results of AFCAPS/TexCAPS. Air Force/Texas Coronary Atherosclerosis Prevention Study. *JAMA* 279(20);1998:1615–22.

Duarte TT, and Spencer CT. Personalized proteomics: The future of precision medicine. *Proteomes* 4(4);2016.

Einstein AJ, Moser KW, Thompson RC, Cerqueira MD, and Henzlova MJ. Radiation dose to patients from cardiac diagnostic imaging. *Circulation* 116(11);2007:1290–1305.

Einstein AJ, Weiner SD, Bernheim A, Kulon M, Bokhari S, Johnson LL, Moses JW, and Balter S. Multiple testing, cumulative radiation dose, and clinical indications in patients undergoing myocardial perfusion imaging. *JAMA* 304(19);2010:2137–44.

Elashoff MR, Wingrove JA, Beineke P et al. Development of a blood-based gene expression algorithm for assessment of obstructive coronary artery disease in non-diabetic patients. *BMC Med Genomics* 4(March);2011:26.

Emami H, Takx RAP, Mayrhofer T, Janjua S, Park J, Pursnani A, Tawakol A, Lu MT, Ferencik M, and Hoffmann U. Nonobstructive coronary artery disease by coronary CT angiography improves risk stratification and allocation of statin therapy. *JACC Cardiovascular Imaging* 10(9);2017:1031–8.

Epstein RS, Moyer TP, Aubert RE, O Kane DJ, Xia F, Verbrugge RR, Gage BF, and Teagarden JR. Warfarin genotyping reduces hospitalization rates results from the MM-WES (Medco-Mayo Warfarin Effectiveness Study). *J Am Coll Cardiol* 55(25);2010:2804–12.

Gao L, and Chen Y-D. Application of drug-coated balloon in coronary artery intervention: Challenges and opportunities. *J Geriatr Cardiol* 13(11);2016:906–13.

Giri AK, Khan NM, Grover S et al. Genetic epidemiology of pharmacogenetic variations in CYP2C9, CYP4F2 and VKORC1 genes associated with warfarin dosage in the Indian population. *Pharmacogenomics* 15(10);2014:1337–54.

Heidenreich PA, Trogdon JG, Khavjou OA et al. Forecasting the future of cardiovascular disease in the United States: A policy statement from the American Heart Association. *Circulation* 123(8);2011:933–44.

Herman L, Froelich J, Kanelos D, St Amant R, Yau M, Rhees B, Monane M, and McPherson J. Utility of a genomic-based, personalized medicine test in patients presenting with symptoms suggesting coronary artery disease. *J Am Board Fam Med* 27(2);2014:258–67.

Jeong Y-H, Lee S-W, Choi B-R, Kim I-S, Seo M-K, Kwak CH, Hwang J-Y, and Park S-W. Randomized comparison of adjunctive Cilostazol versus high maintenance dose clopidogrel in patients with high post-treatment platelet reactivity: Results of the ACCEL-RESISTANCE (adjunctive Cilostazol versus high maintenance dose clopidogrel in patients with clopidogrel resistance) randomized study. *J Am Coll Cardiol* 53(13);2009:1101–9.

Johnson JA, and Cavallari LH. Pharmacogenetics and cardiovascular disease—Implications for personalized medicine. *Pharmacol Rev* 65;2013:987–1009.

Johnson JA, Gong L, Whirl-Carrillo M et al. Clinical pharmacogenetics implementation consortium guidelines for CYP2C9 and VKORC1 genotypes and warfarin dosing. *Clin Pharm Ther* 90(4);2011:625–29.

Kessler T, Vilne B, and Schunkert H. The impact of genome-wide association studies on the pathophysiology and therapy of cardiovascular disease. *EMBO Mol Med* 8(7);2016:688–701.

Kimmel SE, French B, Kasner SE et al. A pharmacogenetic versus a clinical algorithm for warfarin dosing. *N Engl J Med* 369(24);2013:2283–93.

Kivimäki M, Nyberg ST, Fransson EI et al. Associations of job strain and lifestyle risk factors with risk of coronary artery disease: A meta-analysis of individual participant data. *CMAJ* 185(9);2013:763–69.

Klinkman MS, Stevens D, and Gorenflo DW. Episodes of care for chest pain: A preliminary report from MIRNET. Michigan research network. *J Fam Pract* 38(4);1994:345–52.

Kohsaka S, and Makaryus AN. Coronary angiography using noninvasive imaging techniques of cardiac CT and MRI. *Curr Cardiol Rev* 4(4);2008:323–30.

Krishnan MN. Coronary heart disease and risk factors in India—On the brink of an epidemic? *Indian Heart J* 64(4);2012:364–67.

Kromhout D, Menotti A, Kesteloot H, and Sans S. Prevention of coronary heart disease by diet and lifestyle: Evidence from prospective cross-cultural, cohort, and intervention studies. *Circulation* 105(7);2002:893–98.

Laper SM, Restrepo NA, and Crawford DC. The challenges in using electronic health records for pharmacogenomics and precision medicine research. *Pac Symp Biocomput* 21;2016:369–80.

Lee M-S, Flammer AJ, Lerman LO, and Lerman A. Personalized medicine in cardiovascular diseases. *Korean Circ J* 42(9);2012:583–91.

Li JH, Joy SV, Haga SB, Orlando LA, Kraus WE, Ginsburg GS, and Voora D. Genetically guided statin therapy on statin perceptions, adherence, and cholesterol lowering: A pilot implementation study in primary care patients. *Person Med J* 4(2);2014:147–62.

Loh JP, and Wakaman R. Paclitaxel drug-coated balloons: A review of current status and emerging applications in native coronary artery De Novo lesions. *JACC. Cardiovasc Interv* 5(10);2012:1001–12.

Marian AJ, Safavi F, Ferlic L, Dunn JK, Gotto AM, and Ballantyne CM. Interactions between angiotensin-I converting enzyme insertion/deletion polymorphism and response of plasma lipids and coronary atherosclerosis to treatment with fluvastatin: The lipoprotein and coronary atherosclerosis study. *J Am Coll Cardiol* 35(1);2000:89–95.

Mateos-Cáceres PJ, García-Méndez A, Farré AL et al. Proteomic analysis of plasma from patients during an acute coronary syndrome. *J Am Coll Cardiol* 44(8);2004:1578–83.

McGrath S, and Ghersi D. Building towards precision medicine: Empowering medical professionals for the next revolution. *BMC Medical Genomics* 9(1);2016:23.

Mehramiz M, Ghasemi F, Esmaily H et al. Interaction between a variant of CDKN2A/B-Gene with lifestyle factors in determining dyslipidemia and estimated cardiovascular risk: A step toward personalized nutrition. *Clin Nutr (Edinburgh, Scotland)* December 2016.

Mehta SR, Yusuf S, Peters RJ et al. Effects of pretreatment with clopidogrel and aspirin followed by long-term therapy in patients undergoing percutaneous coronary intervention: The PCI-CURE study. *Lancet (London, England)* 358(9281);2001:527–33.

Oemrawsingh RM, Akkerhuis KM, Van Vark LC et al. Individualized angiotensin-converting enzyme (ACE)-inhibitor therapy in stable coronary artery disease based on clinical and pharmacogenetic determinants: The PERindopril GENEtic(PERGENE) risk model. *J Am Heart Assoc* 5(3);2016:e002688.

Parguiña AF, Parguiña AF, Grigorian-Shamajian L et al. Proteins involved in platelet signaling are differentially regulated in acute coronary syndrome: A proteomic study. *PLOS ONE* 5(10);2010:e13404.

Parra EJ, Botton MR, Perini JA et al. Genome-wide association study of warfarin maintenance dose in a Brazilian sample. *Pharmacogenomics* 16(11);2015:1253–63.

Patil J. Pharmacogenetics and pharmacogenomics: A brief introduction. *J Pharmacovigilance* 3(3);2015:e139. doi:10.4172/2329-6887.1000e139.

Postmus I, Trompet S, de Craen AJM, Buckley BM, Ford I, Stott DJ, Sattar N, Slagboom PE, Westendorp RGJ, and Jukema JW. CSK9 SNP rs11591147 is associated with low cholesterol levels but not with cognitive performance or noncardiovascular clinical events in an elderly population. *J Lipid Res* 54(2);2013:561–66.

Rhee CM, Bhan I, Alexander EK, and Brunelli SM. Association between iodinated contrast media exposure and incident hyperthyroidism and hypothyroidism. *Arch Intern Med* 172(2);2012:153–59.

Rosenberg S, Froelich J, Kanelos D, St Amant R, Yau M, Rhees B, Monane M, and McPherson J. Multi-center validation of the diagnostic accuracy of a blood-based gene expression test for assessment of obstructive coronary artery disease in non-diabetic patients. *J Am Board Fam Med* 27(2);2014:258–67.

Sabatine MS, Liu E, Morrow DA, Heller E, McCarroll R, Wiegand R, Berriz G, Roth FP, and Gerszten RE. Metabolomic identification of novel biomarkers of myocardial ischemia. *Circulation* 112(25);2005:3868–75.

Saminathan R, Bai J, Sadrolodabaee L, Karthik GM, Singh O, Subramaniyan K, Ching CB, Chen WN, and Chowbay B. VKORC1 pharmacogenetics and pharmacoproteomics in patients on warfarin anticoagulant therapy: Transthyretin precursor as a potential biomarker. *PLOS ONE* 5(12);2010:e15064.

Sayols-Baixeras S, Lluís-Ganella C, Lucas G, and Elosua R. Pathogenesis of coronary artery disease: Focus on genetic risk factors and identification of genetic variants. *Appl Clin Genet* 7;2014:15–32.

Scheinfeldt LB, Schmidlen TJ, Gerry NP, and Christman MF. Challenges in translating GWAS results to clinical care. *Int J Mol Sci* 17(8);2016:1267.

Shah SH, Arnett D, Houser SR, Ginsburg GS, MacRae C, Mital S, Loscalzo J, and Hall JL. Opportunities for the cardiovascular community in the precision medicine initiative. *Circulation* 133(2);2016:226–31.

Shepherd J, Cobbe SM, Ford I, Isles CG, Lorimer AR, MacFarlane PW, McKillop JH, and Packard CJ. Prevention of coronary heart disease with pravastatin in men with hypercholesterolemia. *N Engl J Med* 333(20);1995:1301–7.

Shi C, Yan W, Wang G, Wang F, Li Q, and Lin N. Pharmacogenetics-based versus conventional dosing of warfarin: A meta-analysis of randomized controlled trials. *PLOS ONE* 10(12);2015:e0144511.

Shuldiner AR, O'Connell JR, Bliden KP et al. Association of cytochrome P450 2C19 genotype with the antiplatelet effect and clinical efficacy of clopidogrel therapy. *JAMA* 302(8);2009:849–57.

Śliwczynski A, and Orlewska E. Precision medicine for managing chronic diseases. *Polskie Archiwum Medycyny Wewnetrznej* 126(9);2016:681–87.

Sofi F, Giusti B, Marcucci R, Gori AM, Abbate R, and Gensini GF. Cytochrome P450 2C19*2 polymorphism and cardiovascular recurrences in patients taking clopidogrel: A meta-analysis. *Pharmacogenomics J* 11(3);2011:199–206.

Steinhubl SR, Berger PB, Mann JT, Fry ETA, DeLago A, Wilmer C, Topol EJ, and CREDO Investigators. Early and sustained dual oral antiplatelet therapy following percutaneous coronary intervention: A randomized controlled trial. *JAMA* 288(19);2002:2411–20.

Stratz C, Amann M, Berg DD, Morrow DA, Neumann F-J, and Hochholzer W. Novel biomarkers in cardiovascular disease: Research tools or ready for personalized medicine? *Cardiol Rev* 20(3); 2012:111–17.

Tenconi MT, Romanelli C, Gigli F, Sottocornola F, Laddomada MS, Roggi C, Devoti G, Gardinali P, and Research Group ATS-OB43 of CNR. The relationship between education and risk factors for coronary heart disease: Epidemiological analysis from the nine communities study. *Eur J Epidemiol* 8(6);1992:763–69.

Thomas GS, Voros S, McPherson JA et al. A Blood-based gene expression test for obstructive coronary artery disease tested in symptomatic nondiabetic patients referred for myocardial perfusion imaging the COMPASS study. *Circ Cardiovasc Genet* 6(2);2013:154–62.

Vargas J, Lima JA, Kraus WE, Douglas PS, and Rosenberg S. Use of the Corus® CAD gene expression test for assessment of obstructive coronary artery disease likelihood in symptomatic nondiabetic patients. *PLoS Curr* 5;2013.

Wan Y-F, Ma X-L, Yuan C, Fei L, Yang J, and Zhang J. Impact of daily lifestyle on coronary heart disease. *Exp Ther Med* 10(3);2015:1115–20.

Yusuf S, Zhao F, Mehta SR, Chrolavicius S, Tognoni G, Fox KK, and Clopidogrel in Unstable Angina to Prevent Recurrent Events Trial Investigators. Effects of clopidogrel in addition to aspirin in patients with acute coronary syndromes without ST-segment elevation. *N Engl J Med* 345(7);2001:494–502.

12

Precision medicine in stroke and other related neurological diseases

ANJANA MUNSHI, VANDANA SHARMA, AND SULENA SINGH

INTRODUCTION

The understanding of mechanisms of neuronal alteration and maintenance of their molecular signatures during disease progression is a major requirement for clinically correct diagnosis of neurological disease. Numerous diagnostic investigations, including imaging techniques, are opted by concerned clinicians for prediction and analysis of the disease. Apart from these diagnostic measures, genomic profiling is one of the cornerstones of precision or personalized therapy, which not only forecasts the susceptibility to disease but also predicts the best possible treatment for the individual patient. Many genes, including ATP binding cassette subfamily A member 7 (ABCA7), bridging integrator 1 (BIN1), complement receptor 1 (CR1), phospholipase D3 gene (PLD3), and phosphatidylinositol-binding clathrin assembly protein gene (PICALM), have been revealed to contribute toward the excess burden of deleterious coding mutations in Alzheimer's disease (Ma et al., 2014; Jiang et al., 2014; Tan et al., 2014b; Cacace et al., 2015; Vardarajan et al.,

2015). In the epileptic encephalopathies, trio exome sequencing has identified that genes UDP-N-acetylglucosaminyltransferase subunit (ALG), gamma-aminobutyric acid type a receptor β3 gene (GABRB3), dynamin 1 (DNM1), hyperpolarization activated cyclic nucleotide gated potassium channel 1 (HCN1), glutamate ionotropic receptor NMDA type subunit 2A (GRIN2A), gamma-aminobutyric acid type A receptor alpha1 subunit (GABRA1), G protein subunit alpha O1 (GNAO1), potassium sodium-activated channel subfamily T member 1 (KCNT1), sodium voltage-gated channel alpha subunit 2 (SCN2A), sodium voltage-gated channel alpha subunit 8 (SCN8A), and solute carrier family 35 member A2 (SLC35A2) are associated with epileptogenesis. Many of the proteins encoded by these genes have been found to be associated with synaptic transmission (Epi, 2015).

RNA sequencing (RNA-seq) provides the information regarding transcript structure, including single base-pair resolution of transcript boundaries and boundaries between exons. The differential transcription is examined using RNA-seq in Alzheimer's disease in brain tissue, which provides the information about the protein coding genes, non-coding RNAs, and splicing. Networks produced may contribute to underlying neurological disease mechanisms. Several studies using animal models have observed the participation of the non-coding in the progression of Alzheimer's disease (Tan et al., 2014a). Next-generation sequencing can explain not only the complete molecular signatures of cells by transcriptome analyses but also the cascade of events that induces or maintains such signatures by epigenetic analyses.

In order to investigate the gene–environment interaction, the science of epigenetics is preferred, which also provides the information of disease pathogenesis. Epigenetic changes are a way in which the environment can influence genetics, bringing about potentially pathogenic alterations in the way genes are expressed. Epigenetics refers to changes in gene regulation brought about through modifications to the DNA's packaging proteins or the DNA molecules themselves without changing the underlying sequence (Mazzio and Soliman, 2012). Several epigenetic studies have provided compelling evidence for genome-wide significant associations between differentially methylated regions and neurological diseases, and suggest that the correlation relationship could reflect a causative one. Synergistically, all these findings could lead researchers to focus on rare variants in neurological diseases precisely, and as a consequence of this treatment, regimens for individual patients may now be tailored and customized.

PRECISION GENETICS FOR PRECISION MEDICINE IN STROKE

Genome-wide association studies have identified many subtypes of specific single nucleotide polymorphisms (SNPs) related with stroke (Traylor et al., 2012; Holliday et al., 2012). Subsequent studies have found genetic commonalities between stroke subtypes (Kaul and Munshi, 2012; Kilarski et al., 2014; Rostanski and Marshall, 2016). Whole exome sequencing analysis carried out by the National Heart, Lung, and Blood Institute Exome Sequencing Project analyzed approximately 6000 participants from different cohorts of European and African ancestry. In this study, 365 cases of ischemic stroke (small-vessel and large-vessel subtypes) and around 809 controls belonging to European ancestry were sequenced.

For replication, 47 affected sib pairs concordant for stroke subtype and case-control series of an African American ancestry were sequenced. This was the first large-scale exome wide sequencing study which has identified protein-coding variant in PDE4DIP phosphodiesterase 4D interacting protein (PDE4DIP) (rs1778155) associated with an intracellular signal transduction mechanism and in acyl-CoA thioesterase (ACOT4) (rs35724886) with a fatty acid metabolism. Further confirmation of PDE4DIP was found in affected sib-pair families with large-vessel stroke subtype and in African Americans (Auer et al., 2015).

Although this area in ischemic stroke is in its conceiving stage, once established in clinical practice it may provide more precise treatment targets based on identification of various stroke-related protein-coding variants. Genotyping has already begun to influence clinical practice with association between drug and genetic variants being included in prescription guidelines for various medicines such as clopidogrel, metoprolol, and warfarin (Kaufman et al., 2015). Clinical trials involving gene expression profiling are underway to implement all these research-related developments (NCT02014896). Studies based on animals to identify the role of a single gene that can control 80% variation in arterial collaterals in stroke are underway (Sealock et al., 2014).

PHENOTYPE-BASED PRECISION MEDICINE IN STROKE

Health care decisions are made on the basis of information provided by big data in an attempt to use phenotypic precision to determine targeted treatments for individuals from large data sets, such as administrative claims. For example, the Observational Health Data Sciences and Informatics collaborative is an international, multistakeholder partnership that uses large-scale analytics to answer clinical questions from electronic health records and health insurance claims data. Combining the data from various sources limits biases that are inherent in homogeneously sourced data (Rostanski and Marshall, 2016).

Electronic health records contain huge amounts of disease-related useful information that could potentially be used for building models for predicting onset of diseases and its models for predicting treatment. In a study, a stroke-prediction model applied advanced machine-learning techniques to aggregated health records in patients with atrial fibrillation. The model outperformed the CHADS2 score (a method or rule used for clinical evaluation of stroke patients with nonrheumatic atrial fibrillation) and helped in predicting stroke, thus providing the useful information for clinical interpretation (Shahn et al., 2015). Therefore, both the medical claims and large health record data keep the potential to improve prediction accuracy when compared to individual patient-level prediction models.

Meta-analyses have been used to convert divergent clinical trial results into cohesive clinical algorithms. New and updated statistical methods used by the National Institutes of Health for the standardization of clinical data elements now allows large-scale data analysis across many trials using patient-level data pooling (Saver et al., 2012). The increased number of patients with consistent data creates greater power to examine subgroups not possible in individual trials and enables focus on patient-centered outcome variables. In an exploratory study, clinical and infarct location data was used from 131 patients with cryptogenic stroke combined with genetic profiles derived from patients with a known stroke subtype (Jickling et al., 2012). In this study the patients predicted with cardio embolic etiology were found to have more myocardial infarctions and higher CHA2DS2-VASc scores when compared to those predicted with a large or small vessel etiology.

Acute stroke imaging

Radiological neuroimaging is a noninvasive approach used to observe the physical manifestation of the disease or phenotype in order to understand the neurological diseases. In cases of acute stroke, the use of advanced perfusion and collateral imaging has added another level of phenotypic precision medicine. Newer clinical trials are underway that involve unknown time of stroke onset and delayed initiation of medical treatment and are using perfusion and collateral imaging to provide thrombolytic treatment to a targeted subset of stroke patients. Moreover, a combination of viscoelastic methodologies, including thromboelastography and scanning electron microscopy, have been validated as useful techniques for stroke patients in a

personalized patient-centered regime leading to precise treatment (de Villiers et al., 2015).

The complete application of knowledge provided by exome sequencing, epigenetics, and other genetic advances may offer new insights into the complex regulation of gene expression in stroke. A higher rate of precision in phenotype description through the use of big data and the addition of advanced imaging state data promise to complement advances in genomics to provide precision medicine for acute stroke treatment, ischemic stroke and its subtypes, cryptogenic stroke, and secondary stroke prevention.

PRECISION MEDICINE IN OTHER NEUROLOGICAL DISEASES

Epilepsy

Epilepsy is a neurological disorder characterized by fits and seizures due to abnormal activity and discharge of neurons in brain. The management of epilepsy needs comprehensive care in order to address the efficacy, side effects and quality of life issues (Jin et al., 2016). The mainstay of epilepsy is drug therapy with antiepileptic drugs (AEDs). However, there are almost two-thirds of patients responding to monotherapy or rational polytherapy who achieve complete seizure control without major side effects (Jin et al., 2016). The therapeutic area in epilepsy is affected mainly because one-third of the patients with epilepsy are refractory even in the context of introduction of new pharmacological molecules/medicines, also known as pharmacoresistant epilepsy. Therefore, surgery in a subset of patients is the only option left.

The role of precision medicine comes in diagnosis, treatment, and presurgical evaluation, which requires a multimodality approach where each provides unique, accurate, and complementary information about the individual patient. Voxel-based morphometric magnetic resonance imaging analysis, multimodality techniques, computer-aided subtraction ictal single-photon emission computed tomography (SPECT), and many other novel techniques have improved our ability to identify subtle structural and metabolic lesions causing focal seizure (Jin et al., 2016). Advances in the field of epilepsy surgery and precision medicine have been reviewed by Jin et al. (2016). Using all the updated knowledge and various modalities, the clinical outcome of epileptic patients will definitely improve reducing disease-associated morbidity and mortality. We are on the edge of using the best surgical diagnosis and treatment of medical refractory epilepsy for individual patients thus promising a better health care outcome.

Alzheimer's disease

Alzheimer's disease is a neurodegenerative disorder and is the most common form of dementia or cognitive impairment in ageing societies. Currently available drugs have slight influence on disease severity and progression. Interventions used to prevent or decelerate disease progression would reduce not only the individual suffering and would significantly relieve public health burden (Reitz, 2016). According to an estimate if we delay the onset of Alzheimer's by 5 years this would reduce the prevalence by ~50% (Reitz, 2016). Alzheimer's is a pathologically and clinically heterogeneous neurodegenerative disease. The disease-associated key pathological manifestations in the brain involve intracellular deposits of hyperphosphorylated tau protein in the form of neurofibrillary tangles and extracellular β-amyloid (Aβ) protein in diffuse and neurotic plaques, which are generated via sequential cleavage of the amyloid precursor protein (APP) through β- (BACE1) and ϒ-secretase (Reitz, 2016). The clinical complexity has been reflected in several genetic studies where significant progress has been made in identifying the underlying genetic variants, with mapping of 27 susceptibility loci (Ertekin-Taner, 2007; Harold et al., 2009; Kim et al., 2009; Seshadri et al., 2010; Hollingworth et al., 2009; Naj et al., 2011; Cruchaga et al., 2012; Guerreiro et al., 2013; Jonsson et al., 2013; Lambert et al., 2013; Logue et al., 2014; Tosto et al., 2015).

The susceptibility risk and molecular profiles of patients with Alzheimer's disease have shown vast variation, and therefore, categorizing patients on the basis of different risk or molecular profiles into a single entity might lead to hiding small subgroups potentially responsive to a certain treatment regime and nonresponsive to another (Reitz, 2016). The results obtained from many clinical trials to date are not satisfactory and a first round of ongoing trials including the Dominantly Inherited Alzheimer Network (DIAN), the Alzheimer's Prevention Initiative, and Anti-Amyloid Treatment in Asymptomatic Alzheimer's Disease (A4) trial are now incorporating information of underlying pathological mechanisms, selecting patients considered most likely to respond to the anticipated action of the therapeutic tested. The results of trials are still awaited (Sperling et al., 2011; Moulder et al., 2013).

Used for decades in the management of a few rare diseases and now broadly applied in cancer care, the precision medicine approach is now stepping toward Alzheimer's disease with the use of genetic variants and neuroimaging techniques.

Although successful application of precision medicine into diagnosis and treatment of Alzheimer's disease needs an extensive framework to identify risk groups, and the underlying pathophysiological and molecular processes. This will develop new interventions, with the significant involvement of a huge network of biologists, physicians, technology developers, data scientists, geneticists, patient groups, stakeholders, and policy makers. It is anticipated that this is only the initiation of a broad precision medicine approach targeting the clinical, biological, and molecular complexity of Alzheimer's disease. With the use of precision medicine in the treatment of Alzheimer's disease, we can hope for better diagnosis, and development of safe, effective drugs or drug combination cocktails to improve clinical outcome of the disease.

Parkinson's disease

Parkinson's disease is a neurodegenerative disease, with a complex pathogenesis involving an interaction of several genetic variants, each with specific, however, minor impact, and a number of environmental factors (Kieburtz and Wunderle, 2013). This disease is associated with devastating chronic complications involving motor symptoms such as slowness of movement, muscle rigidity, resting tremor, and, in some cases, posture and gait problems. Nonmotor symptoms include sleep disorders, cognitive dysfunction, and digestive problems. Current treatment of Parkinson's is focused mainly on restoring movement by replacing lost neurotransmitter dopamine, with the help of medicines. However, these medicines have become less effective over time and only target a small aspect of the underlying disease pathology with reduced therapeutic efficacy. Therefore, a convergence of applications of genomics, epigenomics, neuroimaging, and metabolomics, and their accurate interpretation is required to understand the underlying disease pathology and to manage the disease in a best possible way.

Numerous genetic variants and biomarkers have been identified to be associated with the development of Parkinson's disease (Korczyn and Hassin-Baer, 2015). The causative mutation in the α-synuclein was identified almost two decades back, the gene encoding a major component of Lewy bodies and Lewy neurites, supposed to be the hallmarks of Parkinson's disease pathology (Polymeropoulos et al., 1997; Bu et al., 2016). To date, more than 16 loci and 11 associated genes have been recognized, which help in our understanding of the pathogenesis and clinical heterogeneity of disease (Corti et al., 2011). Besides the genomic profiling, molecular imaging allows a window into the pathophysiology of Parkinson's *in vivo*, and measures the severity and progression of the disease (Bu et al., 2016). Positron emission tomography or single-photon emission computed tomography imaging, integrated with other individual information such as genomic profiling, can help the clinician to design new and tailored interventions for individual patients in contrast to traditional methods of prescribing medicine. For instance, molecular imaging in the asymptomatic carriers with gene mutation of Parkinson's disease can inform about the disease progression at its preclinical stage, therefore, predicting the response of neuroprotective treatments at the right time in right individuals in the future at the right time (Bu et al., 2016).

New tracers used in the radio imaging of α-synuclein are under development and are of potential importance in terms of diagnostic, prognostic, and monitoring biomarkers of Parkinson's disease. Other measures that can be of ultimate importance include deep brain stimulation in order to block abnormal neural signals that may lead to Parkinson's disease. The precision medicine in Parkinson's disease has been reviewed in detail by Bu et al. (2016).

The concept of precision medicine is based on individualized evaluation and multidimensional information about omics, molecular imaging, deep brain stimulation surgery, and wearable sensors leading to tailored therapeutic strategies in Parkinson's disease. Precision medicine will become a promising and feasible strategy and will provide more benefit to patients in the disease course of time with better clinical outcome.

HARNESSING THE POWER OF PRECISION MEDICINE IN STROKE AND OTHER NEUROLOGICAL DISEASES/CONCLUSION

The development of cost-effective advanced genomic technologies and the increasing number of targeted therapies has placed clinical practice in a unique and enviable position. Health care providers have to deeply analyze the patient's sample to know the identified genetic variants, altered protein structure fitting/unfitting the drug molecule and then select a targeted therapy to treat the patient's disease individually rather than only going for histological, anatomical, and routine clinical examination. However, there are many issues in the implementation of precision medicine for both the patient and clinician, for example, interpretation of genomic analysis by geneticists, designing targeted therapies for patients with refractory disease, and the prediction of clinical outcomes with the help of broad application of omics in medicine, which are just in infancy. In addition, the complexities of heterogeneous neurological disorders (including stroke) have to be taken into consideration before providing precision medicine.

In spite of these issues, and in light of the breathtaking responses that can have major impact on patients suffering from neurological diseases, a tidal wave in the form of precision medicine has been initiated and supported by many countries (in the United States by the National Institutes of Health). Likewise, pharmaceutical companies and other industry stakeholders are developing novel precise and tailored targeted therapies and diagnostic kits to support precision medicine. This will bypass the major adverse drug reactions, thereby informing about the responsiveness or nonresponsiveness of the treatment. Moreover a customized treatment will be available for individual patients.

Numerous clinical trials are underway that will provide necessary information regarding metrics, such as survival and health care costs, efficiency of techniques used for diagnosis associated with genomics/proteomics, and transcriptomics for designing precision medicine. The near future will accommodate the precision medicine paradigm in the standard of clinical care for patients with neurological diseases including stroke as it has already been initiated in oncology practice.

REFERENCES

Auer PL, Nalls M, Meschia JF et al. Rare and coding region genetic variants associated with risk of ischemic stroke: The NHLBIExome Sequence Project. *JAMA Neurol* 72(7);2015:781 8.

Bu LL, Yang K, Xiong WX, Liu FT, Anderson B, Wang Y, Wang J. Toward precision medicine in Parkinson's disease. *Ann Transl Med* 4(2);2016:26.

Cacace R, Bossche VT, Engelborghs S et al. Rare variants in PLD3 do not affect risk for early-onset Alzheimer disease in a European consortium cohort. *Hum Mutat* 36;2015:1226–35.

Corti O, Lesage S, Brice A. What genetics tells us about the causes and mechanisms of Parkinson's disease. *Physiol Rev* 91;2011:1161–218.

Cruchaga C, Haller G, Chakraverty S et al. Rare variants in APP, PSEN1 and PSEN2 increase risk for AD in late-onset Alzheimer's disease families. *PLOS ONE* 7;2012:e31039.

de Villiers S, Swanepoel A, Bester J et al. Novel diagnostic and monitoring tools in stroke: An individualized patient-centered precision medicine approach. *J Atheroscler Thromb* 23(5);2015. [Epub ahead of print]. doi:10.5551/jat.32748.

Epi PM Consortium. A roadmap for precision medicine in the epilepsies. *Lancet Neurol* 14;2015:1219–28.

Ertekin-Taner N. Genetics of Alzheimer's disease: A centennial review. *Neurol Clin* 25;2007:611–67.

Guerreiro R, Wojtas A, Bras J et al. TREM2 variants in Alzheimer's disease. *N Engl J Med* 368;2013:117–27.

Harold D, Abraham R, Hollingworth P et al. Genome-wide association study identifies variants at CLU and PICALM associated with Alzheimer's disease. *Nat Genet* 41;2009:1088–93.

Holliday EG, Maguire JM, Evans TJ et al. Common variants at 6p21.1 are associated with large artery atherosclerotic stroke. *Nat Genet* 44(10);2012:1147–51.

Hollingworth P, Harold D, Sims R et al. Common variants at ABCA7, MS4A6A/MS4A4E, EPHA1, CD33 and CD2AP are associated with Alzheimer's disease. *Nat Genet* 43;2009:429–35.

Jiang T, Yu JT, Tan MS et al. Genetic variation in PICALM and Alzheimer's disease risk in Han Chinese. *Neurobiol Aging* 35;2014:934.e1–3.

Jickling GC, Stamova B, Ander BP et al. Prediction of cardioembolic, arterial, and lacunar causes of cryptogenic stroke by gene expression and infarct location. *Stroke* 43(8);2012:2036–41.

Jin P, Wu D, Li X, Ren L, Wang Y. Towards precision medicine in epilepsy surgery. *Ann Transl Med* 4(6);2016:104.

Jonsson T, Stefansson H, Steinberg S et al. Variant of TREM2 associated with the risk of Alzheimer's disease. *N Engl J Med* 368;2013:107–16.

Kaufman AL, Spitz J, Jacobs M et al. Evidence for clinical implementation of pharmacogenomics in cardiac drugs. *Mayo Clin Proc* 90(6);2015:716–29.

Kaul S, Munshi A. Genetic basis of stroke: An Indian perspective. *Neurol India* 60;2012:498–03.

Kieburtz K, Wunderle KB. Parkinson's disease: Evidence for environmental risk factors. *Mov Disord* 28;2013:8–13.

Kilarski LL, Achterberg S, Devan WJ et al. Meta-analysis in more than 17,900 cases of ischemic stroke reveals an ovel association at 12q24.12. *Neurology* 83(8);2014:678–85.

Kim M, Suh J, Romano D et al. Potential late-onset Alzheimer's disease-associated mutations in the ADAM10 gene attenuate {alpha}-secretase activity. *Hum Mol Genet* 18;2009:3987–96.

Korczyn AD, Hassin-Baer S. Can the disease course in Parkinson's disease be slowed? *BMC Medicine* 13;2015:295.

Lambert JC, Ibrahim-Verbaas CA, Harold D et al. Meta-analysis of 74,046 individuals identifies 11 new susceptibility loci for Alzheimer's disease. *Nat Genet* 45;2013:1452–8.

Logue MW, Schu M, Vardarajan BN et al. Two rare AKAP9 variants are associated with Alzheimer's disease in African Americans. *Alzheimers Dement* 10;2014:609–618.e11.

Ma XY, Yu JT, Tan MS et al. Missense variants in CR1 are associated with increased risk of Alzheimer' disease in Han Chinese. *Neurobiol Aging* 35;2014:443.e17–21.

Mazzio EA, Soliman KF. Basic concepts of epigenetics: Impact of environmental signals on gene expression. *Epigenetics* 7;2012:119–30.

Moulder KL, Snider BJ, Mills SL et al. Dominantly inherited Alzheimer network: Facilitating research and clinical trials. *Alzheimers Res Ther* 5;2013:48.

Naj AC, Jun G, Beecham GW et al. Common variants at MS4A4/MS4A6E, CD2AP, CD33 and EPHA1 are associated with late-onset Alzheimer's disease. *Nat Genet* 43;2011:436–41.

Polymeropoulos MH, Lavedan C, Leroy E et al. Mutation in the alpha-synuclein gene identified in families with Parkinson's disease. *Science* 276;1997:2045–7.

Reitz C. Toward precision medicine in Alzheimer's disease. *Ann Transl Med* 4(6);2016:107.

Rostanski SK, Marshall RS. Precision medicine for ischemic stroke. *JAMA Neurol* 73(7);2016:773–4.

Saver JL, Warach S, Janis S et al. Standardizing the structure of stroke clinical and epidemiologic research data: The National Institute of Neurological Disorders and Stroke (NINDS) Stroke Common Data Element (CDE) project. *Stroke* 43(4);2012:967–73.

Sealock R, Zhang H, Lucitti JL, Moore SM, Faber JE. Congenic fine-mapping identifies a major causal locus for variation in the native collateral circulation and ischemic injury in brain and lower extremity. *Circ Res* 114(4);2014:660–71.

Seshadri S, Fitzpatrick AL, Ikram MA et al. Genome-wide analysis of genetic loci associated with Alzheimer disease. *JAMA* 303;2010:1832–40.

Shahn Z, Ryan P, Madigan D. Predicting health outcomes from high-dimensional longitudinal health histories using relational random forests. *Stat Analysis Data Mining ASA Data Sci J* 8(2);2015:128–36.

Sperling RA, Aisen PS, Beckett LA et al. Toward defining the preclinical stages of Alzheimer's disease: Recommendations from the National Institute on Aging Alzheimer's Association workgroups on diagnostic guidelines for Alzheimer's disease. *Alzheimer's Dement* 7;2011:280–92.

Tan L, Yu JT, Tan MS et al. xGenome-wide serum microRNA expression profiling identifies serum biomarkers for Alzheimer's disease. *J Alzheimers Dis* 40;2014a:1017–27.

Tan MS, Yu JT, Jiang T et al. Genetic variation in BIN1 gene and Alzheimer's disease risk in Han Chinese individuals. *Neurobiol Aging* 35;2014b:1781.e1–8.

Tosto G, Fu H, Vardarajan BN et al. F-box/LRR-repeat protein 7 is genetically associated with Alzheimer's disease. *Ann Clin Transl Neurol* 2;2015:810–20.

Traylor M, Farrall M, Holliday EG et al. Genetic risk factors for ischaemic stroke and its subtypes (the METASTROKE collaboration): A meta-analysis of genome-wide association studies. *Lancet Neurol* 11(11);2012:951–62.

Vardarajan BN, Ghani M, Kahn A et al. Rare coding mutations identified by sequencing of Alzheimer disease genome-wide association studies loci. *Ann Neurol* 78(3);2015:487–98.

13

Precision medicine in osteoporosis and bone diseases

FATMANUR HACIEVLIYAGIL KAZANCI, FATIH KAZANCI,
M. RAMAZAN YIGITOGLU, AND MEHMET GUNDUZ

INTRODUCTION

Bone

Bone is a mineralized connective tissue that exerts important functions in the body. Despite its inert appearance, bone tissue is continuously remodeled through the coordinated actions of bone cells (Datta et al., 2008). Osteoblasts, bone lining cells, osteocytes, and osteoclasts are the cells of the bone. After bones have completed their longitudinal growth, almost 10% of the bone is remodeled each year for mineral homeostasis and adaptation to mechanical load (Divieti Pajevic, 2013; Hendrickx et al., 2015).

The remodeling process is under the control of local and systemic factors (Florencio-Silva et al., 2015). A disequilibrium between bone resorption and formation (turnover) leads to a deterioration in bone health. An increase in bone turnover where resorption exceeds neoformation contributes to a decrease in bone volume and mineralization, loss of trabeculae, deterioration of trabecular connectivity, and the formation of cavities and perforations, all of which leads to osteoporosis (Hendrickx et al., 2015). The rate of bone loss throughout life is dependent on a person's lifestyle and genetic background, which influence the geometric architecture of the bone.

Osteoporosis

Osteoporosis is the most common metabolic disorder of the elderly (Mitchell and Streeten, 2013). By the definition of the World Health Organization (WHO), osteoporosis is characterized by reduced bone mass, impaired bone quality, and predisposition to fractures, which is diagnosed if the bone mineral density (BMD) measured by dual x-ray absorptiometry (DXA) is more than 2.5 standard deviations (SD) below the sex-matched young adult mean and/or on the base of fractures occurring without significant trauma (NIH Consensus Development Panel on Osteoporosis Prevention, Diagnosis, and Therapy, 2001). Osteoporosis may be classified as either primary or secondary. Primary osteoporosis is bone loss associated with the aging process and secondary osteoporosis is bone loss caused by a variety of chronic diseases, medications, and nutritional deficiencies.

EPIDEMIOLOGY

Osteoporosis affects one-third of women and one out of eight men over the age of 50 (Li et al., 2010). Most patients with osteoporosis are asymptomatic, which makes epidemiological research especially difficult. However, it is estimated that over 200 million people worldwide have osteoporosis and the prevalence of osteoporosis is continuing to expand with the increase in life expectancy (Reginster and Burlet, 2006). Postmenopausal women are associated with the highest risk of morbidity. In the United States and the European Union, about 30% of all postmenopausal women have osteoporosis, and it has been predicted that more than 40% of them will suffer one or more fragility fractures during their remaining lifetime (Melton et al., 1992).

MORBIDITY AND MORTALITY RATE

The main clinical outcome of the disease is bone fracture. Hip, spine, and distal forearm fractures are the most serious fractures associated with mortality, morbidity, and economic cost (Harvey et al., 2010). In the European Union, in 2000, the number of osteoporotic fractures was estimated at 3.79 million and the direct costs of osteoporotic fractures to the health services were estimated at €32 billion (Reginster and Burlet, 2006). In 2004, it was forecast that there were approximately 10 million Americans over the age of 50 years with osteoporosis, and an additional 34 million Americans were at risk of the disease. Osteoporosis is directly responsible for ~1.5 million fractures annually, with an estimated health care cost of $17 billion in the United States alone (Gass and Dawson-Hughes, 2006). In 2014, it is reported that up to 49 million individuals met the WHO osteoporosis criteria in a number of industrialized countries in North America, Europe, Japan, and Australia (Wade

et al., 2014), and osteoporosis-related fracture is estimated to happen every 3 seconds (International Osteoporosis Foundation, 2014).

The global life expectancy is increasing steadily and the number of elderly individuals is rising in every geographic region. By the year 2050, the population of individuals aged 65 years and over is expected to increase from 323 million to 1.555 billion (Harvey et al., 2010). These demographic changes could lead to an increase in the number of osteoporotic fractures occurring worldwide among elderly individuals. The numbers of hip fractures worldwide are projected to increase from 1.7 million in 1990 to 6.3 million in 2050 and the annual cost of hip fractures is forecast to rise to $131.5 billion by 2050 (at a cost of $21,000 per patient) (Johnell, 1997).

Although hip fractures are a small proportion of all osteoporosis-related fractures, they are the most serious of all fractures, causing a high burden of morbidity, mortality, and health care costs. Approximately 1.6 million hip fractures occur worldwide each year and 5%–10% of patients experience a recurrent hip fracture (International Osteoporosis Foundation, 2014). In a systematic review of 22 studies, excess mortality during the first year after a hip fracture ranged from 8.4% to 36.0%, and the risk of mortality following hip fracture was estimated to be at least twice as high as that for age-matched control individuals from the general population (Sattui and Saag, 2014). The risk of death related to hip fracture of a 50-year-old woman is equivalent to her risk of death from breast cancer and 4 times higher than that from endometrial cancer (International Osteoporosis Foundation, 2014). In 2010, the number of deaths causally related to osteoporotic fractures in Europe was 43,000 and almost 80% of these were due to hip or vertebral fractures (Kanis et al., 2012).

Vertebral fractures are one of the most frequent complications of osteoporosis; however, only one-third to one-quarter of these fractures are clinically diagnosed and hospitalization rates are low (Sattui and Saag, 2014). In the European Prospective Osteoporosis Study (EPOS), age-standardized incidence of vertebral fracture was reported as 10.7 per 1,000 person-years in women and 5.7 per 1,000 person-years in men (Felsenberg et al., 2002). The excess mortality 1 year after vertebral fractures varies considerably, ranging from 1.9% to 42.0% (Sattui and Saag, 2014).

DIAGNOSIS

The diagnosis of osteoporosis relies on the quantitative assessment of BMD. The BMD value is the amount of bone mass per unit volume (volumetric density), or per unit area (areal density), and is currently considered the best predictor of osteoporotic fractures. For every standard deviation decrease in BMD, the fracture risk increases about twofold. BMD is expressed in grams per square centimeter (g/cm^2), but values differ between machines and manufacturers. Therefore, T-scores and Z-scores are used. T-scores can be defined as the number of standard deviations from the mean bone density values in normal sex-matched young adults (Hlaing and Compston, 2014). The T-score is used to make a diagnosis of osteoporosis in postmenopausal women and in men age 50 years and older. Z-scores represent the number of standard deviations from the normal mean value for age- and sex-matched control subjects (Mauck and Clarke, 2006). Z-scores are used preferentially to assess bone loss in premenopausal females and in men younger than age 50 years (Mauck and Clarke, 2006). A wide variety of techniques is available to assess BMD, including quantitative ultrasound (QUS), quantitative computed tomography (QCT), digital x-ray radiogrammetry, dual-photon radionuclide absorptiometry, magnetic resonance imaging, and DXA (Kanis et al., 2008). Among them, DXA is the most commonly used and considered as the gold standard for BMD evaluation (Karjalainen et al., 2016). Ionizing radiation, sophisticated instrumentation, and higher costs are the major handicaps of this method. Most guidelines suggest a single BMD assessment at or around 65 years of age (Solomon et al., 2016). However osteoporotic fractures are not always associated with low BMD (Compston, 2015). Some clinical risk factors other than low BMD contribute independently to fracture risk. Therefore, a need has risen for a more comprehensive and effective tool in risk assessment of osteoporotic fractures. For this purpose, the Fracture Risk Assessment Tool (FRAX) developed by the WHO on the basis of data from several international cohorts, incorporates established risk factors and BMD to predict individual 10-year risk of hip or major osteoporotic fracture. FRAX is freely available on the Internet and it can be used either with or without BMD values. FRAX has some limitations such as inability to take into account dose response effects for several of the clinical risk factors and the exclusion of some common nonhip nonvertebral fractures from its output.

Biochemical markers of bone turnover have the potential to serve as useful tools in osteoporosis management, but biological and analytical variability of them hinders their use (Hlaing and Compston, 2014). They are usually used in the monitoring of osteoporosis therapy (Wei et al., 2016), because changes in BMD take longer to occur than bone turnover markers (Hlaing and Compston, 2014).

PREVENTION AND TREATMENT

Resistance and weight-bearing exercises can increase muscle mass and BMD temporarily (Mauck and Clarke, 2006). Some osteoporosis-related fractures are caused by a minor fall, so patients may benefit from balance programs, which may reduce the risk of falls.

Serum vitamin D levels decrease with advancing age because individuals are exposed to less sunlight and the skin capacity for vitamin D production is reduced (Mauck and Clarke, 2006). Consequently, calcium absorption decreases. Therefore, calcium (1000 mg/day) and vitamin D supplementation should be ordered to elderly patients if dietary intake and sun exposure are not sufficient (Mauck and Clarke, 2006). A total calcium intake of 1500 mg per day (through diet and/or supplements) and a total vitamin D intake of 600–800 IU per day is recommended for osteoporotic patients (Black and Rosen, 2016). For patients with osteoporosis, calcium supplementation should be used as an adjunct to other pharmacological interventions rather than as monotherapy. Vitamin D supplementation alone has not been shown to reduce the risk of fractures or increase BMD except in vitamin D–deficient individuals (Mauck and Clarke, 2006; Black and Rosen, 2016).

The main purpose of pharmacological therapy in osteoporosis is to reduce the risk of fracture, not just increase BMD. Pharmacologic agents for the treatment of osteoporosis can be classified as either antiresorptive or anabolic. The antiresorptive agents are bisphosphonates, calcitonin, receptor activator of nuclear factor kappa-B ligand (RANKL) antibody, and selective estrogen receptor modulators (SERM). These agents suppress bone turnover by reducing bone resorption. Teriparatide is the only anabolic drug, which stimulates bone

Table 13.1 Drugs and drug targets for osteoporosis

Drug	Drug target
Bisphosphonates	Farnesyl pyrophosphate
Teriparetide	PTH receptor
Denosumab	RANKL
SERMs	Estrogen receptor
Calcitonin	Calcitonin receptor
Odanacatib	Cathepsin K
Abaloparatide	PTH receptor
Romosozumab	Sclerostin

formation. Drugs and drug targets for osteoporosis are summarized in Table 13.1.

Bisphosphonates bind at the bone mineral surface, where they potently inhibit osteoclast-mediated bone resorption and are used orally or intravenously. Bisphosphonates are the most potent and the most widely prescribed antiresorptive agents for both osteoporosis prevention and treatment (Mauck and Clarke, 2006; Maraka and Kennel, 2015). Atypical femoral fractures and osteonecrosis of the jaw are rare but are more serious adverse effects of bisphosphonates (Maraka and Kennel, 2015).

Calcitonin nasal spray is approved for the treatment of osteoporosis by the Food and Drug Administration (FDA). Calcitonin inhibits bone resorption, but it is not considered as first-line treatment for osteoporosis because its efficacy in fracture reduction is lower than other medications (Chesnut et al., 2000).

Denosumab is a human monoclonal antibody that decreases osteoclasts differentiation by binding to the receptor activator of nuclear factor-$\kappa\beta$ ligand (RANKL). RANKL is upregulated in postmenopausal women as a result of estrogen decline (Laskowski et al., 2016). As with bisphosphonates, atypical femur fractures and osteonecrosis of the jaw have been observed with denosumab treatment (Black and Rosen, 2016).

Estrogen has direct effects on bone tissue leading to inhibition of bone resorption and maintenance of bone formation. The FDA has withdrawn approval of estrogen or hormone therapy for treatment of osteoporosis but has continued approval of their use for prevention of the disease (Cosman et al., 2014).

Raloxifene is a selective estrogen reception modulator (SERM) that selectively interacts with estrogen receptors. In some tissues, such as bone, it acts an estrogen agonist while in others, such as breast and uterus, it has antiestrogen effects. Raloxifene is approved for both the prevention and the treatment of postmenopausal osteoporosis at a dose of 60 mg/d (Black and Rosen, 2016). Recently, the combination of another SERM, bazedoxifene, with estrogen was approved by the FDA for the treatment of menopausal symptoms and the prevention of osteoporosis (Cosman et al., 2014).

Recombinant human parathyroid hormone analogues are potent bone anabolic agents (Bhandari et al., 2016) and teriparetide is FDA-approved for osteoporosis treatment in patients at high risk for fracture. Teriparetide initially increases bone formation but subsequently also resorption, so its benefits are quickly lost (Harslof and Langdahl, 2016). Therefore, it is recommended to use an antiresorptive agent following teriparetide treatment (Black et al., 2005).

A number of combinations of therapies are possible, but to date no studies regarding combination therapy have had sufficient power to demonstrate additive or synergistic effects of different combinations on fracture outcomes.

EMERGING TREATMENTS FOR OSTEOPOROSIS

Osteoclasts secrete a lot of proteases during bone resorption, and cathepsin K is one of these proteases (Harslof and Langdahl, 2016). Odanacatib is a reversible inhibitor of cathepsin K that interferes only with osteoclast activity (Gauthier et al., 2008). Studies showed that its antifracture efficacy is comparable to current treatments (Nakamura et al., 2013; McClung et al., 2014), but it has not been approved by medical authorities yet.

The osteocytes inhibit bone formation by producing sclerostin (van Bezooijen et al., 2004). Romosozumab is an antisclerostin antibody that has been reported to stimulate bone formation as well as inhibit bone resorption (Chapurlat, 2016). However, the antifracture efficacy of this drug remains to be determined in ongoing trials.

Abaloparatide is a novel synthetic analog of human PTH-related protein (Hattersley et al., 2016). Preclinical trials showed that it induces bone formation without stimulating resorption (Leder et al., 2015). Its effect seems to be superior to teriparatide in preventing fracture (Hattersley et al., 2016).

IMPORTANCE OF LIFESTYLE, ENVIRONMENTAL, AND GENETIC FACTORS IN OSTEOPOROSIS

Osteoporosis is a multifactorial disease. Many genetic variations, and lifestyle and environmental factors influence the susceptibility to osteoporosis and its complications. Dietary intake of calcium and vitamin D, physical activity, low body mass index, excessive alcohol use, and smoking are modifiable risk factors of the disease (Ferrari, 2005). Increased age, female gender, white race, and positive family history are the other risk factors that cannot be modified (Evenson and Sanders, 2016). Improved understanding of all these hereditary and nonhereditary factors has directed the medical community to an individualized approach to prevention, diagnosis, and treatment of osteoporosis. The implementation of preventive strategies for osteoporosis and targeted education of genetically susceptible individuals in order to reduce risks through lifestyle modification would be done only by recognition of the genetic and environmental basis of the disease and the interactions between them. If individuals genetically susceptible to disease with potential to interact with modifiable risk factors can be detected and avoidance of those risk factors can be achieved, the risk of osteoporosis and its complications can be minimized. Furthermore, individual genetic and environmental factors could influence response to therapeutic agents used in osteoporosis prevention and treatment. A number of studies showed that efficacy and safety of the antiosteoporotic drugs are highly variable among treated patients, even with a given drug and a given dosage (Marini and Brandi, 2012). All these reasons encourage us to understand the genetic basis of osteoporosis for both identifying novel diagnostic and therapeutic targets and managing osteoporosis more effectively.

GENETICS OF OSTEOPOROSIS

Today it is widely accepted that genetic factors have a great role in etiopathogenesis of osteoporosis. Many different genetic variants and their interaction with environmental factors (diet, exercise, etc.) influence the susceptibility to osteoporosis. According to twin and family studies, 50%–85% of BMD variance is genetically determined (Park et al., 2012; Wagner et al., 2013) and the risk of fragility fractures is heritable by mechanisms partly independent from BMD (Ralston and Uitterlinden, 2010). Muscle strength, body mass index, levels of calciotropic hormones in circulation, and biochemical markers of bone turnover are other heritable traits related to osteoporosis (Ralston and Uitterlinden, 2010). Estrogen deficiency is one of the most important factors in the reduction of BMD in postmenopausal women (Karasik and Cohen-Zinder, 2012) and twin studies showed that age at menopause is genetically determined (Snieder et al., 1998). This information provides further support for the role of genetic factors in determining bone loss in women. In order to expose the genetic determinants of osteoporosis, a number of linkage and candidate gene studies have been conducted (Mafi Golchin et al., 2016).

Linkage studies have identified loci for BMD, but these findings have not been replicated between studies, and meta-analysis indicated no such loci (Ioannidis et al., 2007). This led investigators to focus on candidate gene studies, but again most of these studies failed to meet expectations (nonreplicative results) (Clark and Duncan, 2015). Some of the osteoporosis-related genes investigated in candidate gene studies are shown in Table 13.2. Today, genome-wide association studies (GWAS) have identified common genetic variants of small effect size that contribute to regulation of BMD and fracture risk in the general population (Richards et al., 2012). GWAS scan the entire genome to identify novel genes/genome regions with modest effects on complex diseases such as osteoporosis (Hirschhorn and Daly, 2005). More than 66 loci influencing BMD have been reported after GWASs, some of which are also associated with fracture risk. Postulated bone mineral density genes identified in GWAS and meta-analyses of GWAS are shown in Table 13.3.

Many of these variants map to genes in known pathways related to bone biology, including the focal adhesion signaling pathway, TGF-β signaling pathway, osteogenic differentiation of human mesenchymal stem cells, endochondral ossification pathway, vitamin D endocrine pathway, RANK/RANKL/OPG, and Wnt signaling (Mafi Golchin et al., 2016). *RANK/RANKL/OPG* and Wnt signaling pathways are illustrated in Figures 13.1 and 13.2, respectively.

The other loci have no apparently known biological function related to bone tissue. Despite the unarguably huge breakthroughs of GWAS, only 6% of the heritability of BMD has been explained to

Table 13.2 Some of the osteoporosis-related genes investigated in candidate gene studies

Gene symbol	Gene name	Function	Pathway
COL1A1	Collagen, type I, alpha 1	Synthesis of type I collagen	
CYP19A1	Cytochrome P450, family 19, subfamily A, polypeptide 1	The key enzyme catalyzing the conversion of androgens to estrogens (Aromatase)	Estrogen
DBP	D site of albumin promoter (albumin D-box) binding protein	Vitamin D carrier	Vitamin D
ESR1	Estrogen receptor 1	Estrogen receptor	Estrogen
ESR2	Estrogen receptor 2	Estrogen receptor	Estrogen
LRP5	LDL receptor-related protein 5	Wnt coreceptor	Wnt/β-catenin signaling
SFRP1	Secreted frizzled-related protein 1	Modulation of Wnt signaling	Wnt/β-catenin signaling
SOST	Sclerosteosis	Inhibition of Wnt signaling	Wnt/β-catenin signaling
TNFSF11	Tumor necrosis factor ligand superfamily, member 11 (RANKL)	Regulation of bone turnover	RANKL/RANK/OPG
TNFRSF11A	Tumor necrosis factor receptor superfamily, member 11a, NFκB activator (RANK)	Regulation of bone turnover	RANKL/RANK/OPG
TNFRSF11B	Tumor necrosis factor receptor superfamily, member 11b (OPG)	Regulation of bone turnover	RANKL/RANK/OPG
VDR	Vitamin D receptor	Vitamin D receptor	Vitamin D
WNT10B	Wingless-type MMTv integration site family, member 10B	Regulation of bone turnover	Wnt/β-Catenin signaling

date (Richards et al., 2012). Genetic architecture of osteoporosis is affected from several genes that have small effects, but a few of which have large effects.

EPIGENETIC MECHANISMS

Epigenetics refers to stable and heritable changes in gene expression other than changes in DNA sequence. Many environmental factors interact with genes through epigenetic mechanisms, and, in some cases, epigenetic alterations are transmissible beyond a single generation (Holroyd et al., 2012). There are a few studies investigating the role of epigenetic mechanisms on differentiation of bone cells and bone turnover (Boudin et al., 2016) and the results of these studies have shown the importance of epigenetic mechanisms on bone biology (Holroyd et al., 2012).

Single-gene disorders of bone

Several rare bone diseases have been identified that occur as the result of mutations in single genes.

These diseases have provided important and novel insights into genes and pathways involved in regulation of bone mass, bone cell function, and bone quality. Depending on the function of the genes involved and the nature of the genetic variation, a full spectrum of phenotypes is observed ranging from severe bone loss to extreme bone mass. The diseases mentioned in this section and their phenotypic features are summarized in Table 13.4.

While discussing these monogenic disorders, we will divide them into two subgroups according to BMD values.

Table 13.3 Postulated bone mineral density genes identified in genome-wide association studies and meta-analyses

Gene	Full name	Study design	Also associated with osteoporotic fracture?
ABCF2	ATP-binding cassette, subfamily F	Meta-analysis	
ADAMTS18	ADAM metallopeptidase with thrombospondin type 1 motif, 18	GWAS	Y
ALDH7A1	Aldehyde dehydrogenase 7 family, member A1	GWAS	Y
ANAPC1	Anaphase promoting complex subunit 1	Meta-analysis	
ARHGAP1	Rho GTPase activating protein 1	Meta-analysis	
AXIN1	Axin 1	Meta-analysis	
C7ORF58	Chromosome 7 open reading frame 58	Meta-analysis	
C12ORF23	Chromosome 12 open reading frame 23	Meta-analysis	
C18ORF19	Chromosome 18 open reading frame 19	Meta-analysis	
CDKAL1	CDK5 regulatory subunit associated protein 1-like 1	Meta-analysis	
CLCN7	Chloride channel, voltage-sensitive 7	GWAS	
CLDN14	Claudin 14	Meta-analysis	
CPN1	Carboxypeptidase N, polypeptide 1	Meta-analysis	
CRHR1	Corticotropin releasing hormone receptor 1	Meta-analysis	
CTNNB1	Catenin (cadherin-associated protein), beta 1	Meta-analysis	Y
CYLD	Cylindromatosis (turban tumor syndrome)	Meta-analysis	
DCDC5	Doublecortin domain-containing 5	Meta-analysis	Y
DHH	Desert hedgehog	Meta-analysis	
DNM3	Dynamin 3	Meta-analysis	
DOK6	Docking protein 6	GWAS	
ERC1	ELKS/RAB6 interacting/CAST family member 1	Meta-analysis	
ESR1	Estrogen receptor 1	GWAS	
FAM9B	Family with sequence similarity 9, member B	Meta-analysis	
FLJ42280	Putative uncharacterized protein	Meta-analysis	
FMN2/GREM2	Formin 2/gremlin 2	GWAS	
FOXL1	Forkhead box L1	Meta-analysis	
FUBP3	Far upstream element (FUSE) binding protein 3	Meta-analysis	Y
GALNT3	UDP-N-acetyl-alpha-D-galactosamine:polypeptide N-acetylgalactosaminyltransferase 3	GWAS	
GPATCH1	G patch domain containing 1	Meta-analysis	
GPR177/WLS	Wntless homolog (Drosophila)	Meta-analysis	
HDAC5	Histone deacetylase 5	Meta-analysis	

(Continued)

Table 13.3 (*Continued*) Postulated bone mineral density genes identified in genome-wide association studies and meta-analyses

Gene	Full name	Study design	Also associated with osteoporotic fracture?
IBSP	Integrin-binding sialoprotein	GWAS	
IDUA		Meta-analysis	
IL21R	Interleukin 21 receptor	GWAS	
INSIG2	Alpha L-iduronidase	Meta-analysis	
JAG1	Jagged 1	GWAS	
KCNMA1	Potassium large conductance calcium-activated channel, subfamily M, alpha member 1	Meta-analysis	
LACTB2	Lactamase, beta 2	Meta-analysis	
LEKR1	Leucine, glutamate and lysine rich 1	Meta-analysis	
LIN7C	Lin 7 homolog C	Meta-analysis	
LRP4	Low-density lipoprotein receptor-related protein 4	Meta-analysis	
LRP5	Low-density lipoprotein receptor-related protein 5	GWAS	Y
LRRC4C	Leucine-rich repeat-containing 4C	GWAS	
MARK3	MAP/microtubule affinity-regulating kinase 3	GWAS	
MBL2	Mannose-binding lectin (protein C) 2, soluble	Meta-analysis	Y
MEF2C	Myocyte enhancer factor 2C	GWAS	
MEPE	Matrix extracellular phosphoglycoprotein	Meta-analysis	Y
MPP7	Membrane protein, palmitoylated 7 (MAGUK p55 subfamily member 7)	Meta-analysis	
OSBPL1A	Oxysterol binding protein-like 1A	GWAS	
PTHLH	Parathyroid hormone-like hormone	Meta-analysis	
RAP1A	RAS-related protein RAP1A	GWAS	
PKDCC	Protein kinase domain containing, cytoplasmic homolog	Meta-analysis	
NTAN1	Ribosomal protein S6 kinase, 90 kda, polypeptide 5	Meta-analysis	
RSPO3	R-spondin 3	GWAS	
RUNX2	Runt-related transcription factor 2	Meta-analysis	
SALL1/CYLD	Sal-like 1 (*Drosophila*)	Meta-analysis	
SLC25A13	Solute carrier family 25	Meta-analysis	Y
SMG6	Smg-6 homolog, nonsense mediated MRNA decay factor (*C. elegans*)	Meta-analysis	
SMOC1	SPARC-related modular calcium binding 1	Meta-analysis	
SOST	Sclerostin	GWAS	Y
SOX4	SRY (sex determining region Y)-box 4	GWAS	
SOX6	SRY (sex determining region Y)-box 6	Meta-analysis	

(*Continued*)

Table 13.3 (*Continued*) Postulated bone mineral density genes identified in genome-wide association studies and meta-analyses

Gene	Full name	Study design	Also associated with osteoporotic fracture?
SOX9	SRY-box containing gene 9	Meta-analysis	
SP7	Sp7 transcription factor 7	GWAS	
SPP1	Secreted phosphoprotein 1	GWAS	Y
SPP2	Secreted phosphoprotein 2, 24 kda	GWAS	
SPTBN1	Spectrin, beta, non-erythrocytic 1	Meta-analysis	Y
STARD3Nl	STARD3 N terminal like	Meta-analysis	Y
TBC1D8	TBC1 domain family, member 8 (with GRAM domain)	GWAS	
TGFBR3	Transforming growth factor, beta receptor III	GWAS	Y
TNFSF11/ RANKL	Tumor necrosis factor (ligand) superfamily, member 11	GWAS	
TNFRS-F11A/ RANK	Tumor necrosis factor receptor superfamily, member 11a, NFKB activator	GWAS	Y
TNFRS-F11B/ OPG	Tumor necrosis factor receptor superfamily, member 11b	GWAS	
WNT16	Wingless-type MMTV integration site family, member 16	Meta-analysis	Y
WNT4	Wingless-type MMTV integration site family, member 4	Meta-analysis	Y
XKR9	XK, Kell blood group complex subunit-related family, member 9	Meta-analysis	
ZBTB40	Zinc finger and BTB domain containing 40	Meta-analysis	Y

Source: This table is modified from Liu YJ et al., *J Bone Metab* 21(2);2014:99–116.

MONOGENIC DISEASES WITH HIGH BMD

This group of diseases can occur as a result of impaired bone resorption, increased bone formation, or increased bone turnover.

Impaired bone resorption

Osteopetrosis is characterized by failure of osteoclastic bone resorption. The most common type of the disease is called osteoclast-rich osteopetrosis that occurs as the result of defects in osteoclast function. Many different gene mutations that impair the ability of osteoclasts to resorb bone have been reported in this type of osteopetrosis.

Failure in osteoclast differentiation can cause another type of the disease called osteoclast-poor osteopetrosis. Inactivating mutations in the *TNFRSF11A* gene encoding RANK or in the *TNFSF11* gene encoding RANKL cause osteoclast-poor osteopetrosis in many cases (Delgado-Calle et al., 2012). Due to the importance of the RANK–RANKL interactions in bone resorption, this interplay was accepted as a potential target for osteoporosis treatment. Currently, a RANKL-inhibitor, denosumab, is approved for the treatment of osteoporosis (Makras et al., 2015).

Pycnodysostosis is an autosomal recessive bone disorder caused by disruption of the osteoclastic function. Mutations resulting in loss of function in the cathepsin K (CTSK) gene have been shown in this disease. CTSK is the main protease of the bone resorption, and today odanacatib (CTSK inhibitor) is developed for osteoporosis treatment (McClung et al., 2014).

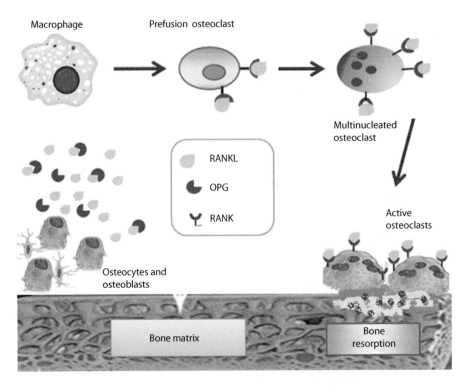

Figure 13.1 RANK/RANKL/OPG pathway in bone turnover. Osteoblasts and osteocytes secrete both RANKL and OPG, which is a soluble decoy receptor for RANKL. In the absence of OPG, RANKL engages RANK present on prefusion osteoclasts. Binding of RANK/RANKL leads to osteoclast maturation and activation, resulting in increased bone resorption.

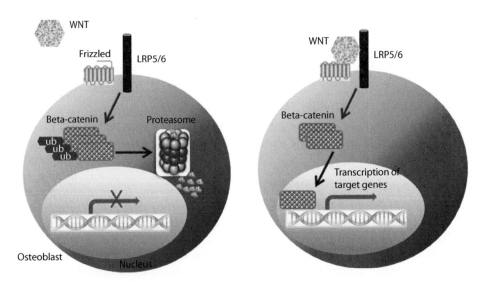

Figure 13.2 Wnt/β-catenin pathway in osteoblasts. In the liganded state on the right side of the figure, Wnt protein binds coreceptors LRP5/6 and Frizzled. This engagement leads to stabilization and accumulation of β-catenin in the cytoplasm and then translocation to the nucleus for transcription of target genes. β-catenin promotes osteoblastic differentiation. In the unliganded state on the left side of the figure, the absence of Wnt ligand binding to the Frizzled receptor and LRP 5/6 coreceptor results in ubiquitination of β-catenin and subsequent degradation in proteasome. Ub, ubiquitin.

Table 13.4 Monogenic bone diseases associated with abnormal bone mass

Disease	Phenotype	Genes
Osteopetrosis	High bone mass, fractures	CLCN7
		TCIRG1
		CATK
		OSTM1
		RANKL (TNFSF11)
		RANK (TNFRSF11A)
Pycnodysostosis	High bone mass	CTSK
Sclerosteosis	High bone mass	SOST
Van Buchem disease	High bone mass	SOST
Craniodiaphyseal dysplasia	High bone mass	SOST
Osteopathia striata	High bone mass	AMER1
Raine syndrome	High bone mass	FAM20C
Camurati–Engelmann disease	High BMD	TGFβ1
Paget's disease of bone	High BMD	SQSTM1
		VCP
Juvenile Paget's disease	High BMD	OPG (TNFRSF11B)
Osteogenesis imperfecta	Low BMD, fractures	COL1A1
		COL1A2
		CRTAP
		LEPRE
		PPIB
Juvenile osteoporosis	Low BMD	WNT1
		LRP5
		PLS3
Osteoporosis-pseudoglioma syndrome	Low bone mass, fractures	LRP5

Increased bone formation

Sclerosteosis occurs due to loss of function mutations in the SOST gene, encoding for sclerostin, an inhibitor of WNT/β-catenin signaling (van Lierop et al., 2011). Van Buchem disease is another autosomal recessive disorder caused by insufficiency of sclerostin. It is caused by a deletion of an enhancer region downstream of SOST, required for adequate sclerostin expression (van Lierop et al., 2013). Another sclerostin-related disorder is craniodiaphyseal dysplasia. It is caused by mutations impairing sclerostin secretion (Kim et al., 2011). After the discovery of molecular pathways of these diseases, sclerostin gained attention as a target to increase BMD in patients with osteoporosis. A monoclonal antibody targeting serum sclerostin, romosozumab, has been developed for osteoporosis treatment (Makras et al., 2015).

Osteopathia striata is a rare, X-linked disorder that is usually mortal for males in the neonatal or fetal stage. Inactivating mutations in the gene coding APC membrane recruitment protein 1 (AMER1), an inhibitor of Wnt signaling, have been shown to give rise to disease (Jenkins et al., 2009). Raine syndrome develops when inactivating mutations occur in the family with sequence similarity 20, member C (FAM20C) gene. This gene encodes a casein kinase that is essential for the osteoblastic differentiation (Ababneh et al., 2013).

Increased bone turnover

Camurati–Engelmann disease is a rare disorder characterized by increased bone turnover. Function mutations in transforming growth factor, beta 1 (TGFb1) have been identified as the genetic cause of the disease (Kinoshita et al., 2000).

Paget's disease of bone (PDB) is caused by an elevated bone turnover rate, leading to excessive production of bone with impaired and disorganized structure.

Mutations in the genes (sequestosome 1 [SQSTM1] and valosin-containing protein [VCP]) encoding for proteins involved in osteoclast differentiation have been found to cause this disorder. It presents with an autosomal dominant mode of inheritance (Delgado-Calle et al., 2012). Inactivating mutations in tumor necrosis factor receptor superfamily member 11b (TNFRSF11B) have been identified as the cause of juvenile Paget's disease (Whyte et al., 2002). This gene encodes a decoy receptor for RANKL, namely, osteoprotegerin (OPG).

MONOGENIC DISEASES WITH DECREASED BMD

Osteogenesis imperfecta is a clinically and genetically heterogeneous group of connective tissue disorders. As well as low BMD and fractures, extraosseus connective tissue symptoms are seen in patients with osteogenesis imperfecta. The International Working Group on Constitutional Disorders of Bone divided the known osteogenesis imperfecta types into five main groups on the basis of the specific clinical features and severity of the disease. Despite the increasing number of known disease-causing genes, the large majority of the cases (~90%) is due to mutations in collagen, type I, alpha 1 (COL1A1); and collagen, type I, alpha 2 (COL1A2) genes (van Dijk et al., 2012).

Juvenile osteoporosis is a group of heritable disorders that present during childhood. Decreased WNT signaling activity because of mutations in WNT1 itself and mutations in LRP5 (which encodes a coreceptor involved in Wnt/β-catenin signaling) have been reported to cause juvenile osteoporosis. Recently, mutations in the plastin 3 (PLS3) gene were identified in patients suffering from an X-linked variant of the disease (van Dijk, 2015).

Osteoporosis-pseudoglioma syndrome is an autosomal recessive disorder that is characterized by severe bone loss and extreme fragility of bones. Loss of function mutations in the LRP5 gene have been reported to cause the disease.

CONCLUSION, CHALLENGES, AND FUTURE DIRECTIONS

Osteoporosis is a highly prevalent bone disorder recognized by low BMD. DXA and FRAX are useful tools for diagnosis of the disease, although they have some limitations. Major morbidity complications of the disease are fractures, especially hip fractures. The osteoporotic fractures are associated with at least a twofold increase in mortality. The risk of death related to hip fracture of a 50-year-old woman is equivalent to her risk of death from breast cancer and 4 times higher than that from endometrial cancer. The number of elderly individuals is rising in every geographic region, and these demographic changes could lead to an increase in the number of osteoporotic fractures occurring worldwide. The number of hip fractures worldwide is forecast to be 6.3 million in 2050 with the annual cost of $131.5 billion. Bone homeostasis involves numerous molecular pathways, and osteoporosis is a complex disease caused by the synergic contribution of several genes. The effect of a single gene on BMD value is relatively small except for rare cases of osteoporosis as explained earlier. Up to 50%–85% of bone loss acceleration is due to genetic factors. Furthermore, environmental factors and lifestyle have important influence on susceptibility of osteoporosis probably through epigenetic mechanisms. Several different methods have been applied to explore the genetic basis of osteoporosis. More than 66 loci have been reported to be associated with BMD in GWAS, but they explain only a small fraction of the total genetic variance in BMD. With the implementation of high-throughput, next-generation DNA sequencing technologies, identification of additional genes can be expected in the near future. For the moment, a number of risk variants within disease-associated genes have been identified as a result of great efforts, and these data allowed novel insights into the pathophysiology of osteoporosis as well as new therapeutic approaches. However, existing knowledge is not enough for providing an explanation for the diagnosis or a tool to assess the genetic risk of osteoporosis in an individual. The undiscovered genetic component likely consists of a combination of many more common variants with increasingly smaller effects and the contributions of rare variants. In the future, some of the missing variance might be revealed by sequencing the exomes (or even whole genomes) of rare variants that have larger effects on BMD than do any of the common variants. Rare variant studies will be potentially useful for two reasons. First, GWAS SNP arrays do not tag variants with minor allele frequencies less than 5%. To assess the low frequency spectrum of allelic variation, direct genotyping of these SNPs is absolutely necessary. Second, rare variants within the exons may negatively influence gene function. In addition, functional studies must also be carried out to explore the likely functional significance of variants identified. Another point to keep in mind is inherited epigenetic modifications and gene-by-gene and gene-by-environmental interactions are significant sources of variation. Although it is clear that epigenetic regulation is involved in the regulation of bone mass, it is unclear which part of the missing heritability can be explained by epigenetic mechanisms. There has been very limited research to date into the genetic determinants of response to treatment with antiosteoporotic agents. With advanced knowledge of the molecules and pathways in bone biology, drug treatments could be optimized according to the patient's individual genetic profile to allow the most beneficial treatment for each patient.

In conclusion, when planning a study related to osteoporosis, sample size should be increased to identify further common variants, next-generation DNA sequencing technologies should be applied to identify rare variants of larger effect size, and functional studies should be done to explore the possible functional significance of variants on BMD.

REFERENCES

Ababneh FK et al. Hereditary deletion of the entire FAM20C gene in a patient with Raine syndrome. *Am J Med Genet A* 161(12);2013:3155–60.

van Bezooijen RL et al. Sclerostin is an osteocyte-expressed negative regulator of bone formation, but not a classical BMP antagonist. *J Exp Med* 199(6);2004:805–14.

Bhandari M et al. Does teriparatide improve femoral neck fracture healing: Results from a randomized placebo-controlled trial. *Clin Orthop Relat Res* 474(5);2016:1234–44.

Black DM et al. One year of alendronate after one year of parathyroid hormone (1–84) for osteoporosis. *N Engl J Med* 353(6);2005:555–65.

Black DM, and Rosen CJ. Clinical practice. Postmenopausal osteoporosis. *N Engl J Med* 374(3);2016:254–62.

Boudin E et al. Genetic control of bone mass. *Mol Cell Endocrinol* 432;2016:3–13.

Chapurlat R. Cathepsin K inhibitors and antisclerostin antibodies. The next treatments for osteoporosis? *Joint Bone Spine* 83(3);2016:254–6.

Chesnut CH, 3rd et al. A randomized trial of nasal spray salmon calcitonin in postmenopausal women with established osteoporosis: The prevent recurrence of osteoporotic fractures study. PROOF Study Group. *Am J Med* 109(4);2000:267–76.

Clark GR, and Duncan EL. The genetics of osteoporosis. *Br Med Bull* 113(1);2015:73–81.

Compston J. FRAX—Where are we now? *Maturitas* 82(3);2015:284–7.

Cosman F et al. Clinician's guide to prevention and treatment of osteoporosis. *Osteoporos Int* 25(10);2014:2359–81.

Datta HK et al. The cell biology of bone metabolism. *J Clin Pathol* 61(5);2008:577–87.

Delgado-Calle J et al. DNA methylation contributes to the regulation of sclerostin expression in human osteocytes. *J Bone Miner Res* 27(4);2012:926–937.

Divieti Pajevic P. Recent progress in osteocyte research. *Endocrinol Metab (Seoul)* 28(4);2013:255–61.

Evenson AL, and Sanders GF. Educational intervention impact on osteoporosis knowledge, health beliefs, self-efficacy, dietary calcium, and vitamin D intakes in young adults. *Orthop Nurs* 35(1);2016:30–6.

Felsenberg D et al. Incidence of vertebral fracture in Europe: Results from the European Prospective Osteoporosis Study (EPOS). *J Bone Miner Res* 17(4);2002:716–24.

Ferrari SL. Osteoporosis: A complex disorder of aging with multiple genetic and environmental determinants. *World Rev Nutr Diet* 95;2005:35–51.

Florencio-Silva R et al. Biology of bone tissue: Structure, function, and factors that influence bone cells. *Biomed Res Int* 2015;2015:421746.

Gass M, and Dawson-Hughes B. Preventing osteoporosis-related fractures: An overview. *Am J Med* 119(4 Suppl 1);2006:S3–11.

Gauthier JY et al. The discovery of odanacatib (MK-0822), a selective inhibitor of cathepsin K. *Bioorg Med Chem Lett* 18(3);2008:923–8.

Harslof T, and Langdahl BL. New horizons in osteoporosis therapies. *Curr Opin Pharmacol* 28;2016:38–42.

Harvey N, Dennison E, and Cooper C. Osteoporosis: Impact on health and economics. *Nat Rev Rheumatol* 6(2);2010:99–105.

Hattersley G et al. Binding selectivity of abaloparatide for PTH-Type-1–receptor conformations and effects on downstream signaling. *Endocrinology* 157(1);2016:141–9.

Hendrickx G, Boudin E, and Van Hul W. A look behind the scenes: The risk and pathogenesis of primary osteoporosis. *Nat Rev Rheumatol* 11(8);2015:462–74.

Hirschhorn JN, and Daly MJ. Genome-wide association studies for common diseases and complex traits. *Nat Rev Genet* 6(2);2005:95–108.

Hlaing TT, and Compston JE. Biochemical markers of bone turnover—uses and limitations. *Ann Clin Biochem* 51(Pt 2);2014:189–202.

Holroyd C et al. Epigenetic influences in the developmental origins of osteoporosis. *Osteoporos Int* 23(2);2012:401–10.

International Osteoporosis Foundation. 2014. Available from http://www.iofbonehealth.org. Accessed March 20, 2016.

Ioannidis JP et al. Meta-analysis of genome-wide scans provides evidence for sex- and site-specific regulation of bone mass. *J Bone Miner Res* 22(2);2007:173–83.

Jenkins ZA et al. Germline mutations in WTX cause a sclerosing skeletal dysplasia but do not predispose to tumorigenesis. *Nat Genet* 41(1);2009:95–100.

Johnell O. The socioeconomic burden of fractures: Today and in the 21st century. *Am J Med* 103(2a);1997:20S–25S, discussion 25S–26S.

Kanis J, Compston J, and Cooper C. The burden of fractures in the European Union in 2010. *Osteoporos Int* 23(Suppl 2);2012:S57.

Kanis JA et al. A reference standard for the description of osteoporosis. *Bone* 42(3);2008:467–75.

Karasik D, and Cohen-Zinder M. Osteoporosis genetics: Year 2011 in review. *Bonekey Rep* 1;2012:114.

Karjalainen JP et al. New method for point-of-care osteoporosis screening and diagnostics. *Osteoporos Int* 27(3);2016:971–7.

Kim SJ et al. Identification of signal peptide domain SOST mutations in autosomal dominant craniodiaphyseal dysplasia. *Human Genetics* 129(5);2011:497–502.

Kinoshita A et al. Domain-specific mutations in TGFB1 result in Camurati-Engelmann disease. *Nat Genet* 26(1);2000:19–20.

Laskowski LK et al. A RANKL wrinkle: Denosumab-induced hypocalcemia. *J Med Toxicol* 12(3);2016:305–8.

Leder BZ et al. Effects of abaloparatide, a human parathyroid hormone-related peptide analog, on bone mineral density in postmenopausal women with osteoporosis. *J Clin Endocrinol Metab* 100(2);2015:697–706.

Li WF et al. Genetics of osteoporosis: Accelerating pace in gene identification and validation. *Hum Genet* 127(3);2010:249–85.

Liu YJ, Zhang L, Papasian CJ, and Deng HW. Genome-wide Association Studies for Osteoporosis: A 2013 Update. *J Bone Metab* 21(2);2014:99–116.

Mafi Golchin M et al. Osteoporosis: A silent disease with complex genetic contribution. *J Genet Genomics* 43(2);2016:49–61.

Makras P, Delaroudis S, and Anastasilakis AD. Novel therapies for osteoporosis. *Metabolism* 64(10);2015:1199–214.

Maraka S, and Kennel KA. Bisphosphonates for the prevention and treatment of osteoporosis. *BMJ* 351;2015:h3783.

Marini F, and Brandi ML. Pharmacogenetics of osteoporosis: What is the evidence? *Curr Osteoporos Rep* 10(3);2012:221–7.

Mauck KF, and Clarke BL. Diagnosis, screening, prevention, and treatment of osteoporosis. *Mayo Clin Proc* 81(5);2006:662–72.

McClung MR et al. Odanacatib anti-fracture efficacy and safety in postmenopausal women with osteoporosis: Results from the phase III long-term odanacatib fracture trial. *Arthritis Rheum* 66(11);2014:S987.

Melton LJ, 3rd et al. Perspective. How many women have osteoporosis? *J Bone Miner Res* 7(9);1992:1005–10.

Mitchell BD, and Streeten EA. Clinical impact of recent genetic discoveries in osteoporosis. *Appl Clin Genet* 6;2013:75–85.

Nakamura T et al. Effect of the cathepsin K inhibitor odanacatib administered once weekly on bone mineral density in Japanese patients with osteoporosis—A double-blind, randomized, dose-finding study. *Osteoporos Int* 25(1);2013:367–76.

NIH Consensus Development Panel on Osteoporosis Prevention, Diagnosis, and Therapy. Osteoporosis prevention, diagnosis, and therapy. *Jama* 285(6);2001:785–95.

Park JH et al. Genetic influence on bone mineral density in Korean twins and families: The healthy twin study. *Osteoporos Int* 23(4);2012:1343–9.

Ralston SH, and Uitterlinden AG. Genetics of osteoporosis. *Endocr Rev* 31(5);2010:629–62.

Reginster J-Y, and Burlet N. Osteoporosis: A still increasing prevalence. *Bone* 38(2, Suppl 1);2006:4–9.

Richards JB, Zheng HF, and Spector TD. Genetics of osteoporosis from genome-wide association studies: Advances and challenges. *Nat Rev Genet* 13(8);2012:576–88.

Sattui SE, and Saag KG. Fracture mortality: Associations with epidemiology and osteoporosis treatment. *Nat Rev Endocrinol* 10(10);2014:592–602.

Snieder H, MacGregor AJ, and Spector TD. Genes control the cessation of a woman's reproductive life: A twin study of hysterectomy and age at menopause. *J Clin Endocrinol Metab* 83(6);1998:1875–80.

Solomon CG, Black DM, and Rosen CJ. Postmenopausal osteoporosis. *N Engl J Med* 374(3);2016:254–62.

van Dijk FS et al. EMQN best practice guidelines for the laboratory diagnosis of osteogenesis imperfecta. *Eur J Hum Genet* 20(1);2012:11–9.

van Dijk FS. Genetics of osteoporosis in children. *Endocr Dev* 28;2015:196–209.

van Lierop AH et al. Patients with sclerosteosis and disease carriers: Human models of the effect of sclerostin on bone turnover. *J Bone Miner Res* 26(12);2011:2804–11.

van Lierop AH et al. Van Buchem disease: Clinical, biochemical, and densitometric features of patients and disease carriers. *J Bone Miner Res* 28(4);2013:848–54.

Wade SW et al. Estimating prevalence of osteoporosis: Examples from industrialized countries. *Arch Osteoporos* 9;2014:182.

Wagner H et al. Genetic influence on bone phenotypes and body composition: A Swedish twin study. *J Bone Miner Metab* 31(6);2013:681–9.

Wei QS et al. Serum osteopontin levels in relation to bone mineral density and bone turnover markers in postmenopausal women. *Scand J Clin Lab Invest* 76(1);2016:33–9.

Whyte MP et al. Osteoprotegerin deficiency and juvenile Paget's disease. *N Engl J Med* 347(3);2002:175–84.

14

Precision medicine in diabetes mellitus

SANDHIYA SELVARAJAN, AKILA SRINIVASAN, NISHANTHI ANANDABASKAR, SADISHKUMAR KAMALANATHAN, AND MELVIN GEORGE

INTRODUCTION

Precision medicine, one of the budding endeavors of modern medicine, is an innovative concept endorsed since 2011 to advocate personalized therapy based on the distinct features of an individual patient (Klonoff, 2015). It varies from personalized medicine by focusing on interindividual genomic characteristics to predict treatment strategies for each patient with the help of tools like genomics, metabolomics, proteomics, and bioinformatics (Collins and Varmus, 2015). Precision medicine embraces three salient components, namely, a specific biomarker, methods of analyzing the biomarker, and treatment based on the biomarker (Figure 14.1). The prime intent of precision medicine is to enhance the success rate of treatment apart from lessening morbidity and mortality (Butz et al., 2013). Succeeding the Precision Medicine Initiative (PMI) by the National Institutes of Health (NIH) to promote the application of precision medicine in various diseases, there has been a sudden upsurge in its application to various diseases including that of diabetes mellitus (Fradkin et al., 2016).

Diabetes mellitus, a metabolic disorder, is characterized by hyperglycemia owing to either insulin resistance or insulin deficiency. This chronic noncommunicable disease has been mounting with high prevalence in both developed and developing countries. The global prevalence of diabetes mellitus among adults has increased from 108 million in 1980 to 422 million in 2014. In 2012, around 2.2 million deaths due to cardiovascular and other complications were attributed to higher blood glucose levels. Likewise, nearly 1.6 million deaths were reported to be due to diabetes mellitus during the year 2015 (World Health Organization, 2016). The prevalence of diabetes mellitus in India is 69.2 million (8.7%) as per the data published in 2015 (World Health Organization, India, 2016).

Diabetes mellitus is classified into various subtypes based on the underlying pathogenesis as listed in Table 14.1 (American Diabetes Association, 2011). Of these subtypes, type 1 diabetes mellitus (T1D) is an autoimmune disorder resulting in pancreatic beta-cell destruction and insulin deficiency. Type 2 diabetes mellitus (T2D) is highly influenced by genetics, environment, and lifestyle, and is associated with a combination of dysfunctional pancreatic beta cells and or insulin resistance. Gestational diabetes mellitus (GDM) associated with glucose intolerance is seen in pregnant women, as the condition is diagnosed for the first time during pregnancy.

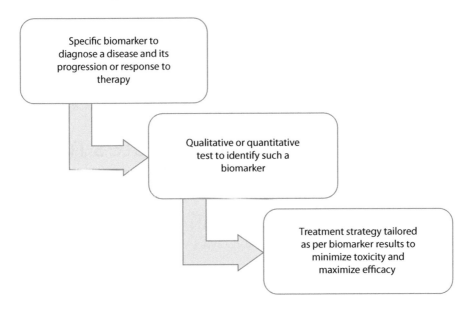

Figure 14.1 Components of precision medicine.

The role of genetics and the genes signifying augmented threat to the occurrence of various types of diabetes mellitus have been identified. GWAS (genome-wide association studies) have identified a wide variety of genetic susceptibility loci for development of type 1 and type 2 diabetes mellitus. The risk of type 1 diabetes is increased from 1 in 300 to 1 in 15 in patients possessing DR3-DQ2/ DR4-DQ8 haplotypes together in their MHC (major histocompatibility complex) region on chromosome 6 (Hartley, 2014). Likewise numerous candidate genes, including peroxisome proliferator-activated receptor gamma (PPARγ2), associated with increased risk of developing type 2 diabetes have also been identified (Abbas et al., 2013). Further, the American Diabetes Association (ADA) states that the children diagnosed with diabetes during the initial 6 months of their life need genetic testing for precise diagnosis of neonatal diabetes, as it is imperative to treat them with sulfonylureas (American Diabetes Association, 2016).

Although a wide array of genetic mutations is implicated in development of diabetes mellitus, none of these genetic profiling tests are currently used in routine screening tests for diagnosis of diabetes mellitus (Gillespie, 2006). In view of the significance of genes involved in instigating this progressive disease, this chapter aims to focus on the applications of precision medicine in various types of diabetes mellitus.

APPLICATION OF PRECISION MEDICINE IN TYPE 1 DIABETES MELLITUS

Type 1 diabetes mellitus is an autoimmune disorder contributing to 5%–10% of the cases of diabetes worldwide (Maahs et al., 2010). It is associated with destruction of pancreatic beta cells resulting in decreased insulin secretion and with course of time this condition progresses to a state of comprehensive insulin deficiency (Gillespie, 2006). Type 1 diabetes mellitus is a genetic disorder and has been found to be associated with nearly 20 human leukocyte antigens (HLA). The first HLA to be identified as the major contributor to the occurrence of familial type 1 diabetes mellitus was found to be located in chromosome 6. The key HLA found to be associated with the prediction of occurrence of type 1 diabetes mellitus are DR4-DQ8, DR3-DQ2 and DR15-DQ (Pociot and McDermott, 2002). On the contrary, it has been found that the presence of HLA like DR15-DQ6 haplotype is highly protective against development of type 1 diabetes mellitus (Price et al., 2001).

Apart from HLA, polymorphisms of insulin genes in chromosome 11 have also been found to confer genetic susceptibility to the development of type 1 diabetes mellitus (Ramos-Lopez et al., 2008). It is also found that polymorphisms in the CTLA4 (cytotoxic T-lymphocyte antigen-4) gene located

Table 14.1 Classification of diabetes mellitus

Classification	Underlying mechanisms
Type 1 diabetes mellitus	Beta cell destruction resulting in insulin deficiency
Type 2 diabetes mellitus	Can vary from predominant insulin resistance with relative insulin deficiency to predominant defect in insulin secretion along with insulin resistance

Other specific types of diabetes	• Genetic defects of beta cell function
	• Genetic defects in insulin action
	• Diseases of exocrine pancreas:
	  • Pancreatitis
	  • Pancreatectomy
	  • Neoplasia
	  • Cystic fibrosis
	  • Hemochromatosis
	  • Fibrocalculous pancreatopathy
	  • Mutations in carboxyl ester lipase
	• Endocrinopathies
	  • Acromegaly
	  • Cushing's syndrome
	  • Glucagonoma
	  • Pheochromocytoma
	  • Hyperthyroidism
	  • Somatostatinoma
	  • Aldosteronoma
	• Drug or chemical induced
	  • Glucocorticoids
	  • Vacor (a rodenticide)
	  • Pentamidine
	  • Nicotinic acid
	  • Diazoxide
	  • Beta adrenergic agonists
	  • Thiazides
	  • Hydantoins
	  • Asparaginase
	  • Alpha interferon
	  • Protease inhibitors
	  • Antipsychotics (atypical and others)
	  • Epinephrine
	• Infections
	  • Congenital rubella
	  • Cytomegalo virus
	  • Coxsackie virus
	• Uncommon forms of immune mediated diabetes
	  • Stiff person syndrome
	  • Anti-insulin receptor antibodies
	• Other genetic syndromes sometimes associated with diabetes
	  • Wolfram's syndrome
	  • Down's syndrome

(Continued)

Table 14.1 (*Continued*) Classification of diabetes mellitus

Classification	Underlying mechanisms
	• Klinefelter syndrome
	• Turner's syndrome
	• Friedrich's ataxia
	• Huntington's chorea
	• Laurence–Moon–Biedl syndrome
	• Mytonic dystrophy
	• Porphyria
	• Prader–Willi syndrome
Gestational diabetes mellitus	Occurs for the first time during pregnancy or is diagnosed during pregnancy

Source: American Diabetes Association, *Diabetes Care* 34(Suppl 1);2011:S62–69.

in chromosome 2q33 are associated with increased risk of type 1 diabetes mellitus (Ueda et al., 2003). Another gene associated with increased risk of type 1 diabetes mellitus is the protein tyrosine phosphatase, a nonreceptor type 22 (PTPN22) gene on chromosome 1p13 that encodes for lymphoid-specific phosphatase (LYP), which prevents spontaneous T cell activation. Similarly polymorphism in the PTPN22 gene leads to development of type 1 diabetes mellitus (Steck et al., 2006). In addition, polymorphisms in IL2RA (CD25), which encodes for the subunit IL-2Rα of the interleukin-2 (IL-2) receptor complex is also implicated in the occurrence of type 1 diabetes mellitus (Vella et al., 2005). Knowledge of these genes and polymorphisms in the occurrence of type 1 diabetes mellitus would pave the way for the early identification of individuals with an increased risk of developing type 1 diabetes mellitus (i.e., even before metabolic derangement occurs). This may help in establishing measures to prevent the occurrence of type 1 diabetes mellitus.

APPLICATION OF PRECISION MEDICINE IN TYPE 2 DIABETES MELLITUS

There has been an escalating trend observed in the global prevalence of type 2 diabetes mellitus in the last decade that could possibly be attributed to the increase in average life span of people (Guariguataa, 2014; World Health Organization, 2017). In recent times considerable measures have been taken to detect all the genes involved in the pathogenesis of type 2 diabetes mellitus. Yet at present only about 5% of patients with type 2 diabetes mellitus have been found to have an identifiable genetic basis. Some of the candidate genes associated with insulin secretion or resistance resulting in type 2 diabetes are shown in Table 14.2 (Sacks and McDonald, 1996; Hani et al., 1999; Barroso et al., 2003; Komurcu-Bayrak, 2012; Brown and Walker, 2016; Magaña-Cerino et al., 2017).

Numerous studies have been done to evaluate the association of various genes with the occurrence of diabetes. It has been found that the K_{ATP} channels present in β-cells have a SUR1 subunit encoded by the ABCC8 gene and a mutation in this gene has been associated with a twofold higher risk for diabetes mellitus with an earlier age of onset in Southwest American Indians (Baier et al., 2015). Apart from K_{ATP} channels, studies have been done to explore the role of PPAR-gamma mutations as a risk factor for the onset of type 2 diabetes mellitus, however, with contradictory results. It has been observed that PPAR-gamma, germ line loss-of-function mutations are associated with an early onset of insulin resistance type 2 diabetes mellitus and hypertension (Barroso et al., 1999). Similarly the presence of proline allele of PPAR-gamma has been shown to be associated with a greater risk for type 2 diabetes. On the contrary, the presence of Pro12Ala polymorphism of PPAR-gamma has been found to be protective against the development of type 2 diabetes mellitus in Caucasians (Altshuler et al., 2000). However, a similar study done by Radha et al. (2006) among Caucasians and South Asians in Texas and South Asians in Chennai showed that the PPAR-gamma Pro12Ala polymorphism provides protection against the occurrence of type 2 diabetes only in Caucasians but not so in South Asians. In addition, the transcription factor 7-like 2 (TCF7L2) gene present in most of the tissues including beta cells of

Table 14.2 Candidate genes associated with type 2 diabetes mellitus

Candidate genes affecting insulin secretion	Candidate genes affecting insulin action	Candidate genes causing insulin resistance
• *SLC2A2* (GLUT2) • MicroRNA (miRNA)-463-3p and ATP-binding cassette A4 (ABCG4) • Insulin promoter factor-1 (IPF-1) gene	• *INSR* (gene coding insulin receptor) • *PIK3R1* (gene encoding p85a regulatory subunit of phosphatidylinositol 3-kinase) • Intronic variant of SNP42 • *SOS1* (son of sevenless homolog 1 in Drosophila) • ABCC8 (sulphonylurea receptor) • KCNJ11 (KIR6.2) • HNF4A • HNF1A	• PPARG • KLF14 • IRS1 • GCKR • FTO • IGF1 • TCF7L2 • NAT2 • ELOVL6 • FABP • RETN • RBP4 • LEP and LEPR • D2 and TSHR

Source: Sacks DB, and McDonald JM, *Am J Clin Pathol* 105(2);1996:149–56; Barroso I et al., *PLoS Biol* 1(1);2003:e20; Brown AE, and Walker M, *Curr Cardiol Rep* 18;2016:75; Magaña-Cerino JM et al., *Mol Genet Genomic Med* 5(1);2017:50–65; Komurcu-Bayrak E., in *Insulin Resistance*, Chap 4, Sarika Arora (Ed.), InTech, 2012, http://cdn.intechopen.com/pdfs/41435/InTech-Impact_of_genetic_polymorphisms_on_insulin_resistance.pdf; Hani EH et al., *J Clin Invest* 104(9);1999:R41–8.

the pancreas has been found to have reduced expression in fat deposits of obese type 2 diabetic patients compared to people with normal blood glucose (Cauchi et al., 2006). A study has identified 6 SNPs (single nucleotide polymorphisms) that are associated with elevated risk for type 2 diabetes mellitus, as shown in Table 14.3 (Zyriax et al., 2013). These findings suggest a robust contribution of genes in the occurrence of type 2 diabetes.

Apart from causing the disease, genes are also shown to have a vital contribution in determining the response to various drugs used in the treatment of type 2 diabetes mellitus. Metformin, the first-line drug used in the treatment for type 2 diabetes mellitus, is known to depict interindividual variations in

response as well as adverse effects depending on the genetic variations in drug metabolizing enzymes. A study that performed an oral glucose tolerance test (OGTT) before and after metformin administration in healthy volunteers to appraise the effect of metformin in the presence of variants of a drug transporter OCT-1 such as OCT1-R61C, -G401S, and -G465R, found that the carriers of decreased function polymorphisms had significantly higher plasma glucose levels compared to those with reference OCT-1 alleles (Shu et al., 2007).

A cohort study carried out on patients with type 2 diabetes found that those on metformin and other drugs that inhibit the function of OCT1 transport were found to be intolerant to metformin. Similarly,

Table 14.3 SNPs associated with elevated risk for type 2 diabetes mellitus

Gene locus	Gene name	Cytogenetic location	SNP
KLF14	Kruppel-like factor 14	7q32.3	rs972283
NOTCH2	Notch 2	1p13-p11	rs10923931
DUSP9	Dual specificity phosphatase 9	Xq28	rs5945326
HHEX	Hematopoietically expressed homeobox	10q24	rs1111875
SLC30A8	Solute carrier family 30 (zinc transporter), member 8	8q24.11	rs13266634
ZBED3	ZBED3 antisense RNA 1	5q13.3	rs4457053

Source: Zyriax B-C, Salazar R, Hoeppner W et al., *PLoS ONE* 8(9);2013:e75807.

in the same study, intolerance to metformin was observed among those patients with reduced function OCT1 alleles compared to patients with normal or one deficient allele with an odds ratio of 2.4 (CI 1.48–3.93, $p < 0.001$) (Dujic et al., 2015). The same study found that those diabetic patients with both reduced function OCT1 alleles and on OCT1 inhibitors were having metformin intolerance 4 times higher than other patients without these risks (Dujic et al., 2015). This clearly envisages the significant role of this drug transporter toward the prediction of therapeutic response to metformin.

Similarly, a study done to evaluate the role of genetic polymorphisms on therapeutic response to sulfonylureas identified the presence of TT in rs12255372 and rs7903146 of the TCF7L2 gene is futile in achieving the HbA1C target (Pearson et al., 2007). Likewise in a study done for 10 weeks with pioglitazone at a dose of 30 mg on type 2 diabetic patients, it has been observed that the presence of S447X variant in lipoprotein lipase (LPL) gene was associated with less response to pioglitazone compared to carriers of S447S genotype with favorable effects on the lipid profile as well as blood pressure (Wang et al., 2007). Moreover the genetic variations have been found to be associated with the occurrence of adverse drug reactions following treatment with pioglitazone. A study instigating the role of genetic variations responsible for the incidence of thiazolidinedione-induced sodium and water retention in type 2 diabetic patients has shown that the AQP2 (aquaporin2) rs296766 T allele and SLC12A1 rs12904216 were associated with occurrences of thiazolidinedione-induced edema (Chang et al., 2011). Apart from the conventional drugs used in diabetes mellitus recently, it has been demonstrated that a variant of α2A adrenergic-receptor (α2AAR) encoding gene resulting in receptor overexpression and decreased insulin secretion among type 2 diabetes patients has been shown to have improved insulin secretion following administration of yohimbine, a α2 adrenergic-receptor blocker (Tang et al., 2014). However, these findings need to be proven in clinical trials and the information can be used for further research in drug discovery and development to bring out new molecules.

GESTATIONAL DIABETES MELLITUS

Intolerance to glucose occurring for the first time or being detected for the first time during pregnancy is known as gestational diabetes mellitus (GDM). The risk factors for GDM include obesity, family history of diabetes and past history of GDM (American Diabetes Association, 2003). The treatment for this condition can vary from dietary modifications to insulin (Seshiah et al., 2004). One of the most common genes implicated in the onset of GDM is glucokinase, located on chromosome 7, which regulates a key enzyme of glucose metabolism. Mutations in this gene have been shown to be associated with GDM in around 5%–10% of Hispanic and Caucasian women (Stoffel et al., 1993). Moreover, this gene is of key interest, as mutations in this can lead to maturity onset diabetes of the young (MODY), which is a form of non-insulin-dependent diabetes mellitus (NIDDM) characterized by mild hyperglycemia appearing in childhood and a milder form of diabetes that in most patients is manageable by dietary modifications alone (Froguel et al., 1993). Yet another gene shown to be associated with GDM is hepatocyte nuclear factor 1-α (HNF1-α). A study conducted in Scandinavian women found that the frequency of L alleles was higher in patients with GDM when compared to controls. Moreover, the dominant model of LL + lL versus ll showed a higher odds ratio signifying the presence of HNF1-α I27L (rs1169288) polymorphism associated with elevated risk of GDM in Scandinavian females (Shaat et al., 2006). Similarly Thr130Ile polymorphism of the transcription factor HNF4α has been found to be playing an important role in GDM, as it was observed that the frequency of minor alleles of this polymorphism was higher in women with GDM compared to healthy postpartum females in a study done in a Mexican population (Monroy et al., 2014). The fact that HNF4α plays a key role in regulating several transcriptomes of liver and pancreatic islets may be a reason for increasing the risk of diabetes by causing dysfunctioning of beta cells (Odom et al., 2004).

ADVANTAGES AND LIMITATIONS OF PRECISION MEDICINE

Whole genome sequencing may result in the detection of genetic variants that are either protective or offensive against the development of a specific disease (Collins and Varmus, 2015). Hence identifying specific genomic biomarkers may help in predicting the risk factor for the occurrence of various diseases including that of diabetes mellitus. The knowledge obtained from genomic research can be used to delay the onset as well as progression of the disease through early intervention (Franks and Poveda, 2017).

However, we should remember that in multifactorial diseases like diabetes and hypertension it is highly challenging to establish the causative genes. One more limitation with genomic studies are that the majority of the SNPs associated with disease are not located in protein coding sites (Shastry, 2002). Moreover, the genetic make upon its own may not be adequate to predict susceptibility to a particular disease, as the disease is predicted by a wide range of environmental factors (Martin et al., 1997). The genotype may not necessarily correlate with the phenotype in many medical conditions, as they do not conform to simple Mendelian inheritance, as these conditions such as susceptibility to diabetes, hypertension, and cancer are known as "complex" traits (Lander and Schork, 1994; Weeks and Lathrop, 1995). Further, most of the genetic studies are carried out in homogeneous populations and thus the external validity of the data generated comes down in real patients who may have other accompanying comorbidities (Beigy, 2015).

Another aspect to be considered is affordability of precision medicine, especially in resource-poor settings. Since specific treatment modalities are targeted at a more specific group of audience based on presence of particular genes or biomarkers, the market for newer drugs becomes smaller as the target population comes down, thereby making development costs of newer drugs unsustainable from such small populations (Aronson, 2015). The accomplishment of precision medicine in the future will be decided upon the success of therapy using the concept and the associated cost (Klonoff, 2015).

CONCLUSION

Precision medicine aims at providing custom-made, personalized therapy to an individual patient based on their genetic changes. Diabetes mellitus being one of the progressively increasing noncommunicable diseases globally with multiple risk factors, measures need to be taken to incorporate the latest advances in medical research to combat the disease at an earlier stage. Precision medicine seems to be one of the promising approaches in the near future to evaluate the multigenic risk factors associated with diabetes mellitus. Based on the knowledge gained using genetic research, precision medicine provides the scope for further research to develop new molecules catering to the needs of personalized therapy in diabetes mellitus.

REFERENCES

Abbas S, Raza ST, Ahmed F et al. Association of genetic polymorphism of PPARγ-2, ACE, MTHFR, FABP-2 and FTO genes in risk prediction of type 2 diabetes mellitus. *J Biomed Sci* 20(1);2013:80.

Altshuler D, Hirschhorn JN, Klannemark M et al. The common PPAR gamma Pro12Ala polymorphism is associated with decreased risk of type 2 diabetes. *Nat Genet* 26;2000:76–80.

American Diabetes Association. Gestational diabetes mellitus. *Diabetes Care* 26(Suppl 1);2003:s103–5.

American diabetes Association. Diagnosis and classification of diabetes mellitus. *Diabetes Care* 34(Suppl 1);2011:S62–69.

American Diabetes Association. Standards of medical care in diabetes. *Diabetes Care* 39(Suppl);2016:S1–12.

Aronson N. Making personalized medicine more affordable. *Ann N Y Acad Sci* 1346(1);2015:81–9.

Baier LJ, Muller YL, Remedi MS et al. ABCC8 R1420H loss-of-function variant in a Southwest American Indian community: Association with increased birth weight and doubled risk of type 2 diabetes. *Diabetes* 64(12);2015:4322–32.

Barroso I, Gurnell M, Crowley VE et al. Dominant negative mutations in human PPAR gamma associated with severe insulin resistance, diabetes mellitus and hypertension. *Nature* 402(6764);1999:880–3.

Barroso I, Luan JA, Middelberg RP et al. Candidate gene association study in type 2 diabetes indicates a role for genes involved in β-cell function as well as insulin action. *PLoS Biol* 1(1);2003:e20.

Beigy M. Generalized resemblance theory of evidence: A proposal for precision/personalized evidence-based medicine. *arXiv preprint*:1512.03825. December 13, 2015.

Brown AE, and Walker M. Genetics of insulin resistance and the metabolic syndrome. *Curr Cardiol Rep* 18;2016:75.

Butz K, Combest A, Steele S. Using precision medicine to improve patient care. 2013 December. Available at: https://www.ppdi.com/~/media/Files/PPDI%20Files/Expert%20Community/Whitepapers/Precision-Medicine-Improve-Patient-Care-Part1.ashx. Accessed November 11, 2017.

Cauchi S, Meyre D, Dina C et al. Transcription factor TCF7L2 genetic study in the French population: Expression in human beta-cells and adipose tissue and strong association with type 2 diabetes. *Diabetes* 55(10);2006:2903–8.

Chang TJ, Liu PH, Liang YC et al. Genetic predisposition and nongenetic risk factors of thiazolidinedione-related edema in patients with type 2 diabetes. *Pharmacogenet Genomics* 21(12);2011:829–36.

Collins FS, and Varmus H. A new initiative on precision medicine. *N Engl J Med* 372(9);2015:793–5.

Dujic T, Zhou K, Donnelly LA et al. Association of organic cation transporter 1 with intolerance to metformin in type 2 diabetes: A GoDARTS study. *Diabetes* 64;2015:1786–93.

Fradkin JE, Hanlon MC, and Rodgers GP. NIH Precision medicine initiative: Implications for diabetes research. *Diabetes Care* 39(7);2016:1080–4.

Franks PW, and Poveda A. Lifestyle and precision diabetes medicine: Will genomics help optimise the prediction, prevention and treatment of type 2 diabetes through lifestyle therapy? *Diabetologia* 60(5);2017:784–792.

Froguel P, Zouali H, Vionnet N et al. Familial hyperglycemia due to mutations in glucokinase: Definition of a subtype of diabetes mellitus. *N Engl J Med* 328(10);1993:697–702.

Gillespie KM. Type 1 diabetes: Pathogenesis and prevention. *CMAJ* 175(2);2006:165–70.

Guariguataa L. Global estimates of diabetes prevalence for 2013 and projections for 2035. *Diabetes Res Clin Pract* 103(2);2014:137–49.

Hani EH, Stoffers DA, Chèvre JC et al. Defective mutations in the insulin promoter factor-1 (IPF-1) gene in late-onset type 2 diabetes mellitus. *J Clin Invest* 104(9);1999:R41–8.

Hartley K. The genomic contribution to diabetes. Briefing note: Diabetes, genomics & public health. PHG Foundation. 2014. Available from http://www.phgfoundation.org/documents/353_1393402422.pdf.

Klonoff DC. Precision medicine for managing diabetes. *J Diabetes Sci Technol* 9(1);2015:3–7.

Komurcu-Bayrak E. Impact of genetic polymorphisms on insulin resistance. In: *Insulin Resistance*, Chap 4, Sarika Arora (Ed.), InTech, 2012. Available from http://cdn.intechopen.com/pdfs/41435/InTech-Impact_of_genetic_polymorphisms_on_insulin_resistance.pdf.

Lander ES, and Schork NJ. Genetic dissection of complex traits. *Science* 265(5181);1994:2037–48.

Maahs DM, West NA, Lawrence JM et al. Chapter 1: Epidemiology of type 1 diabetes. *Endocrinol Metab Clin North Am* 39(3);2010:481–97.

Magaña-Cerino JM, Luna-Arias JP, Labra-Barrios ML et al. Identification and functional analysis of c. 422_423InsT, a novel mutation of the HNF1A gene in a patient with diabetes. *Mol Genet Genomic Med* 5(1);2017:50–65.

Martin N, Boomsma D, and Machin G. A twin-pronged attack on complex traits. *Nat Genet* 17(4);1997:387–92.

Monroy VS, Díaz CA, Trenado LM et al. Thr130Ile polymorphism of HNF4A gene is associated with gestational diabetes mellitus in Mexican population. *J Investig Med* 62(3);2014:632–4.

Odom DT, Zizlsperger N, Gordon DB et al. Control of pancreas and liver gene expression by HNF transcription factors. *Science* 303(5662);2004:1378–81.

Pearson ER, Donnelly LA, Kimber C et al. Variation in TCF7L2 influences therapeutic response to sulfonylureas. *Diabetes* 56;2007:2178–82.

Pociot F, and McDermott MF. Genetics of type 1 diabetes mellitus. *Genes Immun* 3(5);2002:235–49.

Price P, Cheong KYM, Boodhoo A et al. Can MHC class II genes mediate resistance to type 1 diabetes? *Immunol Cell Biol* 79;2001:602–6.

Radha V, Vimaleswaran KS, Babu HN et al. Role of genetic polymorphism peroxisome proliferator-activated receptor-gamma2 Pro12Ala on ethnic susceptibility to diabetes in South-Asian and Caucasian subjects: Evidence for heterogeneity. *Diabetes Care* 29;2006:1046–51.

Ramos-Lopez E, Lange B, Kahles H et al. Insulin gene polymorphisms in type 1 diabetes, Addison's disease and the polyglandular autoimmune syndrome type II. *BMC Medical Genetics* 9;2008:65.

Sacks DB, and McDonald JM. The pathogenesis of type II diabetes mellitus: A polygenic disease. *Am J Clin Pathol* 105(2);1996:149–56.

Seshiah V, Balaji V, Balaji MS et al. Gestational diabetes mellitus in India. *J Assoc Physicians India* 52;2004:707–11.

Shaat N, Karlsson E, Lernmark Å et al. Common variants in MODY genes increase the risk of gestational diabetes mellitus. *Diabetologia* 49(7);2006:1545–51.

Shastry BS. SNP alleles in human disease and evolution. *J Hum Genet* 47(11);2002:561–6.

Shu Y, Sheardown SA, Brown C et al. Effect of genetic variation in the organic cation transporter 1 (OCT1) on metformin action. *J Clin Invest* 117;2007:1422–31.

Steck AK, Liu SY, McFann K et al. Association of the PTPN22/LYP gene with type 1 diabetes. *Pediatr Diabetes* 7;2006:274–8.

Stoffel M, Bell KL, Blackburn CL et al. Identification of glucokinase mutations in subjects with gestational diabetes mellitus. *Diabetes* 42;1993:937–40.

Tang Y, Axelsson AS, Spégel P et al. Genotype-based treatment of type 2 diabetes with an α2A-adrenergic receptor antagonist. *Sci Transl Med* 6;2014:257ra139.

Ueda H, Howson JMM, Esposito L et al. Association of the T-cell regulatory gene CTLA4 with susceptibility to autoimmune disease. *Nature* 423(6939);2003:506–11.

Vella A, Cooper JD, Lowe CE et al. Localization of a type 1 diabetes locus in the IL2RA/CD25 region by use of tag single-nucleotide polymorphisms. *Am J Hum Genet* 76(5);2005:773–9.

Wang G, Wang X, Zhang Q et al. Response to pioglitazone treatment is associated with the lipoprotein lipase S447X variant in subjects with type 2 diabetes mellitus. *Int Journal Clin Pract* 61;2007:552–7.

Weeks DE, and Lathrop GM. Polygenic diseases: Methods for mapping complex disease traits. *Trends Genet* 11;1995:513–9.

World Health Organization, Global report on diabetes, 2016. Available at http://apps.who.int/iris/bitstream/10665/204871/1/9789241565257_eng.pdf. Accessed October 4, 2017.

World Health Organization, Diabetes fact sheet, 2017. Available at http://www.who.int/mediacentre/factsheets/fs312/en/. Accessed October 4, 2017.

World Health Organization, India, 2016. World Health Day. Available at http://www.searo.who.int/india/mediacentre/events/2016/en/. Accessed November 13, 2017.

Zyriax B-C, Salazar R, Hoeppner W et al. The association of genetic markers for type 2 diabetes with prediabetic status—Cross-sectional data of a diabetes prevention Trial. *PLoS ONE* 8(9);2013:e75807.

15

Precision medicine in multiple sclerosis

SHOAIB AHMAD

BACKGROUND: DISEASE PREVALENCE, INCIDENCE, AND MORTALITY RATE (PREFERABLY WORLDWIDE)

Introduction

Multiple sclerosis is a disease affecting the central nervous system. This disease involves the scarring or hardening of patches on nervous tissue of the brain and spinal cord at multiple locations in the body (Multiple Sclerosis Trust [MS Trust], 2017). Multiple sclerosis (MS) is often described as a chronic autoimmune disease. It affects the central nervous system (Lugaresi et al., 2013), including the brain, spinal cord, and the optic nerves. Under normal circumstances, myelin protects the nerve fibers of the CNS involved in the conduction of electrical impulses. In MS, myelin covering is lost or damaged leaving behind the scar tissue (plaques or lesions). In the pathological condition, the impaired or disrupted nerve conduction then produces the characteristic symptoms of MS (Frankel and Jones, 2017).

MS has been classified as a highly heterogeneous disease (Ziemssen et al., 2016). MS is a chronic pathophysiological condition. The general perception of MS being a fatal, infectious, or contagious disease is false (MS Trust, 2017). MS has also been debated/discussed in books (Fletcher, 2000) and challenges in MS have been reviewed (Hohlfeld, 2010).

Incidence

More than 2.3 million people across the globe are affected by MS (Miller and Leary, 2007; Frankel and Jones, 2017). Nearly 400,000 people with MS reside in the United States alone (Markowitz, 2013). Another study has pointed out that MS affects approximately 300,000 U.S. citizens (Leist et al., 2014). Over 100,000 people in the United Kingdom are affected by MS (MS Trust, 2017).

Age considerations

MS is detected in most people when they are in their 20s or 30s (MS Trust, 2017).

Gender considerations

It is known fact that 75% of MS patients are females and 25% are males (MS Trust, 2017).

Ethnic considerations

MS commonly affects persons from ethnic groups (Frankel and Jones, 2017). These groups are listed in Table 15.1.

GENETICS

The forkhead/winged helix gene (FOXP3) is involved in MS (Gholami et al., 2017). Interferon regulatory factor 5 (IRF5) has been validated as the multiple sclerosis risk gene (Vandenbroeck et al., 2011). The SLC9A9 gene is implicated in MS (Esposito et al., 2015). Several genes (Table 15.2) have been associated with MS (Oksenberg and Hauser, 2008).

PATHOLOGICAL BASIS

Neuropathological basis of disease progression has been described in detail (Reynolds et al., 2011).

Table 15.1 Ethnic communities in the United States prone to the risk of MS

Ethnic community	Remarks
African-American	More severe form of MS has been recorded
Asian	—
Hispanic	—
Latino	—
Caucasian	Commonly originating from Northern Europe

Table 15.2 Genes doubted to be associated with MS

Doubtful genes			
APOE	CNTF	HLA-DRB1*1501	MEFV
CCR5	CRYAB	IFNg	OPN
CD24	ESR1	IL1B	PD-1
CD59	GSTM	IL4	TGFB1

Brain atrophy is involved in MS (Chard et al., 2004; Pagani et al., 2005; Tedeschi et al., 2005; Bieniek et al., 2006; Charil et al., 2007; Grassiot et al., 2009; De Stefano et al., 2010; Calabrese et al., 2007; Jones et al., 2013; Pérez-Miralles et al., 2013; Popescu et al., 2013).

Meningeal inflammation has been shown to play an important role in primary progressive multiple sclerosis (Choi et al., 2012). Meningeal B-cell follicles are associated with early onset of MS and its related progression (secondary progressive multiple sclerosis) as reported by Magliozzi et al. (2007).

Invasion of autoreactive T-cells and alterations of the blood-brain barrier (BBB) represent early pathological stages that are characterized by the following events (Smorodchenko et al., 2007):

1. Autoreactive T-cells invasions
2. Changes in blood-brain barrier

MS patients are affected by aberrant activation of helper T lymphocytes. It leads to the demyelination of neurons and manifests in the form of impaired CNS function as well as mobility. Neuronal scarring or sclerosis appears as plaques in MRI scans (Javan et al., 2017).

APPROACHES CURRENTLY USED FOR THE DISEASE DIAGNOSIS AND TREATMENT (MARKERS, GENOMICS, PROTEOMICS, EPIGENOMICS, IMAGING, OR ANY OTHER RELEVANT TECHNOLOGIES/ APPROACHES)

Diagnosis

There is no single test that can detect MS. Red blood cell distribution width is an easy parameter for diagnosis of MS (Hu, 2016). The correlation of the clinical conditions and the test reports can

Table 15.3 Diagnostic techniques for MS

Simple diagnosis	Specialist diagnosis
Medical history (focus on neurological aspects and familial history)	Lumbar puncture (or spinal tap)
Blood tests	Visual evoked potentials (VEP)
Thorough neurological exam	Somatosensory evoked potentials (SEPs)
Magnetic resonance imaging	

often help in early detection of MS usually with the involvement of specialist physicians (Frankel and Jones, 2017). The diagnostic techniques have been depicted in Table 15.3.

Magnetic resonance imaging (MRI) can be used for diagnosis of MS. MRI can be accomplished by two methods (Fisniku et al., 2008):

1. Gadolinium contrast enhancement
2. T2-hyperintense lesion load

As per the National Institute for Health and Care Excellence (NICE) guidelines, diagnosis of MS can be based on the following blood tests (National Institute for Health and Care Excellence, 2017):

1. Complete blood count (Hb, TLC, DLC, etc.)
2. HIV serology (ELISA, RIA)
3. Inflammatory markers (ESR, C-reactive protein)
4. Liver function test (SGOT, SGPT, alkaline phosphatase)
5. Renal function test
6. Serum calcium
7. Serum glucose
8. Thyroid function tests (T3, T4, TSH)
9. Vitamin B12

Optic coherent tomography (OCT) can be used for detecting MS (Villoslada, 2010). OCT can help in determining microcystic macular edema in the retina (Lang et al., 2015). Noninvasive OCT is used in detecting ocular diseases such as nystagmus (Antony et al., 2016). Volumetric assessment of retinal ganglion cell layer by OCT can be used for diagnosing MS (Davies et al., 2011).

BIOMARKERS

Katsavos and Anagnostouli (2013) have given an elaborate description of biomarkers for MS and have classified them into three groups:

1. Genetic/immunogenetic biomarkers
2. Laboratorial biomarkers in body fluids
3. Imaging: biomarkers

Development of biomarkers for MS has been discussed in detail (Bielekova and Martin, 2004). Some of the important biomarkers for MS are listed in Table 15.4.

BIOMARKERS FOR DIAGNOSIS

Established biomarkers and potential markers have been elaborately described by Derfuss (2012) as described in Tables 15.5 and 15.6.

Treatment options for MS

Therapeutic modalities for MS can be typically categorized into two general groups (Frankel and Jones, 2017; also see Table 15.7):

The first group—The disease-modifying medications approved by the Food and Drug Administration (FDA) are used to slow or hamper the disease progression.

Table 15.4 Biomarkers for MS

Biomarker	Reference
Neurofilament light protein	Malmeström et al. (2003)
Glial fibrillary acidic protein	Malmeström et al. (2003)
N-acetylaspartate	Trentini et al. (2014)
Neurofilaments	Trentini et al. (2014)
Circulating T cell	Pender et al. (2003)
Myelin basic protein peptides	Belogurov et al. (2008)
Circulating microRNAs	Vistbakka et al. (2016)
Sphingosine-1-phosphate receptors	Brana et al. (2014)

Table 15.5 Established biomarkers for diagnosis

Valid biomarkers	Biomarkers (without clear evidence)
1. Cerebral spinal fluid-specific oligoclonal bands 2. Intrathecal immunoglobulin production 3. Intrathecal antiviral immunoglobulin production 4. Magnetic resonance imaging 5. Aquaporin 4 antibodies	1. Neutralizing antibodies against beta-interferon 2. Neutralizing antibodies against natalizumab 3. Antibodies against JC virus

Table 15.6 Potential biomarkers

Potential biomarkers (valid)	Potential biomarkers (without clear evidence)
1. CD56 bright natural killer cells 2. Genetics	1. Cytokines/chemokines 2. Myelin oligodendrocyte glycoprotein antibodies 3. Intrathecal/oligoclonal immunoglobulin M production 4. Transcriptomics

Table 15.7 Drugs used in treatment of MS

Category	Drug
Corticosteroids	Methylprednisolone Prednisone
Muscle relaxants	Baclofen Tizanidine Dantrolene Onabotulinumtoxin A
Nerve conduction enhancers	Fampridine
Fatigue reducers	Amantadine Modafinil Armodafinil
Bladder medications	Tolterodine tartrate Oxybutynin Mirabegron Trospium chloride Fesoterodine Solifenacin succinate
Medications for paresthesias	Carbamazepine Amitriptyline Gabapentin Pregabalin Duloxetine hydrochloride
Mood stabilizing medications	Divalproex sodium
Medication for pseudobulbar affect	Combination of dextromethorphan and quinidine

Source: Based on Frankel D, and Jones H, Living with multiple sclerosis [brochure], available at nationalmssociety.org. Accessed on March 11, 2017.

The second group—Includes drugs and techniques that contribute toward improvement or alleviation of symptoms. In fact, symptom management is crucial for living well with MS.

Differences in ethnicity also have the potential to affect the pharmacokinetic profile of the drugs (Wu et al., 2012). Interferon regulatory factor 5 (IRF5) gene can affect IFNβ therapy in MS (Vandenbroeck et al., 2011). IRF5 variants affect clinical outcomes of IFNβ therapy (Vosslamber et al., 2011). Pharmacogenomic analysis of response to IFNβ therapy in MS has been reported (Byun et al., 2008). Anti-John Cunningham (anti-JC) virus antibodies are utilized in the safety monitoring of natalizumab therapy (Bloomgren et al., 2012). Several biomarkers have been identified for MS treatment (Derfuss, 2012) and are classified as follows:

Established biomarkers (with valid evidence)
1. Neutralizing antibodies against natalizumab
2. Antibodies against JC virus

Established biomarkers (with some experimental/clinical evidence)
1. Magnetic resonance imaging
2. Neutralizing antibodies against beta-interferon
3. Aquaporin 4 antibodies

Potential biomarkers (with some experimental/clinical evidence)
1. CD56 bright natural killer cells
2. Cytokines/chemokines
3. Transcriptomics
4. Genetics

WHY GENETIC, ENVIRONMENTAL, LIFESTYLE, AND OTHER FACTORS SHOULD BE CONSIDERED IN PRECISION MEDICINE AND HOW THEY IMPACT ON PREDICTING, PREVENTING, MANAGING, AND TREATING THE DISEASE

Studies have indicated that obesity, serum vitamin D levels, exposure to sunlight, and smoking may impact the disease progression (Ascherio and Munger, 2007). MRI is often used in the MS diagnosis. Patient's age and gender can affect MRI measures (Tintore et al., 2015).

Lifestyle modification strategy

The patients need to develop healthy living habits and take diets that do not cause constipation. Dietary changes, lifestyle modifications, and replenishment of fluid are proven methods to overcome constipation. One needs to undertake exercises for the proper stress management. A bent toward spiritual life may also help. The necessary psychological support (from all family members, friends, and peer group) can also play an important part in the betterment of living with MS (Frankel and Jones, 2017).

Major recommendations in lifestyle modification are as follows:

1. Adequate intake of fluids helps in prevention of bladder complications.
2. One needs to avoid unsafe exposure to harmful UV rays.
3. One is required to ensure proper and adequate amounts of vitamin D.
4. Smoking is to be avoided at all costs. Active as well as passive smoking is to be reduced or eliminated to the best possible extent.
5. Stretching exercises can help reduce spasticity or muscle stiffness. Aerobic exercise and energy-management strategies help in overcoming fatigue, which is one of the most important features of MS. Pelvic floor physical therapy (popularly known as Kegel exercises) can help in overcoming genitourinary problems.
6. Cognitive problems can be managed with rehabilitation/training. Cognitive rehabilitation aids in the matters related to improve memory,

attention, and so forth. Mood changes need talk therapy.

Physical therapies can be used for exercises, endurance, and stamina. Occupational therapies help people achieve maximum independence and optimal functioning. Speech therapy can help in overcoming problems in speech or chewing and swallowing.

WHY PERSONALIZED AND PRECISION MEDICINE ARE NEEDED AND HAVE POTENTIAL IN IMPROVING HEALTH

Precision medicine approach

Personalized treatment has been discussed in MS (Correale, 2011). Personalized treatment for MS is the need of the hour (Derfuss, 2012). Personalized treatment of MS is in focus (Giovannoni and Rhoades, 2012). A combination of clinical features and MRI parameters can help in finding the current state of disease and its progression or its response to medication (Derfuss, 2012). Precision medicine for MS has witnessed a paradigm shift (Dobrota et al., 2014). Shifting paradigms in MS have been debated (Golan et al., 2016).

General consideration with precision medicine

Multidisciplinary care interventions and rehabilitation of MS patients has assumed a greater significance (Skovgaard et al., 2012). Nintendo Wii (a physiotherapy modality) can be used for managing MS (Thomas et al., 2014). Personalized training games can be used in the rehabilitation of MS patients (Octavia and Coninx, 2014). Patient-reported outcomes are an integral part of precision medicine for MS (Lejbkowicz et al., 2012). Self-management and satisfaction play an important role in MS (Giovannoni and Rhoades, 2012). Multiple sclerosis-associated retroviruses specifically need personalized attention (Curtin et al., 2015).

During recent years, precision medicine has become the focus of medical advancement (Willis and Lord, 2015). It has expedited the development of novel therapeutic agents (D'Amico et al., 2016) as listed in Table 15.8.

Table 15.8 Therapeutic agents under investigation

Therapeutic agents	Potential benefit(s)
Adrenocorticotropic hormone	Anti-inflammatory effects
Biotin	Reducing disease progression
Amiloride	Neuroprotective effects
Fluoxetine	Suppression of microglia activation
Riluzole	Antiglutamatergic effects
LINGO-1	Inhibition of oligodendrocyte precursor cells differentiation
Domperidone	Increase in prolactin secretion
Erythropoietin	Neuroprotective abilities
Hematopoetic stem cell transplantation	Resetting of immune system
Ibudilast	Reduction in tumor necrosis factor levels
Idebenone	Antioxidant properties
Lipoic acid	Antioxidant properties
Lithium	Regulation of inflammatory processes
Masitinib	Degranulation of mast cells
MIS416	Modulation of t-cell-mediated autoimmune responses
NeuroVax	Vaccine strategies
Oxcarbazepine	Neuroprotective properties
Simvastatin	Immunomodulatory and neuroprotective effects
Siponimod (BAF312)	Antitumor, antioxidant, and anti-inflammatory benefits
Sunphenon epigallocatechin-3-gallate (EGCg)	Antioxidant, and anti-inflammatory effects

Source: D'Amico E et al., *Int J Mol Sci* 17;2016:1725, doi:10.3390/ijms17101725.

Antibodies also need to be explored for therapeutic potential (D'Amico et al., 2016). Presently two types of monoclonal antibodies are under investigation:

1. Monoclonal antibodies (for natalizumab and alemtuzumab)
2. Anti-CD20 monoclonal antibodies (for rituximab and ocrelizumab)

Challenges and research opportunities (future directions)

Three challenges have been identified for personalized medicine of MS (Gafson et al., 2017):

1. Precision diagnosis
2. Prediction of therapeutic response
3. Personalized monitoring of prediction

Personalized medicine for MS is an uphill and challenging task. Biomarkers of clinical, genetic,

imaging, or immunological nature need to be developed. Academic and industry cooperation is now needed to bring these markers into clinical practice (Derfuss, 2012).

Omics technologies in MS (Table 15.9) include the following techniques (Katsavos and Anagnostouli, 2013):

1. Genomics is used for whole DNA sequencing.
2. Transcriptomics is used for studying RNA sequences. It is commonly exemplified by microarrays and next-generation sequencing.
3. Proteomics is used for investigating distribution of specific proteins.
4. Lipidomics is concerned with recognition of cellular lipid pathways.
5. Metabolomics can be utilized for gaining insight into metabolic pathways implicated in MS.
6. Epigenomics is focused on epigenetic modifications that bring MS susceptibility.

Pharmacogenomics needs to be explored beyond routine genotypic profiling. It can be used

Table 15.9 Applications of omics in MS

Technology	Specific application in MS
Genomics	Recognition of vitamin D receptor element
Transcriptomics	TOB-1 gene downregulation
Proteomics	Ninjurin-1
Lipidomics	CNS lipid epitopes, IgM against myelin lipids
Metabolomics	N-acetyloaspartate in CSF
Epigenomics	Chromatin architecture in MS

to determine metabolism of drugs such as interferons (IFNs). It also implies that this approach can be used to find the drugs of choice when MS patients present with comorbidities (Hegen et al., 2015).

CONCLUSION

Multiple sclerosis is a heterogeneous disease of the central nervous system resulting from demyelination of the nervous tissue. MS is thought to result from a faulty FOXP3 gene. No single diagnostic test exists for MS. Physicians often need to rely on correlation between the clinical features and findings from magnetic resonance imaging. Corticosteroids, interferons, and other immunotherapies need to be supplemented with stem cell therapy/gene therapy. Omics can definitely play a role in the diagnosis and treatment of MS. Better biomarkers can help in early detection of the disease. MS will undergo a paradigm shift with the introduction of personalized or individualized medicine in MS.

ACKNOWLEDGMENTS

The author acknowledges the continuous encouragement and moral support from his teachers at Jamia Hamdard, New Delhi (Dr. Shibli Jameel and Dr. R. Zafar) and his family members (particularly Mrs. Zeba and Dr. Sultan Anjum).

REFERENCES

Antony BJ, Lang A, Swingle EK, Al-Louzi O, Carass A, Solomon S, Calabresi PA, Saidha S, and Prince JL. Simultaneous segmentation of retinal surfaces and microcystic macular edema in SDOCT volumes. *Proc SPIE Int Soc Opt Eng* 2016:9784, doi: 10.1117/12.2214676.

Ascherio A, and Munger KL. Environmental risk factors for multiple sclerosis. Part II: Noninfectious factors. *Ann Neurol* 61;2007:504–13.

Belogurov AA, Kurkova IN, Friboulet A et al. Recognition and degradation of myelin basic protein peptides by serum autoantibodies: Novel biomarker for multiple sclerosis. *J. Immunol* 180;2008:1258–67.

Bielekova B, and Martin R. Development of biomarkers in multiple sclerosis. *Brain* 127;2004:1463–78.

Bieniek M, Altmann DR, and Davies GR. Cord atrophy separates early primary progressive and relapsing remitting multiple sclerosis. *J Neurol Neurosurg Psychiatry* 77;2006:1036–9.

Bloomgren G, Richman S, Hotermans C et al. Risk of natalizumab-associated progressive multifocal leukoencephalopathy. *N Engl J Med* 366;2012:1870–80.

Brana C, Frossard MJ, Pescini Gobert R, Martinier N, Boschert U, and Seabrook TJ. Immunohistochemical detection of sphingosine-1-phosphate receptor 1 and 5 in human multiple sclerosis lesions. *Neuropathol Appl Neurobiol* 40;2014:564–78.

Byun E, Caillier SJ, Montalban X et al. Genomewide pharmacogenomic analysis of the response to interferon beta therapy in multiple sclerosis. *Arch Neurol* 65;2008:337–44.

Calabrese M, Atzori M, Bernardi V et al. Cortical atrophy is relevant in multiple sclerosis at clinical onset. *J Neurol* 254;2007:1212–20.

Chard DT, Griffin CM, and Rashid W. Progressive grey matter atrophy in clinically early relapsing-remitting multiple sclerosis. *Mult Scler* 10;2004:387–91.

Charil A, Dagher A, Lerch JP. Focal cortical atrophy in multiple sclerosis: Relation to lesion load and disability. *NeuroImage* 34;2007:509–17.

Choi SR, Howell OW, Carassiti D, Magliozzi R, Gveric D, Muraro PA, Nicholas R, Roncaroli F, and Reynolds R. Meningeal inflammation plays a role in the pathology of primary progressive multiple sclerosis. *Brain* 135;2012:2925–37.

Correale J. Personalized treatment in multiple sclerosis. *Curr Neurol Neurosci Rep* 11(6);2011:523–5.

Curtin F, Perron H, Faucard R, Porchet H, and Lang AB. Treatment against human endogenous retrovirus: A possible personalized medicine approach for multiple sclerosis. *Mol Diagn Ther* 19(5);2015:255–65.

D'Amico E, Patti F, Zanghì A, and Zappia M. A personalized approach in progressive multiple sclerosis: The current status of disease modifying therapies (DMTs) and future perspectives. *Int J Mol Sci* 17;2016:1725, doi: 10.3390/ijms17101725.

Davies EC, Galetta KM, Sackel DJ, Talman LS, Frohman EM, Calabresi PA, Galetta SL, and Balcer LJ. Retinal ganglion cell layer volumetric assessment by spectral-domain optical coherence tomography in multiple sclerosis: Application of a high-precision manual estimation technique. *J Neuroophthalmol* 31(3);2011:260–4.

Derfuss T. Personalized medicine in multiple sclerosis: Hope or reality? *BMC Medicine* 10;2012:116–20.

Dobrota R, Mihai C, and Distler O. Personalized medicine in systemic sclerosis: Facts and promises. *Curr Rheumatol Rep* 16(6);2014:425.

Esposito F, Sorosina M, Ottoboni L et al. A pharmacogenetic study implicates SLC9a9 in multiple sclerosis disease activity. *Ann Neurol* 78;2015:115–27.

Fisniku LK, Brex PA, Altmann DR et al. Disability and T2 MRI lesions: A 20-year follow-up of patients with relapse onset of multiple sclerosis. *Brain* 131;2008:808–17.

Fletcher P. Do NICE and CHI have no interest in safety? Opinion of the book NICE, CHI and the NHS reforms. Enabling excellence or imposing control? *Adverse Drug React Toxicol Rev* 19(3);2000:167–76.

Frankel D, and Jones H. Living with multiple sclerosis [brochure]. Available at nationalmssociety.org. Accessed on March 11, 2017.

Gafson A, Craner MJ, and Matthews PM. Personalised medicine for multiple sclerosis care. *Multiple Sclerosis J* 23(3);2017:362–9.

Gholami M, Darvish H, Ahmadi H, Rahimi-Aliabadi S, Emamalizadeh B, Amirabadi MRE, Jamshidi J, and Movafagh A. *Iran Red Crescent Med J* 19(1);2017:e34597.

Giovannoni G, and Rhoades RW. Individualizing treatment goals and interventions for people with MS. *Curr Opin Neurol* 25(Suppl);2012:S20–7.

Golan D, Staun-Ram E, and Miller A. Shifting paradigms in multiple sclerosis: From disease-specific, through population-specific toward patient-specific. *Curr Opin Neurol* 29(3);2016:354–61.

Grassiot B, Desgranges B, Eustache F, and Defer G. Quantification and clinical relevance of brain atrophy in multiple sclerosis: A review. *J Neurol* 256;2009:1397–412.

Hegen H, Auer M, and Deisenhammer F. Pharmacokinetic considerations in the treatment of multiple sclerosis with interferon-beta. *Expert Opin Drug Metab Toxicol* 11;2015:1803–19.

Hohlfeld R. "Gimme five": Future challenges in multiple sclerosis. ECTRIMS Lecture 2009. *Mult Scler* 16(1);2010:3–14.

Hu Z. Red blood cell distribution width: A promising index for estimating activity of autoimmune disease. *J Lab Precis Med* 1;2016:4–9.

Javan MR, Jalali Nezhad A, Safa A, Mohammadi MH, and Jamebozorgi K. Personalized medicine toward multiple sclerosis; current challenges and future prospects. *Int J Basic Sci Med* 2(1);2017:11–15.

Jones BC, Nair G, Shea CD, Crainiceanu CM, Cortese IC, and Reich DS. Quantification of multiple-sclerosis-related brain atrophy in two heterogeneous MRI datasets using mixed-effects modeling. *NeuroImage Clin* 3;2013:171–9.

Katsavos S, and Anagnostouli M. Biomarkers in multiple sclerosis: An up-to-date overview. *Multiple Sclerosis Int* 2013, article ID 340508. http://dx.doi.org/10.1155/2013/340508.

Lang A, Carass A, Swingle EK, Al-Louzi O, Bhargava P, Saidha S, Ying HS, Calabresi PA, and Prince JL. Automatic segmentation of microcystic macular edema in OCT. *Biomed Opt Express* 6(1);2015:155–69.

Leist T, Hunter SF, Kantor D, and Markowitz C. Novel therapeutics in multiple sclerosis management: Clinical applications. *Am J Med* 127(1);2014:S2.

Lejbkowicz I, Caspi O, and Miller A. Participatory medicine and patient empowerment towards personalized healthcare in multiple sclerosis. *Expert Rev Neurother* 12(3);2012:343–52.

Lugaresi A, di Ioia M, Travaglini D, Pietrolongo E, Pucci E, and Onofrj M. Risk-benefit considerations in the treatment of relapsing-remitting multiple sclerosis. *Neuropsychiatr Dis Treat* 9;2013:893–914.

Magliozzi R, Howell O, Vora A, Serafini B, Nicholas R, Puopolo M, Reynolds R, and Aloisi F. Meningeal B-cell follicles in secondary progressive multiple sclerosis associate with early onset of disease and severe cortical pathology. *Brain* 130;2007:1089–104.

Malmeström C, Haghighi S, Rosengren L, Andersen O, and Lycke J. Neurofilament light protein and glial fibrillary acidic protein as biological markers in MS. *Neurology* 61;2003:1720–5.

Markowitz CE. Multiple sclerosis update. *Am J Manag Care* 19(16 Suppl);2013:s294–300.

Miller DH, and Leary SM. Primary-progressive multiple sclerosis. *Lancet Neurol* 6;2007:903–12.

Multiple Sclerosis Trust (MS Trust). 2017. What is multiple sclerosis? Poster, available at mstrust. org.uk. Accessed on March 11, 2017.

National Institute for Health and Care Excellence (NICE). 2017. Diagnosing multiple sclerosis. Available at https://pathways.nice.org.uk/pathways/multiple-sclerosis. Accessed on March 11, 2017.

Octavia JR, and Coninx K. Adaptive personalized training games for individual and collaborative rehabilitation of people with multiple sclerosis. *Biomed Res Int* 2014;2014:345728.

Oksenberg J, and Hauser SL. Genetics of multiple sclerosis. In: *Multiple Sclerosis*. Raine CS, McFarland H, Hohlfeld R (Eds). Saunders Elsevier, London; 2008, pp. 214–25.

Pagani E, Rocca MA, and Gallo A. Regional brain atrophy evolves differently in patients with multiple sclerosis according to clinical phenotype. *AJNR* 26;2005:341–6.

Pender MP, Csurhes PA, Wolfe NP, Hooper KD, Good MF, McCombe PA, and Greer JM. Increased circulating T cell reactivity to GM3 and GQ1b gangliosides in primary progressive multiple sclerosis. *J Clin Neurosci* 10;2003:63–6.

Popescu V, Agosta F, Hulst HE et al. Brain atrophy and lesion load predict long term disability in multiple sclerosis. *J Neurol Neurosurg Psychiatry* 84;2013:1082–91.

Pérez-Miralles F, Sastre-Garriga J, Tintoré M et al. Clinical impact of early brain atrophy in clinically isolated syndromes. *Mult Scler* 19;2013:1878–86.

Reynolds R, Roncaroli F, Nicholas R, Radotra B, Gveric D, and Howell O. The neuropathological basis of clinical progression in multiple sclerosis. *Acta Neuropathol* 122;2011:155–70.

Skovgaard L, Djerre L, Hadir N et al. An investigation of multidisciplinary complex health care interventions—steps towards an integrative treatment model in the rehabilitation of people with multiple sclerosis. *BMC Complement Altern Med* 12;2012:50.

Smorodchenko A,Wuerfel J, Pohl EE, Vogt J, Tysiak E, Glumm R, Hendrix S, Nitsch R, Zipp F, and Infante-Duarte C. CNS-irrelevant T-cells enter the brain, cause blood-brain barrier disruption but no glial pathology. *Eur J Neurosci* 2007;26(6):1387–98.

De Stefano N, Giorgio A, and Battaglini M. Assessing brain atrophy rates in a large population of untreated multiple sclerosis subtypes. *Neurology* 74;2010:1868–76.

Tedeschi G, Lavorgna L, and Russo P. Brain atrophy and lesion load in a large population of patients with multiple sclerosis. *Neurology* 65;2005:280–5.

Thomas S, Fazakarley L, Thomas PW, Brenton S, Collyer S, Perring S, Scott R, Galvin K, and Hillier C. Testing the feasibility and acceptability of using the Nintendo Wii in the home to increase activity levels, vitality and well-being in people with multiple sclerosis (Mii-vitaliSe): Protocol for a pilot randomised controlled study. *BMJ Open* 4(5);2014:e005172.

Tintore M, Rovira A, Rio J et al. Defining high, medium and low impact prognostic factors for developing multiple sclerosis. *Brain* 138;2015:1863–74.

Trentini A, Comabella M, Tintoré M et al. N-acetylaspartate and neurofilaments as biomarkers of axonal damage in patients with progressive forms of multiple sclerosis. *J Neurol* 261;2014:2338–43.

Vandenbroeck K, Alloza I, Swaminathan B et al. Validation of IRF5 as multiple sclerosis risk gene: Putative role in interferon beta therapy and human herpes virus-6 infection. *Genes Immun* 12;2011:40–45.

Villoslada P. Biomarkers for multiple sclerosis. *Drug News Perspect* 23(9);2010:585–95.

Vistbakka J, Elovaara I, Lehtimäki T, and Hagman S. Circulating microRNAs as biomarkers in progressive multiple sclerosis. *Mult Scler J* 23(3);2016:403–12.

Vosslamber S, van der Voort LF, van den Elskamp IJ et al. Interferon regulatory factor 5 gene variants and pharmacological and clinical outcome of Interferonbeta therapy in multiple sclerosis. *Genes Immun* 12;2011:466–72.

Willis JC, and Lord GM. Immune biomarkers: The promises and pitfalls of personalized medicine. *Nat Rev Immunol* 15;2015:323–29.

Wu K, Mercier F, David OJ et al. Population pharmacokinetics of fingolimod phosphate in healthy participants. *J Clin Pharmacol* 52;2012:1054–68.

Ziemssen T, Kern R, and Thomas K. Multiple sclerosis: Clinical profiling and data collection as prerequisite for personalized medicine approach. *BMC Neurol* 16;2016:124.

16

Precision medicine in asthma and chronic obstructive pulmonary disease

SHOAIB AHMAD

INTRODUCTION

Asthma and chronic obstructive pulmonary disease (COPD) are respiratory diseases. They are more common in developed countries (Agusti et al., 2016a, 2016b).

These two diseases are growing at an alarming level. Though the number of hospitalizations and mortalities are on a downward trend, the loss of human life and/or functionality and economic burden compels us to look for better vistas for the clinical management of asthma and COPD.

Clinical and immunological biomarkers can be used for prognosis, diagnosis, and prediction of therapeutic response for asthma as well as COPD. Biomarkers can also be used for drug targeting and identifying the patients who can respond to novel therapeutics (Heaney and McGarvey, 2017).

Asthma and COPD may share biological mechanisms. There are similarities in clinical, functional, and diagnostic features. A precision medicine strategy for asthma and COPD has been proposed (Agusti et al., 2016a, 2016b).

PRECISION MEDICINE IN ASTHMA

Introduction

Asthma is a common disease characterized by chronic inflammation of the airway. It mainly affects the respiratory system and a number of symptoms are produced including wheezing, breath shortness, chest tightness, and coughing. These symptoms vary over time. The intensity of the symptoms also undergoes a change as the expiratory airflow changes (Global Initiative for Asthma [GINA], 2015).

EPIDEMIOLOGY

Epidemiology of adult asthma is a challenge (Toren, 1999). Asthma has been reported to affect 1%–18% of the people in different countries (GINA, 2015). Asthma affects 300 million people in the world, and another 100 million people may be added by 2025 (American Thoracic Society [ATS], 2015). Annually, 250,000 deaths across the globe are attributed to asthma (Bateman et al., 2008). Asthma affects 8% (18.7 million) of the adult population in the United States (Blackwell et al., 2014). As per the estimates of 2014, the United States has more than 25 million people affected with asthma. Out of this, 7 million are children (National Heart, Lung, and Blood Institute [NHLBI], 2014). Asthma costs $56 billion on an annual basis. This burden includes medical costs, lost schooldays and workdays, and early deaths (ATS, 2015). The total number of people in Europe affected with severe asthma in 2015 was estimated to be 1.5 million (Chung et al., 2014). The economic burden of asthma in Europe was predicted at €19.3 billion in 2015 (Demoly et al., 2009; Aubier et al., 2015). Asthma has become a challenge for the Western world, particularly the United States and Europe (Table 16.1).

Table 16.1 Prevalence of asthma in United States and Europe

Country	Affected population
United States	25 million
United Kingdom	4.67 million
Germany	3.27 million
France	2.28 million
Italy	2.26 million
Spain	1.58 million

Source: National Heart, Lung, and Blood Institute (NHLBI), August 4, 2014, http://www.nhlbi.nih.gov/health/health-topics/topics/asthma/; Aubier M et al., The Uncovering Asthma Report, Marlborough, MA: Boston Scientific, 2015.

Efforts have been made to locate sex-specific evidence in asthma (Moerman et al., 2009). Asthma in the elderly population has been discussed in detail (Gillman and Douglass, 2012). The Behavioral Risk Factor Surveillance System (BRFSS) can be used for estimating the prevalence of asthma amongst specific races (Goodman, 2010). Strategies have been devised for multiple imputation in longitudinal epidemiological studies (Spratt et al., 2010). Incidence of asthma and COPD is increasing in societies with Western lifestyles (Braun and Tschernig, 2006).

MOLECULAR PATHOGENESIS

Bronchoconstriction is a characteristic symptom of COPD and asthma (Seehase et al., 2011). Erythrocyte glutathione peroxidase activity levels are elevated in the case of asthma (Tho and Candlish, 1987). IL-16 is a proinflammatory cytokine involved in the development of asthma. It is responsible for recruitment of CD4+ T cells to sites of asthma (Nicoll et al., 1999). IL-13 is a Th2 cytokine acting as an important mediator of airway inflammation and leading to asthma lesions. IL-13 can be estimated using the Erenna immunoassay system (St Ledger et al., 2009). TNFalpha and IL-13 are involved in activation of oxytocin receptors. This activity in airway smooth muscle cells is implicated in asthma development (Amrani et al., 2010). 20-HETE has been reported to play a pivotal role in mediation of acute ozone-induced airway hyperresponsiveness (AHR; Cooper et al., 2010). Phosphoinositide 3-kinase gamma (PI3Kgamma) is involved in the pathogenesis of asthma. PI3Kgamma modulates calcium oscillations and hence regulates contraction of the airway smooth muscles. PI3Kgamma inhibitors can help overcome AHR (Jiang et al., 2010). Prostaglandin D(2) is a noninvasive biomarker of asthma and it has been found to cause contractions in peripheral lung tissue (Larsson et al., 2011). Trichostatin A (TSA) is an inhibitor of histone deacetylase (HDAC), which produces an inhibitory effect on airway hyperresponsiveness (Banerjee et al., 2012).

APPROACHES CURRENTLY USED FOR ASTHMA DIAGNOSIS AND TREATMENT

Guidelines on grading quality of evidence and subsequent recommendations in clinical practice have been reported (Brozek et al., 2009). The NHLBI has published Expert Panel Report 3 titled "Guidelines for the Diagnosis and Management of Asthma" (Hahn, 2009). Concerns have been raised on over-reliance on instrumental approach for disease identification (Baser, 2009). A scheme for diagnosis (Table 16.2) has been suggested (GINA, 2015).

Generalization of longitudinal annual FEV1 decline (LLD) has been suggested. LLD is actually based on spirometry precision (Hnizdo et al., 2007). The American Thoracic Society (ATS) curves are considered sufficient for testing spirometer efficacy (Lefebvre et al., 2014). The peak flow meters are handy diagnostic tools with some variability in comparison to the Fleisch pneumotachograph. It clearly indicated that value scales of such flow meters required adjustments (van Schayck et al., 1990).

CLASSIFICATION OF ASTHMA PHENOTYPES

Recognizable patterns of demographic, clinical, and pathological characteristics found in asthma patients are often termed as "asthma phenotypes" (Table 16.3) (GINA, 2015).

Table 16.2 Scheme for diagnosis of asthma

Diagnosis steps
History and family history
Physical examination
Radiological examination
Lung function testing
Bronchial provocation tests
Allergy tests
Exhaled nitric oxide
Differential diagnosis

Table 16.3 Well-recognized asthma phenotypes

Phenotype	Linkage
Allergic asthma	Allergies
Nonallergic asthma	No allergies
Late-onset asthma	Adult life
Asthma with fixed airflow limitation	Long-standing asthma
Asthma with obesity	Obesity

Source: Global Initiative for Asthma (GINA), Pocket guide for asthma management and prevention, 2015, www.ginasthma.org.

NEW APPROACHES IN ASTHMA PHENOTYPING

Phenotyping has been a much debated topic in asthma research. Several clinical phenotypes of asthma were reported (Bel, 2004). The need to shift to endotyping instead of phenotyping has been suggested (Agache, 2013). Severe asthma phenotypes have been identified (Moore et al., 2010). A cluster analysis approach has been used for asthma phenotyping (Haldar et al., 2008). Wheezing phenotypes have been associated with childhood lung function and airway responsiveness (Henderson et al., 2008). Latent class analysis has been used for distinction between phenotypes of childhood wheezing and coughing (Spycher et al., 2008). Wheezing phenotypes in young preschool children have been reported (Just et al., 2013).

Asthma treatment

Asthma is always in discussion for internal medicine specialists (John et al., 2013). Atopic march also affects asthma (Ker and Hartert, 2009).

ASTHMA MEDICATION

Asthma cannot be completely cured. It can very well be controlled using medication and lifestyle modification. Currently, there are two types of treatments available for asthma. The first treatment is concerned with regular medication and it is termed maintenance treatment. The second one is concerned with emergency management and termed as rescue treatment (ATS, 2015). Most asthmatic patients need medicine that can effectively reduce airway inflammation and also be helpful in preventing the onset of severe asthma. Most of these medications are, however, ineffective during an asthma attack (ATS, 2015).

The maintenance medicines include the following treatments:

1. Inhaled corticosteroids (most common and nonhabit-forming substances)
2. Leukotriene modifiers
3. Long-acting beta agonists
4. Long-acting anticholinergics
5. Theophylline
6. Shots

The rescue medicines include the following treatments:

1. Short-acting beta2 agonists (help in opening the airway by respiratory muscle relaxant action)
2. Ipratropium

Eosinophils can be used as biomarkers of therapeutic response (Bagnasco et al., 2016).

BIOLOGICAL AGENTS

Biologics are expensive therapeutic options. Presently, 21 biologics are under development for use by patients with respiratory diseases including asthma. Only one biologic agent—omalizumab—is approved by the U.S. Food and Drug Administration (FDA) for asthma. Omalizumab is indicated for severe, persistent allergic asthma, which is ordinarily not controlled by inhalational corticosteroids. Another biologic, mepolizumab, is under consideration for FDA approval for use in patients with severe eosinophilic asthma. Another third biologic, lebrikizumab, is in clinical trials for reducing airway inflammation (ATS, 2015).

NOVEL PHARMACOTHERAPEUTIC APPROACHES

New pharmacological targets are being identified for asthma and COPD (Braun and Tschernig, 2006). Several animal models have been proposed for the study of asthma (Mullane and Williams, 2014). Steroids have been found to cause complete reversal of albuterol-induced beta(2)-adrenergic receptor tolerance in human subjects (Cooper and Panettieri, 2008).

Novel pharmacotherapeutic approaches include:

a. Surfactant proteins—Surfactant protein D (SP-D) has been found to inhibit early airway response in *Aspergillus fumigatus*-treated mice. SP-D also reduces bronchial hyperresponsiveness, eosinophilia, and T-helper type 2 cytokines in allergic asthma (Erpenbeck et al., 2006).
b. PI3Kgamma inhibitors—Treatment with a PI3Kgamma inhibitor can reduce AHR (Jiang et al., 2012).
c. Histamine and serotonin antagonists—A new compound (C-027) has been found to inhibit IgE-mediated bronchoconstriction. C-027 has been found to have action as a histamine and serotonin antagonist in airways (Cooper et al., 2011).
d. Potassium channel activators—KCNQ (Kv7) potassium channel activators are nowadays debated as bronchodilators (Brueggemann et al., 2014).
e. Precision-guided immunotherapy—Cytotoxic T lymphocyte antigen 4 (CTLA4) Ig-modified dendritic cells have beneficial effects on asthma. CCR7 helps in dendritic cells homing. Both molecules may be combined together for precision-guided immunotherapy (Chen et al., 2014).
f. Phosphodiesterase-4 inhibitors—Roflumilast is an upcoming drug for curing asthma (Ilango et al., 2013).

Nowadays, anti-IL-13 strategies make use of anrukinzumab, lebrikizumab, and tralokinumab, whereas anti-IL-4 strategies include pascolizumab, pitakinra, and dupilumab (Bagnasco et al., 2016).

BRONCHIAL THERMOPLASTY

Bronchial thermoplasty (BT) is only intended for severe asthma adult patients within a research environment (ATS, 2015). BT has proved to be of use in reduction in airway smooth muscle (ASM). BT uses radiofrequency energy to reduce the mass of ASM. The BT modality relies on reducing the mass of ASM by thermal injury and helps in reducing the severity of asthma. The Alair™ Bronchial Thermoplasty System approved by the FDA in 2010 for conditional use in asthma has been used for BT (BlueCross BlueShield Association, 2014).

Genetics of asthma

Asthma is linked to markers on chromosome 12 (Wilkinson et al., 1998). Fetal growth and the development of childhood asthma have a clear-cut relationship (Leadbitter et al., 1999). Several genes and loci have been identified for asthma susceptibility (Guilleminault et al., 2017).

Genes for asthma susceptibility are as follows:

1. ORMDL3
2. GSDMB

Identified susceptibility loci are as follows:

1. HLA-DRA on 6p21
2. HLA-DQB1 on 6p21
3. IL13 on 5q31
4. IL1RL1/IL18RL1 on 2q12
5. IL33 on chromosome 9p24
6. WDR36/TSLP on 5q22

Efforts have been made to utilize genomics in research on health disparities (Fullerton et al., 2012). Asthma endotyping is now in discussion (Agache, 2013). Ancestry may also have been a factor in the development of asthma (Ortega and Meyers, 2014). Epigenetics is now considered to explore a link between prenatal/early-life exposure and susceptibility for asthma (de Planell-Saguer et al., 2014). Six new loci have been found to be associated with forced vital capacity. These loci are listed in Table 16.4 (Loth et al., 2014).

Why personalized and precision medicine are needed and have potential in improving health

BIOMARKERS

Zedan et al. (2016) have established criteria for asthma biomarkers:

1. Ability to identify poor symptoms perceivers
2. Ability to predict disease activity and outcomes

3. Ability to identify nonresponders asthma phenotypes
4. Ability to characterize clinical asthma phenotypes

Multiple markers (Table 16.5) for type 2 inflammation in asthma have been identified (Fricker et al., 2017).

Blood eosinophils as biomarkers have acceptable accuracy (Bayes and Cowan, 2016). Eosinophil counts and serum IgE are considered most reliable biomarkers. Lung function and nasal cytology can be used as ancillary biomarkers (Ciprandi et al., 2017). Blood eosinophils are good biomarkers for T-helper type 2 cell-high asthma therapies. It can also be used to identify greater resistance to steroids (Stokes and Casale, 2016). Blood eosinophils and periostin are important biomarkers in clinical trials of type-2 cytokine inhibitors (Peters et al., 2016). Periostin has potential to be used as a biomarker for T-helper-2 (Th2)-driven asthma (Jeanblanc et al., 2017). Nineteen endogenous urinary biomarkers have been identified for possible use in asthma/COPD (Khamis et al., 2017). Genotyping has led to establishing rs572527200 as a SNP marker for asthma (Ponomarenko et al., 2015). Allergic sensitization and blood eosinophil counts may find use in managing early asthma in young children (Anderson et al., 2017). It is very unfortunate that biomarkers are not utilized in clinical practice. They are also not included in current guidelines on asthma (Agache and Rogozea, 2017).

Asthma pharmacogenetics and individual genetic profiling have been suggested for personalized medicine. It can help in creating tailor-made treatment options for individualized patients. This approach will also ensure that better therapeutic outcomes are achieved and the heath burden is relieved or minimized (Ortega et al., 2015).

Table 16.4 Six new loci associated with forced vital capacity

Position of locus (in or near)
EFEMP1
BMP6
MIR129-2-HSD17B12
PRDM11
WWOX
KCNJ2

Source: Loth DW et al., *Nat Genet* 46(7);2014:669–77.

Table 16.5 Markers for type 2 inflammation in asthma

Markers
Eosinophils in sputum
Eosinophils in blood
Exhaled NO
Serum IgE
Periostin

There is an important and immediate need for identifying medicinally important pharmacogenomics loci for asthma treatment. We are far away from making this reality directly turn relevant and work for a majority of patients. An effective combination of network modeling, functional validation, and integrative omics can make asthma pharmacogenomics affordable and clinically relevant (Park et al., 2015). CCR7 helps in CTLA4 Ig-modified dendritic cells homing. CTLA4 Ig and CCR7 combined together have potential for precision-guided immunotherapy (Chen et al., 2014).

Challenges and research opportunities

As per Gauthier et al. (2015) findings, areas of further research in respiratory precision medicine include:

1. Identification of additional clinical and molecular phenotypes
2. Identification of additional clinical and molecular phenotypes
3. Validation of predictive biomarkers
4. Identification of new areas for possible therapeutic interventions

It is hoped that epigenetics will help in regulating genes responsible for inflammation in aspirin-exacerbated asthma. DNA methylation and histone protein modification methods can be used for this purpose. Developing noninvasive clinical biomarkers for this type of asthma is a challenge (Dahlin and Weiss, 2016).

Several markers (such as eosinophils, IgE, FENO, and periostin) have potential to predict anti-IL-13 and anti-IL-4 therapeutic strategies (Bagnasco et al., 2016). Molecular endotyping using airway gene expression profiling can help in understanding disease mechanisms. It can also aid in determination of potential targets for novel treatments (Wesolowska-Andersen and Seibold, 2015).

Transcriptomics have a potential for endotyping. A combination of FENO/urinary bromotyrosine can be used in predicting response to the steroids (Bayes and Cowan, 2016). Another challenging task is personalized medicine and a biomarker-based approach children suffering from severe asthma (Bozzetto et al., 2015). Phenotype/

endotype based and mechanism-specific therapy is needed for severe persistent asthma (Lang, 2015).

PRECISION MEDICINE FOR CHRONIC OBSTRUCTIVE PULMONARY DISEASE (COPD)

Introduction

Chronic obstructive pulmonary disease (COPD) is a progressive respiratory disease. COPD results in worsening obstruction to airflow in the lungs over a considerable period of time. COPD also includes chronic bronchitis and emphysema. In case of chronic bronchitis, airway linings face constant irritation and inflammation. As a result, the lining starts getting thickened. This thickening is followed by development of thick mucus in the airways. Both of these pathophysiological developments lead to difficulty in breathing. In cases of emphysema, the walls between the air sacs of the lungs are damaged, and, consequently, surface area available for the gaseous exchange is reduced. Clinically, it has been observed that most COPD patients suffer from both chronic bronchitis as well as emphysema (NHLBI, 2015a, 2015c; Sunovion Pharmaceuticals Inc., 2015). COPD has been identified as the disease requiring high precision in family practice for successful management (van Doorn-Klomberg et al., 2013).

OCCURRENCE

According to the estimates of World Health Organization (WHO), 64 million people across the world suffer from COPD (Woolcock Institute of Medical Research, 2015). Fore example, around 1 in every 13 Australians over the age of 40 years has COPD (Woolcock Institute of Medical Research, 2015) and India has been warned of COPD emergence (Jindal, 2006; Ladhani, 2016). COPD kills more than 3 million people annually (Salvi and Agrawal, 2012; Suthar et al., 2015). COPD by 2020 is expected to be the third leading cause of death (Jindal et al., 2006).

COPD has affected the American population to a significant extent, where 15.7 million adults have been diagnosed with COPD (Wheaton et al., 2015), and an estimated 12 million adults remain undiagnosed (NHLBI, 2015c). In the United States, COPD causes over 120,000 deaths per year and is the third leading cause of death (Table 16.6; NHLBI, 2015c).

Table 16.6 COPD in United States

Parameter	Result
Affected world population	67 million
Affected U.S. population	15.7 million
Undiagnosed U.S. population	12 million
Mortality in United States	0.12 million
Economic burden in United States	$29.5 billion

The U.S. Centers for Disease Control and Prevention (CDC) has issued Public Health Strategies for preventing COPD (CDC, 2011).

Approaches currently used for the disease diagnosis and treatment

PATHOGENESIS OF COPD

Pathogenesis of COPD includes the following theories (Vijayan, 2013):

1. Proteinase–antiproteinase hypothesis
2. Immunological mechanisms
3. Oxidant–antioxidant balance
4. Systemic inflammation
5. Apoptosis and ineffective repair

Natural innate cytokine responds to immune-modulators (Switalla et al., 2010). Lipopolysaccharides (LPS) elicit a range of immunological effects and are involved in COPD. Single nucleotide polymorphisms have been linked with susceptibility to COPD. These polymorphisms take place in the TSPYL-4 and NT5DC1 genes (Guo et al., 2012). Desmosine (DES) and isodesmosine have the potential to be used as putative biomarkers for COPD (Viglio et al., 2014). Afamin, a member of the albumin gene family, cannot be used for determining COPD progression (Dieplinger et al., 2013). Six new loci are associated with forced vital capacity. These loci have been identified as EFEMP1, BMP6, MIR129-2-HSD17B12, PRDM11, WWOX, and KCNJ2. These loci were found in African-American, Korean, Chinese, and Hispanic subjects (Loth et al., 2014).

DIAGNOSIS

There is no single specific or confirmatory diagnostic test for COPD. To be honest, the accurate diagnosis depends on the clinical judgment of the physicians. This judgment relies on a combination of history, physical examination, and confirmed airflow obstruction using postbronchodilator spirometry. Patients over the age of 35 presenting with complaints of exertion-related breathlessness, chronic cough, sputum production, frequent winter "bronchitis," or wheezing need to be properly diagnosed for COPD (Derbyshire Joint Area Prescribing Committee, 2015). Diagnostic testing may require a chest x-ray or CT (CAT scan), blood tests, or sputum sample testing. In some cases, cardiac ultrasound and overnight sleep studies may also be needed for accurate diagnosis (Woolcock Institute of Medical Research, 2015).

The diagnosis of COPD is primarily based on a physical examination by a trained general practitioner or a specialist with appropriate training. Many times, the patient's individual medical and family history is also helpful. Lung function testing constitutes the main diagnostic criteria for COPD. Spirometer conducted by a physician is often the primary diagnostic criteria. Spirometry reliably measures the total amount of exhaled air. Spirometry can also be used in detecting COPD in early stages much before the onset of characteristic symptoms (NHLBI, 2015b).

CLASSIFICATION OF DIAGNOSTIC TECHNIQUES

Diagnostic techniques used for COPD are summarized in Table 16.7.

COPD PHENOTYPES

Phenotyping is also relevant in COPD research. COPD phenotypes have been identified (Marsh et al., 2008). The need for treating COPD by clinical phenotypes has been highlighted (Miravitlles et al., 2013). Asthma has been identified as a risk factor for COPD (Silva et al., 2004). Clinical phenotypes of COPD and asthma have been discussed (Carolan and Sutherland, 2013). Table 16.8 shows the COPD phenotypes (Turner et al., 2015).

Therapeutic options for the treatment of COPD

Therapeutic options are classified into the following broad categories (Delmer, 2015):

1. Pulmonary rehabilitation
2. Vaccinations
3. Oxygen therapy

Table 16.7 Diagnostic techniques for COPD

Type	Example of diagnostic technique
Radiological techniques	Chest x-ray, nuclear medicine, real-time sonography, color Doppler ultrasonography (USG), CT and MRI
Traditional pulmonary function tests	Cardiogenic oscillation, negative expiratory pressure, exhaled temperature and exhaled breath condensate, fractional exhaled nitric oxide
Newer Pulmonary function tests	Pulmonary arterial pressure, inductive plethysmography, oximetry-cutaneous capnography, pulse oximetry, computed tomography
Sophisticated assays	B-type natriuretic peptide assay, presage st2 assay
Invasive techniques	RBM biopsy, inflammatory cell counts

4. Pharmacotherapy
5. Exacerbation management
6. Exacerbation pharmacotherapy
7. Monitoring and follow-up
8. Overcoming the barriers associated with COPD treatment
9. Educational strategies

The details of these modalities or approaches are discussed next.

PULMONARY REHABILITATION

Studies have indicated the fruitfulness of nicotine replacement therapy and use of drugs such as varenicline, bupropion, and nortriptyline (Delmer, 2015). Pulmonary rehabilitation for COPD is an accepted practice (Stein et al., 2009).

VACCINATIONS

Vaccinations such as influenza and pneumococcal polysaccharide are also useful in treatment of COPD. The influenza vaccine contains killed or attenuated viruses. It is administered once a year. Pneumococcal polysaccharide vaccines are suitable for adults aged 65 years or more. It can also be given to younger adults having comorbid conditions (Delmer, 2015).

OXYGEN THERAPY

Oxygen therapy involves long-term administration of oxygen. If administered for usually more than 15 hours a day, it can increase the survival rate in patients who have faced or are expected to have chronic respiratory failure or severe hypoxemia associated with resting stage. This therapy also keeps the progressive hypoxemia under check (Delmer, 2015).

PHARMACOTHERAPY

The pharmacotherapy is aimed at achieving one or more of the following targets (Delmer, 2015):

1. Reduction of symptoms
2. Frequency of exacerbations
3. Intensity of exacerbations
4. Prevent disease progression
5. Increase exercise tolerance
6. Improvement of general health
7. Reduction in mortality rates

Common drugs used in the treatment of COPD are summarized in Table 16.9 (Adil et al., 2015).

EXACERBATION MANAGEMENT

Exacerbation management is aimed at minimizing the impact as well occurrence of the exacerbations. Bronchodilators, systemic corticosteroids, and antibiotics may also be of great help. Adjunct therapy and adequate respiratory support may also be desired in certain cases (Delmer, 2015).

EXACERBATION PHARMACOTHERAPY

Short-acting β-agonist alone or in combination with a short-acting anticholinergic agent can

Table 16.8 Phenotypes of COPD

COPD phenotype
High IgE phenotype
Reversibility to β2-agonists phenotype
Type 1 respiratory failure phenotype
Type 2 respiratory failure phenotype
Upper zone emphysema phenotype
α1-antitrypsin deficiency phenotype

Source: Turner AM et al., Eur Respir Rev 24;2015:283–98.

Table 16.9 Common drugs used in the treatment of COPD

Category	Drug
β₂-agonists	Salbutamol
	Salmetrol
Antibiotics	Penicillin
	Cephalosporin
	Fluoroquinolone
	Nitro-imidazole
	Macrolide
	Aminoglycoside
Inhaled corticosteroids	Budesonide
Anticholinergics	Ipratropium bromide
	Scopolamine
Xanthine derivatives	Theophylline/etophylline
Mucolytics	Ambroxol
	Acetyl cysteine
	Bromhexine/terbutaline/ guaifenesin
Oral corticosteroids	Hydrocortisone
	Methyl prednisolone
Antihistamines	Levocetirizine
	Chlorpheniramine
	Loratidine
Leukotriene antagonist	Montelukast

Source: Adil MS et al., *J Clin Diagn Res* 9(11);2015: FC05–8.

be used for exacerbation management in COPD patients. For example, 40 mg of prednisone per day for 5 days can be given by oral route. In case of bacterial infections in COPD patients confirmed by sputum cultures, aminopenicillin with or without clavulanic acid, macrolide, or tetracycline treatment for 5 to 10 days can be used (Delmer, 2015).

MONITORING AND FOLLOW-UP

Once improvement results from pharmacologic or nonpharmacological interventions, spirometry can be done on an annual basis. Smoking habits and cessation needs constant monitoring. Indirect exposure to smoking also needs to be eliminated or minimized to the lowest level possible. Therapeutic regimen, patient compliance, and exacerbation need close monitoring (Delmer, 2015).

Genetic linkage

Several genes (Table 16.10) have been identified in smokers that make them susceptible to COPD (Agusti et al., 2016a, 2016b).

Novel approaches in the management of COPD

1. Dietary changes—Inflammatory processes in patients with COPD is involved in proteolysis and subsequent destruction of elastic fibers. A study has confirmed that foods with high levels of antioxidants may have a protective effect in the lung (Varraso et al., 2015).
2. Exercise—It has been experimentally shown in the clinical studies that a treatment method involving deep-breathing exercise with oxygen inhalation led to improvement of lung function in COPD patients. The superiority of the results was confirmed when the results of this particular group were compared with the results of the patients on deep-breathing exercise or oxygen inhalation alone. This approach has ample scope for use in clinical situations (Liu et al., 2015).
3. Prednisone treatment—Prednisone 30–50 mg daily for 5 days is prescribed for acute moderate or severe exacerbations in COPD patients. Previously, 10 to 14 days of oral prednisone treatment was recommended. It has been clinically observed that 5 days of prednisone treatment has a dual advantage: It effectively treats the current exacerbations and prevents subsequent exacerbations (Leuppi et al., 2013; Walters et al., 2014; Abramson et al., 2014).
4. Antibiotic therapy—Long-term antibiotics are effective for treating frequent exacerbations

Table 16.10 COPD susceptibility genes in smokers

COPD susceptibility genes
FAM13A
HHIP
CHRNA3
CHRNA5
IREB2
Region on chromosome 19

of COPD. However, they should be prescribed only after evaluating the risk of antibiotic resistance. Long-term macrolides prescribing should not be used as a mere rule of thumb. Rather, it should be decided on a case-to-case basis by a respiratory physician (Bryant, 2015).

5. Plant-based therapy—Peucedani Radix (*Qian-hu*) used in COPD in Chinese medicine contains angular-type pyranocoumarins (APs), which can be measured by solid phase extraction-chiral LC-MS/MS (Song et al., 2014).

6. Rehabilitation with music therapy—Music therapy has been tried in patients with COPD at the Louis Armstrong Center of Music and Medicine—Mount Sinai Beth Israel (MSBI). Patients receiving music therapy along with standard rehabilitation exhibited better improvement in symptoms as compared to the patients on standard rehabilitation alone (Canga et al., 2015; Mount Sinai Hospital, 2015).

7. Cognitive-behavioral therapy—It has been clinically proven that cognitive-behavioral therapy has no significant effect on symptoms of anxiety or depression in COPD patients (Health Quality Ontario, 2015).

BIOMARKERS

Several physiological, cellular, proteomic, and genetic markers are available for COPD (Rooney and Sethi, 2015). Agusti et al. (2016a, 2016b) have classified a number of biomarkers (Table 16.11) for COPD.

EPIGENETIC BIOMARKERS

Epigenetic biomarkers can also be used in COPD. Such markers include histone methylation and miRNAs (Agusti et al., 2016a, 2016b). Eosinophils and type 2 inflammatory markers can be used to identify asthma-COPD overlap syndrome (ACOS) molecular phenotype (Christenson, 2016).

Challenges and research opportunities

GENE EXPRESSION ANALYSIS

Currently none of the available therapies can help in restricting the progression of COPD. Therefore, an urgent need exists for the development of new and effective treatments for COPD. There is an urgency to identify molecular targets for COPD treatment. Here, techniques such as gene expression profiling, serial analysis of gene expression, or microarrays may be utilized. Several research teams have mapped comparative gene expression in lung tissues of COPD patients, and it has been found that differentially regulated genes are associated with disease progression. Such studies have identified molecular mechanisms of COPD and also suggested new targets for COPD treatments (Chen et al., 2008).

CONCLUSIONS

Asthma and chronic obstructive pulmonary disease (COPD) have a link with the so-called Westernization and change in lifestyle. Both asthma and COPD involve changes in the airways and lungs leading to loss of quality of life, thus necessitating the standard of medical care. There is an urgent need to include the omics technologies in the diagnosis, molecular pathogenesis, phenotyping, management, and evolving guided-precision therapy for asthma and COPD.

Table 16.11 Classification of biomarkers for COPD

Cellular biomarkers	Blood protein biomarkers	Gene studies	Sputum transcriptomics	Serum metabolomics
Sputum neutrophils	Fibrinogen	Smoking history	Airflow limitation	Serum profile
Circulating WBC	CC16	COPD susceptibility	Blood biomarkers	Exhaled breath
	SP-D	HHIP	Emphysema	condensate
	CCL18 (PARC)	COPD subtypes	COPD susceptibility	pH
	sRAGE	Emphysema		Adenosine/
	Inflammome	Cachexia		purines
	Adipokines	Blood biomarkers		
	Vitamin D			

ACKNOWLEDGMENTS

The author wishes to thank all his teachers at Jamia Hamdard, New Delhi and also his family members. Special thanks are also due to Dr. D. Barh for suggestions and constructive criticism. The author also acknowledges the support from his wife (Mrs. Zeba) and daughters (Shifa and Azka) who sacrificed their quality time.

REFERENCES

Abramson M, Frith P, Yang I et al. COPD-X concise guide for primary care. 2014. Available at www.copdx.org.au. Accessed January 2015.

Adil MS, Khan MA, Khan MN, Sultan I, Khan MA, Ali SA, and Farooqui A. EMPADE study: Evaluation of medical prescriptions and adverse drug events in COPD patients admitted to intensive care unit. *J Clin Diagn Res* 9(11);2015:FC05–8.

Agache, IO. From phenotypes to endotypes to asthma treatment. *Curr Opin Allergy Clin Immunol* 13(3);2013:249–56.

Agache I, and Rogozea L. Asthma biomarkers: Do they bring precision medicine closer to the clinic? *Allergy Asthma Immunol Res* 9(6);2017:466–76.

Agusti A, Bel E, Thomas M et al. Treatable traits: Toward precision medicine of chronic airway diseases. *Eur Respir J* 47(2);2016a:410–9.

Agusti A, Gea J, and Faner R. Biomarkers, the control panel and personalized COPD medicine. *Respirology* 21;2016b:24–33.

Amrani Y, Syed F, Huang C, Li K, Liu V, Jain D, Keslacy S. Expression and activation of the oxytocin receptor in airway smooth muscle cells: Regulation by TNFalpha and IL-13. *Respir Res* 11;2010:104.

Anderson WC 3rd, Apter AJ, Dutmer CM et al. Advances in asthma in 2016: Designing individualized approaches to management. *J Allergy Clin Immunol* 140(3);2017:671–80.

American Thoracic Society (ATS). Patient resources for asthma. thoracic.org/patients/patient-resources/topic-specific/asthma.php, 2015.

Aubier M, Herth FJ, and Niven R. *The Uncovering Asthma Report*. Marlbourough, MA: Boston Scientific; 2015.

Bagnasco D, Ferrando M, Varricchi G et al. A critical evaluation of anti-IL-13 and anti-IL-4 strategies in severe asthma. *Int Arch Allergy Immunol* 170(2);2016:122–31.

Banerjee A, Trivedi CM, Damera G, Jiang M, Jester W, Hoshi T, Epstein JA, and Panettieri Jr RA. Trichostatin A abrogates airway constriction, but not inflammation, in murine and human asthma models. *Am J Respir Cell Mol Biol* 46(2);2012:132–8.

Baser, O. Too much ado about instrumental variable approach: Is the cure worse than the disease? *Value Health* 12(8);2009:1201–9.

Bateman ED, Hurd SS, Barnes PJ et al. Global strategy for asthma management and prevention: GINA executive summary. *Eur Respir J* 31;2008:143–78.

Bayes HK, and Cowan DC. Biomarkers and asthma management: An update. *Curr Opin Allergy Clin Immunol* 16(3);2016:210–7.

Bel EH. Clinical phenotypes of asthma. *Curr Opin Pulm Med* 10;2004:44–50.

Blackwell DL, Lucas JW, and Clarke TC. Summary health statistics for U.S. adults: National Health Interview Survey, 2012. *Vital Health Stat* 10(260);Feb 2014:1–61.

BlueCross BlueShield Association. Bronchial thermoplasty for treatment of inadequately controlled severe asthma. *TEC Assessments* 29(12);2014:1–32.

Bozzetto S, Carraro S, Zanconato S et al. Severe asthma in childhood: Diagnostic and management challenges. *Curr Opin Pulm Med* 21(1);2015:16–21.

Braun A, and Tschernig T. Animal models of asthma: Innovative methods of lung research and new pharmacological targets. *Exp Toxicol Pathol* 57(Suppl 2);2006:3–4.

Brozek JL, Akl EA, Alonso-Coello P et al. Grading quality of evidence and strength of recommendations in clinical practice guidelines. Part 1 of 3. An overview of the GRADE approach and grading quality of evidence about interventions. *Allergy* 64(5);2009:669–77.

Brueggemann LI, Haick JM, Neuburg S, Tate S, Randhawa D, Cribbs LL, and Byron KL. KCNQ (Kv7) potassium channel activators as bronchodilators: Combination with a beta2-adrenergic agonist enhances relaxation of rat airways. *Am J Physiol Lung Cell Mol Physiol* 306(6);2014:L476–86.

Bryant L. Long-term antibiotics to reduce COPD exacerbations: Pros and cons. *Prim Health Care* 7(1);2015:78–80.

Canga B, Azoulay R, Raskin J et al. AIR: Advances in Respiration—Music therapy in the treatment of chronic pulmonary disease. *Respir Med* 109(12);2015:1532–9.

Carolan BJ, and Sutherland ER. Clinical phenotypes of chronic obstructive pulmonary disease and asthma: Recent advances. *J Allergy Clin Immunol* 131;2013:627–34.

Centers for Disease Control and Prevention (CDC). *Public Health Strategic Framework for COPD Prevention*. Atlanta, GA: Centers for Disease Control and Prevention; 2011.

Chen ZH, Kim HP, Ryter SW, and Choi AMK. Identifying targets for COPD treatment through gene expression analyses. *Int J Chron Obstruct Pulmon Dis* 3(3);2008:359–70.

Chen Y, Wang Y, and Fu Z. T lymphocyte antigen 4-modified dendritic cell therapy for asthmatic mice guided by the CCR7 chemokine receptor. *Int J Mol Sci* 15(9);2014:15304–19.

Ciprandi G, Tosca MA, Silvestri M et al. Inflammatory biomarkers for asthma endotyping and consequent personalized therapy. *Expert Rev Clin Immunol* 13(7);2017:715–21.

Christenson SA. The reemergence of the asthma-COPD overlap syndrome: Characterizing a syndrome in the precision medicine era. *Curr Allergy Asthma Rep* 16(11);2016:81.

Chung KF, Wensel SE, Brozek JL et al. International ERS/ATS guidelines on definition, evaluation and treatment of severe asthma. *Eur Respir J* 43;2014:343–73.

Cooper PR, Mesaros AC, Zhang J, Christmas P, Stark CM, Douaidy K, Mittelman MA, Soberman RJ, Blair IA, and Panettieri RA. 20-HETE mediates ozone-induced, neutrophil-independent airway hyper-responsiveness in mice. *PLOS ONE* 5(4);2010:e10235.

Cooper PR, and Panettieri Jr RA. Steroids completely reverse albuterol-induced beta(2)-adrenergic receptor tolerance in human small airways. *J Allergy Clin Immunol* 122(4);2008:734–40.

Cooper PR, Zhang J, Damera G, Hoshi T, Zopf DA, and Panettieri Jr RA. C-027 inhibits IgE-mediated passive sensitization bronchoconstriction and acts as a histamine and serotonin antagonist in human airways. *Allergy Asthma Proc* 32(5);2011:359–65.

Dahlin A, and Weiss ST. Genetic and epigenetic components of aspirin-exacerbated respiratory disease. *Immunol Allergy Clin North Am* 36(4);2016:765–89.

de Planell-Saguer M, Lovinsky-Desir S, and Miller RL. Epigenetic regulation: The interface between prenatal and early-life exposure and asthma susceptibility. *Environ Mol Mutagen* 55(3);2014:231–43.

Derbyshire Joint Area Prescribing Committee. Chronic obstructive pulmonary disease (COPD) management, June 2015.

Delmer TL. The gold standard—Understanding and treating COPD, Universal Activity No.: 0798-0000-15-017-L01-PandT, released May 14, 2015.

Demoly P, Paggiaro P, Plaza V et al. Prevalence of asthma control among adults in France, Germany, Italy, Spain and the UK. *Eur Respir Rev* 18;2009:112, 105–12.

Dieplinger B, Egger M, Gabriel C, Poelz W, Morandell E, Seeber B, Kronenberg F, Haltmayer M, Mueller T, and Dieplinger H. Analytical characterization and clinical evaluation of an enzyme-linked immunosorbent assay for measurement of afamin in human plasma. *Clin Chim Acta* 425;2013:236–41.

Erpenbeck VJ, Ziegert M, Cavalet-Blanco D et al. Surfactant protein D inhibits early airway response in Aspergillus fumigatus-sensitized mice. *Clin Exp Allergy* 36(7);2006:930–40.

Fricker M, Heaney LG, and Upham JW. Can biomarkers help us hit targets in difficult-to-treat asthma? *Respirology* 22(3);2017:430–42.

Fullerton SM, Knerr S, and Burke W. Finding a place for genomics in health disparities research. *Public Health Genomics* 15(3–4);2012:156–63.

Gauthier M, Ray A, and Wenzel SE. Evolving concepts of asthma. *Am J Respir Crit Care Med* 192(6);2015:660–8.

Gillman A, and Douglass JA. Asthma in the elderly. *Asia Pac Allergy* 2(2);2012:101–8.

Global Initiative for Asthma (GINA). Pocket guide for asthma management and prevention. 2015. Available at www.ginasthma.org.

Goodman MS. Comparison of small-area analysis techniques for estimating prevalence by race. *Prev Chronic Dis* 7(2);2010:A33.

Guilleminault L, Ouksel H, Belleguic C et al. Personalised medicine in asthma: From curative to preventive medicine. *Eur Respir Rev* 26;2017:160010.

Guo Y, Gong Y, Shi G, Yang K, Pan C, Li M, Li Q, Cheng Q, Dai R, Fan L, and Wan H. Single-nucleotide polymorphisms in the TSPYL-4 and NT5DC1 genes are associated with susceptibility to chronic obstructive pulmonary disease. *Mol Med Rep* 6(3);2012:631–8.

Hahn DL. Importance of evidence grading for guideline implementation: The example of asthma. *Ann Fam Med* 7(4);2009:364–9.

Haldar P, Pavord ID, Shaw DE et al. Cluster analysis and clinical asthma phenotypes. *Am J Respir Crit Care Med* 178;2008:218–24.

Health Quality Ontario. *Cognitive-Behavioural Therapy for Anxiety and Depression in Patients with COPD: A Rapid Review.* Toronto: Health Quality Ontario; 2015 February. 21 p. Available at: http://www.hqontario.ca/evidence/evidence-process/episodes-of-care#community-copd.

Heaney LG, and McGarvey LP. Personalised medicine for asthma and chronic obstructive pulmonary disease. *Respiration* 93(3);2017:153–61.

Henderson J, Granell R, Heron J et al. Associations of wheezing phenotypes in the first 6 years of life with atopy, lung function and airway responsiveness in mid-childhood. *Thorax* 63;2008:974–80.

Hnizdo E, Sircar K, Yan T, Harber P, Fleming J, and Glindmeyer HW. Limits of longitudinal decline for the interpretation of annual changes in FEV1 in individuals. *Occup Environ Med* 64(10);2007:701–7.

Ilango K, Rajanandh MG, and Nageswari AD. Roflumilast: An upcoming drug for curing asthma and COPD. *IJPRT* 5(2);2013:130–5.

Jeanblanc NM, Hemken PM, Datwyler MJ et al. Development of a new ARCHITECT automated periostin immunoassay. *Clin Chim Acta* 464;2017:228–35.

Jiang H, Abel PW, Toews ML, Deng C, Casale TB, Xie Y, and Tu Y. Phosphoinositide 3-kinase gamma regulates airway smooth muscle contraction by modulating calcium oscillations. *J Pharmacol Exp Ther* 334(3);2010:703–9.

Jiang H, Xie Y, Abel PW, Toews ML, Townley RG, Casale TB, and Tu Y. Targeting phosphoinositide 3-kinase gamma in airway smooth muscle cells to suppress interleukin-13-induced mouse airway hyperresponsiveness. *J Pharmacol Exp Ther* 342(2);2012:305–11.

Jindal SK. Emergence of chronic obstructive pulmonary disease as an epidemic in India. *Indian J Med Res* 124;2006a:619–30.

Jindal SK, Aggarwal AN, Chaudhry K et al. A multicentric study on epidemiology of chronic obstructive pulmonary disease and its relationship with tobacco smoking and environmental tobacco smoke exposure. *Indian J Chest Dis Allied Sci* 48;2006b:23–9.

John G, Darbellay P, Drepper M, Spechbach H, Fosenbauer MB, Perrier A, and Carballo S. [Hospital based internal medicine in 2012]. *Rev Med Suisse* 9(370);2013:193–6, 198.

Just J, Saint-Pierre P, Gouvis-Echraghi R et al. Wheeze phenotypes in young children have different courses during the preschool period. *Ann Allergy Asthma Immunol* 111;2013:256–61.e1.

Ker J, and Hartert TV. The atopic march: What's the evidence? *Ann Allergy Asthma Immunol* 103(4);2009:282–9.

Khamis MM, Adamko DJ, and El-Aneed A. Development of a validated LC- MS/MS method for the quantification of 19 endogenous asthma/copd potential urinary biomarkers. *Anal Chim Acta* 989;2017:45–58.

Ladhani S. COPD—An Indian perspective, *J Lung Pulm Respir Res* 3(1);2016:00070.

Lang DM. Severe asthma: Epidemiology, burden of illness, and heterogeneity. *Allergy Asthma Proc* 36(6);2015:418–24.

Larsson AK, Hagfjard A, Dahlen SE, and Adner M. Prostaglandin D(2) induces contractions through activation of TP receptors in peripheral lung tissue from the guinea pig. *Eur J Pharmacol* 669(1–3);2011:136–42.

Leadbitter P, Pearce N, Cheng S, Sears MR, Holdaway MD, Flannery EM, Herbison GP, and Beasley R. Relationship between fetal growth and the development of asthma and atopy in childhood. *Thorax* 54(10);1999:905–10.

Lefebvre Q, Vandergoten T, Derom E, Marchandise E, and Liistro G. Testing spirometers: Are the standard curves of the american thoracic society sufficient? *Respir Care* 59(12);2014:1895–904.

Leuppi JD, Schuetz P, Bingisser R et al. Short-term vs conventional glucocorticoid therapy in acute exacerbations of chronic obstructive pulmonary disease: The REDUCE randomized clinical trial. *JAMA* 309;2013:2223–31.

Liu YQ, Yan LX, Zhang LY, Song QH, and Xu RM. Conspicuous effect on treatment of mild-to-moderate COPD by combining deep-breathing exercise with oxygen inhalation. *Int J Clin Exp Med* 8(6);2015:9918–24.

Loth DW, Artigas MS, Gharib SA et al. Genome-wide association analysis identifies six new loci associated with forced vital capacity. *Nat Genet* 46(7);2014:669–77.

Marsh SE, Travers J, Weatherall M et al. Proportional classifications of COPD phenotypes. *Thorax* 63;2008:761–7.

Miravitlles M, Soler-Cataluna JJ, Calle M, and Soriano JB. Treatment of COPD by clinical phenotypes: Putting old evidence into clinical practice. *Eur Respir J* 41;2013:1252–6.

Moerman CJ, Deurenberg R, and Haafkens JA. Locating sex-specific evidence on clinical questions in MEDLINE: A search filter for use on OvidSP. *BMC Med Res Methodol* 9;2009:25.

Moore WC, Meyers DA, Wenzel SE et al. Identification of asthma phenotypes using cluster analysis in the Severe Asthma Research Program. *Am J Respir Crit Care Med* 181;2010:315–23.

Mount Sinai Hospital. Music therapy increases effectiveness of pulmonary rehabilitation for COPD patients. 2015, December 23. Available at https://medicalxpress.com/news/2015-12-musictherapy-effectiveness-pulmonary-copd.html. Accessed February 3, 2016.

Mullane K, and Williams M. Animal models of asthma: Reprise or reboot? *Biochem Pharmacol* 87(1);2014:131–9.

National Heart, Lung, and Blood Institute (NHLBI). 2014 August 4. Available at http://www.nhlbi.nih.gov/health/health-topics/topics/asthma/.

National Heart, Lung, and Blood Institute (NHLBI). Best practices in care: Six key messages for managing patients with asthma. 2015a. Available at http://www.nhlbi.nih.gov/files/docs/guidelines/asthma_qrg.pdf.

National Heart, Lung, and Blood Institute. How is COPD diagnosed? 2015b. Available at http://www.nhlbi.nih.gov/health/health-topics/topics/copd/diagnosis.html. Accessed April 20, 2015.

National Heart, Lung and Blood Institute (NHLBI). What is COPD? 2015c. Available at http://www.nhlbi.nih.gov/health/health-topics/topics/copd/. Accessed April 20, 2015.

Nicoll J, Cruikshank WW, Brazer W, Liu Y, Center DM, and Kornfeld H. Identification of domains in IL-16 critical for biological activity. *J Immunol* 163(4);1999:1827–32.

Ortega VE, and Meyers DA. Implications of population structure and ancestry on asthma genetic studies. *Curr Opin Allergy Clin Immunol* 14(5);2014:381–9.

Ortega VE, Meyers DA, and Bleecker ER. Asthma pharmacogenetics and the development of genetic profiles for personalized medicine. *Pharmgenomics Pers Med* 8;2015:9–22.

Park HW, Tantisira KG, and Weiss ST. Pharmacogenomics in asthma therapy: Where are we and where do we go? *Annu Rev Pharmacol Toxicol* 55;2015:129–47.

Peters MC, Nguyen ML, and Dunican EM. Biomarkers of airway type-2 inflammation and integrating complex phenotypes to endotypes in asthma. *Curr Allergy Asthma Rep* 16(10);2016:71.

Ponomarenko M, Rasskazov D, Arkova O, Ponomarenko P, Suslov V, Savinkova L, and Kolchanov N. How to use SNP_TATA_comparator to find a significant change in gene expression caused by the regulatory SNP of this gene's promoter via a change in affinity of the tata-binding protein for this promoter. *Biomed Res Int* 2015;2015: article ID 359835.

Rooney C, and Sethi T. Biomarkers for precision medicine in airways disease. *Ann NY Acad Sci* 1346(1);2015:18–32.

Salvi S, and Agrawal A. India needs a national COPD prevention and control programme. *JAPI* 60;2012:4–5.

Seehase S, Schleputz M, Switalla S et al. Bronchoconstriction in nonhuman primates:

A species comparison. *J Appl Physiol (1985)* 111(3);2011:791–8.

Silva GE, Sherrill DL, Guerra S, and Barbee RA. Asthma as a risk factor for COPD in a longitudinal study. *Chest* 126;2004:59–65.

Song Y, Jing W, Yang F, Shi Z, Yao M, Yan R, and Wang Y. Simultaneously enantiospecific determination of (+)-trans-khellactone, (+/–)-praeruptorin A, (+/–)-praeruptorin B, (+)-praeruptorin E, and their metabolites, (+/–)-cis-khellactone, in rat plasma using online solid phase extraction-chiral LC-MS/MS. *J Pharm Biomed Anal* 88;2014:269–77.

Spratt M, Carpenter J, Sterne JA, Carlin JB, Heron J, Henderson J, and Tilling K. Strategies for multiple imputation in longitudinal studies. *Am J Epidemiol* 172(4);2010:478–87.

Spycher BD, Silverman M, Brooke AM, Minder CE, and Kuehni CE. Distinguishing phenotypes of childhood wheeze and cough using latent class analysis. *Eur Respir J* 31;2008:974–81.

St Ledger K, Agee SJ, Kasaian MT, Forlow SB, Durn BL, Minyard J, Lu QA, Todd J, Vesterqvist O, and Burczynski ME. Analytical validation of a highly sensitive microparticle-based immunoassay for the quantitation of IL-13 in human serum using the Erenna immunoassay system. *J Immunol Methods* 350(1–2);2009:161–70.

Stein K, Dyer M, Milne R, Round A, Ratcliffe J, and Brazier J. The precision of health state valuation by members of the general public using the standard gamble. *Qual Life Res* 18(4);2009:509–18.

Stokes JR, and Casale TB. Characterization of asthma endotypes: Implications for therapy. *Ann Allergy Asthma Immunol* 117(2);2016:121–5.

Sunovion Pharmaceuticals Inc Chronic obstructive pulmonary disease (COPD) fact sheet. 2015. (BRO058-15 7/15).

Suthar NN, Patel KL, and Shah J. A study on awareness of chronic obstructive pulmonary disease (COPD) among Smokers. *Ntl J of Community Med* 6(4);2015:547–53.

Switalla S, Lauenstein L, Prenzler F et al. Natural innate cytokine response to immunomodulators and adjuvants in human precision-cut lung slices. *Toxicol Appl Pharmacol* 246(3);2010:107–15.

Tho LL, and Candlish JK. Superoxide dismutase and glutathione peroxidase activities in erythrocytes as indices of oxygen loading in disease: A survey of one hundred cases. *Biochem Med Metab Biol* 38(1);1987:74–80.

Toren K. Challenges for the new century in the epidemiology of adult asthma. *Scand J Work Environ Health* 25(6);1999:558–63.

Turner AM, Tamasi L, Schleich F, Hoxha M, Horvath I, Louis R, and Barnes N. Clinically relevant subgroups in COPD and asthma. *Eur Respir Rev* 24;2015:283–98.

van Doorn-Klomberg AL, Braspenning JC, Feskens RC, Bouma M, Campbell SM, and Reeves D. Precision of individual and composite performance scores: The ideal number of indicators in an indicator set. *Med Care* 51(1);2013:115–21.

van Schayck CP, Dompeling E, van Weel C, Folgering H, and van den Hoogen HJ. Accuracy and reproducibility of the assess peak flow meter. *Eur Respir J* 3(3);1990:338–41.

Varraso R, Chiuve SE, Fung TT et al. Alternate Healthy Eating Index 2010 and risk of chronic obstructive pulmonary disease among US women and men: Prospective study. *BMJ* 350;2015:h286.

Viglio S, Stolk J, Luisetti M, Ferrari F, Piccinini P, and Iadarola P. From micellar electrokinetic chromatography to liquid chromatography-mass spectrometry: Revisiting the way of analyzing human fluids for the search of desmosines, putative biomarkers of chronic obstructive pulmonary disease. *Electrophoresis* 35(1);2014:109–18.

Vijayan VK. Chronic obstructive pulmonary disease, *Indian J Med Res* 137;2013:251–69.

Walters JA, Tan DJ, White CJ et al. Different durations of corticosteroid therapy for exacerbations of chronic obstructive pulmonary disease. *Cochrane Database Syst Rev* 12;2014:CD006897.

Wesolowska-Andersen A, and Seibold MA. Airway molecular endotypes of asthma: Dissecting the heterogeneity. *Curr Opin Allergy Clin Immunol* 15(2);2015:163–8.

Wheaton, AG, Cunnningham, TJ, Ford, ES, and Croft, JB. Employment and activity limitations among adults with chronic obstructive pulmonary disease—United States, 2013. *MMWR*

64(11);2015, March 27:289–95. Available at http://www.cdc.gov/mmwr/.

Wilkinson J, Grimley S, Collins A, Thomas NS, Holgate ST, and Morton N. Linkage of asthma to markers on chromosome 12 in a sample of 240 families using quantitative phenotype scores. *Genomics* 53(3);1998:251–9.

Woolcock Institute of Medical Research. COPD: Chronic Obstructive Pulmonary Disease, Version 1. 2015. Australia.

Zedan MM, Osman AM, Laimon WN et al. Airway inflammatory biomarker: Could it tailor the right medications for the right asthmatic patient? *Iran J Immunol* 13(2);2016:70–88.

17

Customized DNA–directed precision nutrition to balance the brain reward circuitry and reduce addictive behaviors

KENNETH BLUM, MARCELO FEBO, ERIC R. BRAVERMAN,
MONA LI, LYLE FRIED, ROGER WAITE, ZSOLT DEMOTROVICS,
WILLIAM B. DOWNS, DEBMALYA BARH, BRUCE STEINBERG,
THOMAS McLAUGHLIN, AND RAJENDRA D. BADGAIYAN

INTRODUCTION

In 1969 we used precision to land on the moon, so why can't we use precision medicine to stop the opiate/opioid epidemic in America? The genomics era has brought forth new hope for the future of medicine and, in particular, a better understanding of psychiatry. We now understand the important roles of DNA, polymorphic associations, and brain reward circuitry in a new perspective on addictive behaviors. We present here a review of neurogenetics and nutrigenomics as potential keys that provide more accurate genetic diagnoses to improve dopamine D2 agonist therapy and dopaminergic activation. Through many experiments, we are ready to propose a novel approach that challenges the recovery field to incorporate genomic-based tools in treatment programs at addiction clinics. We propose using the Genetic Addiction Risk Score (GARS)™ for appropriate reward deficiency syndrome (RDS) diagnosis, the Comprehensive Analysis of Reported Drugs CARD™ to determine both compliance and abstinence during treatment, natural D2 agonistic therapy (KB220PAM™), and eventually mRNA (patent pending) to determine pre- and postcandidate gene expressions in RDS. We are, therefore, proposing a paradigm shift we have called "Reward Deficiency Solutions System (RDSS)™" (Blum et al., 2014a), most recently referred to as "Precision Behavioral Management" (PBM).

Have you ever wondered why so many people in America and across the globe are falling victim to the chain of addictive behaviors, part of the worst epidemic in the history of the word? The answer is in part both genetic and environmental (epigenetic). Recently researchers found that epigenetic effects on the chromatin structure of our DNA are a legacy that passes from generation to generation.

Scientists like Stephen Hawking suggest that we are made up of self-assembled molecules generated over 14 billion years. More interesting is that we as *Homo sapiens* differ in our DNA by only 0.5%. The latest research demonstrates that each person has on average about 60 new mutations in comparison to their parents. Even more remarkable, the human brain contains billions of neurons working in concert to provide us the gift of "well-being" free of mental disease and stress. The number of neurons in the brain varies dramatically from species to species. One estimate (published in 2012) puts the human brain at about 85 billion neurons and approximately 85 trillion synapses. It turns out that 20% of our entire body's energy is budgeted to keep our brain working normally. The differences between individual humans are approximately the 4.25 billion neurons and 4.25 trillion synapses that make us unique.

This difference affects the 7.4 billion humans that roam our earth working and living together to achieve some degree of productivity and happiness. However, in the twenty-first century humans face daily reminders of horrific diseases, like cancer, that develop because of genetic and epigenetic differences and lead to fatalities and mental impairments that influence billions of neurons and trillions of synapses. This molecular rearrangement of our genome makes each of us unique. For example, how dopamine functions in our reward system may also be unique (Rena et al., 2001). One example, among other gene variations involving brain reward, is that genetic differences account for the presence of attention deficit/hyperactivity disorder (ADHD), a subtype of reward deficiency syndrome (RDS), in approximately 8%–12% of children in the United States and 4% of adults worldwide (Blum et al., 2008a). You also may be surprised to know that, at birth, an estimated 100 million people in the United States carry a gene form (allele) of just one genetic variation that involves brain dopamine D2 receptors. We know that carriers of the allele DRD2*Taq* A1 have 30%–40% lower D2 receptors in the brain (Noble et al., 1991). So what does this mean regarding our romance with getting high—"turning on" and "turning off" with potent psychoactive drugs (e.g., alcohol, cocaine, and opiates)—and resultant addiction and fatalities seen in our kids?

In 1990, the first association of a variant (A1) on the dopamine D2 receptor gene (DRD2) and severe alcoholism was discovered and published by Blum and Noble et al. in *JAMA*. Later experiments showed that individuals who carry this variant have 30%–40% lower dopamine receptors than DRD2 A2 carriers (Noble et al., 1991). Being born with this single gene variation (DRD2 A1 form), which causes low dopamine receptors, sets an individual up to have a high addiction risk (vulnerability) to any substance or behavior that stimulates the neuronal release of dopamine. In fact, in 1996, Blum's laboratory used a mathematical model (called the Bayesian theorem) to predict that an individual born with the A1 allele (variant) has a 74.4% risk of developing an RDS behavior like an addiction (Blum et al., 1995). People with that allele will have an initial acute response to using a psychoactive drug or experiencing pathological gambling, or whatever behavior stimulates enough neuronal dopamine for them to feel normal well-being possibly for the first time. Unfortunately, chronic consumption/experiences lead to epigenetic changes that further reduce dopamine receptor numbers and a stronger need to abuse can lead to unwanted uncontrollable behaviors and even narcotic overdose followed by death.

How were the genes involved in reward found? The chemical messengers (neurotransmitters) in the brain are like keys that turn on various functions of genes. The neurotransmitters that participate in evoking pleasurable feelings in the reward circuitry work in a cascading fashion throughout the brain. These interactions (the brain reward cascade [BRC]) may be viewed as activities of subsystems of a larger system with simultaneous cascades. The goal is the generation of feelings of well-being by the eventual release of just the right amount of dopamine at the reward site. In this scenario, there are at least seven major neurotransmitters and their pathways involved: *Serotonin, endocannabinoids, endorphin (enkephalin), GABA, glutamine, acetylcholine, and dopamine.* There are thousands of published studies about these reward genes and pathways that influence the function of these named neurotransmitters (Blum and Kozlowski, 1990). This research involved the identification of gene (DNA) variations or alleles that individuals are born with and epigenetic (environmental RNA) changes that may alter the healthy, intended function of DNA.

Dysfunctional DNA is due to what is referred to as *single nucleotide polymorphisms*, frequently called SNPs (pronounced "snips"). SNPs are the most ubiquitous of genetic variations. Every SNP represents a single nucleotide and they can therefore replace cytosine with thymine within a stretch of DNA. SNPs regularly occur throughout individuals' DNA; on average they can be found approximately every 300 nucleotides, translating to nearly 10 million SNPs in the human genome. They are commonly found at DNA points between genes and can act as biological markers of genes specific to various diseases. When SNPs occur within or near a gene (in a regulatory region), they alter the gene's function. If these SNPS show up in the brain reward cascade-set of genes (Blum et al., 2014c), the neurotransmission will be dysfunctional resulting in a loss of dopamine regulation or balance (homeostasis). Too little dopamine will at birth predispose people to "want" or "like" psychoactive drugs or even behaviors like hypersexuality and gambling. Compromised DNA with risk variations (alleles) can predispose them to become victims to the chain of addictive behaviors.

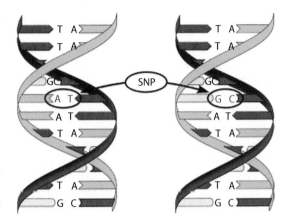

Following 28 years of extensive research from many scientists worldwide, a panel of 11 reward gene risk variants called the Genetic Addiction Risk Score (GARS™) has been developed. When GARS was compared to the Addiction Severity Index (ASI) used in many clinical settings, it was found to predict the severity of both alcohol and drug dependency (Blum, 2015) significantly.

Currently the GARS test is commercially available and is the property of Geneus Health LLC., located in San Antonio, Texas. The GARS test is now available and for the first time, clinicians and parents will be able to assess the vulnerability of their patients and, most important, of their children to chemical dependency and RDS behaviors like addiction, ADHD, and autism spectrum disorders. The common thread across all these risk gene variants is that they lead to a low dopamine (hypodopaminergia) function or deficit. There are arguments against testing such as fear of labeling and arguments for testing like knowing the risk. The real issue or challenge, however, is what can be done if risk alleles are found. It is understandable that when there is one gene–one disease (OGOD) illnesses like in Huntington's disease, and when treatment is unavailable, and prevention remains a problem, why know the risk?

Have we found a safe gentle nonaddictive solution that will provide the brain a means to balance the neurotransmitters involved in the BRC culminating in true dopamine homeostasis?

Despite variant genes, and epigenetic and environmental insults, holistic approaches like mindfulness, exercise, spirituality, and particularly amino acid therapy (KB220PAM formulations) have been shown to reduce relapse and increase brain dopamine homeostasis (Blum et al., 2012).

We suggest that clinicians will now be able to genetically test our children for unwanted reward gene risk variants that predispose them to dopamine deficiency and lack of reward and risk for addiction, but possibly even use this genetic testing to circumvent RDS behaviors (Blum et al., 2014b).

Genetic risk for substance abuse and other RDS behaviors can be identified by the GARS test and explain why some individuals are vulnerable and others not. With continued research, genetic and epigenetic dopamine deficiencies can be treated, relapse reduced, and we can free ourselves from the clutches of powerful addictive behaviors and bring balance and happiness to our lives.

DNA CUSTOMIZATION IS A REALITY

The advent of DNA customization breaks through the traditional paradigm of one-size-fits-all dietary supplementation (and even drug therapy) and can present a promising new, more accurate and beneficial standard of nutrition. Individuals can now receive supplementation of nutraceutical products that are designed for them and customized to meet their unique genetically determined needs. No more trial-and-error guesswork is necessary. This new model provides nutritional supplements that offer significantly greater value to the individual than conventional one-size-fits-all formulations.

The introduction of DNA-customized technology represents a fundamental change in dietary supplementation. As history demonstrates, in addition to availing hope and opportunity, change can also be met with skepticism, doubt, and even resistance. For various reasons, many people just don't want to change the status quo, especially those with a heavily vested interest in conventional dogma. Yet there are others, who, seeing the possibilities, want to partake in the exciting new paradigm and market inferior versions of the technology, albeit with anecdotal data. They offer the promise of DNA customization with woefully inadequate technology capabilities and tools.

Practical applications

The field of science out of which DNA customization has emerged is called nutrigenomics. Nutrigenomics is the study of the relationships between nutrition, the response of genes, and their combined effects on overall health and behavior.

However, not all nutrigenomic technologies are the same. To better evaluate the benefits of DNA customization, a few questions need to be explored and answered.

- What is the importance and value of DNA-customized nutrition (why use it)?
- What exactly is DNA-customized nutrition?
- What criteria need to be satisfied for DNA-customized nutrition to be worthwhile?

WHAT IS THE IMPORTANCE AND VALUE OF DNA-CUSTOMIZED NUTRITION (WHY USE IT)?

As mentioned earlier, although we all have basic similarities that may appear to make us pretty much the same, or at least make us need the same things in general, the fact is that *there are many distinct differences in humans that can alter basic individual needs.* Some of our most obvious differences include gender, age, race, body types, body sizes, tastes and appetites, physical and mental capabilities and activities, talents and skills, eye and hair color, and blood types. Other less obvious differences can influence our bodies and behaviors just as profoundly. Some of these differences include, among others, variations in

- Metabolic processes (the ability to produce and regulate energy)
- Ability to tolerate stress
- Emotions
- Behaviors
- Lifestyle (i.e., food, drink, and activity choices)
- Immune system

The wide range of variations in these differences creates a variety of differences in our individual nutritional needs. *These differences result from genetic differences and are responsible for cultural diversity, as a root cause* (that is, polymorphisms in genotypes and their gene expressions). An example of this is the fact that alcohol affects Asians and Native Americans quite differently than Caucasians. Genotypes represent specific gene variants. An individual is not and does not have only one genotype but a vast number of genotypes. In order to effectively utilize DNA-customized formulations, it is necessary to know the many genotypes of an individual across multiple biological and metabolic pathways.

In the earliest years of dietary supplementation, vitamin and mineral preparations were based on the lowest common denominator, that is, what minimum amount of nutrients seemed to best address the most basic and common needs of the "masses." Many supplement companies followed the Recommended Dietary Allowance (RDA) guidelines to meet the basement of biological necessity. The inadequacies of that minimalist model from those early days through to the present day became evident. This realization has led to elementary types of "category customization" to address the different nutritional needs caused by those factors cited earlier (gender, age, athletics, blood type, etc.). Even with this simple type of customization, a relative "sameness" of the formulas is still maintained within those groups (e.g., all women get the same type of women's formula).

While addressing some "general" differences within the categories, these conventional formulas were, and still are, one-size-fits-all, that is, for example, young women need more iron and young men need more zinc. One-size-fits-all formulas do not meet an individual's specific needs and can inadvertently create some nutritional imbalances.

Moreover, there are apparent flaws and inadequacies of common lifestyles which, when combined with a one-size-fits-all supplement, are ineffective in stopping the increasing levels of chronic pain and mental disorders, such as obesity, cardiovascular disease, hypertension, Alzheimer's, ADHD, and the current opiate/opioid epidemic. Importantly, certain groups of people were, and are, more prone (i.e., predisposed) to suffer a higher incidence and severity of certain types of conditions. Examples include Caucasian and Asian women being more prone to osteoporosis and people of African descent being prone to more severe cardiovascular disease, diabetes, and obesity. This fact is underlined by the enormous quantity and range of pharmaceutical drugs created to address these disorders in different ways and the plethora of adverse side effects that can occur due to differences in how patients react to the drugs. Attempts to find natural remedies to address these ill-health conditions gave rise to an ever-increasing selection of one-size-fits-all "condition-specific" dietary supplement formulas as well.

Consumers' obsession with finding better ways to improve their health has resulted in a meteoric rise in the quantity and range of dietary choices and alternatives, especially in dietary supplements. This is evidenced by the wide availability of a mind-numbing array of dietary supplements from which to choose the "one that's best for me"—clear evidence of an intrinsic desire for customization and personalization. The search for the best supplement to achieve a health objective (e.g., weight loss or increased athletic performance) or address an ill-health condition (e.g., lower cholesterol, resolve digestive disorders, address depression) is almost endless.

The problem is that one often observes "what worked for somebody else" and then thinks that product will apply to oneself. We often also take the advice or recommendation of "informed" store personnel to make supplement choices. In some cases, the product will work. But, in many cases, they barely work or don't work at all. Moreover, sometimes the product effect backfires, promoting negative results or reactions, thus confirming evidence that we have genetic differences. Nevertheless, even dietary supplement companies try to get in on the popular trends and formulate products very similar to popular-selling versions on the market, perpetuating the one-size-fits-all and works-one-way formula fads. Innovation seems to be confined to the next best miracle fruit extract or the like.

With the discovery of DNA and the advent of the human genome project, an entirely new dimension of and opportunity for nutritional customization was born in the field of nutrigenomics. An important fact must now be punctuated: One size doesn't fit all for your clothes or your health needs. Another important factor crucial to the success of a DNA-customized nutraceutical product, previously mentioned, is whether the genes being targeted for nutraceutical customization are the most appropriate to provide the greatest benefit. In fact, DNA customization eliminates the trial-and-error guesswork of finding the best product formulation for the individual, ensuring the greatest and most sustainable benefit. This is easier said than done, but remains a laudable goal for the future of general medicine and, in this context, addiction medicine. Blum's group has already published on this possibility earlier about KB220 variants in obesity (Blum et al., 2008d, 2008e, 2012b, 2014a; Plomin et al., 1994).

WHAT EXACTLY IS A DNA-CUSTOMIZED NUTRACEUTICAL?

DNA analysis and nutraceutical customization is a real possibility now but will always require additional focused research for development of advanced products. Following is an overview of nutrigenomic DNA customization process:

1. Find the right science-based ingredients that influence therapeutically important pathways—a crucial factor.
2. Existing research must validate the dose-dependent mechanisms of action (MOA) and beneficial effects of those ingredients.
3. Determine if, and by how much, those ingredients influence gene expression.
4. Dose-dependent results are reviewed to establish the baseline dose needed to achieve beneficial gene-expression effects in a reasonable portion of the population.
5. Research is then either conducted or, if available, reviewed to determine a range of dose-dependent gene-expression results.
6. Dose-dependent gene expression results are then compared to results in the MOA and efficacy research.
7. Ingredients selected through this process become "DNA qualified" and are used in DNA-customized formulations.
8. Research is then reviewed to determine which genotypes exhibit the kind of gene expressions noted in the gene expression studies.
9. Algorithms are then established for DNA-customized dose-dependent applications for each DNA-qualified ingredient based on the genotypes for each pathway.
10. Then research is reviewed or conducted on select populations of genotype-qualified candidates to confirm and/or better define and catalog the efficacy parameters.
11. Once this process is completed, genuine DNA-qualified ingredients can be customized for the gene polymorphisms of an individual.
12. Specific genes of an individual are analyzed, via analysis of a buccal swab.
13. Genotypes for crucial pathways are determined and ingredient potencies adjusted to accommodate gene-expression variations of the individual.
14. Risk alleles of a number of important physiological pathways are selected on the basis of a common thread (i.e., hypodopaminergic trait/state). Each risk allele must have been associated with a particular risk compared to controls in many published studies.

WHAT CRITERIA NEED TO BE SATISFIED FOR DNA-CUSTOMIZED NUTRITION TO BE WORTHWHILE?

Any one part of your body cannot be well, unless you care for the whole body.

—Attributed to Plato

It is clearly beneficial to address foundational health factors that affect the whole body in conjunction with other, more focused targets (synergistic symbiosis). For example, to address weight loss without first considering genetic factors and the controlling involvement of the brain is a formula for failure, repeatedly demonstrated with conventional methods. The shortcoming of most conventional therapeutic tactics, even nutraceutical ones, is that they rely solely on narrow, compartmentalized, and/or downstream symptomatic targets (Rana et al., 2016). Some examples of these types of benefit targets include:

- Anti-inflammatory
- Appetite suppression
- Blocking carb absorption
- Blocking fat absorption
- Diuretics
- Stimulants
- Calming and soothing
- Libido enhancement
- Cholesterol-lowering
- Blood sugar lowering

Conventional nutraceutical formulations that address these types of "benefit" targets routinely ignore the gene-expression-based mechanisms of underlying "upstream" causes promoting their "downstream" conditions. They are generally targeted at mechanisms of action that achieve superficial or temporary symptomatic relief at best.

There could be many "causes" for a supposed malady and its symptoms. For instance, a depressed individual might try taking St. John's wort (Apaydin et al., 2016) to feel better (trial-and-error). Importantly, there are many causes and types of depression. What works for one cause and type might not work for another or even make other types worse. Knowing the genotypes of genes in certain pathways in the brain would enable a DNA-customized formula to promote healthy brain function and reduced guesswork.

Another example of this premise is the use of fat blockers to reduce fat calorie absorption and promote "weight" loss. The fact is that, in the end, calorie blocking and deprivation work against the body's genetic survival instructions. The initial effect could be to reduce fat calorie absorption, resulting in some initial weight loss. But, such deprivation tactics create a survival crisis, triggering a genetically mandated protection response. The longer-term result of this effect is to lower the basal metabolic rate (rate of calorie burning), increase fat storage, and increase cravings and food consumption to ensure survival, which promotes greater weight regain (colloquially referred to as yo-yo weight gain) (Casazza et al., 2015).

Many conventional nutraceutical products, especially condition-specific formulas, target symptomatic relief. In contrast, by exerting a nutraceutical influence on gene expressions, regarding precision medicine, DNA-customized nutraceutical products must address the whole body, and, importantly, the brain's control of all bodily functions (neuro-nutrigenomics). The result corrects the effects of specific gene variants (polymorphisms), normalizing target tissue function. Moreover, DNA-customized formulas should be symbiotic, improve the overall health and function of the body in a synergistic manner, and satisfy the criteria required to provide efficient and sustainable outcomes. Those criteria are explained next.

Criterion 1: The biological effect should be "body friendly" (i.e., biologically and immunologically compatible)

DNA-customized nutrition is satisfying a need of the body. Therefore, the body should not have to protect, reject, or retaliate against the effects the

formula's ingredients have on the body. For example, pharmaceutical drugs, stimulants, fat blockers, and mega-dose nutrient overloading (pharmaco-logic-like) effects, including all stimulants, suppressants, and calorie deprivation tactics could fight against what has been termed "body friendly."

Criterion 2: Gene polymorphisms identified and targeted for nutrigenomic effects need to exert a significant influence on the origin of a problem and the source of the targeted outcome

In simple terms, get to the genetic source. For example, in contrast to single-effect stimulants and calorie-deprivation tactics, weight loss needs to be the result of correcting biosystem inadequacies. At least five major pathways need to be simultaneously addressed:

- Energy metabolism (production and regulation)
- Reward and craving management in the brain
- Neuroendocrine management
- Stress management
- Immune system function

Criterion 3: The nutraceutical benefit should not only correct the target tissue structure and function, it should make a significant contribution to overall health

To develop DNA-customized and nutrient-specific formulas for RDS (Downs et al., 2009) and all substance and nonsubstance-seeking addictive behaviors, all ingredients and their interactions must be considered as they work in concert to balance the BRC (Blum et al., 2016a) that will provide an effect on overall health of an individual.

Criterion 4: The beneficial effects should be obvious, different, transformational, and sustainable for most patients

In vitro studies can simulate a powerful mechanism of action in cell cultures. However, when human studies are conducted on the same product to find a specific benefit (i.e., blocking fat absorption to induce weight loss) the outcomes are usually skewed across a range of benefits. The results could look like the following:

- 30% of the people received strong benefits (also a possible placebo effect)
- 30% received minor benefits
- 15% received no benefits
- 15% actually experienced reverse effects, that is, increased weight gain

According to a traditional perspective, this outcome would still be considered positive and have significant medical value. However, in this example, these results suggest that as much as 70% of the patients will experience some level of disappointment. This is the primary reason that many people stop taking a weight loss product within 3 to 6 months. This also disregards the yo-yo rebound weight gain effect (the failure effect), which appears to happen in at least 95% of the cases with time. The need for precision medicine will come of age especially when genetic testing could provide meaningful results and medical benefits.

Conventional methods do not comply with all of the four important criteria

First, *traditional methods might not be "body friendly."* In many cases products impose a reduced fat and calorie intake (blocking) on the body. The initial effect could be weight loss (especially in the 30% of people). However, the long-term effect will be the result of the body's attempt to protect itself from the "famine" or "be thrifty" (Joffe and Zimmet, 1998) upshot. This effect might be significantly greater in certain individuals in the 70% group. The body will ultimately try to protect against reduced nutrient absorption with increased fat storage, a natural survival mechanism.

Second, *"fat-blocking" methods do not address multiple metabolic or genetic causes of the problem.* Just blocking fat absorption is not addressing the many other issues related to obesity, whereby many genes and subsequent polymorphisms play important roles in the induction of aberrant cravings, enhancing brown fat synthesis (Polyák et al., 2016), and inducing metabolic anomalies (Higginson et al., 2016).

Third, *conventional methods do not make a significant contribution to overall health.* It is indisputable that weight loss is associated with many health benefits. However, eventual reduced health and amplified yo-yo weight fluctuation is highly probable after significant weight decrease due to the aforementioned first and second criteria, and the fact that absorption of healthy lipids like Omega-3s and fat-soluble vitamins are blocked in the process.

Fourth, *conventional methods do not provide obvious, different, and positive transformational effects that are sustainable for most patients.* We must keep in mind that most dietary supplement research, especially for weight loss products, evaluates results over an 8- to 12-week period. This is a very short-term window in a long-term process. Considering the reality of the weight gain–rebound effect, this is an inadequate time to evaluate more telling longer-term effects. In fact, history reveals that conventional weight loss programs almost without fail have indeed failed. If the products work at all, conventional calorie-deprivation weight loss tactics (including fat blockers), without exception, provide results for only a temporary period. Ultimately, conventional methods lead to the regaining of weight. Given this, and due to the facts cited in the first three criteria, for the vast majority of people, the results are not obvious, not different from other similar-type products, not transformational, and not sustainable.

A major reason that human research results are skewed is due to the wide variation in genetic factors. Identifying these factors enables the use of the right therapeutic modalities in the appropriate amounts to achieve the greatest benefits in the most people for the longest time. When done correctly, the incredible value of DNA customization will be indisputable.

GENETIC TESTING: FACT OR FICTION

Although some physicians may think that by using 23andMe genomic testing they could adequately determine a person's risk for RDS behaviors, they are mistaken. In the first place, the FDA did not allow 23andMe to provide any interpretive data to an individual related to medical health issues and in fact threatened removal of a such a misleading test. While 23andme obtained new clearance and approval on a number of gene-based predisposition tests, the de novo approval does not cover RDS or addiction. The following information will help inform the medical community about cautions concerning genetic testing in general.

Concerning molecular genetic testing, there are three types of current interest. They are *pharmacogenetics*, primarily evaluating metabolizing enzymes for high and low metabolizers with for example opiates; *Genetic Addiction Risk Score*, to determine through a panel of reward gene polymorphisms stratification risk or vulnerability to all RDS behaviors including pain tolerance; and *pharmacogenomics*, personalized addiction medicine based on genotyping an individual and targeting specific gene loci.

Pharmacogenetic testing

Various alleles in the P450 system are currently utilized in pain medicine clinics to evaluate metabolic concerns to help identify high and low metabolizers. For the most part, this has not translated to significant clinical utility, but may have some relevance regarding buprenorphine/naloxone treatments. When used in conjunction with GARS this could provide valuable information (De Fazio et al., 2008).

Genetic Addiction Risk Score (GARS)

We now know that nature (genes), nurture (environment), and behavioral outcomes result from a combination of 50% genes and 50% epigenetics. Thus, molecular genetic or DNA testing is critical, especially in linking aberrant behaviors to any individual.

Blum's laboratory proposed that disruptions from gene variations (polymorphisms) or the environment (epigenetics) within this brain reward cascade result in aberrant addictive behaviors or RDS. Despite the international research to discover specific or potential gene clusters from high-density SNP arrays, many attempts have proven inconclusive. However, more recently, Palmer et al. (2015) observed that 25%–36% of the variance in vulnerability to substance dependence is attributed to common single nucleotide polymorphisms. The effect is additive since such common single nucleotide polymorphisms are shared across the indicators of comorbid drug problems. Recent research has also revealed that specific candidate gene variants account for risk prediction. Earlier work from Blum's laboratory,

adopting a Bayesian approach, determined a positive predictive value (PPV) for the DRD2 A1 variant (fewer D2 receptors) of 74% (Blum et al., 1995). This PPV demonstrated that children with this polymorphism have a very high risk of addiction disorders involving drugs, food, or other aberrant behaviors in their future. Since these findings, global laboratories, including the National Institute on Drug Abuse (NIDA) and the National Institute on Alcohol Abuse and Alcoholism (NIAAA), have not only confirmed this early work (Xu et al., 2004) but also extended the scope of many other candidate genes, in particular genes and second in the reward circuitry of the brain (Goldman, 1995).

For instance, Moeller et al. (2013) suggested that drug cues contribute to relapse, and their neurogenetic results have identified the DAT1R 9R-allele as a vulnerability allele for relapse especially during early abstinence (i.e., detoxification). The DAT1R 9R-allele influences the fast acting transport of dopamine sequestered from the synapse leading to a hypodopaminergic trait.

It is practical to employ genetic testing to discover reward circuitry gene polymorphisms, especially those related to dopaminergic pathways and opioid receptors. By determining genetic vulnerabilities and risks, treatment outcomes can vastly improve. Understanding reward circuitry involvement in buprenorphine effects and genotypes provide more insight into personalizing patients' clinical experience during opioid replacement therapy (Blum et al., 2014b).

Blum's genetic risk score represents a panel of known reward genes and associated risk polymorphisms providing genetic risk for addiction and other behaviors including medical monitoring and clinical outcome response.

Pharmacogenomics: Customized addiction medicine

Furthering this work, Blum and Gerald Kozlowski identified the brain reward cascade (BRC; Blum et al., 2016b). BRC as a concept acts as a blueprint explaining neurotransmitter interactions in the reward circuit. In addition, respective reward genes that regulate such neurotransmitters ultimately control the amount of dopamine released into both the reward site and other brain regions.

The integrity of resting state functional connectivity is crucial for normal homeostatic function.

Zhang et al. (2015) showed that heroin addicts exhibit reduced connectivity between the dorsal anterior cingulate cortex (dACC) and rostral anterior cingulate cortex (rACC), as well as reduced connectivity between the subcallosal anterior cingulate cortex (sACC) and dACC. The heroin addicts' variations in functional connectivity of three subregions of the ACC indicate that these three subregions, along with other key brain areas (such as the dorsal striatum, putamen, orbital frontal cortex, dorsal striatum, cerebellum, amygdala) potentially modulate heroin addiction. Blum's laboratory and Zhang's group (Blum et al., 2015b) showed that in abstinent heroin addicts, KB220Z™ (a putative dopamine D2 agonist) increased BOLD activation in caudate-accumbens-dopaminergic pathways, compared to a placebo following one-hour acute administration. In addition, KB220Z™ also induced the reduction of resting state activity in the putamen. In the second phase of this pilot study, all 10 abstinent heroin-dependent subjects had three brain regions of interest (ROIs), which were shown to be significantly activated from the resting state by KB220Z™ compared to the placebo (P < 0.05). Increased functional connectivity was also observed in a system involving the dorsal anterior cingulate, medial frontal gyrus, nucleus accumbens, posterior cingulate, occipital cortical areas, and cerebellum.

Employing DNA-based testing, successful development of polymorphic gene testing enables customized (personalized) antiobesity compounds, exemplifying the future of personalized addiction medicine utilizing Geneus Health's Genetic Addiction Risk Score (GARS).

lacking panels of other candidate genes and outcome studies. Another example is that a well known Pharmacogenetic company claims that having one polymorphism in the mu opioid receptor leads to opioid dependence. One gene alone cannot provide the necessary information to make such a claim because it is more complex. Other misleading claims suggest that the patient results are compared to population controls. Our review of their "disease-free" controls revealed substantial flaws, without controlling for any RDS behaviors (Chen et al., 2005). For their claims to be true, they would have to utilize what has been termed "super controls." In short, population controls may carry many unseen RDS behaviors that must be identified so that the control would be RDS free. Also the severity of the phenotype in question must be considered as well otherwise spurious results will ensue.

To reiterate, otherwise, adopting genetic testing will lead to spurious and inaccurate results. Some of these companies have selected genes for testing that may be involved in risky behavior, but they are not using the proper variant in their tests, or they use rare variants that do not truly prove addiction risk. Indeed, Mayer and Höllt (2006) correctly stated that the "vast number of non-coding, intronic or promoter polymorphisms in the opioid receptors may influence addictive behavior, but these polymorphisms are far less studied, and their physiological significance remains to be demonstrated." Such companies have never investigated the scientific integrity and predictability of their genetic full-panel tests, let alone actual addiction risk or any associated behaviors.

Fiction

Although there is a slew of promising experiments for RDS behaviors involving candidate gene associations, there are also some negative outcomes (Doremus-Fitzwater and Spear, 2016).

Currently, a few companies have entered into genetic testing in the addiction and pain medication industry to offer what they claim to be "personalized care." However, many of these companies have not fully understood the science behind the application. They often make exaggerated commercial claims using Blum's original work, stating that their genetic test is 74% predictive (Blum et al., 1995). This is obviously false as they used a single gene (DRD2) to support their predictability claim,

Facts

Although we, the authors, may have a personal bias due to the many years that the Blum laboratory has dedicated to developing an accurate genetic test to predict true liability/risk for RDS and associated behaviors, we will attempt to explain why the successful development of the first GARS in conjunction with the Institute of Behavioral Genetics, University of Colorado, Boulder has the real potential of impacting the entire addiction and pain medicine field.

To develop GARS™ we first selected (1) ten reward candidate genes (DRD1, 2, 3, 4; DAT1; serotonin transporter, COMT, MAO, GABA, Mu opiate receptor) and (2) a number of SNPs and point mutations that influence the net release of dopamine at the brain reward site. The SNPs and point mutations were chosen to reflect a hypodopaminergic trait. To validate our methods we partnered with the developers of the Addiction Severity Index-Media Version (ASI-MV), a test mandated in 18 states that determines both alcohol and drug severity risk scores (Butler et al., 2015).

In the process, we collaborated with seven diverse treatment centers across the United States, obtaining a total of 393 subjects that we genotyped using the GARS panel. All data were genotyped and analyzed at the Institute for Behavioral Genetics. For 273 subjects, we discovered a marked association between the summed score of GARS panel risk alleles (variant forms) and both the drug (P < 0.05) and ASI-MV alcohol (p < 0.004) severity indices.

In effect, the correlation indicated the higher the number of risk alleles, the stronger the prediction of alcohol or drug use severity. Notably, family or domestic problems, psychological issues, and medicalization significantly correlated as well. Interestingly, though, a change in any specific SNP resulted in no correlation. This signifies the importance of selected GARS panels, as any deviation will produce false results similar to scientifically vapid commercial tests. Weighting each allele by increasing its score power also resulted in lost significance suggesting the importance of a clustering of genes implicated in the BRC rather than any one gene polymorphism by itself.

BENEFITS OF GARS

We found that a 10-gene panel driven by risk alleles for hypodopaminergic state, consisting of 11 SNPs, was significantly associated with the level of ASI-MV alcohol and drug risk severity score, supporting clinical relevance and utility.

To our knowledge, this is the first and only correlation of a panel of genes, with established polymorphisms that reflect the brain reward cascade with the ASI-MV alcohol and drug risk severity score ever accomplished to date. While other studies are required to further confirm and extend the GARS test to include other genes and polymorphisms that associate with hypodopaminergic trait, these results provide clinicians with a noninvasive genetic test.

Unlike 23andMe genomic testing, other than finding ancestral heritage, GARS can be used to improve clinical interactions and decision making:

- Attenuation of guilt and denial
- Corroboration of family geneograms
- Risk severity–based decisions about appropriate therapies including pain medications and risk for addiction liability
- Appropriate level of care placement, for example, in-patient, out-patient, intensive out-patient, residential
- Length of stay in treatment
- Genetic severity–based relapse and recovery liability and vulnerability
- Pharmacogenetic medical monitoring for better clinical outcomes, for example, the A1 allele of the DRD2 gene reduces the binding to delta receptors in the brain, reducing naltrexone's clinical effectiveness

HAVE WE CREATED DNA-CUSTOMIZED NUTRITION?

In light of these early studies and a scarcity of relevant research, we decided to design a study evaluating the process of DNA customization of a nutritional solution for both wellness and weight management. We review the results of a number of studies (Blum et al., 2008c, 2008d, 2008e) in which Blum's laboratory genotyped 1058 subjects who were administered KB220z (formerly LG9939, Recompozize, Genotrim), a complex neuroadaptagen nutraceutical-dl-phenylalanine, along with chromium, l-tyrosine, and other select amino acids and adaptogens, based on polymorphic outcomes. The customized formulas involved a minimum of 175 SNPs covering 16 genes important to the BRC and, most importantly, involved in "dopamine homeostasis." Within a small subset, simple t-tests comparing parameters before and after 80 days on the nutraceutical were performed.

The clinical outcomes of the studies were weight loss (p < 0.008), appetite suppression (p < 0.004), sugar craving reduction (p < 0.008), snack reduction (p < 0.005), reduction of late night binging (p < 0.007), increased energy (p < 0.004), increased perception of overeating (p < 0.02), enhanced quality of sleep (p < 0.02), and increased happiness (p < 0.02). Polymorphic correlates were also obtained for various genes (PPAR gamma 2, MTHFR, 5-HT2a, and DRD2) with positive clinical parameters tested in this study. Notably, in all the outcomes and gene polymorphisms, a significant Pearson correlation only occurred for the

DRD2 gene polymorphism (allele A1) with days on treatment ($r = 0.42$, $p = 0.045$). This two fold increase is a very important genotype for compliance in treatment.

In addition, Blum's group systematically evaluated the impact of polymorphisms of these five candidate genes as important targets for the development of a DNA-customized nutraceutical KB220z to combat obesity with particular emphasis on body recomposition as measured by body mass index (BMI). A total of 21 individuals were evaluated in a preliminary investigational study of KB220z.

We based the experiment on the results of each subjects' buccal swab genotyping and a customized nutraceutical formula was provided specific to the function of measured gene polymorphisms of the five gene candidates assessed. Subsequently, at the beginning and every 2 weeks each subject completed a modified Blum-Downs OPAQuE Scale™ [Overweight Patient Assessment Questionnaire]. The involved alleles included the DRD2 A1, MTHFR C 677 T, 5HT2a 1438G/A, PPAR-γPro12Ala, and Leptin Ob1875 <208 bp. Pre and post hoc analysis revealed a large difference between the starting BMI and the BMI after an average of 41 days (28–70 d) of KB220z intake in 21 individuals. The pre-BMI was averaged at 31.2 (weight/Ht^2) compared to the post-BMI of 30.4 (weight/Ht^2), with significance value $P < 0.034$ (one-tailed). Similarly the preweight in pounds was 183.52 compared to the postweight of 179 pounds, with a significance value of $P < (0.047)$. Trends for reduction of late night snacking, carbohydrate craving reduction, reduction of stress, and reduction of waist circumference were also observed. In the 41-day period, a weight loss trend was seen as well: 71.4% of subjects lost weight. Fifteen out of 21 subjects lost weight with a z score of 2.4 and significance value of $P < (0.02)$ and this group (53%) overall lost on average over 2.5% of their starting weight.

Other preliminary conclusions requiring extensive investigation, using a path analysis (noncustomized KB220z), include notable associations regarding antiobesity-related behaviors. In a 1-year cross-sectional open trial study of 24 unscreened subjects, administration of oral KB220z variant lead to substantial benefits: stress reduction, sleep enhancement, increase in energy level, generalized well-being, improvement in mental focus/memory, reduction in cravings (sweets/carbs), improvement in blood sugar levels, reduction in food consumption, loss of inches around waist, weight loss, reduction in blood pressure, exercise performance improvement, reduction in drug-seeking behavior, reduction in hyperactivity, and reduction in cholesterol levels.

GENETIC TESTING FOR CUSTOMIZATION

We therefore propose that utilizing this natural dopaminergic-activating (potentially DNA-customized) approach over time leads to neuronal DA release at the NAc, causing a proliferation of D2 receptors (Downs et al., 2013). We are enthusiastic about previous results utilizing KB220Z in neuroimaging and qEEG studies (Blum et al., 2010; Miller et al., 2010) to demonstrate both enhanced resting state functional connectivity in abstinent heroin addicts (Blum et al., 2015b) and regulation of widespread theta activity in the cingulate gyrus of abstinent psychostimulant abusers (Blum et al., 2010), similar to effects observed in alcoholics and heroin addicts (Miller et al., 2010).

Other support for anticraving effects of dopaminergic activation in humans can be found in numerous peer-reviewed published clinical trials including randomized double-blind placebo-controlled studies. In fact, decreased alcohol and cocaine craving behavior was seen with animal gene therapy using cDNA vectors of the DRD2 gene implanted into the NAc (Myers and Robinson, 1999; Thanos et al., 2001, 2005, 2008).

We the authors are cognizant that dopaminergic activation in the long-term instead of blocking dopamine (Blum et al., 2008b) should be utilized to treat not only alcohol, cocaine, and nicotine cravings, but glucose craving and other known behavioral addictions (e.g., gambling, hypersexuality [Blum et al., 2015a]). Thus the coupling of genetic antecedents such as GARS and nutrition may be a very viable alternative approach for the treatment of all RDS behaviors (Blum et al., 2008c, 2008d, 2008e).

Additional research from Blum's laboratory developed a theoretical modeling study, which evaluated health and economic implications of a nutrigenomic product for weight loss. Blum's group (Meshkin and Blum, 2007) based a nutrigenomic economic model on linking (1) published study data on product efficacy of ingredients, (2) validated clinical assessments already used in health economics data, and (3) data on condition prevalence and macroeconomic costs of illness. The model demonstrates that a DNA-customized nutraceutical positively decreases the cost of illness at the macroeconomic and microeconomic level, based on cost-effectiveness and cost-benefit analysis. This model can thus forecast important prognostic health economic implications of a nutrigenomic intervention to demonstrate a theoretical model of nutrigenomic economics as it relates to obesity.

Reiterations of fifteen variant formulae allowed Blum's group (Blum et al., 2008c, 2008d, 2008e) to use polymorphic targets of various reward genes (serotonergic, opioidergic, GABAergic, and dopaminergic) to customize KB220 (neuroadaptogen-amino-acid therapy [NAAT]) with specific algorithms. For 1000 obese subjects in the Netherlands, a small subset was administered different KB220 formulae customized according to individuals' respective DNA polymorphisms, which translated to significant decreases in both BMI and weight in pounds.

Blum's laboratory developed GARS along with Brett Haberstick, Andrew Smolen, and others. When 10 genes were selected with appropriate variants, it appeared that there was a statistically notable association between the ASI-MV alcohol and drug severity scores and GARS. This was found true for 273 patients at seven treatment centers. This correlation sets the framework for early clinical diagnosis and identification of addictive predispositions, showing personalized medicine as a nutrigenomic solution for RDS behaviors (Blum et al., 2014a).

Blum et al. (2015b) reported that a well-known variant of KB220PAM actually increased resting state functional connectivity in abstinent heroin addicts. In fact, the enhanced activation of resting state functional connectivity was observed in a network involving the dorsal anterior cingulate, medial frontal gyrus, nucleus accumbens, posterior cingulate, occipital cortical areas, and cerebellum.

In other published work at the University of Florida, Febo, Blum, and others found that KB220Z considerably stimulates (above placebo) seed regions of interest, such as the hippocampus, left nucleus accumbens, anterior thalamic nuclei, cingulate gyrus, prelimbic and infralimbic loci. The reaction caused by KB220Z establishes substantial functional connectivity and increased brain volume recruitment, as well as enhanced dopaminergic functionality across brain reward circuitry. This robust and selective response is clearly clinically significant (Febo et al., 2017).

CONCLUSION

We, the authors, would like to now propose Precision Behavioral Management (PBM), a protocol that promotes early diagnostic identification and stratification of risk alleles through the use of GARS, thereby allowing nutrigenomic targeting of these risk alleles through the alteration of neuronutrient amino acid therapy ingredients as an algorithmic function of polymorphic DNA SNPS. A novel approach of this can produce the first ever nutrigenomic solution for addiction, pain, and other RDS addictive behaviors (see Figure 17.1). Welcome to the new era of genomic addiction medicine (Blum et al., 2015c; Blum and Badgaiyan, 2013). So maybe just like the famous moon walk, by using precision medicine we will be able to differentiate our cultural diversity here in America and globally and develop DNA-directed antiopiate/opioid precision medicine that promotes "dopamine

Figure 17.1 Schematic of DNA-customization application for a nutrigenomic solution to RDS.

homeostasis" instead of continued opioid locked-in addiction with associated zombie-like behaviors (Hill et al., 2013). Utilizing these simple principles based on empirical genetic evidence concerning various RDS behaviors (Blum et al., 1990, 1996, 2011; Levey et al., 2014; Farris et al., 2014; Ducci and Goldman, 2012; Gyollai et al., 2014; Yan et al., 2014; Guo et al., 2016), we are now poised to overcome the deviating unwelcomed opiate/opioid American second epidemic.

REFERENCES

Apaydin EA, Maher AR, Shanman R, Booth MS, Miles JN, Sorbero ME, Hempel S. A systematic review of St. John's wort for major depressive disorder. *Syst Rev* 5(1);2016:148.

Blum K. "Dopamine Resistance" in Brain Reward Circuitry as a function of Genetic Addiction Risk Score (GARS) polymorphisms in RDS: Synaptamine Complex Variant (kb220z) induced "dopamine sensitivity" as a Pro-Recovery Agent. Presented at the *Reward Deficiency Syndrome (RDS) Inaugural Conference*. San Francisco, November 16–18, 2015.

Blum K, Badgaiyan RD. Addiction research and therapy in the 21st century: Providing a forum for evidence-based addiction medicine. *J Addict Res Ther* 4;2013:1000e117.

Blum K, Badgaiyan RD, Gold MS. Hypersexuality addiction and withdrawal: Phenomenology, neurogenetics and epigenetics. *Cureus* 7(10);2015a:e348.

Blum K, Chen AL, Braverman ER. Attention-deficit-hyperactivity disorder and reward deficiency syndrome. *Neuropsychiatr Dis Treat* 4(5);2008a:893–918.

Blum K, Chen AL, Chen TJ et al. Activation instead of blocking mesolimbic dopaminergic reward circuitry is a preferred modality in the long term treatment of reward deficiency syndrome (RDS): A commentary. *Theor Biol Med Model* 2008b;5:24.

Blum K, Chen AL, Chen TJ et al. LG839: anti-obesity effects and polymorphic gene correlates of reward deficiency syndrome. *Adv Ther* 25(9);2008c:894–913.

Blum K, Chen ALC, Chen TLC et al. Dopamine D2 receptor Taq A1 allele predicts treatment compliance of LG839 in a subset analysis of a pilot study in The Netherlands. *Gene Therapy Mol Biol* 12(1);2008d:129–40.

Blum K, Chen TJ, Morse S et al. Overcoming qEEG abnormalities and reward gene deficits during protracted abstinence in male psychostimulant and polydrug abusers utilizing putative dopamine D₂ agonist therapy: part 2. *Postgrad Med* 122(6);2010:214–26.

Blum K, Chen AL, Oscar-Berman M et al. Generational association studies of dopaminergic genes in reward deficiency syndrome (RDS) subjects: Selecting appropriate phenotypes for reward dependence behaviors. *Int J Environ Res Public Health* 8(12);2011:4425–59.

Blum K, Chen TJH, Williams L et al. A short term pilot open label study to evaluate efficacy and safety of LG839, a customized DNA directed nutraceutical in obesity: Exploring nutrigenomics. *Gene Ther Mol Biol* 12(2);2008e:371–82.

Blum K, Febo M, Badgaiyan RD. Fifty years in the development of a glutaminergic-dopaminergic optimization complex (KB220) to balance brain reward circuitry in reward deficiency syndrome: A pictorial. *Austin Addict Sci* 1(2);2016a: pii: 1006.

Blum K, Febo M, Badgaiyan RD, Demetrovics Z, Simpatico T, Fahlke C, Oscar-Berman M, Li M, Dushaj K, Gold MS. Common neurogenetic diagnosis and meso-limbic manipulation of hypodopaminergic function in Reward Deficiency Syndrome (RDS): Changing the recovery landscape. *Curr Neuropharmacol* 15(1);2016b:184–94.

Blum K, Febo M, McLaughlin T, Cronjé FJ, Han D, Gold SM. Hatching the behavioral addiction egg: Reward Deficiency Solution System (RDSS)™ as a function of dopaminergic neurogenetics and brain functional connectivity linking all addictions under a common rubric. *J Behav Addict* 3(3);2014a:149–56.

Blum K, Gardner E, Oscar-Berman M, Gold M. "Liking" and "wanting" linked to Reward Deficiency Syndrome (RDS): Hypothesizing differential responsivity in brain reward circuitry. *Curr Pharm Des* 18(1);2012:113–8.

Blum K, Giordano J, Hauser M et al. Coupling Genetic Addiction Risk Score (GARS™), Comprehensive Analysis of Reported Drugs (CARD™) dopamine agonist therapy (KB220z™): Reward Deficiency Solution

System (RDSS). *Abstract accepted to European Psychiatry Congress*, Vienna, Austria, 2014b March 18–20.

Blum K, Kozlowski GP. Ethanol and neuromodulator influences. A cascade model of reward. In: *Alcohol and Behaviour: Basic and Clinical Aspects Progress in Alcohol Research*. Ollat H, Parvez S, Parvez H (Eds), VSP International Science Publishers, Utrecht, Netherlands, 1990, pp. 131–50.

Blum K, Liu Y, Wang W et al. rsfMRI effects of KB220ZTM on neural pathways in reward circuitry of abstinent genotyped heroin addicts. *Postgrad Med* 127(2);2015b:232–41.

Blum K, Noble EP, Sheridan PJ, Montgomery A, Ritchie T, Jagadeeswaran P, Nogami H, Briggs AH, Cohn JB. Allelic association of human dopamine D2 receptor gene in alcoholism. *JAMA* 263(15);1990:2055–60.

Blum K, Oscar-Berman M, Demetrovics Z et al. Genetic Addiction Risk Score (GARS): Molecular neurogenetic evidence for predisposition to Reward Deficiency Syndrome (RDS). *Mol Neurobiol* 50(3);2014c:765–96.

Blum K, Oscar-Berman M, Stuller E et al. Neurogenetics and nutrigenomics of neuro-nutrient therapy for Reward Deficiency Syndrome (RDS): Clinical ramifications as a function of molecular neurobiological mechanisms. *J Addict Res Ther* 3(5);2012b:139.

Blum K, Sheridan PJ, Wood RC et al. The D2 dopamine receptor gene as a determinant of reward deficiency syndrome. *J R Soc Med* 89(7);1996:396–400.

Blum K, Simpatico T, Badgaiyan RD et al. Coupling neurogenetics (GARS™) and a nutrigenomic based dopaminergic agonist to treat Reward Deficiency Syndrome (RDS): Targeting polymorphic reward genes for carbohydrate addiction algorithms. *J Reward Defic Syndr* 1(2);2015c:75–80.

Blum K, Wood RC, Braverman ER et al. The D2 dopamine receptor gene as a predictor of compulsive disease: Bayes' theorem. *Funct Neurol* 10(1);1995:37–44.

Butler SF, McNaughton EC, Black RA. Tapentadol abuse potential: A post marketing evaluation using a sample of individuals evaluated for substance abuse treatment. *Pain Med* 16(1);2015:119–30.

Casazza K, Brown A, Astrup A et al. Weighing the evidence of common beliefs in obesity research. *Crit Rev Food Sci Nutr* 55(14);2015:2014–53.

Chen TJ, Blum K, Mathews D et al. Are dopaminergic genes involved in a predisposition to pathological aggression? Hypothesizing the importance of "super normal controls" in psychiatric genetic research of complex behavioral disorders. *Med Hypotheses* 65(4);2005:703–7.

De Fazio S, Gallelli L, De Siena A, De Sarro G, Scordo MG. Role of CYP3A5 in abnormal clearance of methadone. *Ann Pharmacother* 42(6);2008:893–7.

Doremus-Fitzwater TL, Spear LP. Reward-centricity and attenuated aversions: An adolescent phenotype emerging from studies in laboratory animals. *Neurosci Biobehav Rev* 70;2016:121–34.

Downs B, Oscar-Berman M, Waite R et al. Have we hatched the addiction egg: Reward Deficiency Syndrome Solution System™ *J Genet Syndr Gene Ther* 4(136);2013:14318.

Downs BW, Chen AL, Chen TJ et al. Nutrigenomic targeting of carbohydrate craving behavior: Can we manage obesity and aberrant craving behaviors with neurochemical pathway manipulation by Immunological Compatible Substances (nutrients) using a Genetic Positioning System (GPS) Map? *Med Hypotheses* 73(3);2009:427–34.

Ducci F, Goldman D. The genetic basis of addictive disorders. *Psychiatr Clin North Am* 35(2);2012:495–519.

Farris SP, Arasappan D, Hunicke-Smith S et al. Transcriptome organization for chronic alcohol abuse in human brain. *Mol Psychiatry* 20(11);2014:1438–47.

Febo M, Blum K, Badgaiyan RD et al. Enhanced functional connectivity and volume between cognitive and reward centers of naïve rodent brain produced by pro-dopaminergic agent KB220Z. *PLoS One* 12(4);2017:e0174774.

Goldman D. Candidate genes in alcoholism. *Clin Neurosci* 3(2);1995:174–81.

Guo P, Li Y, Eslamfam S et al. Discovery of novel genes mediating glucose and lipid metabolisms. *Curr Protein Pept Sci* 18(6);2016:609–18.

Gyollai A, Griffiths MD, Barta C et al. The genetics of problem and pathological gambling:

A systematic review. *Curr Pharm Des* 20(25);2014:3993–9.

Higginson AD, McNamara JM, Houston AI. Fatness and fitness: Exposing the logic of evolutionary explanations for obesity. *Proc Biol Sci* 283(1822);2016: pii: 20152443.

Hill E, Han D, Dumouchel P et al. Long term Suboxone™ emotional reactivity as measured by automatic detection in speech. *PLOS ONE* 8(7);2013:e69043.

Joffe B, Zimmet P. The thrifty genotype in type 2 diabetes: An unfinished symphony moving to its finale? *Endocrine* 9(2);1998:139–41.

Levey DF, Le-Niculescu H, Frank J et al. Genetic risk prediction and neurobiological understanding of alcoholism. *Transl Psychiatry* 4;2014:e391.

Mayer P, Höllt V. Pharmacogenetics of opioid receptors and addiction. *Pharmacogenet Genomics* 16(1);2006:1–7.

Meshkin B, Blum K. Folate nutrigenetics: A convergence of dietary folate metabolism, folic acid supplementation, and folate antagonist pharmacogenetics. *Drug Metab Lett* 1(1);2007:55–60.

Miller DK, Bowirrat A, Manka M et al. Acute intravenous synaptamine complex variant KB220™ "normalizes" neurological dysregulation in patients during protracted abstinence from alcohol and opiates as observed using quantitative electroencephalographic and genetic analysis for reward polymorphisms: Part 1, pilot study with 2 case reports. *Postgrad Med* 122(6);2010:188–213.

Moeller SJ, Parvaz MA, Shumay E et al. Gene x abstinence effects on drug cue reactivity in addiction: multimodal evidence. *J Neurosci* 33(24);2013:10027–36.

Myers RD and Robinson DE. Mmu and D2 receptor antisense oligonucleotides injected in nucleus accumbens suppress high alcohol intake in genetic drinking HEP rats. *Alcohol* 118(2–3);1999:225–33.

Noble EP, Blum K, Ritchie T et al. Allelic association of the D2 dopamine receptor gene with receptor-binding characteristics in alcoholism. *JAMA Psychiatry* 48(7);1991:648–54.

Palmer RH, Brick L, Nugent NR et al. Examining the role of common genetic variants on alcohol, tobacco, cannabis and illicit drug dependence: Genetics of vulnerability to drug dependence. *Addiction* 110(3);2015:530–7.

Plomin R, Owen MJ, McGuffin P. The genetic basis of complex human behaviors. *Science* 264(5166);1994:1733–9.

Polyák Á, Winkler Z, Kuti D et al. Brown adipose tissue in obesity: Fractalkine-receptor dependent immune cell recruitment affects metabolic-related gene expression. *Biochim Biophys Acta* 1861(11);2016:1614–22.

Rana S, Kumar S, Rathore N et al. Nutrigenomics and its impact on life style associated metabolic diseases. *Curr Genomics* 17(3);2016:261–78.

Rena G, Begg F, Ross A et al. Molecular cloning, genomic positioning, promoter identification, and characterization of the novel cyclic amp-specific phosphodiesterase PDE4A10. *Mol Pharmacol* 59(5);2001:996–1011.

Thanos PK, Michaelides M, Umegaki H and Volkow ND. D2R DNA transfer into the nucleus accumbens attenuates cocaine self-administration in rats. *Synapse* 62(7);2008:481–6.

Thanos PK, Rivera SN, Weaver K, Grandy DK, Rubinstein M, Umegaki H, Wang GJ, Hitzemann R, Volkow ND. Dopamine D2R DNA transfer in dopamine D2 receptor-deficient mice: Effects on ethanol drinking. *Life Sci* 77(2);2005:130–9.

Thanos PK, Volkow ND, Freimuth P, Umegaki H, Ikari H, Roth G, Ingram DK, Hitzemann R. Overexpression of dopamine D2 receptors reduces alcohol self-administration. *J Neurochem* 78(5);2001:1094–103.

Xu K, Lichtermann D, Lipsky RH et al. Association of specific haplotypes of D2 dopamine receptor gene with vulnerability to heroin dependence in 2 distinct populations. *JAMA Psychiatry* 61(6);2004:597–606.

Yan J, Aliev F, Webb BT et al. Using genetic information from candidate gene and genome-wide association studies in risk prediction for alcohol dependence. *Addict Biol* 19(4);2014:708–21.

Zhang Y, Gong J, Xie C et al. Alterations in brain connectivity in three sub-regions of the anterior cingulate cortex in heroin-dependent individuals: Evidence from resting state fMRI. *Neuroscience* 284;2015:998–1010.

Index